Lecture Notes of
the Unione Matematica Italiana

Luc Tartar

The General Theory of Homogenization

A Personalized Introduction

Springer

UMI

Luc Tartar
Carnegie Mellon University
Department of Mathematical Sciences
Pittsburgh, PA, 15213
U.S.A.
tartar@cmu.edu

ISSN 1862-9113
ISBN 978-3-642-05194-4 e-ISBN 978-3-642-05195-1
DOI 10.1007/978-3-642-05195-1
Springer Heidelberg Dordrecht London New York

Library of Congress Control Number: 2009941069

Mathematics Subject Classification (2000): 35J99, 35K99, 35L99, 35S99, 74Q05, 74Q10, 74Q15, 74Q20, 76A99

Cover design: SPi Publisher Services

Printed on acid-free paper

Springer is part of Springer Science+Business Media (www.springer.com)

Dedicated to Sergio SPAGNOLO

Helped with the insight of Ennio DE GIORGI, he was the first in the late 1960s to give a mathematical definition concerning homogenization in the context of the convergence of Green kernels: G-convergence.

to François MURAT

Starting from his discovery of a case of nonexistence of solutions for an optimization problem, in the spirit of the earlier work of Laurence YOUNG, which was not known in Paris, we started collaborating in the early 1970s and rediscovered homogenization in the context of optimal design problems, leading to a slightly more general framework: H-convergence and compensated compactness.

to Évariste SANCHEZ-PALENCIA

It was his work on asymptotic methods for periodically modulated media in the early 1970s that helped me understand that my joint work with François MURAT was related to questions in continuum mechanics, and this gave me at last a mathematical way to understand what I was taught in continuum mechanics and physics at École Polytechnique, concerning the relations between microscopic, mesoscopic, and macroscopic levels, without using any probabilistic ideas!

to Lucia

to my children
Laure, Michaël, André, Marta

to my grandchildren
Lilian, Lisa

and to my wife
Laurence

Preface

In 1993, from 27 June to 1 July, I gave ten lectures for a CBMS–NSF conference, organized by Maria SCHONBEK at UCSC,[1] Santa Cruz, CA. As I was asked to write lecture notes, I wrote the parts concerning homogenization and *compensated compactness* in the following years, but I barely started writing the part concerning *H-measures*.

In the fall of 1997, facing an increase in aggressiveness against me, I decided to put that project on hold, and I devised a new strategy to write lecture notes for the graduate courses that I was going to teach at CMU (*Carnegie Mellon University*),[2,3] Pittsburgh, PA. After doing so for the courses that I taught in the spring of 1999 and in the spring of 2000, I made the texts available on the web page of CNA (Center for Nonlinear Analysis at CMU). For the graduate course that I taught in the fall of 2001, I still needed to write the last four lectures, but I also prepared the last version of my CBMS–NSF course, from the summer of 1996, to make it also available on the web page of CNA, so that those who received a copy of various chapters would not be the only ones to know the content of those chapters that I wrote.

This led to a sharp increase of aggressiveness against me, so after putting my project on hold, I learned to live again in a hostile environment.

---

[1] Maria Elena SCHONBEK, Argentinean-born mathematician. She worked at Northwestern University, Evanston, IL, at VPISU (Virginia Polytechnic Institute and State University), Blacksburg, VA, at University of Rhode Island, Kingston, RI, at Duke University, Durham, NC, and she now works at UCSC (University of California at Santa Cruz), Santa Cruz, CA.

[2] Andrew CARNEGIE, Scottish-born businessman and philanthropist, 1835–1919. Besides endowing a technical school in Pittsburgh, PA, which became Carnegie Tech (Carnegie Institute of Technology) and then CMU (*Carnegie Mellon University*) after it merged in 1967 with the Mellon Institute of Industrial Research, he funded about three thousand public libraries, and those in United States are named *Carnegie libraries*.

[3] Andrew William MELLON, American financier and philanthropist, 1855–1937. He founded the Mellon Institute of Industrial Research in Pittsburgh, PA, which merged in 1967 with Carnegie Tech (Carnegie Institute of Technology) to form CMU (*Carnegie Mellon University*).

In the summer of 2002, I started revising my first two lecture notes by adding information about the persons whom I mention in the text, and for doing this I used footnotes, despite a warning by KNUTH [45][4] that footnotes tend to be distracting, but as he added "Yet Gibbon's *Decline and Fall* would not have been the same without footnotes,"[5] I decided not to restrain myself. I cannot say if my excessive use of footnotes resembles that of GIBBON, as I have not yet read *The History of the Decline and Fall of the Roman Empire* [34], but I wonder if the recent organized attacks on the western academic systems are following some of the reasons that GIBBON proposed for explaining the decline and the collapse of the mighty Roman empire.

Where should I publish my lecture notes once written? I found the answer in October 2002 at a conference at Accademia dei Lincei in Roma (Rome), Italy, when my good friends Carlo SBORDONE and Franco BREZZI mentioned their plan[6,7] to have a series of lecture notes at UMI (*Unione Matematica Italiana*), published by Springer.[8]

I submitted my first lecture notes for publication in the summer of 2004, but I took a long time before making the requested corrections, and they appeared only in August 2006 as volume 1 of the UMI Lecture Notes series [116], *An Introduction to Navier–Stokes Equation and Oceanography.*[9,10]

I submitted my second lecture notes for publication in August 2006, and they appeared in June 2007 as volume 3 of the UMI Lecture Notes series [117], *An Introduction to Sobolev Spaces and Interpolation Spaces.*[11]

I submitted my third lecture notes for publication in January 2007 and they appeared in March 2008 as volume 6 of the UMI Lecture Notes series [119], *From Hyperbolic Systems to Kinetic Theory, A Personalized Quest.*

---

[4] Donald Ervin KNUTH, American mathematician, born in 1938. He worked at Caltech (California Institute of Technology), Pasadena, CA, and at Stanford University, Stanford, CA.

[5] Edward GIBBON, English historian, 1817–1877.

[6] Carlo SBORDONE, Italian mathematician, born in 1948. He works at Università degli Studi di Napoli Federico II, Napoli (Naples), Italy. He was president of UMI (Unione Matematica Italiana) from 2000 to 2006.

[7] Franco BREZZI, Italian mathematician, born in 1945. He works at Università degli Studi di Pavia, Pavia, Italy. He became president of UMI (Unione Matematica Italiana) in 2006.

[8] Julius SPRINGER, German publisher, 1817–1877.

[9] Claude Louis Marie Henri NAVIER, French mathematician, 1785–1836. He worked in Paris, France.

[10] Sir George Gabriel STOKES, Irish-born mathematician, 1819–1903. He worked in London and in Cambridge, England, holding the Lucasian chair (1849–1903).

[11] Sergei L'vovich SOBOLEV, Russian mathematician, 1908–1989. He worked in Leningrad, in Moscow, and in Novosibirsk, Russia. There is now a Sobolev Institute of Mathematics of the Siberian branch of the Russian Academy of Sciences, Novosibirsk, Russia. I first met Sergei SOBOLEV when I was a student, in Paris in 1969, and conversed with him in French, which he spoke perfectly (all educated Europeans at the beginning of the twentieth century learned French).

In the summer of 2007, it was time for me to think again about my CBMS–NSF course. Because I already wrote lecture notes on how homogenization appears in optimal shape design [111] for lectures given during a CIME–CIM summer school, organized by Arrigo CELLINA and António ORNELAS,[12,13] in Tróia, Portugal, in June 1998, I wrote this book in a different way, describing how my ideas in homogenization were introduced during my quest for understanding more about continuum mechanics and physics, so that chapters follow a loose chronological order.

As in my preceding lecture notes, I use footnotes for giving some biographical information about people related to what I mention, and in the text I use the first name of those whom I met. In my third lecture notes, I started putting at the end of each chapter the additional footnotes that are not directly related to the text but expand on some information found in previous footnotes; in this book, instead of presenting them in the order where the names appeared, I organized the additional footnotes in alphabetical order.

When one misses the footnote containing the information about someone, a chapter of biographical information at the end of the book permits one to find where the desired footnote is.

I may be wrong about some information that I give in footnotes, and I hope to be told about my mistakes, and that is true about everything that I wrote in the book, of course!

I want to thank my good friends Carlo SBORDONE and Franco BREZZI for their support, in general, and for the particular question of the publication of my lecture notes in a series of *Unione Matematica Italiana.*

I want to thank *Carnegie Mellon University* for according me a sabbatical period in the fall of 2007, and *Politecnico di Milano* for its hospitality during that time, at it was of great help for concentrating on my writing programme.

I want to thank *Université Pierre et Marie Curie* for a 1 month invitation at *Laboratoire Jacques-Louis Lions*,[14,15] in May/June 2008, as it was during

---

[12] Arrigo CELLINA, Italian mathematician, born in 1941. He works at Università di Milano Bicocca, Milano (Milan), Italy.

[13] António COSTA DE ORNELAS GONÇALVES, Portuguese mathematician, born in 1951. He works in Évora, Portugal.

[14] Pierre CURIE, French physicist, 1859–1906, and his wife Marie SKŁODOWSKA-CURIE, Polish-born physicist, 1867–1934, received the Nobel Prize in Physics in 1903 in recognition of the extraordinary services they have rendered by their joint research on the radiation phenomena discovered by Professor Henri BECQUEREL, jointly with Henri BECQUEREL. Marie SKŁODOWSKA-CURIE also received the Nobel Prize in Chemistry in 1911 in recognition of her services to the advancement of chemistry by the discovery of the elements radium and polonium, by the isolation of radium, and the study of the nature and compounds of this remarkable element. They worked in Paris, France. Université Paris VI, Paris, is named after them, UPMC (Université Pierre et Marie Curie).

[15] Jacques-Louis LIONS, French mathematician, 1928–2001. He received the Japan Prize in 1991. He worked in Nancy and in Paris, France; he held a chair (analyze

this period that I wrote the last chapters of the book. I want to thank François MURAT[16] for his hospitality during my visits to Paris for almost 20 years and for his unfailing friendship for almost 40 years.

I could not publish my first three lecture notes and start the preparation of this fourth book without the support of Lucia OSTONI. I want to thank her for much more than providing the warmest possible atmosphere during my stays in Milano, because she gave me the stability that I lacked so much during a large portion of the last 30 years, so that I now feel safer for resuming my research, whose main goal is to give a sounder mathematical foundation to twentieth century continuum mechanics and physics.

Milano, June 2008

Luc TARTAR
Correspondant de l'Académie des Sciences, Paris
Membro Straniero dell'Istituto Lombardo Accademia di Scienze e Lettere, Milano
University Professor of Mathematics, Department of Mathematical Sciences, Carnegie Mellon University, Pittsburgh, PA 15213-3890, USA

PS: (Pittsburgh, August 2008) Although I finished writing the book at the end of June, while I was in Milano attending the last meeting of Instituto Lombardo before the summer, I still had to check the chapter on notation and to create an index, and while doing that, I realized that I should explain my choices in a better way, in particular the subject of Chap. 1.

My general goal is to understand in a better way the continuum mechanics and the physics of the twentieth century, that is, the questions where small scales appear, plasticity and turbulence on the one hand, atomic physics and phase transitions on the other, and I think that the General Theory of Homogenization (GTH) as I developed it is crucial for starting in the right direction, but as there are a few dogmas to change, if not to discard completely, in continuum mechanics and in physics, I need to explain why the difficulties are similar to those that appeared in religions, where the deadlocks still remain.

---

mathématique des systèmes et de leur contrôle, 1973–1998) at Collège de France, Paris. The laboratory dedicated to functional analysis and numerical analysis which he initiated, funded by CNRS (Center National de la Recherche Scientifique) and UPMC (Université Pierre et Marie Curie), is now named after him, LJLL (Laboratoire Jacques-Louis Lions). He was my teacher at École Polytechnique in Paris in 1966–1967; I did research under his direction until my thesis in 1971.

[16] François MURAT, French mathematician, born in 1947. He works at CNRS (Centre National de la Recherche Scientifique) and UPMC (Université Pierre et Marie Curie), in LJLL (Laboratoire Jacques-Louis Lions), Paris, France.

Describing my family background and my studies is a way to answer the question that should be asked in the future: among those who realized at the end of the twentieth century that some of the dogmas in continuum mechanics and physics had to be discarded as wrong and counter-productive, what explains how they could start thinking differently? Should I say that I do not know who else but myself fits in this category? I expect that by telling this story, more will be able to follow a path similar to mine in the future, that is, there will be more mathematicians interested in the other sciences than mathematics!

Because I use the words parables and gospels in the first sentence of Chap. 1, some may stop reading the book, but in the second sentence I explain why parables are like general theorems, and by the end of the second footnote at the bottom of the first page, one will already learn that I am no longer a Christian, so that any misunderstanding about my intentions should result from the prejudices of the reader against religions, which is not a scientific attitude, and at the end of the book it should be clear that many "scientists" behaved in the recent past like religious fundamentalists.

What I advocate is for all to use their brain in a critical way!

Additional footnotes: BECQUEREL,[17] DUKE,[18] Federico II,[19] LUCAS H.,[20] NOBEL,[21] STANFORD.[22]

Detailed Description of Contents

a.b: refers to Corollary, Definition, Lemma, or Theorem # b in Chap. # a, while (a.b) refers to Eq. # b in Chap. # a.

Chapter 1: Why Do I Write?
About my sense of duty.

Chapter 2: A Personalized Overview of Homogenization I
About my understanding of homogenization in the 1970s.

[17] Antoine Henri BECQUEREL, French physicist, 1852–1908. He received the Nobel Prize in Physics in 1903, in recognition of the extraordinary services he has rendered by his discovery of spontaneous radioactivity, jointly with Pierre CURIE and Marie SKŁODOWSKA-CURIE. He worked in Paris, France.

[18] Washington DUKE, American industrialist, 1820–1905. Duke University, Durham, NC, is named after him.

[19] Friedrich VON HOHENSTAUFEN, German king, 1194–1250. Holy Roman Emperor, as Friedrich II, 1220–1250. He founded the first European state university in 1224, in Napoli (Naples), Italy, where he is known as Federico secondo, and Università degli Studi di Napoli is named after him.

[20] Henry LUCAS, English clergyman and philanthropist, 1610–1663. The Lucasian chair in Cambridge, England, is named after him.

[21] Alfred Bernhard NOBEL, Swedish industrialist and philanthropist, 1833–1896. He created a fund to be used as awards for people whose work most benefited humanity.

[22] Leland STANFORD, American businessman, 1824–1893. Stanford University is named after him (as is the city of Stanford, CA, where it is located).

Chapter 3: A Personalized Overview of Homogenization II
About my understanding of homogenization after 1980.

Chapter 4: An Academic Question of Jacques-Louis Lions
Studying in Paris in the late 1960s, the question of J.-L. Lions (4.1)–(4.3), the counter-example of Murat (4.4)–(4.6); 4.1: the basic one-dimensional homogenization lemma (4.7)–(4.9), a natural relaxation problem (4.10)–(4.14); 4.2: characterization of sequential weak $\star$ limits (4.15)–(4.18), around the ideas of L.C. Young.

Chapter 5: A Useful Generalization by François Murat
Research and development, technical ability, a two-dimensional problem (5.1); 5.1: layering in x_1 for the special case in $\mathbb{R}^2$ (5.2)–(5.4); 5.2: layering in x_1 for the symmetric elliptic case in $\mathbb{R}^N$ (5.5)–(5.12); 5.3: layering in x_1 for the not necessarily symmetric or elliptic case in $\mathbb{R}^N$ (5.13)–(5.16).

Chapter 6: Homogenization of an Elliptic Equation
Distinguishing the G-convergence of Spagnolo, the H-convergence of Murat and myself, and the Γ-convergence of De Giorgi; 6.1: G-convergence (6.5) and (6.6), the work of Spagnolo (6.1)–(6.4) and (6.7)–(6.10), V-ellipticity and norm (6.11); 6.2: abstract weak convergence of $(T^m)^{-1}$ (6.12); 6.3: $\mathcal{M}(\alpha, \beta; \Omega)$ (6.13) and (6.14); 6.4: H-convergence; 6.5: $\mathcal{M}(\alpha, \beta; \Omega)$ is sequentially closed for H-convergence (6.15)–(6.22), computing a convex hull for obtaining bounds (6.32)–(6.29); 6.6: a lower semi-continuity result in the symmetric case (6.30) and (6.31); 6.7: lower and upper bounds in the symmetric case (6.32).

Chapter 7: The Div–Curl Lemma
7.1: A case where coefficients are products (7.1)–(7.3); 7.2: the div–curl lemma (7.4)–(7.8); 7.3: a counter-example for $\int_\omega (E^n, D^n)\,dx$ (7.9)–(7.14), a generalization of Robbin using the Hodge theorem; 7.4: the necessity of (E, D) (7.15)–(7.18), a generalization of Murat to the L^p setting (7.19)–(7.21), a generalization of Hanouzet and Joly, a fake generalization.

Chapter 8: Physical Implications of Homogenization
About conjectures and theorems, my approach to different scales based on weak convergences, what internal energy is, and the defects of the second principle of thermodynamics, homogenization of first-order equations is important for turbulence and quantum mechanics, the nonexistent "particles" of quantum mechanics and the defects of the Boltzmann equation, the div–curl lemma in electrostatics, errors about effective coefficients in the literature, the div–curl lemma in electricity, and in equipartition of energy (8.1)–(8.4).

Chapter 9: A Framework with Differential Forms
9.1: The generalization of Robbin using the Hodge theorem, electrostatics with differential forms (9.1)–(9.5), the Maxwell–Heaviside equation with differential forms (9.6)–(9.20), a question about the Lorentz force and the motion of charged particles (9.21)–(9.23).

Chapter 10: Properties of H-Convergence

The danger of applying a general theory to too many examples, the ability with abstract concepts, my method of oscillating test functions (10.1)–(10.6); 10.1: the uniform bound for A^{eff} (10.7)–(10.12); 10.2: transposition in H-convergence (10.13)–(10.18); 10.3: independence from boundary conditions (10.19)–(10.21); 10.4: convergence up to the boundary for some variational inequalities (10.22)–(10.35); 10.5: local character of H-convergence (10.36)–(10.42); 10.6: a result of De Giorgi and Spagnolo (10.43)–(10.47); 10.7: preserving order by H-convergence (10.48)–(10.51); a counter-example of Marcellini (10.52)–(10.55); 10.8: perturbation of $\mathcal{M}(\alpha,\beta;\Omega)$ (10.56) and (10.57); 10.9: estimating $||A^{\mathrm{eff}} - B^{\mathrm{eff}}||$ for perturbations (10.58)–(10.68); 10.10: C^k and analytic dependence upon a parameter.

Chapter 11: Homogenization of Monotone Operators

An analogue of V-ellipticity for monotone operators (11.1)–(11.3); 11.1: the class $\mathcal{M}on(\alpha,\beta;\Omega)$ (11.4); 11.2: homogenization for $\mathcal{M}on(\alpha,\beta;\Omega)$ (11.5)–(11.17), a nonlinear analogue of symmetry (11.18); 11.3: homogenization of k-monotone and cyclically monotone operators in $\mathcal{M}on(\alpha,\beta;\Omega)$ (11.19)–(11.22); 11.4: an analogue of a result of De Giorgi and Spagnolo (11.23)–(11.34); 11.5: lower and upper bounds (11.35)–(11.42).

Chapter 12: Homogenization of Laminated Materials

12.1: The general one-dimensional case (12.1)–(12.9), an interpretation from electricity, using physical models in mathematics and drawings in geometry; 12.2: an application of the div–curl lemma (12.10); 12.3: sequences not oscillating in (x,e) (12.11) and (12.12), proofs of 12.2 and of 12.3 (12.13)–(12.19), a hyperbolic situation (12.20), my general method for laminated materials (12.21)–(12.33); 12.4: correctors for laminated materials (12.34)–(12.40).

Chapter 13: Correctors in Linear Homogenization

13.1: The general correctors (13.1)–(13.13); 13.2: a first type of lower-order terms (13.14)–(13.17); 13.3: a second type of lower-order terms (13.18)–(13.28); 13.4: some weak limits are better than expected (13.29)–(13.41); 13.5: correctors for 13.3 (13.42)–(13.49); 13.6: a third type of lower-order terms (13.50)–(13.52), the case of periodic data (13.53)–(13.55); 13.7: correctors in the periodic case (13.56)–(13.61).

Chapter 14: Correctors in Nonlinear Homogenization

Remarks on nonlinear elasticity; 14.1: the formula for laminated materials in $\mathcal{M}on(\alpha,\beta;\Omega)$ (14.1)–(14.11); 14.2: correctors for laminated materials in $\mathcal{M}on(\alpha,\beta;\Omega)$ (14.12)–(14.15); 14.3: correctors for $\mathcal{M}on(\alpha,\beta;\Omega)$ (14.16)–(14.27); 14.4: weak limits of $|grad(u_m)|^2$ (14.28)–(14.31), an application (14.32)–(14.35), the formula for laminated materials in nonlinear elasticity (14.36)–(14.40).

Chapter 15: Holes with Dirichlet Conditions
15.1: Homogenization for holes with Dirichlet conditions and data bounded in $L^2(\Omega)$ (15.1)–(15.10); 15.2: constants in the Poincaré inequality (15.11)–(15.13); 15.3: Homogenization for holes with Dirichlet conditions and data bounded in $H^{-1}(\Omega)$ (15.14)–(15.18); 15.4: a lemma involving the volume of the hole in a period cell (15.19)–(15.29); 15.5: the convergence of a rescaled sequence in the periodic case (15.30)–(15.40); 15.6: correctors in the periodic case (15.41)–(15.54), the convergence of the Stokes equation to the Darcy law according to Ené and Sanchez-Palencia.

Chapter 16: Holes with Neumann Conditions
Hypotheses on the holes (16.1)–(16.4); 16.1: using the extensions to prove convergence (16.5)–(16.9); 16.2: passing to the limit in a variational equation (16.10) and (16.11); 16.3: homogenization for holes with Neumann conditions (16.12)–(16.29), remarks on the periodic case (16.30)–(16.33).

Chapter 17: Compensated Compactness
The evolution of the ideas of Murat and myself; 17.1: a necessary condition for sequential weak lower semi-continuity (17.1)–(17.12); 17.2: a necessary condition for sequential weak continuity (17.13) and (17.14); 17.3: quadratic forms satisfying $Q(\lambda) \geq 0$ for all $\lambda \in \Lambda$ (17.15)–(17.38); 17.4: quadratic forms satisfying $Q(\lambda) = 0$ for all $\lambda \in \Lambda$ (17.39) and (17.40), examples (17.41)–(17.43), the general characteristic set $\mathcal{V}$ (17.44); 17.5: necessary conditions of higher-order (17.45)–(17.50); 17.6: a condition motivated by a result of Šverák (17.51) and (17.52).

Chapter 18: A Lemma for Studying Boundary Layers
The importance of asking questions, setting of the problem asked by J.-L. Lions (18.1)–(18.9); 18.1: my generalization of the Lax–Milgram lemma (18.10)–(18.14), my construction of M (18.15)–(18.18); 18.2: applying my abstract lemma (18.19)–(18.30); 18.3: my more general approach based on the Lax–Milgram lemma (18.31)–(18.45).

Chapter 19: A Model in Hydrodynamics
Explaining my model (19.1) and (19.2); 19.1: homogenization of my model (19.3)–(19.26), the case $div(w^n) = 0$ (19.27)–(19.29), a hint about H-measures, defects of kinetic theory.

Chapter 20: Problems in Dimension $N = 2$
Characterizing mixtures of two isotropic conductors (20.1)–(20.3), a preceding interaction between mathematics and physics, distinguishing conjectures and theorems, an observation of J. Keller (20.4)–(20.7); 20.1: $\frac{(A^n)^T}{\det(A^n)}$ H-converges to $\frac{(A^{\rm eff})^T}{\det(A^{\rm eff})}$ (20.8); 20.2: $\det(A^n) = \kappa$ implies $\det(A^{\rm eff}) = \kappa$ (20.9); 20.3: $\tau_P(M) = (-c\,R_{\pi/2} + d\,M)(a\,I + b, R_{\pi/2}M)^{-1}$ defines a group homomorphism if $\det(P) = a\,d - b\,c > 0$ (20.10) and (20.11); 20.4: $\tau_P(A^n)$ H-converges to $\tau_P(A^{\rm eff})$ (20.12)–(20.18), the Beltrami equation (20.19); 20.5:

writing the Beltrami equation as a system (20.20)–(20.24); 20.6: a characterization of symmetric M with $M_{1,1}, M_{2,2} > 0$, $\det(M) = 1$ and $\text{Trace}(M) \geq 2$ (20.25) and (20.26); 20.7: the formula for laminated materials uses an inversion (20.27)–(20.32); 20.8: closed discs inside the closed unit disc are stable by H-convergence (20.33)–(20.36), the conjecture of Mortola and Steffé.

Chapter 21: Bounds on Effective Coefficients
Using symmetries; 21.1: change of variable (21.1)–(21.6), equations which are not frame indifferent; 21.2: $\mathcal{B}(\theta)$, $\mathcal{H}(\theta)$, $\mathcal{K}(\theta)$ (21.7)–(21.10), the intuition about defining $\mathcal{K}(\theta)$; 21.3: basic estimates (21.11)–(21.13); 21.4: generating bounds using correctors (21.14)–(21.23), a choice of functionals based on compensated compactness (21.24)–(21.27); 21.5: a result in linear algebra (21.28); 21.6: a general lower bound (21.29)–(21.32); 21.7: a general upper bound (21.33)–(21.36), more general functionals (21.37)–(21.39); 21.8: mixtures of two isotropic conductors (21.40)–(21.48), the conjectured bounds of Hashin and Shtrikman, a result of Francfort and Murat and myself.

Chapter 22: Functions Attached to Geometries
Some approaches are not homogenization; 22.2: same geometries for two materials M^1, M^2 and defining $F(\cdot, M^1, M^2)$ (22.1)–(22.5), special cases (22.6) and (22.7); 22.2: numerical range (22.8); 22.3: its convexity (22.9)–(22.11); 22.4: the cases of $F(\cdot, M^1, M^2)$ and $\big(F(\cdot, M^1, M^2)\big)^{-1}$ (22.12)–(22.14); 22.5: a more precise result (22.15) and (22.16), transposed and complex conjugate (22.17); 22.6: using order (22.18); 22.7: Pick functions and Herglotz functions; 22.8: Herglotz functions (22.19)–(22.24); 22.9: Pick functions (22.25)–(22.29), using the constraints $g(1) = 1$ and $g'(1) = 1 - \theta$ (22.30)–(22.33); 22.10: generalizing 20.1 (22.34); 22.11: $\frac{F^T}{det(F)}$ for $N = 2$ (22.35) and (22.36), the reiteration formula in the simple case (22.37); 22.12: the reiteration formula in the general case (22.38)–(22.41), remarks about percolation.

Chapter 23: Memory Effects
Observations of physical phenomena and conjectures about equations used as models, why the second principle is wrong, why an experiment of spectroscopy is related to effective equations with nonlocal effects, a too-general question (23.1), my simplified model (23.2)–(23.4); 23.1: the Laplace transform of the kernel (23.5)–(23.7); 23.2: solving (23.5) by convolutions (23.8)–(23.15), looking in the correct family of equations in the linear case, my first approach using only convolutions (23.16)–(23.21); 23.3: my solution using Pick functions (23.22)–(23.25); 23.4: characterizing the Radon measure defining the kernel (23.26)–(23.31), a possible origin of irreversibility, the case where the kernel is a finite combination of exponentials (23.32)–(23.35), a different way to write the effective equation (23.36) and (23.37).

Chapter 24: Other Nonlocal Effects
Time-dependent coefficients and a nonlinear equation (24.1)–(24.4), an approach by perturbation (24.5)–(24.9); 24.1: the expansion of the kernel

(24.10)–(24.31), a degenerate elliptic problem (24.32)–(24.34); 24.2: its effective equation (24.35)–(24.44), a different way to write the effective equation (24.45) and (24.46), a study by Amirat, Hamdache, and Ziani (24.47) and (24.48); 24.3: their effective equation (24.49)–(24.56), their different way to write the effective equation (24.57) and (24.58), models for a porous medium, a nonlinear model (24.59)–(24.62), its perturbation expansion (24.63)–(24.77), analogy with Feynman diagrams, a truncated expansion (24.78)–(24.80).

Chapter 25: The Hashin–Shtrikman Construction
25.1: Equivalent media (25.1) and (25.2); 25.2: the Hashin–Shtrikman coated spheres (25.3)–(25.8); 25.3: using a Vitali covering (25.9)–(25.17); 25.4: coated spheres give optimal bounds (25.18)–(25.21); 25.5: $g''(1)$ for binary mixtures (25.22)–(25.24), a remark of Bergman on cubic symmetry; 25.6: optimal values of Pick functions for Taylor expansion at order 2 (25.25)–(25.31), giving the Hashin–Shtrikman bounds in "dimension" d (25.32) and (25.33); 25.7: bounds for a Pick function g when $z\,g\left(\frac{1}{z}\right)$ is a Pick function (25.34)–(25.39), a remark of Milton for ternary mixtures; 25.8: a Riccati equation for general coated spheres (25.40)–(25.44), properties of Riccati equations (25.45)–(25.48); 25.9: generalizing the remark of Milton to arbitrary proportions (25.49)–(25.58); 25.10: the optimal radial construction (25.59)–(25.63).

Chapter 26: Confocal Ellipsoids and Spheres
Confocal ellipsoids (26.1); 26.1: derivatives of an implicit function (26.2)–(26.8); 26.2: particular solutions for isotropic materials in the confocal ellipsoids geometry (26.9)–(26.15); 26.3: a Riccati equation for general confocal ellipsoids (26.16)–(26.20); 26.4: its explicit solution for a binary mixture in the coated ellipsoid case (26.21)–(26.25), why I use old methods of explicit solutions, the difficulty of learning some fields of mathematics by lack of scientific behavior of the specialists, some of the useless fashions that I witnessed, the defect of not mentioning the names of those who had the ideas and of advertising things which are wrong; 26.5: the Schulgasser construction for the radial case (26.26)–(26.30); 26.6: extension by Francfort and myself to the confocal ellipsoids case (26.31)–(26.37); 26.7: solving (26.26); 26.8: solving (26.31), (26.38)–(26.40), can the Schulgasser construction improve 25.10? (26.41); 26.9: it does not (26.42)–(26.51), a two-dimensional case of Gutiérrez, Murat, Weiske and myself (26.52)–(26.62); 26.10: a corresponding Riccati equation (26.63) and (26.64), discussion of the result (26.65)–(26.71), a result of Francfort and myself about the natural character of confocal ellipsoids (26.72)–(26.84).

Chapter 27: Laminations Again, and Again
27.1: A formula for laminated materials (27.1); 27.2: a result of Braidy and Pouilloux (27.2)–(27.5), disadvantage of being shown a line of proof, my writing the general formula for laminations as a differential equation (27.6)–(27.13); 27.3: my formula for repeating laminations (27.14)–(27.16); 27.4: identifying a term in (27.14) and (27.17), my use of relaxation techniques for ordinary differential equations (27.18)–(27.20); 27.5: my direct method

(27.21)–(27.28); 27.6: the inverse after adding a rank one matrix (27.29); 27.7: my formula for laminating m materials in one direction (27.30)–(27.36).

Chapter 28: Wave Front Sets, H-Measures
Singular support of L. Schwartz, wave front set of Hörmander (28.1)–(28.6), propagation of microlocal regularity is not propagation of singularities, oscillations, and concentration effects (studied in a microlocal way) are more important in continuum mechanics and physics than singularities, were H-measures known before I introduced them?, the intuition about H-measures (28.7)–(28.11), $\mathbb{S}^{N-1}$ is a simple way to talk about a quotient space; 28.1: operators M_b and P_a (28.13) and (28.14), using the Plancherel formula (28.15) and (28.16); 28.2: a first commutation lemma (28.17) and (28.18); 28.3: a homogeneous of degree 0; 28.4: $a(s\,\xi, s^2\tau) = a(\xi,\tau)$ for $s > 0$ (28.19)–(28.25), using results of Coifman, Rochberg, and Weiss; 28.5: existence of scalar H-measures (28.26)–(28.29), vectorial H-measures (28.30), scalar first-order equation (28.31) and (28.32); 28.7: the localization principle (28.33)–(28.38); 28.8: scalar first-order equation (28.39) and (28.40); 28.9: gradients (28.41)–(28.43); 28.10: wave equation (28.44)–(28.48); 28.11: compensated compactness with variable coefficients (28.49)–(28.53); 28.12: symbols (28.54) and (28.55), examples (28.56)–(28.58); 28.13: symbol of a product (28.59); 28.14: weak $\star$ limit of $S_1 U_k^m \overline{S_2 U_\ell^m}$ (28.60), periodically modulated sequences (28.61) and (28.62); 28.15: the H-measure it defines (28.63)–(28.66); 28.16: the H-measure for a concentration effect at a point (28.67)–(28.69); 28.18: the necessity of some convergences in $H^{-1}_{loc}(\Omega)$ strong (28.70)–(28.73).

Chapter 29: Small-Amplitude Homogenization
Two approximations from Landau and Lifshitz (29.1) and (29.2), my interpretation using small-amplitude homogenization (29.3)–(29.11); 29.1: the correction in γ^2 uses H-measures (29.12)–(29.26), the injectivity of a mapping, my model of Chap. 19 (29.27) and (29.28); 29.2: expressing $M^{\rm eff}$ with H-measures (29.29)–(29.38); 29.3: density in x of the projection of H-measures for sequences in L^p (29.39) and (29.40); 29.4: application to the Taylor expansion of $F(\cdot, M^1, M^2)$ on the diagonal (29.41) and (29.42); 29.5: H-measures associated to characteristic functions (29.43) and (29.44).

Chapter 30: H-Measures and Bounds on Effective Coefficients
Description of the general method (30.1)–(30.12) and (30.13)–(30.17); 30.1: notation $\langle\langle\mu, Q(x,\xi,U)\rangle\rangle$; 30.2: a lower bound (30.18)–(30.24); 30.3: a consequence (30.25)–(30.28), the case of binary mixtures (30.29) and (30.30); 30.4: an upper bound (30.31)–(30.43); 30.5: a consequence (30.44)–(30.47), the case of binary mixtures (30.48) and (30.49).

Chapter 31: H-Measures and Propagation Effects
How conserved quantities hide at mesoscopic level, an error of thermodynamics, how waves carry conserved quantities around; 31.1: a second commutation lemma (31.1)–(31.8); 31.2: the Poisson bracket (31.9); 31.3: the

second commutation lemma with standard operators (31.10), an improved regularity hypothesis uses a result of Calderón, differences between localization and propagation (31.11)–(31.14); 31.4: the scalar first-order hyperbolic case (31.15)–(31.27), how to generalize the result to more general systems (31.28)–(31.36); 31.5: the scalar wave equation (31.37)–(31.48), differences between propagation of H-measures and geometrical optics, a question of smoothness of the coefficients (31.49), the question of initial data, transformation of H-measures under local diffeomorphisms (31.50)–(31.53).

Chapter 32: Variants of H-Measures

My idea for introducing one characteristic length (32.1) and (32.2); 32.1: it gives H-measures independent of x_{N+1}, the idea of semi-classical measures of P. Gérard (32.3); 32.2: semi-classical measures for one-dimensional oscillations (32.4)–(32.6); 32.3: and its H-measures (32.7) and (32.8); 32.4: a commutation lemma (32.9)–(32.11); 32.5: two compactifications; 32.6: H-measure on the compactification (32.12)–(32.14), a mistake of P.-L. Lions and Paul, the Wigner transform (32.15)–(32.17), the idea of P.-L. Lions and Paul using the Wigner transform (32.18), my approach with P. Gérard using two-point correlations (32.19)–(32.33), more general equations for the localization principle (32.34); 32.7: the localization principle away from 0 (32.35)–(32.42); 32.8: its implication at ∞ (32.43) and (32.44); 32.9: another form of the localization principle at ∞ (32.45)–(32.50), a computation with P. Gérard on a sequence with two characteristic lengths (32.51)–(32.60), an intuitive explanation with beats, puzzling facts about spectroscopy, the approach of P. Gérard for deriving equations for the two-point correlation measures for the Schrödinger equation (32.61)–(32.68), for the heat equation (32.69)–(32.74), a computation with P. Gérard for k-point correlation measures for the heat equation (32.75)–(32.77), the case of variable coefficients (32.78)–(32.83), a computation of P. Gérard on how the Lorentz force appears from the Dirac equation with a large mass term, my research programme.

Chapter 33: Relations Between Young Measures and H-Measures

Why Young measures cannot see differential equations and cannot characterize microstructures; 33.1: laminating m materials in one direction at order γ^2 (33.1) and (33.2); 33.2: H-measures associated to characteristic functions (33.3)–(33.6), a model from micromagnetism, mixing r materials (33.7)–(33.10); 33.3: a first type of construction (33.11)–(33.18); 33.4: a second type of construction (33.19)–(33.39), an analogy with matrices of inertia (33.40)–(33.46); 33.5: admissible decompositions (33.47) and (33.48); 33.6: a third type of construction (33.49)–(33.54); 33.7: H-measures constructed by lamination (33.55)–(33.63); 33.8: a generalization (33.64); 33.9: a first type of construction (33.65); 33.10: a second type of construction (33.66)–(33.70); 33.11: a third type of construction (33.71)–(33.78); 33.12: sequences corresponding to a given Young measure and satisfying some particular differential system.

Chapter 34: Conclusion

My early difficulties about reading and writing, splitting some chapters into two parts, remarks on homogenization in optimal design, adapted microstructures for heat conduction and elasticity, remarks about three-point correlations, the difficulty of discovering useful generalizations, why periodicity assumptions are not so useful, when does the frequency of light play a role, the geometrical theory of diffraction (GTD) of Keller, about Bloch waves and the Bragg law for X-ray diffraction, about concentration effects, beyond partial differential equations and GTH.

35: Biographical Information

Basic biographical information for people whose name is associated with something mentioned in the book.

36: Abbreviations and Mathematical Notation

References

Index

Contents

Chapter 1
Why Do I Write?

I often quote the parable of talents from the gospels.[1] Parables are like general theorems, and they can be transmitted by people who do not necessarily understand all the various applications of the teaching: if after stating a general theorem one gives an example, the weak students only understand the example while the bright students foresee that the theorem applies to many situations. The gospels repeatedly show that the disciples of Jesus of Nazareth did not understand what the parables were about,[2] as they often asked for examples. The parables of talents which appear in Matthew 25:14–30 and Luke 19:12–27 differ, but the scenario is that a master left for a long trip and gave various amounts, five talents, two talents, one talent, to three of his servants, and when he came back he asked them to report about what they did. The servant who received five talents made them fructify and earned five more, the servant who received two talents earned two more, and they

---

[1] The four evangelists, Matthew, Mark, Luke, and John, are not very well known. Matthew supposedly was a tax collector, chosen by Jesus to be one of his 12 disciples. Mark supposedly was a hellenist, converted by Peter. Luke supposedly was a physician, converted by Paul. John supposedly was a disciple of John the Baptist, who became one of the 12 disciples of Jesus.

[2] Jesus of Nazareth is believed by Christians to be the (unique) son of God, and the messiah whom Jews were waiting for, hence its title Christ, which probably has that meaning in Greek. Of course, I consider that he was only human, and I often refer to him as the Teacher. According to the gospels, he practised meditation past the point where one can do miracles, but without using that power for a personal advantage. He was executed by the Romans, probably because some of his followers believed him to be the messiah whom Jews were waiting for, and whom they expected to put an end to the Roman occupation. Oriental religions mention that after death the bodies of people who are extremely advanced on the spiritual path may shrink, and even dissolve completely, sometimes leaving hair and nails, an effect called "rainbow body"; could it be the reason why the body of Jesus could not be found?

L. Tartar, *The General Theory of Homogenization*, Lecture Notes of the Unione Matematica Italiana 7, DOI 10.1007/978-3-642-05195-1_1,

were both praised, with the same words.[3] The servant who received one talent said that he was afraid to lose it, and he buried it into the ground, so that he only gave back the initial amount, and he was punished. The versions in the gospels were probably distorted from an original teaching,[4] which I believe is reported in an apocryphal gospel, which has a fourth servant who also received one talent, and this servant tried to make it fructify but he lost it; however, in the end this servant was not punished, and again it was the servant who did not try to use his talent who was punished. Obviously, either the disciples of Jesus or those to whom they told the story could not understand why the servant who lost his talent was not punished, so they took him out of the parable, probably because they thought that the parable was about money, but that interpretation using money is a dull one, and cannot be the meaning intended by the Teacher, of course!

Although the talent in the parable was a unit of money (probably like a pound of silver), I interpret it as a gift for something useful, like mathematics, and my interpretation is that we are not the creators of our brains, and anyone who received a very efficient brain is bound to be successful and he/she should not be proud about that, but anyone who misuses his/her talent should be castigated, and that applies to the very bright mathematicians who do not attack difficult problems and settle for more elementary ones (for them), in order to be praised for solving many of these easy problems, instead of trying

---

[3] My father pointed out to me that the praises are identical, so that the servant who was given five talents and earned five more was not considered more worthy than the servant who was given two talents and earned two more!

[4] I once told my father, who was a Protestant minister, that I did not think that Jesus existed, and that it did not matter because only his teachings are important, but he disagreed, because he believed in resurrection. Many years after, in reading magazines published by BAS (Biblical Archaeology Society), Washington, DC, either *BAR* (*Biblical Archaeology Review*) or *BR* (*Bible Review*), I learned that the first version of the gospel of Mark, which is the earliest of the four gospels, ended after the women found the tomb empty, and a sequel was written a few centuries after (obviously for making it conform to the other gospels, which were written afterwards, and talked about resurrection), and I checked that my father knew about that. For me it is the sign that the gospel of Mark was written before the dogma of resurrection was invented, and propagated by Paul, who I think was the real inventor of Christianity. However, I finally thought that Jesus existed, arguing that the reporting of parables by the evangelists show a superior Teacher who was trying to transmit a deep message to uneducated students, and I see the fact that the evangelists transmitted us the information that the disciples of Jesus did not understand his teachings as a sign that they did not invent the whole story. Actually, if some teachings must be transmitted orally, and without too much distortion, by people who do not really understand what the teachings are about, then one is bound to invent using parables for transmitting the teachings. Of course, distortions occurred later, and I imagine that the original version of the gospel of Mark contained what the Teacher taught and how he died, without trying to interpret how his body could dissolve.

to open new doors. There should be no shame for failing to open a closed door behind which something very interesting is supposed to be found, but one should be able to show that one made efforts in reasonable directions; one should also give advice to those who also plan some attempts, explaining what was tried before, and possibly why it did not work.

I tried to follow these general principles, and as *I was lucky to study in Paris in the mid 1960s, in a special scientific environment* that is almost impossible to reproduce nowadays, I feel the need to *explain what I was taught and what knowledge I added by my own research work*, not so much because it is my own but because it should help the young researchers for avoiding the long and useless meanderings that many others are still following.[5] Also, *I witnessed the behavior of a few famous mathematicians*, and I met many in person, which was the initial reason why I wanted to share some biographical information about them, but then I tried to find biographical information concerning those whom I did not meet, usually for the obvious reason that they lived in a different period. I obtained much of my information by searching the Internet, but not everything comes from such a reliable source as MacTutor, http://www-history.mcs.st-andrews.ac.uk/, the web site from University of St Andrews, St Andrews, Scotland, which is dedicated to history of mathematics, and some of my information coming from other sources could be slightly wrong, and I expect every interested reader to tell me about my mistakes. I am not interested in the actual citizenship of the people whom I mention, but sometimes they were born in a different country than the one where they worked, and my point is to show that exchanges between countries and continents play a role in the creation and dissemination of knowledge. My hope is that this biographical information will help give a more global picture about how *science progresses by the work of many, coming from different times, different places, and different cultures.*[6]

---

[5] There could be psychological reasons why many continue on a path which was already shown to be wrong, but a few have political reasons to mislead these researchers whom I am trying to educate.

[6] With the help of MacTutor (http://www-history.mcs.st-andrews.ac.uk/), one can look at mathematicians from the past (including some astronomers, and some philosophers), and one finds that 75% of the (86) names of people born before 500 are Greek (and 12% are Chinese, and 8% are Indian), that 70% of the (7) names of people born between 500 and 750 are Indian (and 15% are Chinese, and 15% are European), that 80% of the (36) names of people born between 750 and 1,000 are Arabic (and 20% are Indian), that 44% of the (32) names of people born between 1,000 and 1,250 are European (and 25% are Arabic, 19% are Chinese, and 12% are Indian), and that 69% of the (46) names of people born between 1,250 and 1,500 are European (and 15% are Arabic, and 12% are Indian). It is an interesting fact that there are no Greek names after 500, and that from 1,400 to 1,500 almost all names are European. One may deduce that the development of mathematics (or more generally of all sciences) was not independent of economical, political, and religious factors in the past. Therefore, one should stay alert about counteracting the bad tendencies which are observed nowadays.

I consider mathematics as a part of a big puzzle, certainly quite an important piece of science, and I learned about the interplay of various scientific fields of research, a little more than most mathematicians, and this did not just happen by chance.

In the mid 1960s, I succeeded at exams which gave me the possibility to study either at *École Normale Supérieure* or at *École Polytechnique*, Paris, France. I wanted to do something useful, and no one told me that *mathematics can be useful for something else than teaching mathematics*, but I thought that engineers were doing useful things, and this idea led me to choose to study at École Polytechnique, which is not actually an engineering school, as I only understood much later. I did not know what the work of an engineer is, and no one in my family knew about that either, so I took my decision alone; after 1 year, Laurent SCHWARTZ gave an evening talk,[7] on the *role and duties of scientists*, and he mentioned that engineers do a lot of administration, and that made me understand that I needed to change my orientation, and I decided to do research in mathematics, possibly with an applied twist, in agreement with my original choice. After studying at École Polytechnique, it became clear that this choice gave me an enormous advantage on the majority of mathematicians, because of what I studied outside mathematics.

During the first year I learned about classical mechanics, which is an eighteenth century point of view of mechanics, based on ordinary differential equations; during the second year I learned about continuum mechanics, which is a nineteenth century point of view of mechanics, based on partial differential equations; I did hear a little about a twentieth century point of view of mechanics, which included questions about turbulence and plasticity, the latter being the research topic of the teacher, Jean MANDEL,[8] but a few years after I discovered that the mathematical tools for that point of view did not exist yet, and my research work (after my thesis) transformed into developing a new mathematical approach for that.

Studying analysis with Laurent SCHWARTZ [86], and numerical analysis with Jacques-Louis LIONS, who became my thesis advisor, was the best preparation for hearing about all the mathematical tools in partial differential equations which were used for understanding continuum mechanics and

---

[7] Laurent SCHWARTZ, French mathematician, 1915–2002. He received the Fields Medal in 1950 for his work in functional analysis. He worked in Nancy, in Paris, at École Polytechnique, which was first in Paris (when he was my teacher in 1965–1966 [86]), and then in Palaiseau, and at Université Paris 7 (Denis Diderot), Paris, France.

[8] Jean MANDEL, French mathematician, 1907–1982. He worked in Saint-Étienne and in Paris, France. He was my teacher for the course of continuum mechanics at École Polytechnique in 1966–1967 in Paris [58].

physics [52] (although neither of them was really interested in mechanics or physics), before the introduction of the ideas that I started developing in the mid 1970s.

I also learned about classical physics, special relativity, quantum mechanics and statistical physics, but with teachers who often gave the impression that they did not know how to disentangle mathematics and physics, and I thought later that it could be the result of an infamous classification by COMTE,[9] a French philosopher, who studied at École Polytechnique for 1 year, and obviously valued abstraction so much that he put mathematics above all other sciences,[10] before astronomy,[11] physics, chemistry, biology, in this order. One needs different abilities for becoming a good mathematician, a good physicist, a good chemist, or a good biologist, and it is not wise to disparage others because they possess an ability that one has not, so I find quite silly, if not completely ridiculous, to imagine a linear order between various fields, inside or outside science, whatever its definition is.[12] Nowadays, there are many people who lack the abilities for mathematics, like the sense of abstraction for example, and they would choose another field more suited to their interests and abilities, were it not for this unnatural attraction created by the Comte classification, or other silly reasons.

My teacher in probability was not good, and as the teacher of statistical physics gave me a bad impression too, I was bound to distrust any probabilistic model for linking different phenomena, and I was glad to discover in the early 1970s that I could avoid probabilities altogether for relating what happens at different scales, and use various types of weak convergence instead; finding this was not only due to some joint work that I did with François MURAT [93], generalizing some earlier work of Sergio SPAGNOLO [89, 90],[13] helped with the insight of Ennio DE GIORGI [22],[14] but also to some particular applications of Évariste SANCHEZ-PALENCIA [81, 82],[15] which helped

---

[9] Auguste COMTE, French philosopher, 1798–1857. He worked in Paris, France.

[10] Mathematics is one of the sciences, and the sentence "mathematics and science" was probably coined by experts in sabotage.

[11] This explains why nowadays, many of those who chose to study physics because they thought that they were not good enough for studying mathematics end up in astrophysics.

[12] It is precisely that mistake which makes weaker people in one group believe that they are worth much more than stronger people in another group, a disease which grew too much in our times, and which is called *racism*!

[13] Sergio SPAGNOLO, Italian mathematician, born in 1941. He works at Università degli Studi di Pisa, Pisa, Italy.

[14] Ennio DE GIORGI, Italian mathematician, 1928–1996. He received the Wolf Prize in 1990, for his innovating ideas and fundamental achievements in partial differential equations and calculus of variations, jointly with Ilya PIATETSKI-SHAPIRO. He worked at Scuola Normale Superiore, Pisa, Italy.

[15] Enrique Évariste SANCHEZ-PALENCIA, Spanish-born mathematician, born in 1941. He works at CNRS (Centre National de la Recherche Scientifique) and UPMC

me understand this new point of view, and this could only happen because I was interested in understanding continuum mechanics and physics, of course! Peter LAX later observed that the idea that some numerical schemes only converge in a weak topology was used before,[16] by VON NEUMANN,[17] but it does not seem that VON NEUMANN thought of changing the way one looks at physics, in the manner that I developed.[18]

Understanding better a subject is both an intellectual advantage and a social disadvantage, because one quickly finds oneself isolated, among a majority who prefers to continue being wrong and lying about it. In 1984, Jean LERAY told me about suffering because one understands more than others,[19] and later I found in a book by Clifford TRUESDELL,[20] which he offered me, a quote of PLANCK,[21] who also described this difficulty: "A new scientific

(Université Pierre et Marie Curie), Paris, France. I knew him under the French form of his first name, Henri, but he now uses his second name, Évariste.

[16] Peter David LAX, Hungarian-born mathematician, born in 1926. He received the Wolf Prize in 1987, for his outstanding contributions to many areas of analysis and applied mathematics, jointly with Kiyoshi ITO. He received the Abel Prize in 2005 for his groundbreaking contributions to the theory and application of partial differential equations and to the computation of their solutions. He works at NYU (New York University), New York, NY.

[17] János (John) VON NEUMANN, Hungarian-born mathematician, 1903–1957. He worked in Berlin, in Hamburg, Germany, and at IAS (Institute for Advanced Study), Princeton, NJ.

[18] I read that VON NEUMANN wrote in a letter in 1935 that he did not believe anymore in the mathematical framework that he devised for quantum mechanics. As he did not make this point known to all, he did not think of changing the way how one looks at physics, and he bears the responsibility that the silly rules of quantum mechanics transformed into dogmas!

[19] Jean LERAY, French mathematician, 1906–1998. He received the Wolf Prize in 1979, for pioneering work on the development and application of topological methods to the study of differential equations, jointly with André WEIL. He worked in Nancy, France, in a prisoner of war camp in Austria (1940–1945), and in Paris, France; he held a chair (théorie des équations différentielles et fonctionnelles, 1947–1978) at Collège de France, Paris.

[20] Clifford Ambrose TRUESDELL III, American mathematician, 1919–2000. He worked at Indiana University, Bloomington, IN, and at Johns Hopkins University, Baltimore, MD.

[21] Max Karl Ernst Ludwig PLANCK, German physicist, 1858–1947. He received the Nobel Prize in Physics in 1918, in recognition of the services he rendered to the advancement of physics by his discovery of energy quanta. He worked in Kiel and in Berlin, Germany. There is a Max Planck Society for the Advancement of the Sciences, which promotes research in many institutes, mostly in Germany (I spent my sabbatical year 1997–1998 at the Max Planck Institute for Mathematics in the Sciences in Leipzig, Germany).

truth does not triumph by convincing its opponents and making them see the light, but rather because its opponents finally die, and a new generation grows up that is familiar with it." For many years I wondered why so many mathematicians pretended to work on problems of mechanics, and said things known to be false by anyone who studied a little; I was not very good at communicating, and I stayed silent, but I felt that many around me showed a mixture of incompetence and intellectually dishonest behavior.

Why pretend that the world is described by ordinary differential equations, as if one did not study partial differential equations? For example, why use the term mechanics for designating classical mechanics, which is an eighteenth century point of view based on ordinary differential equations, and not continuum mechanics, which is a nineteenth century point of view based on partial differential equations, or ignore the twentieth century point of view that goes *beyond partial differential equations*, as I explained during the last 30 years?

Why be interested in studying the asymptotic behavior of equations without saying that so many known effects were neglected in the models used that their time of validity is known to be quite limited?

Why pretend that physical systems minimize their potential energy, as if one did not know the first principle of thermodynamics, that energy is conserved (when one counts all its various forms)? Why ignore the second principle of thermodynamics, despite its defects? Why not say that thermodynamics is not about dynamics but about equilibria, and that equations of state derived from equilibrium might well create havoc if one pretends that they are valid all the time? Why not discuss the defects of the Boltzmann equation,[22] and observe that it was obtained by postulating an irreversible behavior, and thus cannot help one understand how irreversibility occurs?

Why not observe that the rules of quantum mechanics could only be invented by people unaware of partial differential equations, and unable to distinguish between the point of view of NEWTON,[23] where there are forces acting at a distance, and the point of view of H. POINCARÉ,[24] which EINSTEIN did not seem to understand,[25] where there are none? Why not

[22] Ludwig BOLTZMANN, Austrian physicist, 1844–1906. He worked in Graz and Vienna, Austria, in Leipzig, Germany, and then again in Vienna.

[23] Sir Isaac NEWTON, English mathematician, 1643–1727. He worked in Cambridge, England, holding the Lucasian chair (1669–1701). There is an Isaac Newton Institute for Mathematical Sciences in Cambridge, England.

[24] Jules Henri POINCARÉ, French mathematician, 1854–1912. He worked in Paris, France. There is an Institut Henri Poincaré (IHP), dedicated to mathematics and theoretical physics, part of UPMC (Université Pierre et Marie Curie), Paris.

[25] Albert EINSTEIN, German-born physicist, 1879–1955. He received the Nobel Prize in Physics in 1921, for his services to theoretical physics, and especially for his discovery of the law of the photoelectric effect. He worked in Bern, in Zürich, Switzerland, in Prague, now capital of the Czech Republic, at ETH (Eidgenössische Technische Hochschule), Zürich, Switzerland, in Berlin, Germany, and at IAS (Institute for

observe that those who pretend to see things traveling faster than the speed of light c play with equations whose best derivation is to let c tend to ∞ in a more realistic physical description? Why not observe that the difficulties between waves and particles disappear when one understands that there are only waves satisfying partial differential equations, and that as there are no particles the question of understanding where they are is meaningless?

Is it not becoming obvious that one needs to go *beyond partial differential equations*, which I explained for about 30 years, so why are there so many people who keep thinking in terms of ordinary differential equations?

I was only aware of a few of these questions in the mid 1970s, when I developed my new approach to continuum mechanics which mixed ideas from homogenization and from *compensated compactness*, first described in my Peccot lectures at the beginning of 1977,[26] at Collège de France, in Paris. For what concerned questions of physics, the situation was more delicate, because I could not believe the classical presentations, usually obscured by an excessive amount of probabilities, so often used for masking the fact that one does not know much yet about the phenomena that one pretends to study. In 1977, I understood why the second principle of thermodynamics needed improvement, by embedding the question into a more general homogenization problem, and in 1980, I understood that the appearance of nonlocal effects by homogenization is probably behind the strange rules of spontaneous absorption and emission which physicists invented, and the key for understanding turbulence, but in the summer of 1982 I still did not see how to extend my ideas to quantum mechanics and statistical mechanics, and it was due to the help of Robert DAUTRAY,[27] that I could improve my understanding of physics. On one hand, he offered me a position at CEA (Commissariat à l'Énergie Atomique), so that I could leave Université Paris Sud, Orsay, France; on the other hand, I benefited from his advice about what to read, and this helped me understand how a part of physics could be described in the same spirit as my previous research programme, and this is how I understood about *H-measures* and their variants [105]. I am very grateful to Robert DAUTRAY for that, because *physics is a very difficult subject for a mathematically oriented mind*, as physicists' statements usually lack precision, and by following the advice of a very competent person, one learns that there

Advanced Study), Princeton, NJ. The Max Planck Institute for Gravitational Physics in Potsdam, Germany, is named after him, the Albert Einstein Institute.

[26] Claude Antoine PECCOT, French child prodigy, 1856–1876.

[27] Ignace Robert DAUTRAY (KOUCHELEVITZ), French physicist, born in 1928. It is thanks to him that I worked at CEA (Commissariat à l'Énergie Atomique) from 1982 to 1987, and that my understanding of physics improved.

are a few important questions which need to be understood in a better way, and one gains an invaluable amount of time in not having to identify these questions by oneself. After that, one observes that most mathematicians who think that they understand physics are only playing one of the many games which physicists invented, without a good physicist telling them that they should not build too much on a game which is not good physics at all;[28] even some games with a long life, like quantum mechanics or statistical physics, survive mostly because they were transformed into dogmas, which makes them difficult to discard, but their defects are too obvious to be ignored.[29]

A few years ago, I heard a talk by a physicist which showed how difficult it is for a mathematician to assess the value of what physicists say; this one, who put a lot of humor in his presentation, chose a suggestive title, "before the big-bang," and at the end I asked him a question, mentioning that temperature is an equilibrium concept, and wondering if he thought that in the first few milliseconds just after the big-bang (which he believed in), matter was in equilibrium at temperatures of a few million (or billion) degrees, and he answered yes!

I read an article by POISSON from 1807,[30] where he pointed out that the speed of sound could not have been computed before by using the available data about compressibility of air, because the usual relation where the pressure is proportional to the density of mass gives an incorrect value for the speed of sound, and maybe NEWTON already knew this discrepancy; instead, POISSON used a law $p = c\,\varrho^\gamma$, which was proposed by LAPLACE,[31] probably for heuristic reasons. One explains now that *the propagation of a wave is too fast a phenomenon for heat to flow* so that the process is adiabatic (isentropic). This was a source of error for a few mathematicians, starting with RIEMANN,[32] who worked too much with the equations of isentropic gases, as if adiabatic changes were the rule, but it seems to me that some physicists are as deluded as some mathematicians if they believe that matter reaches instantaneously its equilibrium at a temperature of million

---

[28] There is a parable about that, which talks about building a house on the sand.

[29] I first learned about religions because my father was a Protestant minister, but after rejecting the idea of God for intuitive reasons when I was 12 or 13, I became interested in religions in order to make up my mind about GOD (see fn. 34, p. 10). It was only much later, after fighting against vote-rigging in Orsay, and observing the powerful allies of my political opponents and their methods of destruction, that I realized how much one can learn from the mistakes of the past concerning religions, as it helps understand how some of the actual chaos in science was generated.

[30] Siméon Denis POISSON, French mathematician, 1781–1840. He worked in Paris, France.

[31] Pierre-Simon LAPLACE, French mathematician, 1749–1827. He was made count in 1806 by Napoléon I, and marquis in 1817 by Louis XVIII. He worked in Paris, France.

[32] Georg Friedrich Bernhard RIEMANN, German mathematician, 1826–1866. He worked at Georg-August-Universität, Göttingen, Germany.

degrees, whatever this means.[33] One should remember that physicists misled the "scientific" community and the funding agencies for almost 50 years by claiming that they were going to control fusion, but now they estimate that they might succeed in the second part of the twenty first century; of course, they are quite careful not to say explicitly that one important reason is that *one must still discover the properties of matter at temperatures of a few million degrees*!

It was not so difficult for me to discover what is wrong with a few laws believed by physicists, and I think that being educated as a Calvinist and losing my faith in God by the age of 13 helped form my character in a useful way for science,[34] in that I cannot lie and I cannot accept any dogma without criticizing it, and will preferably tear it to pieces and wonder why some people believe it. However, becoming a mathematician implies that one must know the hypotheses and postulates that one makes: having postulated that God does not exist, I needed to check that particular dogma of mine.

From a mathematical point of view, it is impossible to decide if the world was created or has existed forever: in his course at École Polytechnique, in 1965–1966, Laurent SCHWARTZ pointed out that one cannot decide if the universe that we live in is an orientable manifold or not,[35] as orientability is a global property and our information on the universe is local, and the same argument shows that *one cannot decide if the universe was created or not.* Many western pseudo-scientists were so brainwashed by the creation theory in the Bible, that they think that they must reject it by adhering to the big-bang theory, without realizing that both the creationist approach and the big-bang theory are flawed, and as one cannot decide if the universe we live in was created or not, why not imagine that there could be quite a number of universes, some having always existed and some being created in a finite past, where the same particular event occurs, like that of a French mathematician preparing his fourth book, on homogenization.[36]

From a mathematical point of view, it is impossible to decide if one or many gods exist without giving mathematical definitions of divine beings and proving their properties, but again western pseudo-scientists were so brainwashed by the Bible, that they think that they must oppose those who believe that God exists and that the Bible redactors were inspired by God,

---

33 I was told recently that this physicist does not believe that matter was in equilibrium, so that either he did not understand my question, or he answered it in the spirit of his talk, as a joke.

34 I use God to refer to the deity venerated by Jews, Christians, and Moslems, whom I believe to be just a literary character created in the seventh century BCE. I use GOD as a conjecture for a notion too transcendent to be perceived by ordinary beings, like the one Ramakrishna seemed to refer to, with a name that I do not recall [57].

35 Obviously, Laurent SCHWARTZ postulated that the universe is a manifold!

36 There is no obvious reason why the book should be finished, or that the finished books should be the same in all the realizations of that event.

by advocating flawed theories like that of the big-bang. Archaeology gave us so much information on the past that the anachronisms in the Bible suggest a redaction around the seventh century BCE,[37] probably by gluing the oral traditions of various remnants of tribes together, but even if one becomes convinced that Abraham never existed except as a literary character used for the purpose of creating a unified theory, based on a real Abram who founded a tribe bearing his name, and was remembered for immolating his first son to his god, it would not say much about the existence or nonexistence of GOD! Indeed, despite the flaws in the Bible, which imply that the beliefs of Jews, Christians, and Moslems are quite questionable, GOD might well exist, obviously in a different manner than that described in the texts of the three Abrahamic religions, where a lot of human defects were projected onto God. Perhaps these observations should suggest that putting an order between religions is as silly as putting an order among the sciences.

What I find more important than arguing about questions on which we cannot gather much information, like if the universe was created or not, is to assess the mathematical laws that govern the universe, and some which are used now are certainly wrong, but physicists often refuse to listen to the hints that something is amiss, even if it comes from a Nobel laureate, like ALFVÉN,[38] who pointed out that a few observations in the cosmos must have an electromagnetic explanation instead of a gravitational one. Those who defend the dogma of gravitation prefer to invent dark matter, dark forces, and dark energy for continuing to pretend that gravitation is the main factor, although they are still unable to tell us what mass is!

In December 1984, I wrote a few letters to Laurent SCHWARTZ where I described the incidence of vote-rigging that I opposed in Orsay, and in one of these letters I wrote that I thought that gravitation is not an independent force. My analysis, which I did not mention in the letter, and which Laurent SCHWARTZ never enquired about, was that in the Dirac equation,[39] the mass term should not be introduced,[40] because I expected that such a term could appear by itself through a homogenization effect, similar to that studied in

---

[37] BCE = before common era, CE = common era.

[38] Hannes Olof Gösta ALFVÉN, Swedish-born physicist, 1908–1995. He received the Nobel Prize in Physics in 1970, for fundamental work and discoveries in magneto-hydrodynamics with fruitful applications in different parts of plasma physics, jointly with Louis NÉEL. He worked in Uppsala and Stockholm, Sweden, at UCSD (University of California San Diego), La Jolla, CA, and at USC (University of Southern California), Los Angeles, CA.

[39] Paul Adrien Maurice DIRAC, English physicist, 1902–1984. He received the Nobel Prize in Physics in 1933, jointly with Erwin SCHRÖDINGER, for the discovery of new productive forms of atomic theory. He worked in Cambridge, England, holding the Lucasian chair (1932–1969).

[40] It is not scientific to change a term in an equation in order that it fits with something that one observed.

elliptic situations by Doina CIORANESCU and François MURAT,[41] and *mass would then be explained as pure electromagnetic energy stored inside "particles,"* which would not necessarily be "electrons" or "positrons"; of course, the matter field $\psi \in \mathbb{C}^4$ in the Dirac equation is coupled with the Maxwell equation,[42] which I proposed to call the Maxwell–Heaviside equation,[43] where ϱ and j are sesqui-linear in ψ,[44] and the scalar potential V and vector potential A appear in the equation for ψ, and the Planck constant h appears for coupling the matter field ψ with V and A, so that h always appears when one studies interaction between light and matter. Shortly after, I learned of a proposal by BOSTICK of a toroidal structure of the electron [9],[45] published in January 1985, also with the mass being the stored electromagnetic energy, but he completed the Maxwell equation by using the de Broglie wavelength for the electron.[46]

It seems, indeed, that there are mathematical laws governing the universe, but they should use no probabilities, as this idea comes from a mistake in logic often made by physicists, which I call pseudo-logic: if a game A creates a result B and one observes something like B, physicists too often believe that it proves that nature plays game A; on the contrary, students in mathematics fail their exams if they think that A implies B is the same as B implies A! As will be seen in this course, homogenization in hyperbolic situations may lead to nonlocal effects appearing in the effective equations,[47] and it is useful to observe that my proofs, as well as those of my students and their collaborators, use no probabilities. However, one may, a posteriori, invent probabilistic games whose outputs are the kind of effective equations with nonlocal effects which were obtained, but it is a mistake in logic, very similar

---

[41] Doina POP-CIORANESCU, Romanian-born mathematician. She works at CNRS (Centre National de la Recherche Scientifique) in LJLL (Laboratoire Jacques-Louis Lions) at UPMC (Université Pierre et Marie Curie), Paris, France.

[42] James CLERK MAXWELL, Scottish physicist, 1831–1879. He worked in Aberdeen, Scotland, in London and in Cambridge, England, where he held the first Cavendish professorship of physics (1871–1879).

[43] Oliver HEAVISIDE, English engineer, 1850–1925. He worked as a telegrapher, in Denmark, in Newcastle upon Tyne, England, and then did research on his own, living in the South of England. We owe to him the simplified version of the *Maxwell equation* using vector calculus, as he replaced by a set of 4 equations in 2 variables what MAXWELL had written as a set of 20 equations in 20 variables, so that I prefer to call it the *Maxwell–Heaviside equation.*

[44] Sesqui is a prefix meaning one and a half, and sesqui-linear is the complex analogue of bilinear: one has linearity in one variable, but anti-linearity in the other variable, which is counted for half in respect to linearity.

[45] Winston Harper BOSTICK, American physicist, 1916–1991. He worked at Stevens Institute of Technology, Hoboken, NJ.

[46] Prince Louis Victor Pierre Raymond DE BROGLIE, 7th duke, French physicist, 1892–1987. He received the Nobel Prize in Physics in 1929, for his discovery of the wave nature of electrons. He worked in Paris, France.

[47] Only simple cases were understood, and a general theory is still missing.

to pseudo-logic, to deduce that there are probabilities creating these laws. If a proof of a theorem uses a mathematical theory $\mathcal{C}$, it is quite silly to believe that all possible proofs of that theorem must use theory $\mathcal{C}$; actually, a classical activity of mathematicians is precisely to look for simple proofs, or just other proofs, often because some of them may lead to generalizations which the first proof does not provide. If a first guess of physicists used a probabilistic argument, it only has a chronological importance, and there is no reason to prefer probabilistic arguments because they worked once.

Actually, one important reason why my derivation of an effective equation containing a nonlocal effect is better than deriving the equation from a probabilistic game is that one does not need new laws, and my method of proof shows that the extended law is but a consequence of the old one. Of course, the probabilistic games have the defect that one must know what the result is for discovering a probabilistic game that creates the observed effect. It does not look very scientific to me; actually, I was taught in high school that one should not put in the hypotheses what one wants to find in the conclusion!

I once read that an anthropologist saw the witch-doctor of the tribe that he was observing prepare a strange mixture and let it ferment for some time before using it for curing a particular disease.[48] It was amazing that there were some good results, because our western medical cure used antibiotics, so the anthropologist sent the potion of the witch-doctor to be analyzed, and indeed something developed, akin to penicillin. Among the many ingredients used by the witch-doctor, only a few were useful for the right fermentation to take place, but not knowing which, he needed to prepare his potion in exactly the same conditions that worked once. Obviously, modern pseudo-scientists also use this technique of repeating arguments that worked, despite the fact that they do not make any sense, so is it so different from witchcraft?

Seeing then the chaotic situation which resulted from the search for the mathematical laws followed by nature, I think that it is the role of mathematicians interested in other sciences to create a little order, and for doing that I found that the mathematical theory of homogenization, as I developed it with François MURAT,[49] contains many ideas which should help putting a lot of things on a sounder basis.

A few people pretended that some of my results were already known, and it could be possible, but more likely those who made these claims did not understand what mathematics is about, and if they knew that there cannot be a mathematical result before there is a definition of what one is looking for, they would probably say that some results were conjectured, instead of proven. It is not my method of work to read much, and I did not try to read what physicists, engineers, or "applied mathematicians" wrote in order

[48] It is not clear to me if the story that I read is true, and it was possibly invented as a kind of parable, so it is what the story teaches which is important.

[49] One should be aware of some wrong uses of the term homogenization by my opponents, who specialized in fake mechanics and fake physics.

to discover who were the first to make the correct conjectures, and if their sketches of "proofs" made any sense in view of the mathematical definitions that were introduced later, by Sergio SPAGNOLO in the late 1960s [89, 90], or by François MURAT and myself in the early 1970s [71, 93]. If my subtitle for these lecture notes is *A Personalized Introduction*, it is because I mostly present what I did, alone or with some collaborators, for the development of the subject, and besides the mathematical results, I want to explain the importance for a better development of continuum mechanics and physics, by identifying the questions of homogenization which were not dealt with correctly before.

Those who will disagree with my understanding of physics may be right, but they should observe that this is a mathematical theory, with results which are proven, and if they feel that this mathematical theory is not the right one for physics they should describe in precise terms why, so that mathematicians interested in other sciences, like myself, can think about their proposal, but if their argument is that it is not the way physicists think, then it has no scientific value, because a million physicists can be wrong, if they follow a wrong dogma![50]

My strong opposition to the incidence of vote-rigging in Orsay had the unfortunate effect that the friends of my political opponents showed a growing tendency to attribute my ideas to others; although I found it hurtful, I could easily imagine why some people would behave in this way because of their political orientation, but I was extremely puzzled, and hurt much more, by the same behavior from people who had not shown such a political orientation: either they hid it, or they had a different one but took advantage of the situation. In 1984, Jean LERAY told me about his own difficulties of that kind, more than 30 years earlier, and he said that it was a good sign that people stole my ideas, as it showed that I had some new ideas of my own, while one could not say the same about those who misbehaved by incorrectly attributing my ideas.

After a while I thought that a bigger problem was not so much that my ideas were attributed to others, but that those given credit for my ideas did not even understand them, which could be a reason why they distorted these ideas. However, could it be also that my ideas were distorted on purpose by others, and should I agree with the advice of KIPLING in his poem *If*?[51]

If you can bear to hear the truth you've spoken
Twisted by knaves to make a trap for fools, ...

---

[50] *Democracy has not much place in science*, and one does not vote for deciding if a result is true, as the result of a vote would only tell how many conjecture the result to be true, and how many conjecture the result to be false, and I expect that in most cases one would forget to count how many conjecture the result to be undecidable!

[51] Joseph Rudyard KIPLING, Indian-born British author, 1865–1936.

I consider that inducing students and researchers willingly in error is the worst sin of a teacher, and I wondered about the behavior of those who seemed to be doing exactly that, because my duty is to help the researchers who were led astray, by telling them the truth. Of course, my religious background tells me also to have compassion for those who did not receive a brain very adapted to scientific work,[52] and I realize how difficult their lives must be, fearing that more and more people would observe that they talked about topics that they did not understand; I suppose that their desire to steal ideas is for becoming highly considered by the dull crowd, according to the saying *au pays des aveugles les borgnes sont rois*,[53] as they lack the brain for being considered in the company of the bright few? At least I should try not to mention their names explicitly.[54] I should teach about what I understood of the interaction between mathematics, mechanics and physics,[55] for the benefit of everyone, including those who stole some results from me or others, and in some sense these people should seriously study the courses that I write, so as not to appear too deluded in front of the new generation of students, who will hopefully understand a little better what I am teaching.

In describing others' ideas I often add why I could not have a particular idea myself, either because it is not a good idea for what I tried to do, or because it is a good idea but I thought in a different way than the person

[52] Of course, I disagree with the actual trend of bowing to the propaganda of hiring not so competent people coming from various under-represented groups. It was always the tradition in academic systems (apart from those in countries under dictatorship!) to look for bright people, which is a quite rare commodity! I am in favor of free education for all, which seems the best way to avoid people being manipulated by politicians and religious leaders, and to find the children with bright minds who are born in various under-represented groups.

[53] In the land of the blind, the one-eyed man is king.

[54] However, I think that being a good researcher implies being a good detective, and every good student should then deduce after a while who are the ones who should not be credited for ideas that they use but are not theirs.

[55] Chemistry and biology should be added later on. It seems to me that there are difficult questions resembling homogenization in these disciplines, but for the moment there are still a few abstract concepts that need to be understood, and I see chemistry as a domain of possible applications once the general theory will be developed a little more. Although I never studied biology, I can discern that a few mathematicians who are pretending to work in biology are doing a bad job, which is not surprising in view of their failure to do a good job on questions of mechanics and physics before, where the mathematical framework is much clearer.

who found it. My religious upbringing forbids me to lie or to steal, and if I misattribute an idea it could be because I was not told whose idea it is, or because I read about a wrong attribution (although I do not read much), or because I discovered some results which were known before, a process which Ennio DE GIORGI once described as *Chi cerca trova, chi ricerca ritrova.*[56] In this case, I noticed that my memory first reminds me about how and why I discovered the result and only after comes the reminiscence that others proved the same result, possibly earlier, and my memory is not always as clear about their names. If I made mistakes in attributing anything, I would like to know; as one says in France, *Errare humanum est, sed perseverare diabolicum,*[57] so only repeated mistakes should be considered a real fault.

Actually, I think that all my ideas were very simple, and it is just because many bright specialists of partial differential equations were not interested in understanding continuum mechanics and physics, that I had these ideas before them. For example, my proof of existence of *H-measures* starts in a way quite reminiscent of some computations done by Lars HÖRMANDER [40],[58] but curiously many do not seem to see what is different between his point of view and mine, although the difference is great. I have no doubt that if Lars HÖRMANDER made the effort to imagine that there is something interesting in continuum mechanics or physics, and if he understood that I knew something about that, he could show interest in hearing about the mathematical questions to solve in this approach; certainly, he could be successful where I failed.[59]

[56] It is a play on words on what I knew in French as *Qui cherche trouve*, which comes from the gospels: "Ask and it will be given to you; seek and you will find; knock and the door will be opened to you" (Matthew 7:7, Luke 11:9).

[57] To err is human, but to continue erring is diabolical.

[58] Lars HÖRMANDER, Swedish mathematician, born in 1931. He received the Fields Medal in 1962 for his work on partial differential equations. He received the Wolf Prize in 1988, for fundamental work in modern analysis, in particular, the application of pseudo-differential and Fourier integral operators to linear partial differential equations, jointly with Friedrich HIRZEBRUCH. He worked in Stockholm, Sweden, at Stanford University, Stanford, CA, at IAS (Institute for Advanced Study), Princeton, NJ, and in Lund, Sweden.

[59] Unlike Jean LERAY, Lars HÖRMANDER understood nothing about human problems in 1984, and although he invited me to come to the Mittag-Leffler Institute, he pretended not to understand why I needed to be sure that I was not going to meet there anyone from the group of my political opponents from Orsay, who by their *racist* behavior led me to a nervous breakdown and to the verge of suicide.

I like to quote the example of the Caliph AL MA'MUN,[60] who created in his capital Baghdad a special institution, the *house of wisdom*, where AL KHWARIZMI worked,[61] with the goal of translating Greek philosophical and mathematical works into Arabic;[62] thanks to him, some Greek philosophical texts survived through their translation into Arabic, because at that time Europe was going through the dark ages, and the interest for learning either did not exist or did not include these old works by pagans.

Science is not adapted to any particular culture or limited by geographical boundaries, and *inside science the role of mathematics is crucial*, and not much quantitative analysis can be done without some form of mathematics, but mathematics has another advantage for those living in regions in need, that one can start practising it without waiting for the economic situation to change. *Political and economic factors are important for the development of knowledge*, and its transmission to future generations, and I think about those whose work is made difficult if not almost impossible because of some disastrous economic or political situation around them; I wish that in the difficult times that may be ahead, the importance of education will not be forgotten, and that those who have the political power in their hands will understand the need for everyone to live in decent conditions, and that includes clean air to breathe, pure water to drink, enough food for the body, and also a lot of knowledge for the spirit.[63]

---

[60] Abu al-'Abbas 'abd Allah AL MA'MUN ibn Harun, 7th Caliph of the Abbasid dynasty, 786–833. He ruled over the Moslem world from Baghdad, now capital of Iraq.

[61] Abu Ja'far Muhammad ibn Musa AL KHWARIZMI (or better KHAWARIZMI as I was told), "Iraqi" mathematician, 780–850. He worked in Baghdad, then the capital of the Moslem world, now capital of Iraq, but it is not known where he was from, and even the term "Arab" would be misleading because he certainly spoke Arabic, but he was probably not from Arabia. The word *algebra* was derived from the title of his treatise *Hisab al-jabr w'al-muqabala*, and the term *algorithm* was coined from his name.

[62] My father told me that the translations were made in two steps, with Christian scholars translating first from Greek to Syriac, and with Moslem scholars then translating from Syriac to Arabic.

[63] This is much more important than obtaining the right to vote, which only serves to being manipulated by politicians, unless everyone received education and was taught how the institutions function at every level, which is a bare minimum of what every citizen deserves to know.

I hope that the description of the ideas that I introduced, the explanations of the various reasons which led me to introduce them, and the discussions of the various questions that one should address in order to go further will help students everywhere. I wish all students a productive scientific career, hoping that a few of them will be eager to continue the work and transmit an improved knowledge to another generation of students. Having been raised in a religious environment, and finding myself gifted with an inquisitive brain, I apply the same spirit of research to all questions that I encounter, be it religion or science, or anything else, and it is my form of worship to try to use my knowledge of mathematics for understanding how the world around us functions at various levels, be it questions of continuum mechanics or physics (as my understanding of chemistry is a little weak and my understanding of biology nonexistent), or questions which are not a part of classical science. I hope that some of my readers will be interested in following me in that trend, and in order to learn about studying some questions for a while even though one may not see how they could be used for one's own goal, it is useful to meditate about the following suggestion: *Learn everything, and you will see afterward that nothing is useless*, which was the motto of Hugo of St Victor.[64,65]

[64] Hugo VON BLANKENBURG, German-born theologian, 1096–1141. He worked at the monastery of Saint Victor in Paris, France.

[65] I often heard people say about some famous scientists from the past, that luck played an important role in their discovery, but the truth must be that they would miss the importance of the new hint that occurred if they did not know beforehand all the aspects of their problem. Those who present chance as an important factor in discovery probably wish that every esoteric subject that they like be considered important and funded, but that is not at all what the quoted motto is about.

Additional footnotes: ABEL,[66] AL 'ABBAS,[67] BONAPARTE/Napoléon I,[68] CAVENDISH,[69] CORNELL,[70] DIDEROT,[71] FIELDS,[72] FOURIER J.-B.,[73] Georg-August/George II,[74] Friedrich HIRZEBRUCH,[75] HOPKINS,[76] ITO,[77] John the

---

[66] Niels Henrik ABEL, Norwegian mathematician, 1802–1829. The Abel Prize is named after him.

[67] AL 'ABBAS ibn 'abd al-Muttalib, uncle of MUHAMMAD, 566–652. The Abbasid Caliphs later claimed the caliphate because he was their ancestor.

[68] Napoléon BONAPARTE (Napoleone BUONAPARTE), French general, 1769–1821. He became Premier Consul after his coup d'état in 1799, was elected Consul à vie in 1802, and he proclaimed himself emperor in 1804, under the name Napoléon I (1804–1814, and 100 days in 1815).

[69] Henry CAVENDISH, English physicist and chemist (born in Nice, not yet in France then), 1731–1810. He lived in London, England. He founded the Cavendish professorship of physics at Cambridge, England.

[70] Ezra CORNELL, American philanthropist, 1807–1874. Cornell University, Ithaca, NY, is named after him.

[71] Denis DIDEROT, French philosopher and author, 1713–1784. He worked in Paris, France, and he was the editor-in-chief of the Encyclopédie. Université Paris 7, Paris, is named after him.

[72] John Charles FIELDS, Canadian mathematician, 1863–1932. He worked in Meadville, PA, and in Toronto, Ontario. The Fields Medal is named after him.

[73] Jean-Baptiste Joseph FOURIER, French mathematician, 1768–1830. He worked in Auxerre, in Paris, France, accompanied BONAPARTE in Egypt, was prefect in Grenoble, France, until the fall of Napoléon I, and worked in Paris again. Université de Grenoble I, Grenoble, is named after him, and the Institut Fourier is its department of mathematics.

[74] Georg Augustus, 1683–1760. Duke of Brunswick-Lüneburg (Hanover), he became king of Great Britain and Ireland in 1727, under the name of George II. He founded Georg-August-Universität in Göttingen, Germany, in 1734.

[75] Friedrich HIRZEBRUCH, German mathematician, born in 1927. He received the Wolf Prize in 1988, for outstanding work combining topology, algebraic and differential geometry, and algebraic number theory; and for his stimulation of mathematical cooperation and research, jointly with Lars HÖRMANDER. He worked in Bonn, Germany.

[76] Johns HOPKINS, American financier and philanthropist, 1795–1873. Johns Hopkins University, Baltimore, MD, is named after him.

[77] Kiyoshi ITO, Japanese mathematician, born in 1915. He received the Wolf Prize in 1987, for his fundamental contributions to pure and applied probability theory, especially the creation of the stochastic differential and integral calculus, jointly with Peter LAX. He worked in Kyoto, Japan, but also at Aarhus University, Aarhus, Denmark (1966–1969) and at Cornell University, Ithaca, NY (1969–1975).

Baptist,[78] Louis XVIII,[79] M.,[80] MITTAG-LEFFLER,[81] MUHAMMAD,[82] Napoléon I = BONAPARTE, NÉEL,[83] Paul (apostle),[84] Peter (apostle),[85]

---

[78] John the Baptist, Jewish preacher and ascetic, 30. According to the gospels, he was a cousin of Jesus of Nazareth.

[79] Louis Stanislas Xavier de France, 1755–1824, count of Provence, was king of France from 1814 to 1824, under the name of Louis XVIII.

[80] M. (Mahendranath GUPTA), Bengali teacher, 1854–1932. Disciple of Ramakrishna, he called himself M., and was the author of *The Gospel of Ramakrishna* [57].

[81] Magnus Gösta MITTAG-LEFFLER, Swedish mathematician, 1846–1927. He worked in Stockholm, Sweden. The Mittag-Leffler Institute in Stockholm is named after him.

[82] MUHAMMAD, Arab mystic and legislator, 570–632. He was the prophet of Islam.

[83] Louis Eugène Félix NÉEL, French physicist, 1904–2000. He received the Nobel Prize in Physics in 1970, for fundamental work and discoveries concerning antiferromagnetism and ferrimagnetism which have led to important applications in solid state physics, jointly with Hannes ALFVÉN. He worked in Strasbourg, and in Grenoble, France.

[84] Paul (Saul) of Tarsus (in actual Turkey), apostle, founder of Christianity, 10–67.

[85] Peter (Simon, or Cephas), one of the 12 disciples of Jesus, first Pope, 64.

PIATETSKI-SHAPIRO,[86] Ramakrishna,[87] SCHRÖDINGER,[88] STEVENS,[89] Georges TARTAR (my father),[90] Vivekananda,[91] WEIL A.,[92] WOLF.[93]

[86] Ilya PIATETSKI-SHAPIRO, Russian-born mathematician, born in 1929. He received the Wolf Prize in 1990, for his fundamental contributions in the fields of homogeneous complex domains, discrete groups, representation theory and automorphic forms, jointly with Ennio DE GIORGI. He worked at Tel-Aviv University, Tel Aviv, Israel.

[87] Sri Ramakrishna Paramahamsa (Gadadhar CHATTOPADHYAY), Bengali mystic, 1836–1886. Some of his teachings were transmitted by one of his students known as M. [57], and by his main student, Swami Vivekananda.

[88] Erwin Rudolf Josef Alexander SCHRÖDINGER, Austrian-born physicist, 1887–1961. He received the Nobel Prize in Physics in 1933, jointly with Paul Adrien Maurice DIRAC, for the discovery of new productive forms of atomic theory. He worked in Vienna, Austria, in Jena and in Stuttgart, Germany, in Breslau (then in Germany, now Wrocław, Poland), in Zürich, Switzerland, in Berlin, Germany, in Oxford, England, in Graz, Austria, and in Dublin, Ireland.

[89] Edwin Augustus STEVENS, American engineer and philanthropist, 1795–1868. The Stevens Institute of Technology, Hoboken, NJ, is named after him.

[90] Georges Elias TARTAR, Syrian-born Protestant minister, 1913–2003. After coming to Paris, France, in 1935 (while Syria was a French Protectorate), he worked as a tailor, he studied Protestant theology, he was a missionary in Aleppo, Syria, he was a minister for ERF (Église Réformée de France), the French Calvinist church, in France and in Algeria, he taught Arabic in Algeria, he worked at the French embassy in Amman, Jordan, in parallel with his main project: to tell Moslems about how Jesus and his mother are described in the Coran, in ways far superior to anyone else, apart from God (who, being the same God revered by Jews and Christians as the Coran mentions, cannot then have an untranslatable name, and Allah is just his name in Arabic!).

[91] Swami Vivekananda (Narendranath DUTTA), Bengali philosopher and monk, 1863–1902. Chief disciple of Ramakrishna, he was a spiritual leader of the philosophies of Vedanta and Yoga.

[92] André WEIL, French-born mathematician, 1906–1998. He received the Wolf Prize in 1979, for his inspired introduction of algebro-geometry methods to the theory of numbers, jointly with Jean LERAY. He worked in Aligarh, India, in Haverford, PA, in Swarthmore, PA, in São Paulo, Brazil, in Chicago, IL, and at IAS (Institute for Advanced Study), Princeton, NJ.

[93] Ricardo WOLF, German-born inventor, diplomat and philanthropist, 1887–1981. He emigrated to Cuba before World War I; from 1961 to 1973 he was Cuban Ambassador to Israel, where he stayed afterwards. The Wolf Foundation was established in 1976 with his wife, Francisca SUBIRANA-WOLF, 1900–1981, "to promote science and art for the benefit of mankind."

Chapter 2
A Personalized Overview of Homogenization I

Most of the important developments of physics during the twentieth century were concerned with describing relations between different scales, and it is quite interesting for a competent outsider to observe how physicists dealt with the extraordinary *challenge of discovering what happens at mesoscopic and microscopic scales*, often from observations made at a macroscopic level.

Such a challenge can hardly be addressed without making mistakes of the pseudo-logic type, where physicists naively believe that nature plays game A, because they guess that playing such a game gives a result resembling a partial observation B. Mathematicians should warn against such mistakes in logic, and propose help in finding a reasonable framework for that unnatural question of looking for an equation when one already knows the solution.[1] It is not the fact that physicists use wild guesses about what really happens which is the problem, but the fact that they forget to tell students that they must find better guesses, and their worst mistake, of course, is to invent dogmas to make it more difficult for future students to say that a part of the rules that they learned is nonsense!

One easy way for physicists to avoid a few mistakes is to observe that it is time to stop thinking in terms of ordinary differential equations and to start thinking instead in terms of partial differential equations, but they act as if they have not understood the difference between eighteenth century classical mechanics and nineteenth century continuum mechanics. They should observe that twentieth century mechanics and physics require going a step

[1] When I studied the appearance of nonlocal effects by homogenization of hyperbolic equations [104], I first characterized a weak limit of solutions of a sequence of partial differential equations, and then the difficulty was to find a natural class of equations in which to search for an effective equation that it satisfies. As will be seen in Chaps. 23 and 24, my answer is only valid in special linear cases, and more general situations are not understood.

L. Tartar, *The General Theory of Homogenization*, Lecture Notes of the Unione Matematica Italiana 7, DOI 10.1007/978-3-642-05195-1_2,

further because of the presence of various small scales, like for turbulence and plasticity on the mechanics side, and for questions of atomic physics (involving "particles" which do not exist) on the physics side. Of course, mathematicians should have pointed out such obvious observations a long time before I did, but one problem is that very few mathematicians care about other sciences than mathematics.

I shall describe the way I became aware that homogenization is important for a better understanding of continuum mechanics and physics. Because I write that physicists encountered situations of homogenization, which they have not handled well, one may think that I do not appreciate what physicists did in the past, but for doing my job of developing better mathematical tools, I used the intuition coming from some of the wild guesses made by good physicists before. Physicists usually work at the level which mathematicians call intuition, and it is inherent to their job that they say things which do not make much sense to mathematicians, one difficulty being that the equations of physics must be discovered; despite what physicists think, this is far from being done.

I know that physicists do a different job than mathematicians, and I am not criticizing that,[2] but I am telling mathematicians that there is interesting work for them to do,[3] in developing other simplifying concepts for explaining what physics is about. If I write about a new theory to be developed in the continuation of what I did, *beyond partial differential equations*, it is because better mathematical tools are not only necessary for solving some of the equations that were proposed, but also for writing the new types of equations which are necessary for twentieth century mechanics and physics.[4]

The term homogenization was used in nuclear engineering when Ivo BABUŠKA introduced it in the mathematical literature in the early 1970s [4],[5] one of his examples being to compute temperatures and stresses in the core of a nuclear reactor; I heard about his work from Carl DE BOOR,[6] after giving a talk about my joint work with François MURAT [93], at the beginning of the academic year 1974–1975, which I spent at UW, Madison, WI.

---

[2] However, I suggest that some "physicists" with a Comte complex do a bad job of trying to do mathematics instead of physics, because they usually fail from a mathematician's point of view, as well as from the point of view of real physicists, who do not suffer from a Comte complex.

[3] Among those who may dislike what I say are some "applied mathematicians", because I criticize as nonsense some of the models which they use.

[4] In my analysis, we live at a time similar to that of NEWTON, who invented infinitesimal calculus for expressing the laws of classical mechanics, and the differential equations which are implied!

[5] Ivo M. BABUŠKA, Czech-born mathematician, born in 1926. He worked at Charles University, Prague, Czech Republic, at UMD (University of Maryland), College Park, MD, and he works now at University of Texas, Austin, TX.

[6] Carl DE BOOR, German-born mathematician, born in 1937. He worked at UW (University of Wisconsin), Madison, WI.

Ivo BABUŠKA was working in the context of periodic media, as in the earlier work of Évariste SANCHEZ-PALENCIA [81, 82], which helped me understand that my work with François MURAT (for understanding an academic question of optimization proposed by Jacques-Louis LIONS [51], described in Chap. 4) was related to questions of effective properties of mixtures, and this helped me develop a new mathematical approach for what I was taught at École Polytechnique in my courses in continuum mechanics [58] and in physics, about questions involving various scales.

My new point of view has the advantage of avoiding the use of probabilities. I already understood that the assumptions of randomness which are introduced in partial differential equations are most of the time a way to hide the fact that one does not understand so much about the question that one studies. Unless one adds a sentence saying that "for the moment, since there are a few things that one does not understand, one will use probabilities", and point out that "it might also be that one worked with equations which are not good enough for describing the effects that one would like to study", one tends to accredit the point of view that one cannot find the laws that nature follows, which is a highly unscientific position, akin to desertion, since the role of scientists is precisely to find the laws that nature follows for various questions.[7]

One should notice that neither the theory of G-convergence, developed by Sergio SPAGNOLO in the late 1960s [89,90], nor the theory of H-convergence, developed by François MURAT and myself in the early 1970s [71,93], have any assumption of a long-range order like for the periodic case. It is also useful to observe that the motivations of Évariste SANCHEZ-PALENCIA and of Ivo BABUŠKA for studying periodic situations were different.

Évariste SANCHEZ-PALENCIA considered mixtures of materials showing a periodic geometry, for questions of diffusion, of heat or of electricity, or questions of linearized elasticity [81,82]. After postulating an asymptotic expansion, he used variational methods for identifying the first term of that expansion when one lets the period length ε tend to 0, which describes the effective equations, which he found to have the same form as the initial equations, but with a constitutive relation corresponding to a general anisotropic medium,[8] even when the materials used are all isotropic.

---

[7] If one separates these questions into various fields, biology, chemistry, mathematics, mechanics, and physics (in alphabetical order), it is mostly because one does not know how to form efficient researchers in general, and one certainly does not know how to form researchers who would understand enough about questions ranging over more than one of these fields.

[8] It seems that some people guessed the first term wrongly, proposing to solve a problem on one period with Neumann conditions, while the correct solution, which Évariste SANCHEZ-PALENCIA deduced from his postulated asymptotic expansion, uses periodic boundary conditions. I suppose that problems with periodic boundary conditions were not found so natural to some people interested in continuum mechanics.

Évariste SANCHEZ-PALENCIA also studied problems which showed the appearance of linear memory effects by homogenization, again in a periodic geometry, and in these cases the effective equations found when one lets the period length ε tend to 0 have a slightly different form than the initial equations [83]. One situation concerned visco-elastic effects, obtained for the effective equation describing the evolution of a mixture of an elastic solid (in the approximation of linearized elasticity) and a liquid (in the approximation of the Stokes equation).[9] Another situation concerned the dependence of the effective dielectric permittivity of a mixture upon frequency, an effect which results from using the Ohm law.[10]

With Horia ENÉ,[11] Évariste SANCHEZ-PALENCIA considered an homogenization approach for explaining the Darcy law for flows in porous media,[12] as the effective equation for the rescaled velocity of a liquid (in the approximation of the Stokes equation) flowing inside a rigid solid showing a periodic geometry, and this case shows then an effective equation having a quite different form than the original equation [26].

Évariste SANCHEZ-PALENCIA was then identifying which equations to use at a macroscopic scale, when one knows the partial differential equations governing a mesoscopic scale, the periodic geometry being a simplifying assumption, which permits one to use the technique of asymptotic expansions.

Around the same time, I heard Alain BAMBERGER mention the work of Georges MATHERON on porous media,[13,14] and when I met him in the mid 1980s at some talks at IFP, Rueil-Malmaison, France, he claimed to derive the Darcy law in the late 1960s; however, Georges MATHERON's "derivation" used probabilistic methods, and since probabilists often impose the models that they like without caring if their assumptions are compatible with what one understands about the partial differential equations of continuum mechanics, I do not know if someone proved his approach to be correct.

---

[9] Évariste SANCHEZ-PALENCIA suggested that the movement of the solid sets the fluid in motion, and that motion is dissipative, but some kinetic energy can be stored in the fluid and recovered later, at least in part, hence a memory effect. He suggested that it gives a qualitative explanation of the visco-elastic behaviour of concrete, whose properties actually change with age, perhaps due to a slow drying process.

[10] Georg Simon OHM, German mathematician, 1789–1854. He taught in various schools before working in München (Munich), Germany.

[11] Horia ENÉ, Romanian mathematician and politician, born in 1941. He works in Bucharest, Romania.

[12] Henry Philibert Gaspard DARCY, French engineer, 1803–1858. He worked in Dijon, France.

[13] Alain BAMBERGER, French mathematician. He worked at École Polytechnique, Palaiseau, France, and then became an administrator at IFP (Institut Français du Pétrole), and at École Polytechnique.

[14] Georges MATHERON, French mathematician, 1930–2000. He worked at École des Mines, in Nancy, Paris, and Fontainebleau, France.

In the spring of 1974, I found that a book by LANDAU & LIFSHITZ contained a puzzling section on the conductivity of mixtures [47],[15,16] and I wondered if physicists knew what they were talking about! Assuming that two different conductors (of respective conductivities α and β) are ground into fine powders, that one mixes them thoroughly (with respective proportions θ and $1-\theta$), that one shakes the mixture well, and that one compresses it, they asked about the conductivity of the resulting mixture, and they performed a strange calculation which leads to a curious formula. From a mathematical point of view, the conductivity of a mixture in dimension $N \geq 2$ does not depend only upon the proportions used, and it is strange that they looked for a formula, but I only found in June 1980 the exact interval where the conductivity of an (isotropic) mixture is found: it is given by the Hashin–Shtrikman bounds [38].[17,18] One must notice, however, that there is no mathematical meaning for grinding, shaking, and compressing, so that it is difficult to assert what the practical result could be.[19] They did not mention that a proportion must be small, so that their formula is obviously wrong for $\theta = \frac{1}{2}$, by lack of symmetry in α and β. Curiously, they quoted no experimental measurements for assessing the practical value of their argument!

Ivo BABUŠKA considered engineering applications where a period of macroscopic size is repeated a large number of times, like for the core of a nuclear reactor, where the cells have an hexagonal cross-section with a cylindrical hole in their centre for lowering bars of uranium [4]. For security reasons, one must check that the temperature in the reactor is under control,[20] and it is important to be able to carry out precise numerical simulations about what could happen in the case of an accident.

---

[15] Lev Davidovich LANDAU, Russian physicist, 1908–1968. He received the Nobel Prize in Physics in 1962, for his pioneering theories for condensed matter, especially liquid helium. He worked in Moscow, Russia.

[16] Evgenii Mikhailovich LIFSHITZ, Russian physicist, 1915–1985. He worked in Moscow, Russia.

[17] Zvi HASHIN, Israeli physicist. He worked in Tel Aviv, Israel.

[18] Shmuel SHTRIKMAN, Belarusian-born physicist, 1930–2003. He worked at the Weizmann Institute of Science, Rehovot, Israel.

[19] If one could take into account the cost of creating the mixture, it might be related to creating a mixture of low cost.

[20] The heat generated by the nuclear reactions is transported away by a first fluid in a closed primary circuit, which must have no leaks, since this fluid becomes radioactive. Heat carried away by the first fluid is transmitted by conduction to a second fluid in a secondary circuit, which does not become radioactive, and this second fluid is used for creating the vapour for the turbines which generate electricity. If a leak happened in the primary circuit, there would be a contamination by radioactive products, of course, but a more dangerous problem would be caused by an important leak, because the cooling of the reactor could become insufficient, so that the temperature of the reactor would increase too much, and if the bars of uranium were not removed quickly enough, one could arrive at a catastrophic meltdown.

The approach of Ivo BABUŠKA leads to the same kind of asymptotic expansion that Évariste SANCHEZ-PALENCIA used before, but the reason for a periodic geometry is different: it is the design chosen by engineers, and although the size of the period is not small, the first term of the asymptotic expansion is interesting, because it leads to an efficient numerical method.

The treatment of Évariste SANCHEZ-PALENCIA was not entirely mathematical, since his intuition about the problems suggested that an asymptotic expansion holds, and after that his proof was valid for identifying the first term. His effective equation was correct, and for a diffusion equation a complete proof followed from the work of Sergio SPAGNOLO, who was invited in May 1975 to talk about G-convergence [91] at a conference at UMD, College Park, MD, organized by Ivo BABUŠKA, who himself reported on the formulas published in the literature [5], the earlier being introduced by POISSON. For a simple plane periodic situation, he found that none of them was correct when the conductivities were far apart.[21]

During the preceding year which I spent at UW, Madison, WI, I simplified my joint work with François MURAT, which he later called H-convergence [71], and the basic idea was rediscovered a few years after by Leon SIMON [88],[22] and it applies to more general equations, like linearized elasticity, which could not be obtained so easily by the approach of Sergio SPAGNOLO [89], because he used a regularity result of MEYERS [62].[23,24]

In May 1975, I learned from Ivo BABUŠKA about the importance of amplifying factors in elasticity, and it was the reason why I introduced correctors later that year (without periodicity assumptions), described in Chap. 13, but I only realized many years after that the defects of the linearization in elasticity are amplified by the multiplication of interfaces. Later that year, I proved a basic theorem of homogenization of monotone operators, described

[21] I was expecting that it would be the case, independently of the precise problem, where the period was a square containing a circular inclusion half the size in area, but I was quite surprised that someone in the audience expressed the strange idea that more than one answer could be valid!

[22] Leon Melvyn SIMON, Australian-born mathematician, born in 1945. He worked at Stanford University, Stanford, CA, at UMN (University of Minnesota), Minneapolis, MN, at University of Melbourne, Melbourne, and at ANU (Australian National University), Canberra, Australia, and again at Stanford University.

[23] Norman George MEYERS, American mathematician. He works at UMN (University of Minnesota), Minneapolis, MN.

[24] In April 1974, at a meeting in Roma (Rome), Italy, Ennio DE GIORGI told me that Meyers' regularity result is not related to the maximum principle, but I only learned a few years after about a proof, which uses the Calderón–Zygmund theorem and interpolation, in a lecture of Jacques-Louis LIONS, who used the idea for linearized elasticity using isotropic materials; that the proof for general linearized elastic materials works was only checked many years after.

in Chap. 11, and 2 years after, at a conference in Rio de Janeiro, Brazil, in August 1977, I discussed the difficulties that I encountered for creating a theory of homogenization valid for nonlinear elasticity [96]. In the late 1980s, when Owen RICHMOND told me that he thought that one needs a theory with higher-order derivatives,[25] I did not understand what he meant; in some sense he was right, although I shall not advocate higher-order derivatives but nonlinear nonlocal effects [107], described in Chap. 24, which are not well understood yet: if there is not yet a theory of homogenization valid for nonlinear elasticity,[26] it is because one has not found the form of the effective equations of that more general theory which follows mathematically, and which nature uses in place of that simpler nonlinear elasticity approach which mathematicians believed in.[27]

Because Jacques-Louis LIONS was convinced of the interest of questions in periodic homogenization by Ivo BABUŠKA, he tried to prove the correctness of the asymptotic expansions with two collaborators, Alain BENSOUSSAN and George PAPANICOLAOU [6],[28,29] but I think that they only succeeded in the case of Dirichlet conditions,[30] before I showed my improved method to

---

[25] Owen RICHMOND, American mathematician, 1928–2001. He worked at ALCOA (Aluminum Company of America), Alcoa Center, PA.

[26] There are deluded people who think that Γ-convergence answered the question. Were they victims of saboteurs, who made them adopt the idiotic belief that every material which contains energy is elastic?

[27] In 1995, at meeting at IMA (Institute for Mathematics and its Applications), UMN, Minneapolis, MN, I was surprised to hear an American engineer boast that he proved something that mathematicians had not, before stating something wrong! I pointed out twice that his statement was wrong, with no reaction from the participants who pretended to be knowledgeable about homogenization, except John WILLIS, who came to me later to point out one of the two errors that I would point out to anyone interested. I then asked my student Sergio GUTIÉRREZ to check some limitations imposed on materials for which one can define a theory of homogenization in linearized elasticity [36,37]; his result shows that one cannot develop a theory of homogenization under strict *strong ellipticity* (related to the physical Legendre–Hadamard condition which renders the system hyperbolic), and that one almost needs the *very strong ellipticity* condition, which permits one to use the *Lax–Milgram lemma*. One should be careful not to deduce that very strong ellipticity has a physical meaning, because linearized elasticity is not a physical theory, since it is not frame-indifferent!

[28] Alain BENSOUSSAN, Tunisian-born mathematician, born in 1940. He worked at Université Paris IX-Dauphine, Paris, France, and he works now at University of Texas at Dallas, Richardson, TX.

[29] George C. PAPANICOLAOU, Greek-born mathematician, born in 1943. He worked at NYU (New York University), New York, NY, and he works now at Stanford University, Stanford, CA.

[30] Johann Peter Gustav LEJEUNE DIRICHLET, German mathematician, 1805–1859. He worked in Breslau (then in Germany, now Wrocław, Poland), in Berlin, and at Georg-August-Universität, Göttingen, Germany.

Jacques-Louis LIONS, after his talk at a conference organized in September 1975 in Luminy, near Marseille, France.[31] He mentioned it in his article for the proceedings of the conference [54], but after that he never mentioned again that it was my method, although he used it extensively in his courses (at Collège de France, in Paris), so that I once asked him why he did not attribute my method to me, and his answer was "everybody knows that it is your method"!

In 1975, I also studied a limiting case, when $\alpha = 0$, corresponding to using an insulator, interpreted as holes in the domain, with the homogeneous Neumann condition imposed on their boundary,[32] described in Chap. 16, without periodicity assumptions, but imposing some regularity condition for ensuring the construction of an extension inside the hole. In December 1975, Ivo BABUŠKA came for a conference in Versailles, France, and I told him about my result, and I said that I expected the non-homogeneous Neumann condition as well as the Fourier condition to lead to similar results,[33] but he asked me what scaling of the coefficients I planned to use, and he showed me that my choice was not physical.

I noticed before that the solutions tend to 0 if one uses the homogeneous Dirichlet condition on the boundaries of the holes, described in Chap. 15, so after finding no interest to that case, I felt in the same way, that I needed to improve my intuitive understanding of continuum mechanics, when I found that Évariste SANCHEZ-PALENCIA was rightly asking two natural questions, of identifying what is the rate at which the solutions tend to 0, and of identifying the weak limit of a rescaled sequence.

It was precisely that kind of programme which Évariste SANCHEZ-PALENCIA followed with Horia ENÉ for deriving the Darcy law out of the Stokes equation in a "periodic porous medium" [26], and I only looked at applying my method to this question because Jacques-Louis LIONS said that he was unable to apply my method of extension in the holes, the difficulty being to control the extension of the "pressure".[34] I looked at the question, in a periodic setting, and I found correct estimates, but for a geometry which Évariste

---

[31] He actually heard my talk at a conference that he organized at IRIA (Institut de Recherche en Informatique et Automatique) in Rocquencourt, France, in June 1974, and I was surprised that after he started to work on the subject, he did not ask François MURAT about the details of our joint work [93] (because I was working at UW (University of Wisconsin), Madison, WI, for 1974–1975). Could it be that he did not realize that it was the same subject?

[32] Franz Ernst NEUMANN, German mathematician, 1798–1895. He worked in Königsberg, then in Germany, now Kaliningrad, Russia.

[33] The condition used by FOURIER is also called a Robin condition, but I could not discover who this ROBIN was.

[34] Because of the hypothesis $\varrho = \varrho_0$ which gives incompressibility, what one calls the (reduced) "pressure" $\frac{p}{\varrho_0}$ is only defined up to an additive constant (depending on time), and it is not necessarily ≥ 0 like a real pressure!

SANCHEZ-PALENCIA pointed out to be unrealistic,[35] and I explained the result in the spring of 1979 at a workshop in Bandol, France; more general periodic geometries were only covered a few years after, by Grégoire ALLAIRE.[36]

I understood one advantage of a mathematical theory of homogenization, that from an experimental identification of which equation holds in a simple homogeneous medium, like an isotropic one, together with which transmission condition holds at a general interface between two simple media, the mathematician can discover a general variational framework and identify all possible limits, and can thus predict the form of the equation valid for a general medium, like an anisotropic one, even though such media might not be observed yet by experimentalists. My first goal was to develop mathematical tools for carrying such identifications, but in the first few years, the opposition of a few persons surprised me.

There was some opposition from mathematicians: once, I explained the point of view that one should try to understand more about the physical meaning of the equations that one is studying, and Jacques-Louis LIONS defended the opposite position, that it is not strictly necessary to do it; after that our paths separated.

In this chapter, I have not yet mentioned the Italian pioneers who were the first to obtain mathematical results in homogenization, Sergio SPAGNOLO and Ennio DE GIORGI, with Antonio MARINO,[37] because they were only concerned with mathematical questions, and Chaps. 2 and 3 are about my point of view that homogenization is important in continuum mechanics and physics, but a negative tendency appeared, to advocate Γ-convergence for questions of fake mechanics, so that the name of Ennio DE GIORGI is used like a shield. Such nonsense is becoming increasingly popular for political reasons, among people who do not know the first principle, that energy is conserved, or that time exists, so that they believe in a fake mechanics/physics where materials minimize their potential energy instantaneously, and when they realize that evolution problems exist, they are deluded enough to only use gradient flows![38] I wrote [113] in order to push this group to acquire more common sense.

There was some opposition from specialists of continuum mechanics, particularly those who did not introduce ideas of their own for homogenization

---

[35] In my construction, the solid parts from different periods were disconnected.

[36] Grégoire ALLAIRE, French mathematician, born in 1963. He worked at CEA (Commissariat à l'Énergie Atomique) in Saclay, at UPMC (Université Pierre et Marie Curie), Paris, and he works now at École Polytechnique, Palaiseau, France.

[37] Antonio MARINO, Italian mathematician, born in 1939. He works at Università di Pisa, Pisa, Italy.

[38] If they were good enough to be students in Paris in the late 1960s, they would observe that the best students are expected to try to tackle hyperbolic problems for a while, and that those who settle for elliptic or parabolic equations at least know that reality is elsewhere!

questions, and who used my general approach on many examples, without emphasizing that François MURAT and myself developed a general method, not restricted to periodic situations. Once, I helped someone by explaining a point that he mistakenly thought easy, but he did not acknowledge that I was the first to solve a more general question (since I used holes which were not necessarily distributed in a periodic way), and that he could not write his article without my help. At the end of my talk in Bandol in the spring of 1979, I was the focus of a virulent attack, which I thought quite mistaken, since it targeted mathematicians who are not interested in continuum mechanics, and in my talk I described mathematical tools for proving that the Darcy law follows from the Stokes equation (in a particular geometry), as asserted by Horia ENÉ and Évariste SANCHEZ-PALENCIA![39]

In the spring of 1975, I gave a talk at NYU, New York, NY, but I did not meet Joe KELLER,[40] and I first had a discussion with him a few months after,[41] at UW, Madison, WI. I had difficulties in understanding what Joe KELLER told me about homogenization for hyperbolic problems, because I said that when coefficients only depend upon x,[42] the homogenization of a wave equation is straightforward, but he disagreed, and I did not understand why. This was an earlier example of my lack of intuition for some questions in continuum mechanics, of course, but I was not aware at the time that a problem occurs if one uses a sequence of initial data with energy concentrated in wavelengths comparable to the period size. For the case that I was considering, of fixed initial data (and using no periodicity hypothesis), such an effect does not exist, at least if one only considers a finite time, since waves propagating in the effective medium do not behave correctly at infinity, an effect which can be studied using Bloch waves in the periodic case.[43]

I am unhappy about Bloch waves, which only have a meaning for periodic media, but I have some hope that one will find a mathematical way to carry

---

[39] Unlike others, Évariste SANCHEZ-PALENCIA was always correct in his attributions, and because he knew that it could take years before I would write down my proof, he preferred to write it as an appendix of his book [83], and asked me if I agreed; he even put my name for that appendix, as if I wrote it myself!

[40] Joseph Bishop KELLER, American mathematician, born in 1923. He received the Wolf Prize for 1996/97, for his innovative contributions, in particular to electromagnetic, optical, acoustic wave propagation and to fluid, solid, quantum and statistical mechanics, jointly with Yakov G. SINAI. He worked at NYU (New York University), New York, NY, and at Stanford University, Stanford, CA.

[41] He pointed out that the title of my talk, which was about control in coefficients of partial differential equations, was not conveying the important fact that it was about properties of mixtures.

[42] Sergio SPAGNOLO later studied the case of coefficients depending upon t, and without regularity it is already challenging to prove the existence of a solution.

[43] Felix BLOCH, Swiss-born physicist, 1905–1983. He received the Nobel Prize in Physics in 1952, jointly with Edward Mills PURCELL, for their development of new methods for nuclear magnetic precision measurements and discoveries in connection therewith. He worked at Stanford University, Stanford, CA.

a similar analysis in some non-periodic situations, because there were other questions of homogenization where the periodic situation gave the correct intuition, which could only be transformed into proofs in the late 1980s, after I introduced *H-measures* [105], which were also introduced independently by Patrick GÉRARD for a different purpose.[44]

It was only in 1979 that I thought of studying homogenization of first-order hyperbolic equations (or systems) with oscillating coefficients, a problem obviously crucial for understanding turbulence, which is "known" to be generated by fluctuations of the velocity field. Of course, it is imperative to avoid probabilities for a question like turbulence, unless one wants to destroy the physical reality, and I wonder about the goal of those who coined such a strange term as "Burgers turbulence", because Eberhard HOPF clearly explained in [39] that BURGERS was wrong in [11] to think that turbulence is created by large values of the velocity,[45,46] since Galilean invariance shows that only variations of the velocity field should be taken into account.[47]

I also thought that it could be the key to a question which led physicists to postulate curious rules, before developing strange dogmas, for explaining what is observed in experiments of spectroscopy, because spectroscopy consists in sending a wave in a slightly heterogeneous medium, and my guess was that this question of homogenization leads to equations involving at least some nonlocal effects, as I checked for a simple case [104], described in Chap. 23, and a more interesting situation was studied along the same lines by Youcef AMIRAT,[48] Kamel HAMDACHE,[49] and Hamid ZIANI[50] [1,2], described in Chap. 24. Proving that in a linear situation with translation invariance an

---

[44] Patrick GÉRARD, French mathematician, born in 1961. He works at Université Paris Sud, Orsay, France.

[45] Eberhard Frederich Ferdinand HOPF, Austrian-born mathematician, 1902–1983. He worked at MIT (Massachusetts Institute of Technology), Cambridge, MA, in Leipzig and in München (Munich), Germany, and at Indiana University, Bloomington, IN, where I met him in 1980.

[46] Johannes Martinus BURGERS, Dutch-born mathematician, 1895–1981. He worked at UMD (University of Maryland), College Park, MD.

[47] Galileo GALILEI, Italian mathematician, 1564–1642. He worked in Siena, in Pisa, in Padova (Padua), Italy, and again in Pisa.

[48] Youcef AMIRAT, Algerian-born mathematician, born in 1949. He worked in Alger (Algiers), Algeria, and he works now at Université de Clermont-Ferrand II (Blaise Pascal), Aubière, France.

[49] Kamel HAMDACHE, Algerian-born mathematician, born in 1948. He worked in Alger (Algiers), Algeria, and then in various laboratories of CNRS (Centre National de la Recherche Scientifique), at ENSTA (École Normale Supérieure des Techniques Avancées), Palaiseau, at ENS (École Normale Supérieure), Cachan, in Bordeaux, at Université Paris Nord, Villetaneuse, and he works now at École Polytechnique, Palaiseau, France. He studied for his PhD (1986) under my supervision.

[50] Abdelhamid ZIANI, Algerian-born mathematician, 1948–2004. He worked in Alger (Algiers), Algeria, at École Polytechnique, Palaiseau, and at Université de Nantes, Nantes, France.

effective equation, correctly constrained to belong to a class of convolution operators, is defined in a unique way from the sequence of coefficients, converging only weakly, shows that the laws valid at a macroscopic scale are not always expressed with partial differential equations, even though they are in that class at a mesoscopic scale. Of course, Évariste SANCHEZ-PALENCIA already made such observations [83], but not in the hyperbolic situations that I was interested in. The natural mathematical question became to identify the correct new class of equations to introduce for more general situations, like for time-varying coefficients, studied by Luísa MASCARENHAS [60],[51] and then myself [107], but the main difficulty lies with nonlinear situations, since I do not know a natural framework for dealing with translation invariance [107], or more general group invariance.

Once one proves that an effective equation must contain an added term like a memory effect, some people may want to find a probabilistic game whose output will be the same equation, but there is no scientific reason to prefer a probabilistic proof to a non-probabilistic one. It is a pity that physicists brainwashed their students to believe that there are probabilities in the laws of nature, but the truth must be that one needs effective equations with nonlocal terms, and that is a quite different matter.

Having found a simple linear setting where linear memory effects appear by homogenization, I hoped that solving nonlinear cases could point out natural classes of materials with fading memory, a subject whose foundations were laid by Bernard COLEMAN,[52] Victor MIZEL,[53] and Walter NOLL,[54] who became my colleagues when I moved to CMU, Pittsburgh, PA, in 1987, and Bernard COLEMAN proposed an intuitive explanation about the appearance of a memory term: a weak limit is akin to taking an average and one then averages local evolutions which are not related, because of the hyperbolic nature of the equation which has the property of transporting various informations about the solutions, like oscillations and concentration effects,[55] and the macroscopic information at time t is certainly not enough for predicting

[51] Maria Luísa MARTINS MACEDO FARIA MASCARENHAS, Portuguese mathematician. She works in Lisbon, Portugal. She studied for her PhD (1983) under my supervision.

[52] Bernard David COLEMAN, American mathematician, born in 1930. He worked at the Mellon Institute of Industrial Research, which merged with Carnegie Tech (Carnegie Institute of Technology) to become CMU (*Carnegie Mellon University*), Pittsburgh, PA, where he was my colleague in 1987–1989, and he works now at Rutgers University, Piscataway, NJ.

[53] Victor Julius MIZEL, American mathematician, 1931–2005. He worked at CMU (*Carnegie Mellon University*), Pittsburgh, PA, being my colleague after 1987.

[54] Walter NOLL, German-born mathematician, born in 1925. He worked at CMU (*Carnegie Mellon University*), Pittsburgh, PA, being my colleague after 1987.

[55] In the case of smooth coefficients, something else is transported, which has no interest for questions of physics, i.e. *propagation of microlocal regularity* according to the ideas of Lars HÖRMANDER, and wrongly calling this effect "propagation of

the future values; however, if one keeps the averages for all intermediate times, it gives the feeling that one could filter out some mesoscopic information about the initial data, and then predict something about the future values.

In their book [120], Clifford TRUESDELL and Robert MUNCASTER observe that the Maxwell–Boltzmann kinetic theory cannot be derived from an initial framework of interacting particles feeling instantaneous forces at a distance,[56,57] which is a time-reversible process, so that it is not possible that an observer reversing time could also deduce the Boltzmann equation in the limit, because of the H-theorem of BOLTZMANN; in other words, the introduction of probabilities destroyed some of the physical reality. The time-reversibility is a consequence of a Hamiltonian structure,[58] a particular structure which was actually first discovered for some ordinary differential equations by LAGRANGE.[59,60]

However, if the limiting equation has a memory term, corresponding to an integral from $-\infty$ to t, then an observer reversing time would write the same limiting equation but with an integral from $+\infty$ to t, and these two equations could have the same solutions, so that no irreversibility would appear; irreversibility would then be introduced by getting rid of the integral terms, and keeping only the information at time t. This effect seems typical of situations where a sequence of semi-groups, or even groups, converges only weakly, in which case the limit is not always a semi-group.

Of course, I already found in the late 1970s that the laws of thermodynamics are wrong, and should be improved. That total energy is conserved is not disputed, but using topologies of weak type for describing the relations between mesoscopic and macroscopic levels, one finds that some part of the energy seems lost in taking a weak limit, and it is only that it becomes hidden at a mesoscopic level, so that one naturally identifies that part with

singularities" may be related to an intention to mislead that I observed among some of his followers.

56 Robert Gary MUNCASTER, American mathematician, born in 1948. He works at University of Illinois, Urbana, IL.

57 They do not mention that most instantaneous forces at a distance violate the relativity principle of POINCARÉ.

58 Sir William Rowan HAMILTON, Irish mathematician, 1805–1865. He worked in Dublin, Ireland.

59 Joseph Louis LAGRANGE (Giuseppe Lodovico LAGRANGIA), Italian-born mathematician, 1736–1813. He worked in Torino (Turin) Italy, in Berlin, Germany, and in Paris, France. He was made count in 1808 by Napoléon I.

60 His work was about perturbations in celestial mechanics: starting from a first approximation of an elliptic orbit, it was natural that he wondered how the parameters describing the orbit would change because of the gravitational pull of another planet, and writing the evolution equation for these parameters he was surprised to discover that they share the same "Hamiltonian" form as that in the initial Cartesian coordinates. I was told that he even used the same letter H that one uses now in "Hamiltonian" studies.

the internal energy which thermodynamics speaks about. It was a mistake of those who developed thermodynamics to believe that there is only one form of internal energy, because in order to understand what classical thermodynamics interprets as heat flowing, which is how the energy hidden inside the material moves around, it is crucial to distinguish between the different forms of internal energy, since each form usually flows according to its own rule, and it is not wise then to postulate an equation for the whole internal energy. However, 10 years occurred between the moment when I became aware of that question and the moment where I proved such an individual transport property, using the *H-measures* that I introduced in the late 1980s for questions of small amplitude homogenization [105] (and which Patrick GÉRARD introduced independently for a different reason): for a linear wave equation $\varrho\, u_{tt} - \sum_{i,j}(a_{i,j}u_{x_j})_{x_i} = 0$ with coefficients independent of t and of class C^2,[61] the density of kinetic energy is $\frac{\varrho\,(u_t)^2}{2}$ and the density of potential energy is $\frac{\sum_{i,j} a_{i,j}u_{x_j}u_{x_i}}{2}$, and the internal energy is described by using a nonnegative measure indexed by (x,t) and directions in the dual variable (ξ,τ), supported on the set where $\varrho\,\tau^2 - \sum_{i,j} a_{i,j}\xi_j\xi_i = 0$, which expresses the *equipartition of hidden energy*,[62] between the kinetic energy and the potential energy, and this measure satisfies a first-order partial differential equation, expressing the fact that it is transported along the bicharacteristic rays of $\varrho\,\tau^2 - \sum_{i,j} a_{i,j}\xi_j\xi_i$.

One should not be lured by the fact that the transport equations that I proved to be valid look like some simple equations from kinetic theory, because in kinetic theory one postulates equations of a given form, compatible with physical intuition, while I proved which equations should be used for various classes of hyperbolic systems. Actually, the transport equations that my general approach generates in the case of the scalar wave equation, the Maxwell–Heaviside equation, the linearized elasticity equation, and the Dirac equation, might be interpreted by physicists as describing the movement of "photons", "polarized photons", "phonons", "electrons" and "positrons", but

[61] I only assumed the coefficients to be of class C^1 for proving a transport equation for some H-measure in [105], but Patrick GÉRARD pointed out that uniqueness may not hold for that transport equation unless one assumes a little more regularity for the coefficients.

[62] In 1974, using the *div–curl lemma* that I had just proved with François MURAT, I had already shown this form of equipartition of hidden energy, that the hidden part of $\frac{\varrho\,(u_t)^2}{2}$ and the hidden part of $\frac{\sum_{i,j} a_{i,j}u_{x_j}u_{x_i}}{2}$ are equal, and this result requires no regularity but L^∞ for the coefficients. In 1977, I applied our improved theory of *compensated compactness* to the Maxwell–Heaviside equation, and I deduced that the hidden part of $\frac{(D,E)}{2}$ and the hidden part of $\frac{(B,H)}{2}$ are equal, which in the linear case is equipartition of hidden energy, between the electric part and the magnetic part. One should notice that my proofs of equipartition of hidden energy have nothing to do with the game of counting degrees of freedom which one teaches in thermodynamics courses.

these should be considered idealized "particles" whose quantified properties cannot be understood from my work, which was concerned with linear systems, and they require an improved theory which is not developed yet, and which should be applicable to semi-linear hyperbolic systems, like the coupled Maxwell–Heaviside/Dirac system.
Additional footnotes: Alberto CALDERÓN,[63] Charles IV,[64] DESCARTES,[65] Sergio GUTIÉRREZ,[66] HADAMARD,[67] LEGENDRE,[68] MILGRAM,[69]

[63] Alberto Pedro CALDERÓN, Argentinean-born mathematician, 1920–1998. He received the Wolf Prize in 1989, for his groundbreaking work on singular integral operators and their application to important problems in partial differential equations, jointly with John W. MILNOR. He worked in Buenos Aires, Argentina, at OSU (Ohio State University), Columbus, OH, at MIT (Massachusetts Institute of Technology), Cambridge, MA, and at The University of Chicago, Chicago, IL. I first heard him talk at the Lions–Schwartz seminar in the late 1960s, and I met him in Buenos Aires when I visited Argentina for 2 months in 1973; he kept strong ties with Argentina, as can be witnessed from the large number of mathematicians from Argentina having studied harmonic analysis, and often working now in the United States.

[64] Charles IV of Luxembourg, 1316–1378. German king and king of Bohemia (in 1346) and Holy Roman Emperor (in 1355) as Karl IV. Charles University, which he founded in Prague in 1348, is named after him.

[65] René DESCARTES, French mathematician and philosopher, 1596–1650. Université de Paris 5, Paris, France, is named after him.

[66] Sergio Enrique GUTIÉRREZ, Chilean mathematician, born in 1963. He works at Pontificia Universidad Católica de Chile, Santiago, Chile. He was my PhD student (1997) at CMU (*Carnegie Mellon University*), Pittsburgh, PA.

[67] Jacques Salomon HADAMARD, French mathematician, 1865–1963. He worked in Bordeaux, in Paris, France, and he held a chair (mécanique analytique et mécanique céleste, 1909–1937) at Collège de France, Paris.

[68] Adrien-Marie LEGENDRE, French mathematician, 1752–1833. He worked in Paris, France.

[69] Arthur Norton MILGRAM, American mathematician, 1912–1960. He worked at Syracuse University, Syracuse, NY, and at UMN (University of Minnesota), Minneapolis, MN.

MILNOR,[70] PASCAL,[71] PURCELL E.M.,[72] RUTGERS,[73] Yakov SINAI,[74] WEIZMANN,[75] John WILLIS,[76] Antoni ZYGMUND.[77]

[70] John Willard MILNOR, American mathematician, born in 1931. He received the Fields Medal in 1962 for his work in differential topology. He received the Wolf Prize in 1989, for ingenious and highly original discoveries in geometry, which have opened important new vistas in topology from the algebraic, combinatorial, and differentiable viewpoint, jointly with Alberto CALDERÓN. He worked at Princeton University, Princeton, NJ, and at SUNY (State University of New York) at Stony Brook, NY.

[71] Blaise PASCAL, French mathematician and philosopher, 1623–1662. Université de Clermont-Ferrand II, Aubière, France, is named after him.

[72] Edward Mills PURCELL, American physicist, 1912–1997. He received the Nobel Prize in Physics in 1952, jointly with Felix BLOCH, for their development of new methods for nuclear magnetic precision measurements and discoveries in connection therewith. He worked at MIT (Massachusetts Institute of Technology) and at Harvard University, Cambridge, MA.

[73] Henry RUTGERS, American colonel, 1745–1830. Rutgers University, Piscataway, NJ, is named after him.

[74] Yakov Grigor'evich SINAI, Russian-born mathematician, born in 1935. He received the Wolf Prize for 1996/97, for his fundamental contributions to mathematically rigorous methods in statistical mechanics and the ergodic theory of dynamical systems and their applications in physics, jointly with Joseph B. KELLER. He worked at Moscow State University, Moscow, Russia, and at Princeton University, Princeton, NJ.

[75] Chaim WEIZMANN, Belarusian-born chemist, 1874–1952. He worked in Geneva, Switzerland, and in Manchester, England. He was a Zionist political leader, and he became the first President of Israel, 1949–1952. The Weizmann Institute of Science, Rehovot, Israel, is named after him.

[76] John Raymond WILLIS, English mathematician, born in 1940. He worked in Bath and in Cambridge, England.

[77] Antoni Szczepan ZYGMUND, Polish-born mathematician, 1900–1992. He worked in Warsaw, Poland and in Wilno (then in Poland, now Vilnius, Lithuania), and at The University of Chicago, Chicago, IL.

Chapter 3
A Personalized Overview of Homogenization II

I mentioned earlier that instantaneous forces at a distance are nonphysical, because of the principle of relativity of POINCARÉ, since instantaneity is impossible to define, but distance is not such an easy concept either, if one considers that for stars up to a few light-years or about one parsec away,[1] the distance is measured by parallax, since these nearby stars move slightly with respect to the background in the course of a year, and further away the stars are too far for measuring their distance, but one observed some relation between luminosity and distance for those stars which are near enough, and so one switches to measuring luminosity, and one pretends that one is measuring "distance", and further away one switches to something else by way of another observed relation that one postulates to be always true, so that when astronomers say that the red-shift is proportional to distance, one wonders if they, almost unknowingly, used the red-shift as a measure of their "distance".

Some particular forces at a distance make sense, like forces in $\frac{1}{r^2}$, corresponding to potentials in $\frac{1}{r}$, because they appear naturally when one considers the solution of a partial differential equation with a Laplacian in a domain of $\mathbb{R}^3$, due to the fact that $\frac{1}{4\pi\, r}$ is the elementary solution of $-\Delta$, so that the formula integrating all forces acting on a "particle" is just a way to express the solution with the help of a Green kernel.[2] However, one should pay attention that POINCARÉ also suggested that when a classical argument involves forces acting at a distance it means that there is an underlying hyperbolic system for propagating the information about "particles", and not an elliptic equation, so that one may be looking at a stationary solution of an hyperbolic system, and indeed Laplacians appear when one looks for stationary solutions of the Maxwell–Heaviside equation. Because specialists of fake mechanics pretend that nature minimizes energy, I wrote [113] to explain what was observed or understood about where energy goes, hoping that a few

---

[1] A parsec is the distance at which 280 million kilometres (the size of the trajectory of the earth around the sun) is seen under an angle of one second of arc.

[2] George GREEN, English mathematician, 1793–1841. He was a miller.

L. Tartar, *The General Theory of Homogenization*, Lecture Notes of the Unione Matematica Italiana 7, DOI 10.1007/978-3-642-05195-1_3,

confused mathematicians could understand that they were pushed to make a mistake of pseudo-logic:[3] one may find equations with Laplacians by minimizing a functional, as was known since the *Dirichlet principle*, not so well named (by RIEMANN) because it was used before by GREEN and by GAUSS,[4] but nature does not arrive at stationary solutions by a process of minimization, since *nature uses conservative hyperbolic systems.* If one observes what looks like a stationary solution, it is because the conserved quantities are carried by waves, away in space or away in the dual variables, so that if one thinks that the solution is stationary it is because one does not see the energy bouncing around in the range of high frequencies; in such situations one often hears people say that there is "noise", but what is going on has nothing to do with the probabilities that they impose on the problem.[5]

The potential in $\frac{1}{r}$ is actually just a particular case of the family of potentials in $\frac{e^{-\alpha\,r}}{r}$ for $\alpha \geq 0$, which were used by physicists, for example by YUKAWA for studying the short-range nuclear forces,[6] or in plasma physics with $\alpha = \frac{1}{r_D}$, where the Debye radius r_D depends upon some parameters in the plasma,[7] and it is interesting to observe that $\frac{e^{-\alpha\,r}}{4\pi\,r}$ is the elementary solution of $-\Delta + \alpha^2$. This is also true when α is purely imaginary, and this case appears when one looks at solutions of the wave equation $u_{tt} - c^2 \Delta\,u = f(x)\,e^{i\,\omega\,t}$, and one wants to show that for *large positive times* u looks like $v(x)\,e^{i\,\omega\,t}$ with $-\Delta\,v - \frac{\omega^2}{c^2} v = f$, a question studied as the *limiting amplitude principle* by Cathleen MORAWETZ [65];[8] she also worked

---

[3] Because these specialists of fake mechanics often misattribute results of mathematicians interested in other sciences like me, I also proposed in this article that those who feel the urge to attribute my ideas to others should at least choose someone who introduced new ideas, like Ennio DE GIORGI!

[4] Johann Carl Friedrich GAUSS, German mathematician, 1777–1855. He worked at Georg-August-Universität, Göttingen, Germany.

[5] The solution that one observes is not really independent of t, and it only stays near a stationary solution, but in a distance related to a coarse topology, of a weak type. If the measurements are too precise, it means that they are made in a finer topology, where the solution does not appear to be stationary. In order to resolve that fake "noise", one must first describe what the solution is doing at a small scale, and then one must assert what is really measured at that scale, or one may prefer to change the type of measurement that is made, in order to take into account the new type of information that one has on the real solution.

[6] Hideki YUKAWA, Japanese physicist, 1907–1981. He received the Nobel Prize in Physics in 1949, for his prediction of the existence of mesons on the basis of theoretical work on nuclear forces. He worked in Osaka, Japan.

[7] Petrus (Peter) Josephus Wilhelmus DEBYE, Dutch-born physicist, 1884–1966. He received the Nobel Prize in Chemistry in 1936, for his contributions to our knowledge of molecular structure through his investigations on dipole moments and on the diffraction of X-rays and electrons in gases. He worked at Cornell University, Ithaca, NY.

[8] Cathleen SYNGE-MORAWETZ (daughter of John Lighton SYNGE), Canadian-born mathematician, born in 1923. She works at NYU (New York University), New York, NY.

on this question with Peter LAX and Ralph PHILLIPS [49],[9] whose theory of scattering is obviously relevant for these questions [50], and with James RALSTON and Walter STRAUSS [66].[10,11] Understanding the boundary condition at ∞ for v forces one to wonder if one should use $\frac{e^{i\,\beta\,r}}{4\pi\,r}$ or $\frac{e^{-i\,\beta\,r}}{4\pi\,r}$ (with $\beta = \frac{\omega}{c}$) as the elementary solution of $-\Delta - \frac{\omega^2}{c^2}$, and this choice involves the Sommerfeld radiation condition,[12] which selects the physical outgoing waves $\frac{e^{-i\,\beta\,(r-c\,t)}}{r}$, instead of the incoming waves $\frac{e^{i\,\beta\,(r+c\,t)}}{r}$, which correspond to information coming from ∞ (and serve for large negative times).

The appearance of $\alpha > 0$ is related to a question of homogenization, for domains with tiny holes on which one imposes a Dirichlet condition.[13] It was studied by Jeff RAUCH and Michael TAYLOR in the United States [79],[14,15] and by Evgeny KHRUSLOV and MARCHENKO in the Soviet Union, I think.[16,17] An abstract framework, very similar to that of H-convergence, was developed by Doina CIORANESCU and François MURAT [15, 16], but checking their hypotheses in situations where the maximum principle does not hold is not an easy task; although they start from an equation $-\Delta\,u = f$ and call the extra term $+c(x)\,u$ appearing in the effective equation a "strange term coming from nowhere", there is a simple intuitive explanation for its appearance: assuming u to be smooth, adding a tiny hole ω around a point x_0 on which a Dirichlet condition is imposed changes the solution essentially by $-u(x_0)\,p_\omega$, where p_ω is the capacity potential of ω, and this adds a term $|u(x_0)|^2 c_\omega$ to the energy $\int_\Omega |grad(u)|^2\,dx$, where c_ω is the electrostatic

---

[9] Ralph Saul PHILLIPS, American mathematician, 1913–1998. He worked at USC (University of Southern California), Los Angeles, CA, and at Stanford University, Stanford, CA.

[10] James Vickroy RALSTON Jr., American mathematician, born in 1943. He works at UCLA (University of California at Los Angeles), Los Angeles, CA.

[11] Walter Alexander STRAUSS, American mathematician, born in 1937. He worked at Stanford University, Stanford, CA, and at Brown University, Providence, RI.

[12] Arnold Johannes Wilhelm SOMMERFELD, German physicist, 1868–1951. He worked in Clausthal, Aachen, and München (Munich), Germany.

[13] It is the electrostatic capacity of the holes which is important for the scaling considered, and Jeff RAUCH and Michael TAYLOR observed that the case of thin wires, which leads essentially to a two-dimensional situation, explains that a metal mesh of thin wires has the same effect as a metal sheet for creating a Faraday cage.

[14] Jeffrey Baron RAUCH, American mathematician, born in 1945. He works at University of Michigan, Ann Arbor, MI.

[15] Michael Eugene TAYLOR, American mathematician, born in 1946. He worked at SUNY (State University of New York) at Stony Brook, NY, and at UNC (University of North Carolina), Chapel Hill, NC.

[16] Evgeny Yakovlevich KHRUSLOV, Ukrainian mathematician, born in 1937. He worked in Kharkov, Ukraine.

[17] Vladimir Aleksandrovich MARCHENKO, Ukrainian mathematician, born in 1922. He worked in Kharkov, Ukraine.

capacity of ω,[18] so that if the density of capacity of the holes is c, one expects an effective energy in $\int_\Omega |grad(u)|^2\,dx + \int_\Omega c\,u^2\,dx$. Transforming this intuitive idea into a proof is not so simple, because the potentials p_{ω_j} of distinct holes ω_j do not decay so fast and one must control the interactions between the corrections from the various holes, and Doina CIORANESCU and François MURAT added an hypothesis that the holes were far enough apart; however, because George PAPANICOLAOU and Raghu VARADHAN have a probabilistic version [77] of the same result with a different hypothesis,[19] I guess that there is an improved condition to be discovered.

There is a similar problem concerning a fluid passing through a sieve, and it was studied by Évariste SANCHEZ-PALENCIA, by François MURAT, and by others, but the reason why I find this question important concerns semilinear hyperbolic systems, because I think that it plays an important role for explaining some of the strange rules which physicists devised for "particles" in atomic physics.

Around 1980, I heard a seminar talk by Yvonne CHOQUET-BRUHAT (who used her husband's name in front of hers as was done in her generation),[20] about a global existence result for the coupled Maxwell–Heaviside/Dirac system with small initial data in a fractional Sobolev space. Her somewhat miraculous geometrical proof [14] followed a suggestion of Demetrios CHRISTODOULOU,[21] to use the conformal invariance of the system, only valid if the Dirac part of the equation has no mass term, and for that she used a special conformal transformation constructed by Roger PENROSE.[22] I was surprised that she had not much idea about the physics behind the equations

[18] In dimension $N \geq 3$, for ω bounded with a smooth boundary, p_ω is the solution of $\Delta p_\omega = 0$ in $\mathbb{R}^N \backslash \omega$ with $p_\omega = 1$ on $\partial\omega$ and $p_\omega = 0$ at ∞; then $c_\omega = \int_{\mathbb{R}^N \backslash \omega} |grad(p_\omega)|^2\,dx$.

[19] Sathamangalam Raghu Srinivasa VARADHAN, Indian-born mathematician, born in 1940. He received the Abel Prize in 2007, for his fundamental contributions to probability theory and in particular for creating a unified theory of large deviations. He works at NYU (New York University), New York, NY.

[20] Yvonne BRUHAT-CHOQUET, French mathematician, born in 1923. She worked at UPMC (Université Pierre et Marie Curie), Paris, France.

[21] Demetrios CHRISTODOULOU, Greek-born mathematician, born in 1951. He worked in Athens, Greece, at Syracuse University, Syracuse, NY, at NYU (New York University), New York, NY, at Princeton University, Princeton, NJ, and he works now at ETH (Eidgenössische Technische Hochschule), Zürich, Switzerland.

[22] Sir Roger PENROSE, English mathematician, born in 1931. He received the Wolf Prize (in Physics!) in 1988, jointly with Stephen W. HAWKING, for their brilliant development of the theory of general relativity, in which they have shown the necessity for cosmological singularities and have elucidated the physics of black holes. In this work they have greatly enlarged our understanding of the origin and possible fate of the Universe. [This is the official reason for them receiving the prize, and it does not reflect my opinion, that it results from a huge Comte complex among physicists, and that a lot of what is done concerning gravitation is sheer nonsense!] He worked at Birkbeck College, London, he held the Rouse Ball professorship (1973–1998)

that she worked on, and that she trusted another differential geometer for that,[23] André LICHNEROWICZ,[24] who told her that the Dirac equation with zero mass term models neutrinos.[25]

Two years after, while reading a book suggested by Robert DAUTRAY when I was working at CEA, I learned how DIRAC obtained his equation, *rightly cheating with the dogmas of quantum mechanics*, and being quite creative in *keeping the symmetries of the theory of relativity*, but I disagreed with his last step, when he added a term containing the "mass of the electron", because he wanted a dispersion relation to fit with something previously used for an "electron". I understand that engineers may fit parameters in ad hoc models so that they can control a process for which the equations are not known, but scientists should not put in their hypotheses what is needed for obtaining a conclusion that was observed! However, I disagreed with DIRAC's choice for another reason, because I thought that a (possibly different) mass term could appear by homogenization, carrying the information about the energy stored in some concentration effects, so that *mass would only be concentrated electromagnetic energy inside "particles"*. Later, I read an interesting suggestion of BOSTICK [9], about a toroidal shape for an "electron", and he also thought of its mass being pure electromagnetic energy, and I thought then that the concentration effects which I was thinking about could appear in

---

in Oxford, England, and then became Gresham Professor of Geometry at Gresham College, London, England.

[23] I observed that differential geometers usually learn nothing about physics, and that they do not perceive that most of the physicists whom they meet suffer from a Comte complex.

[24] André LICHNEROWICZ, French mathematician, 1915–1998. He worked in Strasbourg and Paris, France; he held a chair (physique mathématique, 1952–1986) at Collège de France, Paris.

[25] In his course on continuum mechanics at École Polytechnique [58], Jean MANDEL linearized the equation of hydrodynamics in an infinite ocean of fixed depth H (around a zero velocity field), and after observing that disturbances of the surface decompose into one-dimensional sinusoidal waves travelling at a speed $V(H)$, and that it is the top of one of these sinusoidal waves (easily followed with the eye) which travels at this speed, he pointed out that it is a phase velocity and that there is no transport of mass, although there is transport of linear momentum (and a floating cork moves a little when the waves go by, but does not drift). If there is a sharp decrease in depth near a beach, for example due to the presence of a submerged coral reef, the waves from the open sea arrive much faster than the local speed favoured by the waves and that creates the breaking of waves, the delight of surfers, whose art is precisely about using the linear momentum transported by these waves. A neutrino could be a similar kind of wave, transporting angular momentum with no transport of mass, for a semi-linear hyperbolic system like the Maxwell–Heaviside/Dirac system with no mass term; however, this system also describes "electrons" and "positrons" and plenty of other "particles". I wonder if FERMI, who postulated the existence of the neutrino for a question of conservation of angular momentum in a collision between "particles", thought about that analogy, but he was probably not thinking in terms of waves, like most of the physicists who followed him.

knotted structures, describing different "particles", and although this may look like the dream of Kelvin,[26,27] or string theory, it is important to observe that because I am interested in *semi-linear hyperbolic systems with only the speed of light c as characteristic speed*, like the Maxwell–Heaviside/Dirac system, I am taking into account the observations of POINCARÉ in his principle of relativity, which the other approaches do not; up to now, I was not able to extend my theory, for studying oscillations and concentration effects in sequences of solutions of semi-linear hyperbolic systems.

Before looking at turbulence through the homogenization of first-order differential equations with oscillating coefficients, which is not understood yet, but shows appearance of nonlocal effects in some examples, I wanted to study a model with more features from fluid dynamics, like viscosity and pressure. The problem of understanding small viscosity effects, or high Reynolds numbers,[28] being considered too difficult, I found it useful to invent a simpler model retaining as much as possible of the qualitative properties which I was interested in. In 1976, I used the fact that the nonlinear term in the Navier–Stokes equation may be written $u \times curl(-u) + grad\left(\frac{|u|^2}{2}\right)$, and because of an analogy with electromagnetism where terms in $u \times B$ are due to Lorentz forces (so that the induction field B makes charged particles turn),[29]
I decided to replace $curl(-u)$ by a given oscillating field in order to study its effect. Not knowing what to expect, I decided to begin with the stationary case, and to use the method of asymptotic expansions in a periodic setting for

$$-\nu\,\Delta\,u^\varepsilon + u^\varepsilon \times \frac{1}{\varepsilon}b\Big(\frac{x}{\varepsilon}\Big) + grad\,p_\varepsilon = f, div(u^\varepsilon) = 0 \text{ in } \Omega, u^\varepsilon \in H^1_0(\Omega;\mathbb{R}^3), \tag{3.1}$$

and I did the formal computations with Michel FORTIN,[30] who was visiting Orsay that year and was sharing my office; we first noticed that the average of the periodic vector field b must be 0, or the whole fluid would turn very fast, and in that case we derived an equation satisfied by the first term of the formal expansion; I easily proved the result that we obtained, by using my

[26] William THOMSON, Irish-born physicist, 1824–1907. In 1892 he was made Baron Kelvin of Largs, and thereafter known as Lord Kelvin. He worked in Glasgow, Scotland.

[27] Because THOMSON/Kelvin thought that the universe is made of vortices, he and TAIT initiated the mathematical theory of knots.

[28] Osborne REYNOLDS, Irish-born mathematician, 1842–1912. He worked in Manchester, England.

[29] Hendrik Antoon LORENTZ, Dutch physicist, 1853–1928. He received the Nobel Prize in Physics in 1902, jointly with Pieter ZEEMAN, in recognition of the extraordinary service they rendered by their research into the influence of magnetism upon radiation phenomena. He worked in Leiden, The Netherlands. The Institute for Theoretical Physics in Leiden, The Netherlands, is named after him, the Lorentz Institute.

[30] Michel FORTIN, Canadian mathematician. He works at Université Laval, Québec, Québec.

method of oscillating test functions [95]. It is easy to avoid periodicity hypotheses and to consider terms of the form $u^n \times curl(v^n)$ with v^n converging weakly, but I noticed something else when I wrote it down for a meeting at IMA, Minneapolis, MN, in the fall of 1984 [102]: I considered

$$-\nu\,\Delta\,u^n + u^n \times curl(v^\infty + \lambda\,w^n) + grad\,p_n = f, div(u^n) = 0 \text{ in } \Omega, \quad (3.2)$$

with $v^\infty \in L^3(\Omega;\mathbb{R}^3)$ and $w^n \rightharpoonup 0$ in $L^3(\Omega;\mathbb{R}^3)$ weak, and I did not impose boundary conditions but I assumed that $u^n \rightharpoonup u^\infty$ in $H^1(\Omega;\mathbb{R}^3)$ weak,[31] and I showed that there exists a symmetric nonnegative matrix M, depending only upon ν and a subsequence w^m of w^n that one may have to extract, such that

$$u^m \times curl(w^m) \rightharpoonup \lambda\,M\,u^\infty \text{ in } H^{-1}_{loc}(\Omega;\mathbb{R}^3) \text{ weak}, \quad (3.3)$$

$$\nu\,|grad(u^m)|^2 \rightharpoonup \nu\,|grad(u^\infty)|^2 + \lambda^2(M\,u^\infty, u^\infty) \text{ in } \mathcal{M}(\Omega) \text{ weak} \star, \quad (3.4)$$

so that

$$-\nu\,\Delta\,u^\infty + u^\infty \times curl(v^\infty) + \lambda^2 M\,u^\infty + grad\,p_\infty = f, div(u^\infty) = 0 \text{ in } \Omega. \quad (3.5)$$

Although the force $u^m \times curl(w^m)$ does no work since it is perpendicular to the velocity u^m, it induces oscillations in $grad(u^m)$, so that more energy is dissipated by viscosity (per unit of time, since this is a stationary situation), and (3.4) shows the interesting feature that the added dissipation which appears in the limiting equation is not quadratic in $grad(u^\infty)$, but quadratic in u^∞, contrary to a quite general belief about *turbulent viscosity*; however, one should be careful that the model (3.2) is not compatible with Galilean invariance (after one adds a term $\frac{\partial u}{\partial t}$), so that this remark should be put in the right context. George PAPANICOLAOU later mentioned that forces in $M\,u$ are called Brinkman forces,[32] which are usually connected to the drag created by obstacles in the fluid, so that one may think that the oscillations in w^m act like little obstacles creating a drag. This is described in Chap. 19.

What surprised me more was the exact quadratic effect with respect to the strength parameter λ, and the proof gave me the first hint about *H-measures*, which I was too lazy to define correctly at that time. However, I felt that the quadratic effect was explaining some formulas in quantum mechanics (but not all, since a few different things are mixed under this name), and I was

[31] It is a natural requirement in homogenization, that if one wants to speak about the effective properties of a mixture one should obtain a result which is independent of the boundary conditions. If one fails to do this, one is only talking about global properties of the mixture together with its container. Although he did not mention applications to mixtures in continuum mechanics or physics, Sergio SPAGNOLO obtained such independence results in his work on G-convergence.

[32] Henri Coenraad BRINKMAN, Dutch physicist, 1908–1961. He worked in Delft, The Netherlands.

quite disappointed when I discussed this matter with David BERGMAN and Graeme MILTON,[33,34] because they did not see any reason to correct quantum mechanics,[35] so that I waited two years before looking for a precise mathematical definition, in order to prove results of *small-amplitude homogenization* [105],[36] and immediately after I checked that M can be computed from an H-measure μ associated to the sequence w^m; I found that M is $\frac{1}{\nu}$ times a linear combination of fourth-order moments of μ in ξ [105].[37] This is described in Chaps. 28 and 29.

I then checked the evolution problem, in the whole space in order to avoid some difficult estimates for the "pressure",[38] and I found that a similar correction term appears [103]. Ten years after, with Konstantina TRIVISA and Chun LIU,[39,40] we looked for a formula giving the corresponding matrix, and we found the need for a variant of *H-measures* with a parabolic scaling [112].

The models that I considered do not have Galilean invariance, but the analysis of first-order partial differential equations already showed that the appearance of nonlocal effects destroys the possibility of reversing time in a classical manner, so that one should probably ask more general questions about group invariance of effective equations. I remember now an observation

---

33 David J. BERGMAN, Israeli physicist. He works in Tel Aviv, Israel.

34 Graeme Walter MILTON, Australian-born physicist, born in 1956. He worked at NYU (New York University), New York, NY, and he works now at University of Utah, Salt Lake City, UT.

35 Later, Graeme MILTON mentioned to me that one needs three-point correlations for scattering phenomena, like those appearing in the experiments of spectroscopy which I would like to treat by homogenization, so that one needs other mathematical tools than *H-measures* or their variants, which are somewhat related to two-point correlations.

36 My reason was to give a rational explanation for the efficiency of a formula guessed by LANDAU and LIFSHITZ, whose arguments made absolutely no sense, since they "derived" it from another formula which was obviously false in general.

37 Here μ is a tensor, and it was not in connection with the preceding problem that I characterized the fourth-order moments of a scalar nonnegative measure on $\mathbb{S}^2$ with Gilles FRANCFORT and François MURAT [32], since we were interested in the scalar H-measure associated with the fluctuations of the shear modulus in a question of small-amplitude homogenization in linearized elasticity (using only isotropic materials). My motivation was to explain the five-fold symmetry observed in quasi-crystals, which has no relation with Penrose tilings, of course, since it is the result of microstructures in a metallic ribbon changing to evacuate heat and release stress!

38 Later, Wolf VON WAHL pointed out to me that he used a semi-group approach for handling the "pressure" in the case of an open set Ω.

39 Konstantina TRIVISA, Greek-born mathematician. She worked at Northwestern University, Evanston, IL, and she works now at UMD (University of Maryland), College Park, MD. She was a post doctoral associate of CNA (Center for Nonlinear Analysis) at CMU (*Carnegie Mellon University*), Pittsburgh, PA.

40 Chun LIU, Chinese-born mathematician. He works at Penn State (Pennsylvania State University), State College, PA. He was a post doctoral associate of CNA (Center for Nonlinear Analysis) at CMU (*Carnegie Mellon University*), Pittsburgh, PA.

of Joel Robbin,[41] during one of my stays at UW, Madison, WI, probably in 1980 or earlier: he wondered why the invariance by the Lorentz group, valid in a vacuum, is no longer true in the presence of matter. Quite recently, Constantine Dafermos mentioned that Heaviside already considered memory effects in electromagnetism, and I wonder if Heaviside, or Poincaré in his pioneering work on the principle of relativity, were already aware of that question.[42] The main difficulty is to understand what matter is, without postulating too much about its behaviour, and my guess is that one should look at homogenization questions for the Maxwell–Heaviside/Dirac system, but they are mostly open, so that I cannot yet make precise conjectures about how the invariance by the Lorentz group holds for the correct effective equations, which may involve a hierarchy of partial differential equations, a part of that theory *beyond partial differential equations* which I tried to perceive for many years.

One should be prepared for the general idea that small-scale effects do not behave like the large-scale effects that one is used to in continuum mechanics, but do not behave either as was postulated by the laws of thermodynamics. A new thermodynamics must be developed that should permit each form of information hidden at a mesoscopic level to be transported as it wants to, and one should introduce adapted mathematical objects for describing this evolution; it goes without saying, but it is important to repeat it, that one should avoid using probabilities as much as possible, and that one should prove the correct behaviour from the analysis of partial differential equations. Only then will one have a rational explanation for all the terms appearing in the effective equations, but one should also be prepared to see these terms involve new quantities whose evolution will be described by other partial differential equations, like in the hierarchy that I guessed above.[43]

As my initial discovery of homogenization with François Murat concerned questions of optimal design, I considered it important to find the precise set of effective coefficients that could be obtained by mixing various materials.[44] The method which we first used relies on the *div–curl lemma* (actually its simpler form for gradients, easily seen by integration by parts) [92,93], and at the end of 1977 I proposed an improved method based on our more general *compensated compactness* theory [97]. In June 1980, while I was visiting the

[41] Joel William Robbin, American mathematician, born in 1941. He works at UW (University of Wisconsin), Madison, WI.

[42] Constantine Michael Dafermos, Greek-born mathematician, born in 1941. He worked at Cornell University, Ithaca, NY, and he works now at Brown University, Providence, RI.

[43] Although these new equations may look like some equations from kinetic theory, one should be aware that they should not look like the Boltzmann equation, the defects of which I described in [119].

[44] It is a question often referred to as finding the G-closure of a set.

Courant Institute at NYU,[45] I used my method for proving new bounds in the case of mixing two isotropic conductors, when the effective conductivity is isotropic, and George PAPANICOLAOU told me to compare them with the Hashin–Shtrikman bounds: Zvi HASHIN and SHTRIKMAN guessed the correct bounds [38], but their "proof" did not make sense. I understood much later that filling the gap in their argument "requires" the mathematical tool of *H-measures* [105],[46] which I only developed in 1987, twenty-five years after their argument appeared.

It sometimes takes a long time before a mathematician explains a formal argument, like for Laurent SCHWARTZ giving a meaning to the calculus of HEAVISIDE, or (with a shorter delay) to some computations of DIRAC, by his theory of distributions [85], and I think that his work came out of a purely mathematical question, to give a meaning to the Fourier series $\sum_{n\in\mathbb{Z}} c_n e^{2i\,\pi\,n\,x}$ when the complex Fourier coefficients c_n have at most a polynomial growth in n [84]. I do not know if the person who asked this question thought about particular computations made by engineers or physicists, but in my case I developed the theory of *H-measures* [105] for clarifying some formal computations by LANDAU and LIFSHITZ [47], and not the formal argument of Zvi HASHIN and SHTRIKMAN, [38] which I forgot.[47] However, their construction that the bounds are attained by a geometry of coated spheres was already clear to me in 1980, and I would have understood their construction if I read their article after 1975, but not before.[48] I treated the case where the effective conductivity is anisotropic with François MURAT,

[45] Richard COURANT, German-born mathematician, 1888–1972. He worked at Georg-August-Universität, Göttingen, Germany, and at NYU (New York University), New York, NY. The department of mathematics of NYU is named after him, the Courant Institute of Mathematical Sciences.

[46] As a mathematician, I am careful about claiming that a result needs a particular proof, but here it is not about finding a different proof of the statement, which I was the first to obtain in June 1980, but about understanding if one can salvage an incomplete step in the argumentation used in 1962 by Zvi HASHIN and SHTRIKMAN. Finishing the construction of a house does not mean tearing it down and building a new house along different plans drawn by another architect, and it does not mean either explaining how easy it would be to finish the construction if the house was built in wood, when it is not the case; I use this analogy because some people later wrote a proof in a periodic situation (without mentioning my earlier more general proof that they heard), and Zvi HASHIN and SHTRIKMAN could easily handle a periodic situation, but they knew periodicity to be utterly unrealistic, and they used their physicist's intuition for extending the computations to a general (non-periodic) situation. It is precisely a result of my theory of *H-measures* [105] to provide a mathematical framework where some proofs in a general (non-periodic) situation follow the pattern of the proof in a periodic situation, and I did not hear of earlier mathematical results giving this possibility!

[47] I easily forgot the steps of the formal computation of Zvi HASHIN and SHTRIKMAN, because I did not recognize any physical principle behind them, and they did not propose a general method for proving bounds as I did.

[48] Because I needed the simplified approach to my joint work with François MURAT, which I developed during my first stay at UW, Madison, WI, in 1974–1975.

and proving the bounds used the same functionals in my method, and we thought of generalizing the Hashin–Shtrikman coated spheres geometry by using coated confocal ellipsoids; I asked a question to Edward FRAENKEL,[49] who explained to me three-dimensional ellipsoidal coordinates, but we managed to avoid using them. I presented our result at a meeting at NYU, New York, NY, in June 1981, where I first met Graeme MILTON, still a graduate student from Australia, with already an amazing understanding about bounds on effective coefficients, on his way to becoming the best specialist in the world for that question, for which he described his approach in [64]; I only wrote the article two years later, for a conference in honour of Ennio DE GIORGI [101]. This is described in Chaps. 21, 25, and 26.

Because of a widespread tendency among the advocates of fake mechanics to attribute my ideas to others, my method [97] of 1977 for proving bounds on effective coefficients is rarely attributed to me, and it could be because it was (wrongly) called "the method of translations" by Graeme MILTON,[50] who even described it using the term quasiconvexity which is now almost synonymous to fake mechanics,[51] while my method is based on the *compensated compactness* ideas that I developed with François MURAT [72–74, 98], in the spirit of what I thought in connection with real questions of continuum mechanics in. [98] It would not be too much of a problem if an American did not add to the confusion: in the spring of 1983,[52] he asked me if he could

[49] Ludwig Edward FRAENKEL, German-born mathematician, born in 1927. He worked in London, in Cambridge, in Brighton, and in Bath, England.

[50] I told Graeme MILTON that the name "method of translations" does not describe well my method from [97], but I forgot to mention that the name *method of translations* correctly describes a method of Louis NIRENBERG, who used translations for proving the regularity of solutions of elliptic equations, and I learned that name as a graduate student in the late 1960s, in lectures of Jacques-Louis LIONS.

[51] Quasiconvexity was introduced by MORREY for extending a question of calculus of variations to multi-dimensional problems. Nowadays, the term calculus of variations is often used for unrelated questions of optimization or partial differential equations, by naive "mathematicians" who hope to improve in this way the status of what they do; curiously enough, most of those people believe in the fake mechanics/physics principle of minimizing potential energy, instantaneously! It is almost an insult to associate my name with such nonsense.

[52] We were visiting MSRI (Mathematical Sciences Research Institute), Berkeley, CA, and this American already behaved strangely, attributing homogenization only to Leon SIMON in one of his talks, and I wondered why he was not mentioning the work in the late 1960s of the Italian school, Sergio SPAGNOLO and Ennio DE GIORGI, and the work in the early 1970s of the French school, François MURAT and myself. Leon SIMON found independently [88] an abstract framework for H-convergence [71], and the referee of his article mentioned my article for a conference in Japan in the fall of 1976 [94] (before François MURAT coined the term H-convergence), which was then added in the introduction of [88]. Because a few years after, this American was not mentioning Leon SIMON for homogenization anymore, I thought that his real goal was to avoid any mention of my name, and that he realized that [88] gave me priority!

explain to one of his friends in the USSR, the detail of my method,[53] and it was then quite strange that a few years later he would attribute my method to his friend, who also behaved quite strangely afterward, threatening to sue if I repeated what he told Jean-Louis ARMAND when he came to visit him![54] I must say that I do not understand the rationale for resorting to threats, a behaviour which I did not expect to find in "academic circles", as if it was difficult for good researchers to notice that some "mathematicians" do not even understand the methods which they claim as theirs! Anyway, I found a more general method for obtaining bounds on effective coefficients, using my theory of *H-measures* [105], and I describe it in Chap. 30.

Still in the spring of 1983, at a conference at UW, Madison, WI, I heard Michael RENARDY talk about a joint work with Dan JOSEPH,[55,56] on two-phase flows in pipes, either for the effect of adding a little water to the oil in a pipeline, for a better lubrication, or for the extrusion of a mixture of two molten polymers,[57] and the observations led them to a conjecture, which they verified when the cross-section is a disc. Because they were considering Poiseuille flows,[58] reducing the problem to an elliptic equation on the cross-section, this was one of the optimal design problems which I studied with François MURAT, and their conjecture was that there was a classical solution, which I doubted for an arbitrary cross-section. A month after, discussing with Joel SPRUCK in the lounge of the Courant Institute at NYU,[59] he recalled a result of James SERRIN [87],[60] which I heard him present in

---

[53] I described the method in 1977 [97], but there was a question of choice of functionals, which I explained at the meeting in New York, NY, in June 1981; I only wrote it down in detail in the fall of 1983 [101], for a conference in Paris, France, in honour of Ennio DE GIORGI.

[54] Jean-Louis ARMAND, French engineer, born in 1944. He worked at IRCN (Institut de Recherches de la Construction Navale), Paris, France, at UCSB (University of California at Santa Barbara), Santa Barbara, CA, at University of Aix-Marseille II (Université de la Méditerranée), Marseille, France, at AIT (Asian Institute of Technology), Klongluang, Thailand; he works now at the French Embassy in Tokyo, Japan.

[55] Michael RENARDY, German-born mathematician, born in 1955. He worked at UW (University of Wisconsin), Madison, WI, and he works now at VPISU (Virginia Polytechnic Institute and State University), Blacksburg, VA.

[56] Daniel D. JOSEPH, American mathematician, born in 1929. He worked at Illinois Institute of Technology, Chicago, IL, and at UMN (University of Minnesota), Minneapolis, MN.

[57] Michael RENARDY wrote a book [80] with my colleague Bill HRUSA, and with John NOHEL, on similar mathematical questions for visco-elastic materials.

[58] Jean-Louis Marie POISEUILLE, French physician, 1797–1869. He worked in Paris, France.

[59] Joel SPRUCK, American mathematician, born in 1946. He worked at University of Massachusetts, Amherst, MA, and he works now at Johns Hopkins University, Baltimore, MD.

[60] James B. SERRIN, American mathematician, born in 1926. He worked at UMN (University of Minnesota), Minneapolis, MN.

1972 at a conference in Jerusalem, Israel, and he mentioned a much simpler proof by Hans WEINBERGER [122],[61] and these results were helpful for showing that, for a simply connected cross-section, the conjecture of Dan JOSEPH and Michael RENARDY only holds for a circular cross-section [75].[62]

Because a related question is to study the stability of a Poiseuille flow, among non-Poiseuille flows, of course, I wondered if the nonexistence of a Poiseuille flow solving the conjecture of Dan JOSEPH and Michael RENARDY was not the sign that some kind of turbulent flow would appear in parts of the pipe, and I was naturally led to wonder if turbulent flows are not trying to optimize something. However, I discovered later in a book by Dan JOSEPH on stability of fluid motions [42] that this idea was already proposed by BUSSE,[63] but I only found one of his articles, which looked to me like some kind of self-similar microstructure for boundary layers, and I wondered if his idea was not restricted to effects arising near the boundary. I easily imagined that convection could help for creating a flow transporting mass parallel to the boundary, but with a much larger heat flux than diffusion would allow for a Poiseuille-like flow, in a similar way that convection cells appear in the Rayleigh–Bénard instability,[64,65] so that the efficiency of a turbulent flow could be seen by the angle between the direction of transport of mass and the direction of transport of heat; when I asked a question about that to Olivier PIRONNEAU,[66] he confirmed that heat seems to flow in strange ways in turbulent flows.[67]

[61] Hans Felix WEINBERGER, Austrian-born mathematician, born in 1928. He worked at UMD (University of Maryland), College Park, MD, and at UMN (University of Minnesota), Minneapolis, MN. In 1948, Hans WEINBERGER was one of the first students in the graduate programme at Carnegie Tech (Carnegie Institute of Technology), now part of CMU (*Carnegie Mellon University*), Pittsburgh, PA, and he was a PhD student of my late colleague Richard DUFFIN.

[62] When I gave a talk in Minneapolis in the spring of 1984, Hans WEINBERGER pointed out a mistake that I made concerning the regularity of some solutions of a nonlinear partial differential equation, which I corrected before [75] was in print.

[63] Friedrich H. BUSSE, German-born physicist, born in 1936. He worked at UCLA (University of California at Los Angeles), Los Angeles, CA, and in Bayreuth, Germany.

[64] John William STRUTT, 3rd Baron Rayleigh, English physicist, 1842–1919, known as Lord Rayleigh. He received the Nobel Prize in Physics in 1904, for his investigations of the densities of the most important gases and for his discovery of argon in connection with these studies. He worked in Cambridge, England, holding the Cavendish professorship (1879–1884), after MAXWELL.

[65] Henri BÉNARD, French physicist, 1874–1939.

[66] Olivier PIRONNEAU, French mathematician, born in 1945. He worked at Université Paris Nord, Villetaneuse, and he works now in LJLL (Laboratoire Jacques-Louis Lions) at UPMC (Université Pierre et Marie Curie), Paris, France.

[67] I do not consider the law of diffusion of heat of FOURIER, or the law of diffusion of matter of FICK to be physical; they should only be considered as first approximations of more realistic physical laws.

It is easy to imagine situations where one needs to understand more than one type of effective coefficients, the simplest being to consider two kinds of diffusion, of heat and of electricity for example, and both David BERGMAN and Graeme MILTON treated that question by using a special class of functions of one complex variable which send the upper half plane into itself [7], called Pick functions in harmonic analysis,[68] and such an idea seems to go back to PRAGER;[69] they used this approach in order to derive bounds on effective coefficients, and Graeme MILTON observed that some best bounds involve Padé approximants,[70] but this approach has the defect of considering only mixtures which always give isotropic effective conductivities. This hypothesis is not so realistic if there are no symmetries in the patterns used, and I found no difficulty avoiding it and developing a theory for general mixtures of m materials, using the numerical range of a matrix studied by Eduardo ZARANTONELLO [126],[71] but not much is known for the corresponding class of matrix-valued functions which appears, defined for an adapted set of m matrices with complex entries. I describe this in Chap. 22.

One important goal is to develop a mathematical theory for describing the evolution of microstructures, and one must go beyond the first level which I described in the lectures that I gave at the invitation of Robin KNOPS,[72] at Heriot–Watt University,[73,74] Edinburgh, Scotland, in the summer of 1978.

In my lecture notes [98], actually written by Bernard DACOROGNA,[75] the first step is to use *Young measures* for taking into account the (usually non-

---

[68] Georg Alexander PICK, Austrian-born mathematician, 1859–1942. He worked in Prague, now capital of the Czech Republic.

[69] William PRAGER, German-born mathematician, 1903–1980. He worked in Göttingen and in Karlsruhe, Germany, in Istanbul, Turkey, and at Brown University, Providence, RI.

[70] Henri Eugène PADÉ, French mathematician, 1863–1953. He worked in Lille, in Poitiers, and in Bordeaux, France, and then as chancellor in Besançon, in Dijon and in Aix-Marseille, France.

[71] Eduardo Hector ZARANTONELLO, Argentinean mathematician, born in 1918. He worked in La Plata, in Córdoba, in San Juan, and in San Luis y Cuyo, Argentina. When I first met him in 1971, during my first trip to United States, he was working at MRC (Mathematics Research Center) in Madison, WI, and I met him in the early 1980s at the Scuola Normale Superiore in Pisa, Italy, and in the 1990s he was still working, in Mendoza, Argentina.

[72] Robin John KNOPS, English mathematician, born in 1932. He worked in Nottingham, and in Newcastle-upon-Tyne, England, and he works now at Heriot–Watt University, Edinburgh, Scotland.

[73] George HERIOT, Scottish goldsmith and philanthropist, 1563–1624. He left money to found the Heriot Hospital in Edinburgh, Scotland, part of which became the George Heriot School, and part of which merged with the Watt Institution to form Heriot–Watt College, which became Heriot–Watt University in 1966.

[74] James WATT, Scottish engineer, 1736–1819. He worked in Glasgow, Scotland. Heriot–Watt University in Edinburgh, Scotland, is partly named after him.

[75] Bernard DACOROGNA, Egyptian-born mathematician, born in 1953. He works at EPFL (École Polytechnique Fédérale de Lausanne), Lausanne, Switzerland.

linear) pointwise constitutive relations,[76] and although I was the first to introduce *Young measures* in a context of partial differential equations and for questions of continuum mechanics, I used the term *parametrized measures* which I heard in the late 1960s in talks at the Pallu de la Barrière seminar at IRIA,[77,78] Rocquencourt, France. It was not said in Paris that the notion was first introduced in the late 1930s by Laurence YOUNG [123, 124], whom I actually met in the spring of 1971 at UW, Madison, WI, during my first trip to the United States, and his book [125] was not noticed; it was Ron DIPERNA who proposed later to abandon the name parametrized measures and use the better name *Young measures*.[79,80]

In [98] the second step is to use the *compensated compactness* theory that I developed with François MURAT for deducing information on weak limits of quadratic quantities because of the linear partial differential laws which are satisfied [72–74], so that the *Young measures* are constrained by the differential balance equations.

In [98] the third step is to use the new partial differential equations which follow from the constitutive relations and the balance equations, which Peter LAX named "entropies" [48]; one must notice that this step is not dependent upon a possible hyperbolic character of the system considered.[81] My method of [97] for obtaining bounds on effective coefficients only uses the first two steps of [98], and since *H-measures* are just a better way to per-

---

[76] Laurence Chisholm YOUNG, English-born mathematician, 1905–2000. He worked in Cape Town, South Africa, and at UW (University of Wisconsin), Madison, WI, where I first met him in 1971, during my first trip to United States.

[77] Robert PALLU DE LA BARRIÈRE, French mathematician, born in 1922. He worked in Caen and at UPMC (Université Pierre et Marie Curie), Paris, France.

[78] IRIA later became INRIA (Institut National de Recherche en Informatique et Automatique). Informatique is the French term for computer science, and automatique is the French term for control theory.

[79] Ronald John DIPERNA, American mathematician, 1947–1989. He worked at Brown University, Providence, RI, at University of Michigan, Ann Arbor, MI, at UW (University of Wisconsin), Madison, WI, at Duke University, Durham, NC, and at UCB (University of California at Berkeley), Berkeley, CA.

[80] Those who switched back many years after to the obsolete term of parametrized measures may want to show that they were rewriting my text [98], and probably attributing all my ideas to their friends.

[81] Some adepts of fake mechanics preposterously claimed that I only introduced my method for hyperbolic problems, but the truth is that I developed it first because of questions of stationary (nonlinear) elasticity, and there the equation is elliptic and the entropies used are Jacobian determinants. Then, having rendered obsolete the notion of quasiconvexity used by the adepts of fake mechanics, I wanted to study the evolutionary (nonlinear) elasticity, which indeed is hyperbolic. The adepts of fake mechanics usually ignore that time exists, and when they finally use it they seem not to know that the basic equations of continuum mechanics are hyperbolic, and that it is only by a lack of understanding of a better thermodynamics that one introduces diffusion terms which render the equations partially parabolic.

form the second step of *compensated compactness*, the new way to prove bounds that I introduced in [105] is not so different, but it mixes *Young measures* and *H-measures* and the analogue of the third step would be to have a better understanding about the relations between *Young measures* and *H-measures*, because *H-measures* carry some information on the partial differential equations. I obtained some results in this direction with François MURAT, extending what I mentioned in [105], and presented it in Ferrara, Italy, in the fall of 1991 [109],[82] and in Udine, Italy, in the summer of 1994 [110]; our method of construction, presented in Chap. 33, is very similar to that used by Graeme MILTON [63] and Enzo NESI,[83] but two different improvements seem necessary for going forward: on one hand the necessary conditions given by the *compensated compactness method* should be improved, for providing more relations between the *Young measures* and the *H-measures* of a sequence, possibly by defining a more general type of "entropies" [115], and on the other hand the sufficient conditions given by the constructions by multiple laminations should be improved, possibly by defining a more general type of geometrical constructions.[84]

Despite a limited mathematical understanding about how to characterize the set of effective coefficients for a given situation, I can rephrase in the following way my conjecture that turbulent flows are trying to do something optimal: if one considers the set of effective coefficients concerning the transport of mass, the transport of linear momentum, the transport of angular momentum, the transport of energy, including the heat flux, and a few other interesting quantities like pressure and the Cauchy stress tensor,[85] turbulent flows try to create points on the boundary of this (still not so well known) set; in some way, it is not the detail of the flow which is important but the effective parameters that it creates, and I suppose that once the effective set becomes understood better, one will realize that some turbulent flows are neither isotropic nor stationary, so that KOLMOGOROV's ideas do not

[82] I was presenting an application to a problem of micromagnetism [10], and only the results of Antonio DE SIMONE were obtained (independently) around the same time, by a more explicit construction [23].

[83] Vincenzo NESI, Italian mathematician, born in 1959. He works at Università degli Studi di Roma "La Sapienza", Roma (Rome), Italy.

[84] Graeme MILTON described an argument for linearized elasticity, when one mixes more than seven (isotropic) materials, that some effective properties cannot be obtained by multiple laminations. However, because linearized elasticity is not a physical theory, since it is not frame indifferent, I do not think that it is a good training ground for understanding a new type of micro-geometries, for which I think that group invariance plays an important role.

[85] Augustin Louis CAUCHY, French mathematician, 1789–1857. He was made Baron by Charles X. He worked in Paris, France, and between the 1830 revolution and the 1848 revolution in Paris, he worked in Torino (Turin), Italy.

apply,[86] and the description of the evolution on the boundary of the effective set will be important to analyse, and it might become clearer then what other questions mean, like intermittence.

Even the question of identifying the set corresponding to both the effective coefficients for diffusion and the effective coefficients for linearized elasticity is not easy, although I do not like to mix the linearization in elasticity with homogenization, and one will find some results in this direction in the book by Graeme MILTON [64]. I described in [114] my conjectures on two questions where evacuation of heat and release of elastic stress are important, the first one corresponding to the three-fold symmetries shown by snow flakes, which are flat, and the second one corresponding to the five-fold symmetries shown by the quasi-crystals formed in metallic ribbons. Evacuation of heat is important in the first case, because freezing liberates heat (the latent heat) which must then be evacuated, and laminated structures are optimal for creating an important heat flux in the direction of the planes, and it is important in the second case, because the ribbon was heated above the Curie point of the metal for facilitating the creation of a good magnetic configuration by application of an external magnetic field, and the ribbon is then cooled rapidly for keeping the interesting magnetic configuration that was obtained. Elastic effects are important in the first case, because water increases in volume by freezing, and that generates elastic stress, and they are important also in the second case because one draws on the ribbon for cooling it. For explaining the difference in the symmetries observed, I invoke questions of small-amplitude homogenization (for which I introduced *H-measures* in the late 1980s) for linearized elasticity [106], or the construction by Gilles FRANCFORT and François MURAT of linearized elastic materials by multiple laminations in the mid 1980s [30].[87] When mixing two isotropic materials, Gilles FRANCFORT and François MURAT looked for the smallest number of laminations that creates an isotropic material, and for the two-dimensional case they found three laminations in directions at 120 degrees, but for the three-dimensional case they found six directions, pointing towards the vertices of a regular icosahedron (or equivalently the faces of a regular dodecahedron).[88] When I heard

[86] Andrey Nikolayevich KOLMOGOROV, Russian mathematician, 1903–1987. He received the Wolf Prize in 1980, for deep and original discoveries in Fourier analysis, probability theory, ergodic theory and dynamical systems, jointly with Henri CARTAN. He worked at Moscow State University and at the Steklov Institute, Moscow, Russia.

[87] Gilles André FRANCFORT, French mathematician, born in 1957. He worked at LCPC (Laboratoire Central des Ponts et Chaussées), Paris, and he works now at Université Paris XIII (Paris Nord), Villetaneuse, France.

[88] I first suggested ten directions, normal to the faces of a regular icosahedron.

about quasi-crystals, I quickly understood that the games of Penrose tilings played by physicists are irrelevant (and related to the Comte complex of theoretical physicists), because the physical problem is one of evacuating heat and releasing stress, and since a ribbon of 0.1 millimetre thickness contains of the order of a million atoms in the small direction of the ribbon, it is clearly not a two-dimensional problem. I could only make an analysis in terms of *H-measures*, and a characterization of fourth-order moments of a nonnegative measure living on $\mathbb{S}^2$ is the critical question, which I investigated with Gilles FRANCFORT and François MURAT in [32], and the isotropic elastic mixture is not on the boundary of the set of effective parameters (in the approximation of *H-measures*), but some transversely isotropic mixtures are on the boundary, and they can be obtained with five directions of laminations, pointing towards the vertices of a regular pentagon. However, this is just a sufficient condition for obtaining a transversely isotropic linear elastic material, and in order to go further one needs to understand about the evolution of the *H-measures* describing a mixture, and I guess that this may involve describing a new mathematical object.

These intuitive observations show that homogenization is of great importance for questions like phase transitions, and this physical situation involves time,[89] because one starts from a material in a first phase, liquid for example, and one ends up with the same material in another phase, solid for example, and one usually has to deal with the (latent) heat released by the phase transition, and to understand where it goes. My guess is that physicists have not done a good job on this question, and that the notion of latent heat is badly defined, because the heat released should depend upon which path along microstructures one uses. It seems that answering the type of question which I am thinking about will require new mathematical tools to be developed, so that there is some good work to do by mathematicians, with a better understanding of physics as a prize, but one should first understand what homogenization is along the lines that I followed in these two chapters of overview, and I must now explain in the rest of this book what homogenization is.

[89] Specialists of fake mechanics use the term "phase transition" to mean something completely different, where one shuffles around different materials, which might be the same anisotropic material with different orientation: there is no question of latent heat, and no physics, since there is no time variable for expressing how the different pieces move around in order to discover an "optimal configuration"!

Additional footnotes: BALL R.,[90] BIRKBECK,[91] BROWN N.,[92] CARTAN É.,[93] Henri CARTAN,[94] Charles X,[95] CRAFOORD,[96] Antonio DE SIMONE,[97] Richard DUFFIN,[98] FARADAY,[99] FERMI,[100] FICK,[101] FULLER,[102]

---

[90] Walter William Rouse BALL, English mathematician, 1850–1925. He worked in Cambridge, England. The Rouse Ball professorship at Cambridge, England, is named after him.

[91] George BIRKBECK, English physician and philanthropist, 1776–1841. He worked in London, England, and he founded in 1823 the London Mechanics Institute, later to become Birkbeck College, part of University of London.

[92] Nicholas BROWN Jr., American merchant, 1769–1841. Brown University, Providence, RI, is named after him.

[93] Elie Joseph CARTAN, French mathematician, 1869–1951. He worked in Montpellier, in Lyon, in Nancy, and in Paris, France.

[94] Henri Paul CARTAN, French mathematician, 1904–2008. He received the Wolf Prize in 1980 for pioneering work in algebraic topology, complex variables, homological algebra and inspired leadership of a generation of mathematicians, jointly with Andrei N. KOLMOGOROV. He worked in Lille, in Strasbourg, in Paris, and at Université Paris Sud, Orsay, France, retiring in 1975 just before I was hired there. Theorems attributed to CARTAN are often the work of his father É. CARTAN.

[95] Charles-Philippe de France, 1757–1836, count of Artois, duke of Angoulême, pair of France, was king of France from 1824 to 1830 under the name Charles X.

[96] Holger CRAFOORD, Swedish industrialist and philanthropist, 1908–1982. He invented the artificial kidney, and he and his wife (Anna-Greta CRAFOORD, 1914–1994) established the Crafoord Prize in 1980 by a donation to the royal Swedish academy of sciences, to reward and promote basic research in scientific disciplines, outside those of the Nobel Prize, including mathematics, geoscience, bioscience (particularly in relation to ecology and evolution), and astronomy.

[97] Antonio DE SIMONE, Italian mathematician, born in 1962. He worked at MPI (Max Planck Institute), Leipzig, Germany, and he works now at SISSA (Scuola Internazionale Superiore di Studi Avanzati), Trieste, Italy.

[98] Richard James DUFFIN, American mathematician, 1909–1996. He worked at Carnegie Tech (Carnegie Institute of Technology), Pittsburgh, PA, which then became a part of CMU (*Carnegie Mellon University*) after merging with the Mellon Institute of Industrial Research; he was my colleague after 1987.

[99] Michael FARADAY, English chemist and physicist, 1791–1867. He worked in London, England, as Fullerian professor of chemistry at the Royal Institution of Great Britain.

[100] Enrico FERMI, Italian-born physicist, 1901–1954. He received the Nobel Prize in Physics in 1938, for his demonstrations of the existence of new radioactive elements produced by neutron irradiation, and for his related discovery of nuclear reactions brought about by slow neutrons. He worked in Chicago, IL. The FermiLab (Fermi National Accelerator Laboratory) of DoE (Department of Energy), Batavia, IL, is named after him. The "particles" which physicists call fermions are also named after him.

[101] Adolph Eugen FICK, German physiologist/physicist, 1829–1901. He worked in Zürich, Switzerland, and in Würzburg, Germany.

[102] John FULLER, English politician and philanthropist, 1757–1834. He instituted the Fullerian professorship in chemistry and in physiology at the Royal Institution of Great Britain, London, England.

GRESHAM,[103] Stephen HAWKING,[104] Bill HRUSA,[105] JACOBI,[106] John Paul II,[107] LAVAL,[108] MORREY,[109] Louis NIRENBERG,[110] John NOHEL,[111] STEKLOV,[112] SYNGE,[113] TAIT,[114] Wolf VON WAHL,[115] ZEEMAN.[116]

[103] Sir Thomas GRESHAM, English merchant and financier, 1519–1579. He left the money for the foundation of Gresham College, which was established in 1597.

[104] Stephen William HAWKING, English mathematician, born in 1942. He received the Wolf Prize (in Physics!) in 1988, jointly with Roger PENROSE, for their brilliant development of the theory of general relativity, in which they have shown the necessity for cosmological singularities and have elucidated the physics of black holes. In this work they have greatly enlarged our understanding of the origin and possible fate of the Universe. [This is the official reason for them receiving the prize, and it does not reflect my opinion, that it results from a huge Comte complex among physicists, and that a lot of what is done concerning gravitation is sheer nonsense!] He works in Cambridge, England, holding the Lucasian chair (1980–).

[105] William John HRUSA, American mathematician, born in 1955. He works at CMU (*Carnegie Mellon University*), Pittsburgh, PA, where he has been my colleague since 1987.

[106] Carl Gustav Jacob JACOBI, German mathematician, 1804–1851. He worked in Königsberg (then in Germany, now Kaliningrad, Russia) and Berlin, Germany.

[107] John Paul II (Karol Józef WOJTYLA), Polish-born Pope, 1920–2005. He was elected Pope in 1978.

[108] Blessed François DE (MONTMORENCY) LAVAL, French-born bishop, 1623–1708. He was the first Roman Catholic bishop in Canada, archbishop of Québec, Québec. He was beatified in 1980 by Pope John Paul II. Université Laval, Québec, Québec, is named after him.

[109] Charles Bradfield MORREY Jr., American mathematician, 1907–1980. He worked at UCB (University of California at Berkeley), Berkeley, CA.

[110] Louis NIRENBERG, Canadian-born mathematician, born in 1925. He received the Crafoord Prize in 1982. He works at NYU (New York University), New York, NY.

[111] John Adolf NOHEL, Czech-born mathematician, 1924–1999. He worked at Georgia Tech (Georgia Institute of Technology), Atlanta, GA, and at UW (University of Wisconsin), Madison, WI.

[112] Vladimir Andreevich STEKLOV, Russian mathematician, 1864–1926. He worked in Kharkov, and in St Petersburg (then Petrograd, USSR), Russia. The Steklov Institute of Mathematics, Moscow, Russia, is named after him.

[113] John Lighton SYNGE, Irish mathematician, 1897–1995. He worked in Toronto (Ontario), at OSU (Ohio State University), Columbus, OH, and at Carnegie Tech (Carnegie Institute of Technology), now part of CMU (*Carnegie Mellon University*), Pittsburgh, PA, where he was the head of the mathematics department from 1946 to 1948, and in Dublin, Ireland.

[114] Peter Guthrie TAIT, Scottish physicist, 1831–1901. He worked in Edinburgh, Scotland.

[115] Wolf VON WAHL, German mathematician. He works in Bayreuth, Germany.

[116] Pieter ZEEMAN, Dutch physicist, 1865–1943. He received the Nobel Prize in Physics in 1902, jointly with Hendrik LORENTZ, in recognition of the extraordinary service they rendered by their research into the influence of magnetism upon radiation phenomena. He worked in Leiden, and in Amsterdam, The Netherlands.

Chapter 4
An Academic Question of Jacques-Louis Lions

Having graduated from École Polytechnique in 1967, I should have spent the following year doing some kind of military activity,[1] but DE GAULLE,[2] who was re-elected Président de la République in 1965, decided on a national effort toward research, and since I chose to do research in mathematics with Jacques-Louis LIONS as advisor,[3] my only duty for the year 1967–1968 was to obtain a DEA in "numerical analysis," following courses by Jacques-Louis LIONS and by René DE POSSEL.[4] My advisor asked me to follow the Lions–Schwartz seminar at IHP (where I also followed on my own a course by Salah BAOUENDI),[5] and he also asked me to follow some courses and seminars

[1] École Polytechnique has a military status, and French students sign a 3 year contract with the Army, and in those days, when the school was still in Paris, the first 2 years were the scientific studies.

[2] Charles DE GAULLE, French general and statesman, 1890–1970. He was elected Président de la République in 1958, in the style of the 4th Republic, by the two legislative chambers meeting in Versailles. He had a new constitution for France accepted, and he was re-elected Président de la République in 1965, in the style of the 5th Republic, by popular election, and he resigned in 1969.

[3] Among my teachers, Laurent SCHWARTZ wore a label of "pure mathematician" and Jacques-Louis LIONS a label of "applied mathematician," and choosing to study with Jacques-Louis LIONS was compatible with my preceding choice of studying at École Polytechnique instead of École Normale Supérieure. After learning a little more, I realized that my advisor was not really interested in continuum mechanics or physics, or even numerical analysis (which was the course he taught at École Polytechnique), but I could not make a better choice, since there was no French mathematician in the preceding generation with the type of knowledge which I acquired later, due to my interest in continuum mechanics and physics.

[4] Lucien Alexandre Charles René DE POSSEL, French mathematician, 1905–1974. He worked in Marseille, in Clermont-Ferrand, in Besançon, France, in Alger (Algiers) (then in France, now capital of Algeria), and in Paris, France. He was a teacher for my DEA (Diplôme d'Études Approfondies) in numerical analysis, at Institut Blaise Pascal in Paris, in 1967–1968.

[5] Mohammed Salah BAOUENDI, Tunisian-born mathematician, born in 1937. He worked in Paris, France, at Purdue University, West Lafayette, IN, and he works now at UCSD (University of California at San Diego), La Jolla, CA.

L. Tartar, *The General Theory of Homogenization*, Lecture Notes of the Unione Matematica Italiana 7, DOI 10.1007/978-3-642-05195-1_4,

at IRIA, in Rocquencourt,[6] where a group of researchers worked under his direction, and it was in a room reserved for visitors that I first met François MURAT, who studied at École Polytechnique a year after me. In 1969, some new buildings were finished at a place first known as Halle aux Vins,[7] and then called Jussieu like the nearest metro station,[8] which enlarged the Faculté des Sciences, and Jacques-Louis LIONS took possession of a part for a laboratory jointly funded by CNRS and the university,[9] of functional analysis and numerical analysis,[10] housing a group of researchers from Institut Blaise Pascal, and a few other people, like François MURAT and myself, and we shared our first office.

At the end of my studies at École Polytechnique, my advisor had already told me to read a book on control [78]; then he told me to learn Fortran during the summer (but the language taught for the DEA was Algol), and he asked me to read a book on game theory [43], and later a book on differential games [41], but he never asked me any precise question in these directions. In the fall of 1967 he asked me a question which I solved quickly, which gave him the possibility to make fun of me because of my bad style of writing, but he was kind enough to rewrite my result, and although he told me that it was going to be a report of his group at IRIA, he never included it in that series.[11] On my own I proved an abstract version of the *method of translation*

---

[6] DE GAULLE chose nuclear dissuasion for the military defence of France, and as a consequence he withdrew France from the military part of NATO (North Atlantic Treaty Organization); maybe it was the other way around, that he wanted to get France out of the military part of NATO, so that France must fight alone in case of an attack from the east, hence the development of a nuclear force of dissuasion. Anyway, when NATO transferred its offices from France to Belgium, it left some vacant buildings in Rocquencourt (near Versailles), used for IRIA, and in Paris, used for Université Paris IX Dauphine.

[7] Because it was used by the wine merchants, who then moved to Bercy, on the other bank of the river Seine.

[8] Antoine Laurent DE JUSSIEU, French botanist, 1748–1836. He worked in Paris, France.

[9] Shortly after, the University of Paris was split into a few independent universities, and the directors of laboratories in Jussieu chose between becoming a part of Paris 6, now UPMC (Université Pierre et Marie Curie), or of Paris 7 (Denis Diderot).

[10] Jacques-Louis LIONS was usually teaching applications of functional analysis, and he dealt with numerical analysis, optimal control, or continuum mechanics in the same way, starting from an applied field for finding a question that he could transform into a problem in functional analysis, but he never went back to check if the problems that he solved after were good questions for the applied field that he pretended to consider. That no one dared criticize his style was partly due to the fact that most people who knew numerical analysis or control preferred to continue working on finite-dimensional problems, so that they showed other limitations, and the fact that theoreticians in mechanics suffer from a Comte complex and are afraid to criticize mathematicians (a situation that ill-intentioned mathematicians often take advantage of nowadays, for advertising their preferred version of fake mechanics).

[11] However, it served once, after I pointed out a gap in a proof of Alain BENSOUSSAN during a talk he gave at IRIA: I needed to help him complete his proof, and I quickly found how my result filled the gap.

of Louis NIRENBERG for proving regularity, but when I mentioned it to my advisor, he abruptly said that he did not care (an interesting application was to obtain a result of Salah BAOUENDI and Charles GOULAOUIC,[12] by using interpolation of L^2 spaces with weights instead of pseudo-differential operators).[13] He was frankly upset another time, that I solved a question asked by Jacques PLANCHARD,[14] and I would understand the reaction of my advisor if there was a particular question that he asked me to look at and which I did not, but after I gave him a written solution for a question that he asked in his book [51] about the method of dynamic programming, he never made any comment to me, although it was clear that this question still interested him; I was much too shy to ask him anything about that.

Around the time when I moved into my new office, my advisor asked me to generalize one of his results of nonlinear interpolation [53], which I did rather easily, but apart from looking at applications of my results, I waited for my advisor to ask another question,[15] and during this period I tried to help others solving their problems, but it was not immediately that I asked François MURAT what kind of problem he worked on, and since I was surprised by the result that he proved, I worked with him often for understanding more about this "new" field that we were discovering. We were actually just rediscovering it, but at the beginning we were unaware of the previous work of Sergio SPAGNOLO [89,90].

I later thought that it is useful to rediscover results, because one may find a completely new approach, with more possibilities than the previous studies gave, and also because the knowledge of what was previously done is sometimes an obstacle to the discovery of a new strategy for taming an old problem. Of course, one must mention the previous approaches, once one hears about them.

In [51], Jacques-Louis LIONS asked a purely academic question, to minimize the "cost function"

$$J(a) = \int_0^1 |y - z_d|^2 \, dx, \tag{4.1}$$

[12] Charles GOULAOUIC, French mathematician, 1938–1983. He worked in Rennes, at Université Paris-Sud, Orsay, being my colleague from 1975 to 1979, and at École Polytechnique, Palaiseau, France.

[13] I learned the theory of interpolation of Hilbert spaces in a book by my advisor and Enrico MAGENES [55], but later he gave me his article with Jaak PEETRE [56] to read, and there I learned the theory of interpolation of Banach spaces.

[14] Jacques PLANCHARD, French engineer. He worked at EDF (Électricité de France), Clamart, France.

[15] Which would lead me to complete my thesis, defended in April 1971. After that, I once asked my advisor if he could give me a problem to solve, and he answered that I was now on my own!

when the "state" $y \in H^1\big((0,1)\big)$ solves the "equation of state"

$$-\frac{d}{dx}\Big(a\,\frac{dy}{dx}\Big) + a\,y = f \text{ in } (0,1);\ y(0) \text{ and } y(1) \text{ given,} \tag{4.2}$$

and the "control" a lies in the "admissible control set"

$$\mathcal{A}_{ad} = \big\{a \mid a \in L^\infty\big((0,1)\big), \alpha \le a \le \beta \text{ a.e. in } (0,1)\big\}. \tag{4.3}$$

I put between quotes the terms of *control theory* that my advisor often used, because it is just a problem of optimization, and the term control should be used only for cases where the variable is *time* and where one wants to *use at time t a control depending only on the state at time t*,[16] or possibly earlier.

François MURAT found a case where there is no solution [69]: he chose $z_d(x) = 1 + x^2$ for $x \in (0,1)$, he imposed $f = 0$, $y(0) = 1$ and $y(1) = 2$, and he assumed that $0 < \alpha \le \frac{\sqrt{2}-1}{\sqrt{2}}$ and $\beta \ge \frac{\sqrt{2}+1}{\sqrt{2}}$, and then he proved that $\inf_{a\in\mathcal{A}_{ad}} J(a) = 0$, but $J(a) > 0$ for all $a \in \mathcal{A}_{ad}$. He showed easily that $J(a)$ cannot vanish, because it means $y = z_d$, so that a satisfies the homogeneous differential equation $-\frac{d(2x\,a)}{dx} + a\,(1+x^2) = 0$ on $(0,1)$, whose solutions have the form $\frac{C}{\sqrt{x}} exp(\frac{x^2}{4})$ for $C \in \mathbb{R}$, i.e., are unbounded on $(0,1)$, or identically 0. He showed that $J(a^n) \to 0$ for the sequence a^n defined by

$$a^n(x) = \begin{cases} 1 - \sqrt{\frac{1}{2} - \frac{x^2}{6}} \text{ when } x \in \Big(\frac{2k}{2n}, \frac{2k+1}{2n}\Big), k = 0, \ldots, n-1 \\ 1 + \sqrt{\frac{1}{2} - \frac{x^2}{6}} \text{ when } x \in \Big(\frac{2k+1}{2n}, \frac{2k+2}{2n}\Big), k = 0, \ldots, n-1, \end{cases} \tag{4.4}$$

and he observed that

$$\begin{aligned} &a^n \rightharpoonup a^+ \text{ in } L^\infty\big((0,1)\big) \text{ weak} \star, \text{ with } a^+(x) = 1 \text{ for } x \in (0,1) \\ &\frac{1}{a^n} \rightharpoonup \frac{1}{a^-} \text{ in } L^\infty\big((0,1)\big) \text{ weak} \star, \text{ with } a^-(x) = \frac{1}{2} + \frac{x^2}{6} \text{ for } x \in (0,1), \end{aligned} \tag{4.5}$$

so that by Lemma 4.1 below, the sequence y^n converges in $H^1\big((0,1)\big)$ weak, hence uniformly, to $y^\infty \in H^1\big((0,1)\big)$ solution of

$$-\frac{d}{dx}\Big(a^-\,\frac{dy^\infty}{dx}\Big) + a^+\,y^\infty = 0 \text{ in } (0,1);\ y^\infty(0) = 1, y^\infty(1) = 2, \tag{4.6}$$

but since z_d is solution of (4.6), which has a unique solution by the Lax–Milgram lemma, one has $y^\infty = z_d$, and $\lim_{n\to\infty} J(a^n) = \int_0^1 |y^\infty - z_d|^2\,dx = 0$.

---

[16] Such a problem of control is also called *closed loop control*, by opposition to "open loop control," which is just optimization, and not control!

Lemma 4.1. *Let $I = (x_-, x_+)$ be a bounded interval of $\mathbb{R}$, and assume that*

$$\begin{aligned} &\frac{1}{a_n} \rightharpoonup \frac{1}{a_-} \text{ in } L^\infty(I) \text{ weak } \star \text{ as } n \to \infty, \\ &b_n \rightharpoonup b_+ \text{ in } L^\infty(I) \text{ weak } \star \text{ as } n \to \infty, \end{aligned} \tag{4.7}$$

and

$$\begin{aligned} &z_n \rightharpoonup z_\infty \text{ in } H^1(I) \text{ weak as } n \to \infty, \\ &-\frac{d}{dx}\Big(a_n \frac{dz_n}{dx}\Big) + b_n z_n = f_n \to f_\infty \text{ in } H^{-1}(I) \text{ strong as } n \to \infty, \end{aligned} \tag{4.8}$$

as $n \to \infty$. Then, one has

$$\begin{aligned} &a_n \tfrac{dz_n}{dx} \rightharpoonup a_- \tfrac{dz_\infty}{dx} \text{ in } L^2(I) \text{ strong as } n \to \infty, \\ &b_n z_n \rightharpoonup b_+ z_\infty \text{ in } L^2(I) \text{ weak as } n \to \infty, \\ &-\frac{d}{dx}\Big(a_- \frac{dz_\infty}{dx}\Big) + b_+ z_\infty = f_\infty \text{ in } H^{-1}(I). \end{aligned} \tag{4.9}$$

Proof. [17] One has $f_n = \frac{dg_n}{dx}$ with $g_n \to g_\infty$ in $L^2(I)$ strong (and one can impose $\int_I g_n\,dx = 0$), so that $\xi_n = a_n \frac{dz_n}{dx} + g_n$ is bounded in $L^2(I)$ and $\frac{d\xi_n}{dx} = b_n z_n$ is bounded in $L^2(I)$, and one can extract a subsequence ξ_m converging in $H^1(I)$ weak, hence uniformly on I, to ξ_∞. One deduces that $\frac{dz_m}{dx} = \frac{\xi_m - g_m}{a_m}$ converges in $L^2(I)$ weak to $\frac{\xi_\infty - g_\infty}{a_-}$, which is then $\frac{dz_\infty}{dx}$, so that $\xi_\infty = a_- \frac{dz_\infty}{dx} + g_\infty$, and the limit being independent of the subsequence all the sequence ξ_n converges in $L^2(I)$ strong to $a_- \frac{dz_\infty}{dx} + g_\infty$. This shows that $a_n \frac{dz_n}{dx}$ converges in $L^2(I)$ strong to $a_- \frac{dz_\infty}{dx}$, and since z_n converges uniformly on I to z_∞, $b_n z_n$ converges in $L^2(I)$ weak to $b_+ z_\infty$, hence the equation satisfied by z_∞. □

By applying what one called the direct method of the calculus of variations in the past, before functional analysis was well developed,[18] from any minimizing sequence $a_n \in \mathcal{A}_{ad}$ one can extract a subsequence a_m such that

$$a_m \rightharpoonup a_+ \text{ and } \frac{1}{a_m} \rightharpoonup \frac{1}{a_-} \text{ in } L^\infty\big((0,1)\big) \text{ weak } \star, \tag{4.10}$$

and that

$$J(a_m) \to \widetilde{J}(a_-, a_+) = \int_0^1 |y_\infty - z_d|^2\,dx. \tag{4.11}$$

[17] The proof shows that in (4.8)–(4.9) one can replace $L^2(I)$, $H^1(I)$ and $H^{-1}(I)$ by $L^p(I)$, $W^{1,p}(I)$ and $W^{-1,p}(I)$ for $p \ge 1$, but for $p = 1$ using that $W^{-1,1}(I)$ means $\left\{\frac{dg}{dx} \mid g \in L^1(I)\right\}$ and not the dual of $W_0^{1,\infty}(I)$.

[18] Nowadays, one should only use the term calculus of variations for problems of a geometric nature, and not as synonymous with optimization!

where $y_\infty \in H^1\big((0,1)\big)$ is defined by

$$-\frac{d}{dx}\Big(a_-\frac{dy_\infty}{dx}\Big) + a_+ y_\infty = 0 \text{ in } (0,1);\; y_\infty(0) \text{ and } y_\infty(1) \text{ given}, \tag{4.12}$$

and François MURAT and myself were led to the natural question of characterizing all the possible pairs (a_-, a_+) which may appear in (4.10) for sequences from $\mathcal{A}_{ad}$. We proved that the characterization is

$$\alpha \le a_- \le a_+ \le \frac{a_-(\alpha+\beta) - \alpha\,\beta}{a_-} \le \beta \text{ a.e. in } (0,1), \tag{4.13}$$

or equivalently

$$\frac{1}{a_+} \le \frac{1}{a_-} \le \frac{\alpha+\beta-a_+}{\alpha\,\beta} \text{ a.e. in } (0,1), \tag{4.14}$$

and our proof extended easily to the following more general situation.[19]

Lemma 4.2. *For $K \subset \mathbb{R}^p$, let U^n be a sequence of Lebesgue-measurable functions,*[20,21] *from an open set $\Omega \subset \mathbb{R}^N$ into $\mathbb{R}^p$ satisfying*

$$\begin{aligned} &U^n \rightharpoonup U^\infty \textit{ in } L^\infty(\Omega;\mathbb{R}^p) \textit{ weak} \star \\ &U^n(x) \in K \textit{ a.e. } x \in \Omega. \end{aligned} \tag{4.15}$$

If K is bounded, the characterization of all possible limits U^∞ in (4.15) is

$$U^\infty(x) \in \overline{conv(K)}, \textit{ the closed convex hull of } K, \textit{ a.e. } x \in \Omega, \tag{4.16}$$

but the characterization for a general unbounded set K is

$$\begin{aligned} &\textit{there exists } M < \infty \textit{ such that } U^\infty(x) \in \overline{conv(K_M)}, \textit{ a.e. } x \in \Omega, \\ &\textit{where } K_M = \{k \in K \mid ||k|| \le M\}, \textit{ for } M < \infty. \end{aligned} \tag{4.17}$$

Proof. The closed convex hull of K is the intersection of all the closed half spaces which contain K, and a closed half space H_+ has an equation $\{\lambda | \lambda \in \mathbb{R}^p, L(\lambda) \ge 0\}$ for some non-constant affine function L, and if H_+ contains K one has $L(U^n) \ge 0$ a.e. $x \in \Omega$, so that $L(U^\infty) \ge 0$ a.e. $x \in \Omega$, i.e., $U^\infty(x) \in$

[19] In order to obtain (4.13) or (4.14) from Lemma 4.2, one uses a vector U^n having components a_n and $\frac{1}{a_n}$, and for K the piece of hyperbola defined by $U_1\,U_2 = 1$ with $\alpha \le U_1 \le \beta$.

[20] Henri Léon LEBESGUE, French mathematician, 1875–1941. He worked in Rennes, in Poitiers, and in Paris, France, holding a chair at Collège de France (mathématiques, 1921–1941), Paris.

[21] "Lebesgue integration" was discovered 2 years before LEBESGUE by W.H. YOUNG.

H_+ a.e. $x \in \Omega$; the conclusion $U^\infty(x) \in \overline{conv(K)}$ a.e. $x \in \Omega$ follows, if one is careful to write the closed convex hull as a *countable* intersection of closed half spaces containing K. However, since U^n is bounded in $L^\infty(\Omega; \mathbb{R}^p)$ by the Banach–Steinhaus theorem,[22,23] there exists $M < \infty$ such that $U^n(x) \in K_M$ a.e. $x \in \Omega$ for all n, so that $U^\infty(x) \in \overline{conv(K_M)}$ a.e. $x \in \Omega$.

Let $V \in L^\infty(\Omega; \mathbb{R}^p)$ be such that $V(x) \in \overline{conv(K_M)}$ a.e. $x \in \Omega$. For each m, one cuts $\mathbb{R}^p$ into small cubes of size $\frac{1}{m}$ and for each cube intersecting $\overline{conv(K_M)}$ one chooses a point of $conv(K_M)$, the convex hull of K_M, and one creates a function $W^m \in L^\infty(\Omega; \mathbb{R}^p)$ such that $|V - W^m| \le \frac{1}{m}$ a.e. $x \in \Omega$, and W^m takes only a finite number of values in $conv(K_M)$. On a measurable subset ω of Ω where W^m is constant, one wants to construct a sequence of functions converging in $L^\infty(\omega; \mathbb{R}^p)$ weak $\star$ to W^m and taking their values in K_M, and gluing these functions together will create a sequence converging in $L^\infty(\Omega; \mathbb{R}^p)$ weak $\star$ to W^m, so that one can approach V in this topology, since the weak $\star$ topology of $L^\infty(\Omega; \mathbb{R}^p)$ is metrizable on bounded sets.

Let $W^m = \lambda \in conv(K)$ on ω, so that $\lambda = \sum_i \theta^i k^i$, with $k^i \in K$, $\theta^i \ge 0$, $\sum_i \theta^i = 1$, and the sum is finite (with at most $p+1$ terms by the Carathéodory theorem).[24] One now writes ω as a union of disjoint measurable pieces of diameter at most $\frac{1}{n}$, then one partitions each piece E into (disjoint) measurable subsets E^i with $meas(E^i) = \theta^i meas(E)$, and one defines the function Z^n to be equal to k^i on each such E^i. The claim is then that, as n tends to ∞, the sequence Z^n converges in $L^\infty(\omega; \mathbb{R}^p)$ weak $\star$ to λ; Z^n is bounded, since it only takes a finite number of values in K, and it is enough to check that $\int_\omega \varphi\, Z^n\, dx \to \int_\omega \varphi\, \lambda\, dx$ for every continuous function φ with compact support. Since φ is uniformly continuous, one has $|\varphi(x) - \varphi(y)| \le \varepsilon$ when $|x-y| \le \frac{1}{n}$, so that if $e \in E$, one has $|\varphi(e) \int_E Z^n\, dx - \int_E \varphi\, Z^n\, dx| \le \varepsilon\, M\, meas(E)$ and $|\varphi(e) \int_E \lambda\, dx - \int_E \varphi\, \lambda\, dx| \le \varepsilon\, M\, meas(E)$, but since $\int_E Z^n\, dx = \sum_i \int_{E_i} k^i\, dx = \sum_i \theta^i meas(E) k^i = meas(E)\lambda = \int_E \lambda\, dx$, one deduces that $| \int_\omega \varphi\, Z^n\, dx - \int_\omega \varphi\, \lambda\, dx| \le 2\varepsilon\, M\, meas(\omega)$.

In order to construct the sets E^i from E,[25] one notices that, for a non-constant affine function L, the measure of $\{x \in E \mid L(x) \ge t\}$ is a continuous function of t which grows from 0 to $meas(E)$ and one obtains the desired partition of E by cutting E by suitable hyperplanes $L^{-1}(t_i)$. □

---

[22] Stefan BANACH, Polish mathematician, 1892–1945. He worked in Lwów (then in Poland, now Lvov, Ukraine). There is a Stefan Banach International Mathematical Center in Warsaw, Poland.

[23] Hugo Dyonizy STEINHAUS, Polish mathematician, 1887–1972. He worked in Lwów (then in Poland, now Lvov, Ukraine) until 1941, and after 1945 in Wrocław, Poland.

[24] Constantin CARATHÉODORY, German mathematician (of Greek origin), 1873–1950. He worked at Georg-August-Universität, Göttingen, in Bonn, in Hanover, Germany, in Breslau (then in Germany, now Wrocław, Poland), in Berlin, Germany. After World War I, he worked in Athens, Greece and in Smyrna (then in Greece, now Izmir, Turkey), and in München (Munich), Germany.

[25] This step is valid for any set equipped with a *measure without atoms*.

For $A \subset \mathbb{R}^p$, Lemma 4.2 characterizes the sequential closure of the set

$$X(A) = \{U \in L^\infty(\Omega; \mathbb{R}^p) \mid U(x) \in A \text{ a.e. } x \in \Omega\}, \tag{4.18}$$

for the weak $\star$ topology on $L^\infty(\Omega; \mathbb{R}^p)$, since $\bigcup_M X\big(\overline{conv(A_M)}\big) \subset X(A^*)$, with $A^* = \overline{\bigcup_M conv(A_M)}$. As a consequence, the $L^\infty(\Omega; \mathbb{R}^p)$ weak $\star$ closure of $X(A)$ is $X\big(\overline{conv(A)}\big)$, so that for any unbounded set $A \subset \mathbb{R}^p$ such that $A^* \neq \overline{conv(A)}$, the topology induced on $X\big(\overline{conv(A)}\big)$ by the weak $\star$ topology on $L^\infty(\Omega; \mathbb{R}^p)$ is not metrizable.

I learned about convexity methods related to the Hahn–Banach theorem,[26] and I learned that weak convergence is not adapted to nonlinear questions, apart from using compactness arguments for transforming weak convergence in a space into strong convergence in another space,[27] but although a part of our proof used convexity, I was surprised that the simple characterization of the weak limits of sequences from $X(K)$ was missed before, and I decided to ask my advisor about that. The matter looked important enough to me that I could not wait until the next Friday, which was usually the only day of the week when he came to Jussieu, and I overcame my shyness and I called him; he suggested that I ask the question to Ivar EKELAND,[28] whom I knew from the Pallu de la Barrière seminar at IRIA,[29] and I called him too, and he told me that it was implicitly used in the work of CASTAING,[30,31] who based his arguments on the Lyapunoff theorem.[32,33]

Of course, the counter-example of François MURAT is reminiscent of ideas first used by Laurence YOUNG in the late 1930s for showing nonexistence of minimizers for some functionals, which led him to the notion of *Young measures* [123]. His ideas were actually rediscovered a few times for

[26] Hans HAHN, Austrian mathematician, 1879–1934. He worked in Vienna, Austria.

[27] Jacques-Louis LIONS taught about a dichotomy, between the compactness method and the monotonicity method [52], and a few years later, I unified these apparently unrelated parts by my *compensated compactness method* [98].

[28] Ivar EKELAND, French-born mathematician, born in 1944. He worked at Université Paris IX Dauphine, Paris, France, where he was my colleague from 1971 to 1974, and he works now at UBC (University of British Columbia), Vancouver, British Columbia.

[29] Ivar EKELAND's talk, like most talks at the Pallu de la Barrière seminar at IRIA, was related to questions of measurability, which I still have almost no interest for, and since many advocates of fake mechanics seem to specialize now in questions of measurability, I wonder how being interested in such questions was then explained to be related to the goals of IRIA.

[30] Charles CASTAING, French mathematician, born in 1932. He worked at Université des Sciences et Techniques de Languedoc (Montpellier II), Montpellier, France.

[31] I often heard the name of CASTAING mentioned at the Pallu de la Barrière seminar at IRIA, for a measurable selection theorem of multi-valued mappings.

[32] A. LIAPOUNOFF. I could not find much on this mathematician, who wrote a few articles in French around 1940.

[33] When I met Zvi ARTSTEIN in the spring of 1975, he showed me his simple proof of the Lyapunoff theorem and of bang-bang results in control theory [3].

control problems, where he used the term *chattering controls* [125]: similar ideas were used by a group in the USSR, BOLTYANSKII,[34] GAMKRELIDZE,[35] MISHCHENKO,[36] and Lev PONTRYAGIN,[37] for their maximum principle [78], and they compared their work to that of Richard BELLMAN in dynamic programming,[38,39] but they did not define *generalized solutions* like Laurence YOUNG did, for which the classical (first-order) optimality conditions give the maximum principle, or Jack WARGA,[40] who reinvented chattering controls as *relaxed controls*, or GHOUILA-HOURI,[41] who reinvented *Young measures* as *parametrized measures* [35]. Our Lemma 4.2 seems a simple step which was missed before, not quite as general as *Young measures*, but of a more practical and elementary nature.

Our Lemma 4.2 is also useful for defining practical relaxations of minimizing problems, a question which I heard Ivar EKELAND talk about in the early 1970s, but at a much too abstract level to be of any use, since I remember hearing him mention the Stone–Čech compactification,[42,43] while our Lemma 4.2 goes in the opposite direction of using compactifications as small as possible, and it helps create quite explicit relaxations, which are tractable.

For a set Z and a real function f defined on Z, a *relaxation* of the problem of minimizing f on Z is made of a topological space $\widehat{Z}$, an injection j from Z into $\widehat{Z}$ such that $j(Z)$ is dense in $\widehat{Z}$, and a lower semi-continuous function $\widehat{f}$ on $\widehat{Z}$ such that $f(z) = \widehat{f}\big(j(z)\big)$ for all $z \in Z$, and satisfying a compatibility

[34] Vladimir Grigor'evich BOLTYANSKII, Russian mathematician. He worked in Moscow, Russia.

[35] Revaz Valerianovich GAMKRELIDZE, Georgian-born mathematician, born in 1927. He worked in Moscow, Russia.

[36] Evgenii Frolovich MISHCHENKO, Russian mathematician, born in 1922. He worked in Moscow, Russia.

[37] Lev Semenovich PONTRYAGIN, Russian mathematician, 1908–1988. He worked in Moscow, Russia.

[38] Richard Ernest BELLMAN, American mathematician, 1920–1984. He worked at Princeton University, Princeton, NJ, Stanford University, Stanford, CA, at the RAND corporation, Los Angeles, CA, and at USC (University of Southern California), Los Angeles, CA.

[39] The idea of dynamic programming is due to CARATHÉODORY, who introduced it in his studies about the Hamilton–Jacobi equations, long before Richard BELLMAN popularized it.

[40] Jack WARGA, Polish-born mathematician, born in 1922. He worked at Northeastern University, Boston, MA.

[41] Alain GHOUILA-HOURI, French mathematician, 1939–1966.

[42] Marshall Harvey STONE, American mathematician, 1903–1989. He worked at Columbia University, New York, NY, at Yale University, New Haven, CT, at Harvard University, Cambridge, MA, at The University of Chicago, Chicago, IL, and at the University of Massachusetts, Amherst, MA.

[43] Eduard ČECH, Czech mathematician, 1893–1960. He worked at Masaryk University in Brno, and Charles University in Prague, Czech Republic.

condition, which in the metrizable case is that for every $\widehat{z} \in \widehat{Z}$ there is a sequence $z_n \in Z$ such that $j(z_n) \to \widehat{z}$ and $f(z_n) \to \widehat{f}(\widehat{z})$. The interesting case is when one can choose $\widehat{Z}$ compact, or at least such that $\{\widehat{z} \in \widehat{Z} \mid \widehat{f}(\widehat{z}) \leq \lambda\}$ is nonempty and compact for some $\lambda \in \mathbb{R}$, in which case any minimizing sequence z_n of f in Z is such that the accumulation points of $j(z_n)$ in $\widehat{Z}$ are in the set where $\widehat{f}$ attains its minimum on $\widehat{Z}$ (which is nonempty and compact), and serves as generalized minimizers of f.

Nowadays many use the term relaxation for a weaker form of the problem, introduced later by Ennio DE GIORGI (Γ-convergence), but they never use all the power of his definition which lies in the possibility of choosing the topology that one uses, and instead they use the topologies that they know, rarely adapted to their problem, which tends to show that Γ-convergence is useless, while it is just that they do not know how to use it well.

Additional footnotes: Zvi ARTSTEIN,[44] CHISHOLM,[45] HARDINGE,[46] HARVARD,[47] HILBERT,[48] Enrico MAGENES,[49] MASARYK,[50] Jaak PEETRE,[51] PURDUE,[52] YALE,[53] YOUNG W. H.[54]

[44] Zvi ARTSTEIN, Israeli mathematician, born in 1943. He worked at Brown University, Providence, RI, and he works now at the Weizmann Institute of Science, Rehovoth, Israel.

[45] Grace Emily CHISHOLM-YOUNG, English mathematician, 1868–1944. There are many results attributed to her husband W.H. YOUNG which may be joint work with her, since they collaborated extensively.

[46] Sir Charles HARDINGE, 1st Baron HARDINGE of Penshurst, English diplomat, 1858–1944. He was Viceroy and Governor-General of India (1910–1916).

[47] John HARVARD, English clergyman, 1607–1638. Harvard University, Cambridge, MA, is named after him.

[48] David HILBERT, German mathematician, 1862–1943. He worked in Königsberg (then in Germany, now Kaliningrad, Russia) and at Georg-August-Universität, Göttingen, Germany.

[49] Enrico MAGENES, Italian mathematician, born in 1923. He worked at Università di Pavia, Pavia, Italy.

[50] Tomáš MASARYK, Czech philosopher and politician, 1850–1937. He was the first president of Czechoslovakia (1918–1935).

[51] Jaak PEETRE, Estonian-born mathematician, born in 1935. He worked at Lund University, Sweden.

[52] John PURDUE, American industrialist, 1802–1876. Purdue University, West Lafayette, IN, is named after him.

[53] Elihu YALE, American-born English philanthropist, Governor of Fort St George, Madras, India, 1649–1721. Yale University, New Haven, CT, is named after him.

[54] William Henry YOUNG, English mathematician, 1863–1942. He worked in Liverpool, England, in Calcutta, India, holding the first Hardinge professorship (1913–1917), in Aberystwyth, Wales, and in Lausanne, Switzerland. There are many results attributed to him which may be joint work with his wife, Grace CHISHOLM, since they collaborated extensively.

Chapter 5
A Useful Generalization by François Murat

In industry, research and development are quite distinct activities: finding a new area with oil not too deep below the ground level and exploiting an already discovered oil field have not much in common. In academia, too often what one calls research is merely a question of development: most researchers just apply ideas which already exist, to many different situations.[1]

One reason for this behavior is the "publish or perish" philosophy, which pushes researchers to publish uninteresting generalizations, instead of analysing what was already achieved, selecting an interesting challenge, and spending some time working on it. This incentive to publish too much did not yet exist in France around 1970, and François MURAT and myself both worked at CNRS, but although we were underpaid, our situation offered a lot of freedom, with relative stability.[2]

There is another reason why many researchers specialize in not so interesting generalizations without new ideas, which is that their technical expertise is not so good, and instead of publishing their lengthy computations because they took a long time to do them, it would be better if they could add more time for simplifying them and for reducing their articles to their essential features (probably not so new), but they often lack this kind of ability. One reason why our technical level was reasonably good was the preparation that we went through for succeeding in the competition to enter École Polytechnique:[3] after obtaining good grades at the baccalauréat

[1] In disciplines with an experimental component, one seems to use students as slaves for this development, and one tells them that they do research.

[2] One constraint toward École Polytechnique, was to defend our thesis at most 6 years after having finished the school, or we would have to reimburse the cost of our 2 years of study there.

[3] École Polytechnique offered three hundred places to French students, and in our times the competition was reserved to men accepted for military service; women were only allowed to compete in the early 1970s, but they had to enrol in the Army too (nowadays, there are more places offered, and various competitions, and the result of the competitions is usually a class of about 10% women and 90% men). Foreigners were allowed to compete, and they needed to obtain better grades than the last French student admitted, but they paid the cost of their studies.

L. Tartar, *The General Theory of Homogenization*, Lecture Notes of the Unione Matematica Italiana 7, DOI 10.1007/978-3-642-05195-1_5,

(a national exam at the end of high school) and being accepted in special preparation classes of *mathématiques supérieures* and then of *mathématiques spéciales* on the basis of our preceding ability in mathematics (and possibly physics/chemistry), which implied that we were in the top 5% of our class of age for what concerned scientific questions, we needed to acquire during these 2 years of preparation some kind of mathematical dexterity in order to belong to the top 1% of this selected group. There is no reason why such a training would select good researchers, but the goal of École Polytechnique was not to form researchers!

In all education systems one must go through a selection process in order to be admitted as a researcher, and this possibility is only offered in countries with a reasonably good economy, but I never heard about an efficient way to select people who will have new ideas: the first step in the selection is usually to only consider those who were good at learning what others did before, but being good at research and being good at learning old things are obviously quite different, as one can deduce from the fact that books in mathematics are never about how to have new ideas, but always about teaching old things! To have a new idea is like being able to walk against the crowd if one's reason tells that there could be something interesting in the opposite direction, but it is not about being stubborn and rejecting everything, and one must be able to accept criticism and to change one's mind if convinced by reason, but not by intimidation![4]

In the early days, I was not good at asking new questions, but once I was asked a question I would put my mind to finding ways to answer it. After I tried to understand continuum mechanics from a mathematical point of view, I found that there were so many things that did not make sense to a mathematician in what one taught in continuum mechanics or physics, that new questions popped up easily in my mind.[5]

Finding generalizations of something already done is either obvious or difficult, depending upon the state of mind of the person who wonders about it. After discovering with François MURAT the characterization (4.13) and (4.14)

[4] Hence, research becomes difficult in a dictatorial environment, and dictatorship can be economical, political, religious, or more subtle. Becoming a good mathematician almost requires being good at abstraction and seeing similarities between situations that others do not see related (like for parables). When one encounters a mathematician who does not want to answer questions and who treats others as if they belong to a despicable group, one must understand that he/she uses a form of intimidation: such people are usually not good mathematicians, but are forced to adopt this behavior because of the narrowness of their knowledge (often in their field, which they think to be superior), and they acquire such *racist* tendencies as a solution to their fear, that once they answer a question their real level will be perceived to be nothing to be proud of!

[5] I guess then that one must first discover the type of research that one is good at doing, and it would be better if one could avoid being intimidated into researching something different than that, but it may happen that one is good at something which no funding agency is interested in supporting!

of compatible weak limits a_- and a_+ defined by (4.10), I saw without difficulty that our proof suggested Lemma 4.2 as the natural generalization, but I could not see another generalization of what François MURAT did, or what we did together.

François MURAT looked at a problem in a rectangle $\omega = I_1 \times I_2$, and for a sequence a_n satisfying $0 < \alpha \leq a_n \leq \beta < \infty$ a.e. in ω, and for $f \in H^{-1}(\omega)$, he considered the sequence $y_n \in H_0^1(\omega)$ of solutions of

$$-\frac{\partial}{\partial x_1}\Big(a_n \frac{\partial y_n}{\partial x_1}\Big) - \frac{\partial}{\partial x_2}\Big(a_n \frac{\partial y_n}{\partial x_2}\Big) = f \text{ in } \omega = I_1 \times I_2, \tag{5.1}$$

obtained by the *Lax–Milgram lemma*, and he proved the following result.

Lemma 5.1. *For a_n depending only upon x_1, if*

$$a_m \rightharpoonup a_+ \text{ and } \frac{1}{a_m} \rightharpoonup \frac{1}{a_-} \text{ in } L^\infty(I_1) \text{ weak } \star, \tag{5.2}$$

and the solutions of (5.1) satisfy

$$y_m \rightharpoonup y_\infty \text{ in } H_0^1(\omega) \text{ weak.} \tag{5.3}$$

Then,

$$\begin{aligned} a_m \tfrac{\partial y_m}{\partial x_1} &\rightharpoonup a_- \tfrac{\partial y_\infty}{\partial x_1} \text{ in } L^2(\omega) \text{ weak} \\ a_m \tfrac{\partial y_m}{\partial x_2} &\rightharpoonup a_+ \tfrac{\partial y_\infty}{\partial x_2} \text{ in } L^2(\omega) \text{ weak.} \end{aligned} \tag{5.4}$$

Proof. One writes $f = \frac{\partial g_1}{\partial x_1} + \frac{\partial g_2}{\partial x_2}$ and $\xi_n = a_n \frac{\partial y_n}{\partial x_1} + g_1$, so that ξ_n is bounded in $L^2(\omega) = L^2\big(I_1; L^2(I_2)\big)$ and $\frac{d\xi_n}{dx_1}$ is bounded in $L^2\big(I_1; H^{-1}(I_2)\big)$, since it is $-\frac{\partial(a_n y_n + g_2)}{\partial x_2}$ and $a_n y_n + g_2$ is bounded in $L^2(\omega)$. Then, one applies a compactness argument taught by Jacques-Louis LIONS [52],[6] a consequence of the fact that the injection from $L^2(I_2)$ into $H^{-1}(I_2)$ is compact, and one deduces that ξ_n stays in a compact of $L^2\big(I_1; H^{-1}(I_2)\big)$ strong. For a subsequence ξ_ℓ of ξ_m converging to ξ_∞ in $L^2(\omega)$ weak and in $L^2\big(I_1; H^{-1}(I_2)\big)$ strong, and for $\varphi \in H_0^1(\omega)$, one observes that $\frac{\varphi}{a_\ell}$ converges in $L^2\big(I_1; H_0^1(I_2)\big)$ weak to $\frac{\varphi}{a_-}$, so that $\langle \xi_\ell, \frac{\varphi}{a_\ell}\rangle$ converges to $\langle \xi_\infty, \frac{\varphi}{a_-}\rangle$, but the first term is $\int_\omega \big(\frac{\partial y_\ell}{\partial x_1} + \frac{g_1}{a_\ell}\big)\varphi\, dx_1 dx_2$, and the second term is $\int_\omega \frac{\xi_\infty}{a_-}\varphi\, dx_1 dx_2$, so that one deduces that $\xi_\infty = a_- \frac{\partial y_\infty}{\partial x_1} + g_1$ since φ is arbitrary; the weak limit of $a_\ell \frac{\partial y_\ell}{\partial x_1}$ in $L^2(\omega)$ being $a_- \frac{\partial y_\infty}{\partial x_1}$ independently of the subsequence, the limit is valid for the whole sequence indexed by m.

[6] Jacques-Louis LIONS attributed the variant that he taught to Jean-Pierre AUBIN.

Since the injection of $H^1(\omega)$ into $L^2(\omega)$ is compact, $y_m \to y_\infty$ in $L^2(\omega)$ strong, so that $a_m y_m \to a_+ y_\infty$ in $L^2(\omega)$ weak, and since $a\frac{\partial y}{\partial x_2} = \frac{\partial(a\,y)}{\partial x_2}$ in $H^{-1}(\omega)$ for $a \in L^\infty(I_1)$ and $y \in H^1(\omega)$, one deduces that $a_m\frac{\partial y_m}{\partial x_2} = \frac{\partial(a_m y_m)}{\partial x_2} \rightharpoonup \frac{\partial(a_+ y_\infty)}{\partial x_2} = a_+\frac{\partial y_\infty}{\partial x_2}$ in $H^{-1}(\omega)$ weak, hence in $L^2(\omega)$ weak since it is bounded in $L^2(\omega)$. □

François MURAT found a quite useful generalization, which probably escaped me since I was stuck in the point of view of control theory in which I interpreted the first kind of result. From his point of view, the generalization was more natural, since he was not exposed as much as myself to questions in control theory, and playing with variational elliptic equations was the basic training that our advisor gave to most of his students.[7]

Besides giving François MURAT the possibility to construct other minimization problems without a solution [70], his Lemma 5.1 made the connection with partial differential equations, and an obvious challenge occurred for this new framework, that of identifying a natural relaxation problem. We were not aware of Sergio SPAGNOLO's earlier partial characterization,[8] and we would then rediscover his result, by a slightly different method than his. Then, we would hear about the method of asymptotic expansions used by Évariste SANCHEZ-PALENCIA, and it would open our understanding to crucial applications in continuum mechanics. However, a useful step was to understand a little more about explicit formulas, in particular for laminations, i.e., when coefficients depend only upon (x, e) for a unit vector e, and

---

[7] Jacques-Louis LIONS started from functional analysis and the Lax–Milgram lemma, and he considered abstract equations of the form $A\,u = f$, $u' + Au = f$ with $u(0)$ given, and then, adding $A^T = A$, $u'' + Au = f$ with $u(0)$ and $u'(0)$ given; it contained as applications some partial differential equations, of elliptic, parabolic, hyperbolic, or other types, sometimes with an interpretation in continuum mechanics. When I gave my first graduate course in partial differential equations at CMU (*Carnegie Mellon University*), Pittsburgh, PA, I constructed it the other way around, and I started from ordinary differential equations (the tool for eighteenth century classical mechanics), and I taught first-order scalar partial differential equations, and then some coupled linear or semi-linear systems (which are hyperbolic), and by letting a characteristic speed tend to ∞ I obtained Fourier diffusion equations (which are parabolic), so that the students would not be lured by the fake physics of "Brownian" motion, and then I obtained the Laplace/Poisson equation (which is elliptic) by letting time tend to ∞; after that, I dealt with the wave equation, the linearized elasticity equation, the Stokes equation, and the Navier–Stokes equation. This covered the same equations taught by my advisor, but where he used functional analysis as a goal and continuum mechanics as a pretence that he was interested in applications, I used functional analysis as a tool and understanding continuum mechanics as a goal.

[8] When I met Sergio SPAGNOLO in the summer of 1970 in Varenna, Italy, for a CIME course, he knew about my recent results on interpolation, and he asked me if it could imply his result, so he started to tell me what his result was, but just from the fact that he only used L^∞ hypotheses on the coefficients, I told him that my result in interpolation did not apply to his situation, since I assumed more regularity for the coefficients.

in that case the coefficients of the limiting equation, which we later called *homogenized coefficients* and then *effective coefficients*, are obtained by computing a finite list of weak $\star$ limits.

François MURAT observed that his proof of Lemma 5.1 has a local character and extends to any dimension $N \geq 2$, and one may just assume that $y_n \rightharpoonup y_\infty$ in $H^1(\Omega)$ for an open set $\Omega \subset \mathbb{R}^N$, and then one may restrict attention to a box $\omega \subset \Omega$ with one side perpendicular to e_1.

For a sequence of problems of the form $-div\big(A^n(x_1)grad(u)\big) = f$, starting from the case of isotropic matrices $A^n = a^n\, I$, with $0 < \alpha \leq a^n \leq \beta < \infty$ a.e. in Ω, one creates at the limit special cases with A^{eff} diagonal,[9] and if one starts with A^n diagonal one also finds A^{eff} diagonal;[10] if all A^n have their eigenvalues between α and β a.e. in Ω, then A^{eff} inherits the same property. However, if one uses laminations perpendicular to a vector e not parallel to e_1, the limit may not be diagonal in the initial basis, and one needs then to consider rather general symmetric (positive definite) matrices, and François MURAT proved a generalization of his Lemma 5.1.

Lemma 5.2. *If $A^n \in L^\infty\big(\Omega; \mathcal{L}(\mathbb{R}^N; \mathbb{R}^N)\big)$ depends only upon x_1, and*

$$(A^n)^T = A^n \text{ a.e. in } \Omega, (A^n \xi, \xi) \geq \alpha\, |\xi|^2 \text{ for all } \xi \in \mathbb{R}^N \text{ a.e. in } \Omega, \tag{5.5}$$

for some $\alpha > 0$ independent of n, if one extracts a subsequence A^m with

$$\frac{1}{A^m_{1,1}} \rightharpoonup \frac{1}{A^{\mathit{eff}}_{1,1}} \text{ in } L^\infty(\Omega) \text{ weak } \star, \tag{5.6}$$

$$\frac{A^m_{1,i}}{A^m_{1,1}} \rightharpoonup \frac{A^{\mathit{eff}}_{1,i}}{A^{\mathit{eff}}_{11}} \text{ in } L^\infty(\Omega) \text{ weak } \star, \text{ for } i = 2, \ldots, N, \tag{5.7}$$

$$\frac{A^m_{i,1}}{A^m_{1,1}} \rightharpoonup \frac{A^{\mathit{eff}}_{i,1}}{A^{\mathit{eff}}_{11}} \text{ in } L^\infty(\Omega) \text{ weak } \star, \text{ for } i = 2, \ldots, N, \tag{5.8}$$

$$A^m_{i,j} - \frac{A^m_{i,1}A^m_{1,j}}{A^m_{1,1}} \rightharpoonup A^{\mathit{eff}}_{i,j} - \frac{A^{\mathit{eff}}_{i,1}A^{\mathit{eff}}_{1,j}}{A^{\mathit{eff}}_{11}} \text{ in } L^\infty(\Omega) \text{ weak } \star, \text{ for } i,j = 2, \ldots, N, \tag{5.9}$$

then if

$$-div\big(A^m grad(u_m)\big) = f_m \to f_\infty \text{ in } H^{-1}(\Omega) \text{ strong}, \tag{5.10}$$

$$u_m \rightharpoonup u_\infty \text{ in } H^1_0(\Omega) \text{ weak}, \tag{5.11}$$

[9] The corresponding matrices are either isotropic or have two eigenvalues $a_- < a_+$, of multiplicity 1 and $N-1$.

[10] This is because A^n only depends upon x_1.

one deduces that

$$A^m grad(u_m) \rightharpoonup A^{eff} grad(u_\infty) \text{ in } L^2(\Omega;\mathbb{R}^N) \text{ weak.} \tag{5.12}$$

Of course, once $A^{eff}_{1,1}$ is defined by (5.6), thanks to the ellipticity assumption in (5.5), then (5.7) and (5.8) define $A^{eff}_{1,i}$ and $A^{eff}_{i,1}$ for $i \geq 2$, equally thanks to the symmetry assumption in (5.5), and then (5.9) defines $A^{eff}_{i,j}$ for $i,j \geq 2$, which gives a symmetric A^{eff}. Afterward, François MURAT and myself developed a theory of homogenization for nonsymmetric matrices satisfying a uniform ellipticity assumption, and the analogue of Lemma 5.2 holds if one removes the symmetry assumption in (5.5).

In the late 1970s, I was asked a question about laminations in nonlinear elasticity, and I started by simplifying the proof of the explicit formula for laminations under a general linear partial differential equation, using the *div–curl lemma* that François MURAT and myself obtained in the spring of 1974. The statement of the general approach for the preceding setting is Lemma 5.3 below, and one should notice that (5.13) is a much weaker assumption than the ellipticity condition in (5.5). The proofs of Lemmas 5.2 and 5.3 follow from the general proof, which I shall show in Chap. 12.

Lemma 5.3. *If $A^n \in L^\infty\big(\Omega;\mathcal{L}(\mathbb{R}^N;\mathbb{R}^N)\big)$ depends only upon x_1, and*

$$A^m_{1,1} \geq \alpha \text{ a.e. in } \Omega \text{ for some } \alpha > 0 \text{ independent of } m, \tag{5.13}$$

and that (5.6)–(5.9) hold, then if

$$-div\big(A^m grad(u_m)\big) = f_m \to f_\infty \text{ in } H^{-1}_{loc}(\Omega) \text{ strong,} \tag{5.14}$$

$$u_m \rightharpoonup u_\infty \text{ in } H^1_{loc}(\Omega) \text{ weak,} \tag{5.15}$$

then

$$A^m grad(u_m) \rightharpoonup A^{eff} grad(u_\infty) \text{ in } L^2_{loc}(\Omega;\mathbb{R}^N) \text{ weak.} \tag{5.16}$$

Additional footnotes: Jean-Pierre AUBIN,[11] BROWN R.[12]

[11] Jean-Pierre AUBIN, French mathematician, born in 1939. He worked at Université Paris IX-Dauphine, Paris, France.

[12] Robert BROWN, Scottish-born botanist, 1773–1858. He collected specimens in Australia, and then worked in London, England.

Chapter 6
Homogenization of an Elliptic Equation

In the early 1970s, François MURAT and myself were not aware of the theory of G-convergence, the convergence of Green kernels, which Sergio SPAGNOLO developed in the late 1960s, helped with the insight of Ennio DE GIORGI.

It is sometimes useful not to know about previous attempts to solve a problem, so that one may find a slightly different approach, and this is precisely what happened in our case, and we based our analysis on our Lemma 4.2, while Sergio SPAGNOLO used a regularity result of MEYERS (which we were not aware of). Although generalizing this regularity result is a little difficult, the localization property which Sergio SPAGNOLO proved out of it is stronger than the version that we arrived at by our approach.[1]

The title of my Peccot lectures, given in the beginning of 1977 at Collège de France, in Paris, was "*Homogénéisation dans les équations aux dérivées partielles*,"[2] and if I borrowed the term homogenization from Ivo BABUŠKA, I implied no periodicity assumption like those appearing in the engineering applications which were his motivation.[3] Shortly after, François MURAT coined the term H-convergence for describing our approach, where one wants to identify the weak limits of all the terms which appear in a sequence of partial differential equations; this is then more general than G-convergence, where one wants to identify the weak limit of the solutions.

At the beginning of the 1974–1975 academic year, which I spent at UW, Madison, WI, I gave a talk on my joint work with François MURAT, and Carl DE BOOR showed me after that an article by Ivo BABUŠKA, who used periodicity assumptions, like for the earlier work of Évariste

[1] It is important to be fair in describing advantages and defects of various approaches to a question. As I never read much, there are obviously a lot of results which I am not aware of, but that is different from a trend which grew in recent years for political reasons, since there is a group whose members are keen in advocating fake mechanics or physics models, and who intentionally attribute all my ideas to their friends, who obviously do not understand them very well!

[2] It means "Homogenization in partial differential equations."

[3] Michael VOGELIUS mentioned to me that Ivo BABUŠKA borrowed the term homogenization from the nuclear engineering literature.

L. Tartar, *The General Theory of Homogenization*, Lecture Notes of the Unione Matematica Italiana 7, DOI 10.1007/978-3-642-05195-1_6,

SANCHEZ-PALENCIA which helped me understand that my joint work with François MURAT was related to the question of finding macroscopic properties of mixtures,[4] without any probabilistic framework, of course![5] After discussing with Joel ROBBIN, who showed me how to use the framework of differential forms for obtaining a different proof of the *div–curl lemma*, which I proved with François MURAT in the spring of 1974, it became clear why weak convergence is right for some quantities and not for others, and why H-convergence is natural (and it was found, later but independently, by Leon SIMON).

An interesting consequence of homogenization is that the formulas for mixtures postulated in continuum mechanics or physics by the rules of thermodynamics can at best be approximations, because effective properties cannot be deduced in dimension $N \geq 2$ from the knowledge of the *Young measures*, which see the proportions of the materials used. Observing that something is wrong is easy; finding how to correct the known defects is not so simple!

It "explains" the behavior of the advocates of fake mechanics or physics models, who either play obsolete thermodynamics games, or invent rather silly games based on gradient flows; not surprisingly, this group promotes fake homogenization questions, and pretends not to see that Γ-convergence is not homogenization, showing a clear intention to mislead others.

Γ-convergence was developed by Ennio DE GIORGI for generalizing an aspect of G-convergence which he studied with Sergio SPAGNOLO, and for generalizing also a notion of convergence of convex functions introduced by Umberto MOSCO,[6] and I wonder if he was aware of the abstract theory of relaxation which Ivar EKELAND was describing in Paris in the early 1970s. The main defect of Γ-convergence for continuum mechanics or physics is that it misleads people to study minimization questions, so that it cannot handle time, apart from some quite silly versions of gradient flows! Γ-convergence is a concept belonging to topology and functional analysis, while *homogenization* is a part of the beginning of *a nonlinear microlocal theory*, with the theory of *compensated compactness* which I also developed with François MURAT, and the theory of *H-measures* which I developed in the late 1980s.

---

[4] Our initial motivation, by an academic question of optimal design, helped understand which mixtures are likely to be observed in natural mixing phenomena.

[5] Probabilities are used at places where one does not understand what happens, and their use is natural for engineers, who are asked to tame processes for which no one knows which equations to use, but scientists should be careful to explain that probabilities should be pushed further and further away, until one finds the correct equations. It is pure sabotage from a scientific point of view to brainwash students in believing that the laws of nature contain probabilities!

[6] Umberto MOSCO, Italian-born mathematician, born in 1938. He worked at Università di Roma "La Sapienza," Roma (Rome), Italy, and he works now at WPI (Worcester Polytechnic Institute), Worcester, MA.

The basic problem of G-convergence is to consider in a bounded open set Ω of $\mathbb{R}^N$ a sequence of Dirichlet problems

$$-\,div\big(A^n grad(u_n)\big) = f \text{ in } \Omega;\; u_n \in H_0^1(\Omega), \tag{6.1}$$

where, assuming $0 < \alpha \leq \beta < \infty$, one has

$$\text{for all } n, (A^n)^T = A^n, \alpha\,|\xi|^2 \leq (A^n\xi,\xi) \leq \beta\,|\xi|^2 \text{ for all } \xi \in \mathbb{R}^N, \text{ a.e. in } \Omega. \tag{6.2}$$

Of course, the equation in (6.1) is written in the sense of distributions, and $H_0^1(\Omega)$ is the closure in $H^1(\Omega)$ of smooth functions with compact support, equipped with the norm $||grad(u)||_{L^2(\Omega)}$ (because the Poincaré inequality holds), and $f \in H^{-1}(\Omega)$, the dual space of $H_0^1(\Omega)$, equipped with the dual norm. By the Lax–Milgram lemma (or the F. Riesz theorem since one is in a symmetric situation),[7] there exists a unique solution $u_n \in H_0^1(\Omega)$ of (6.1), satisfying

$$||u_n||_{H_0^1(\Omega)} \leq \frac{1}{\alpha}\,||f||_{H^{-1}(\Omega)}. \tag{6.3}$$

If $f \in L^p(\Omega)$ with $p > \frac{N}{2}$ (or $p \geq 1$ for $N = 1$), then the solution u_n is Hölder continuous,[8,9] and can be expressed by an integral formula

$$u_n(x) = \int_\Omega G_n(x,y) f(y)\,dy \text{ for } x \in \Omega, \tag{6.4}$$

with a (nonnegative) Green kernel G_n, for which some regularity estimates are known, so that a subsequence G_m converges, strongly outside the diagonal, to a kernel G_∞, and it is a natural question to ask whether G_∞ is the Green kernel of a similar equation, with coefficient A^{eff}.

Definition 6.1. *One says that* $A^n \in L^\infty\big(\Omega; \mathcal{L}_{sym}(\mathbb{R}^N;\mathbb{R}^N)\big)$ *G-converges to* $A^{\mathit{eff}} \in L^\infty\big(\Omega; \mathcal{L}_{sym}(\mathbb{R}^N;\mathbb{R}^N)\big)$ *if there exists* $0 < \alpha \leq \beta < \infty$ *such that*

$$\alpha\,|\xi|^2 \leq (A^n\xi,\xi), (A^{\mathit{eff}}\xi,\xi) \leq \beta\,|\xi|^2 \text{ for all } \xi \in \mathbb{R}^N, \text{ a.e. in } \Omega, \text{ and} \tag{6.5}$$

---

[7] Frigyes (Frederic) RIESZ, Hungarian mathematician, 1880–1956. He worked in Kolozsvár (then in Hungary, now Cluj-Napoca, Romania), in Szeged and in Budapest, Hungary.

[8] Otto Ludwig HÖLDER, German mathematician, 1859–1937. He worked in Leipzig, Germany.

[9] It is special to scalar second-order equations to have Hölder continuous solutions for nonsmooth coefficients, but the Morrey theorem does not have this restriction, as Ennio DE GIORGI told me in the spring of 1974.

$$\begin{aligned}&\textit{for all } f \in H^{-1}(\Omega), \textit{ the solutions } u_n \in H_0^1(\Omega) \textit{ of } -div\Big(A^n grad(u_n)\Big) = f\\ &\textit{converge in } H_0^1(\Omega) \textit{ weak to the solution } u_\infty \textit{ of } -div\Big(A^{\mathit{eff}} grad(u_\infty)\Big) = f.\end{aligned} \tag{6.6}$$

Sergio SPAGNOLO proved that for any sequence A^n satisfying (6.2), one can extract a subsequence which G-converges to A^{eff}, satisfying also (6.2) (see Lemma 6.2, and for H-convergence, see Theorem 6.5), but in the locally isotropic case, i.e., $A^n = a_n I$, then A^{eff} is not necessarily proportional to I in dimension $N \geq 2$.[10] He found that G-convergence has a local character (for H-convergence, see Lemma 10.5, but with ω open), i.e., if A^m G-converges to A^{eff}, if B^m G-converges to B^{eff}, and if for all m one has $A^m = B^m$ on a *measurable* subset ω of Ω, then $A^{\mathit{eff}} = B^{\mathit{eff}}$ in ω.[11] He also found that other boundary conditions may lead to the same limit A^{eff} (for H-convergence, see Lemma 10.3).

Later on, Ennio DE GIORGI and Sergio SPAGNOLO showed that if A^n G-converges to A^{eff}, and if v_n converges to v_∞ in $H_0^1(\Omega)$ weak then for any measurable subset ω of Ω one has

$$\liminf_n \int_\omega \big(A^n grad(v_n), grad(v_n)\big)\,dx \geq \int_\omega \big(A^{\mathit{eff}} grad(v_\infty), grad(v_\infty)\big)\,dx, \tag{6.7}$$

and that if A^n G-converges to A^{eff} and $u_\infty \in H_0^1(\Omega)$, then there exists a sequence u_n converging to u_∞ in $H_0^1(\Omega)$ weak such that for any measurable subset ω of Ω one has

$$\lim_n \int_\omega \big(A^n grad(u_n), grad(u_n)\big)\,dx = \int_\omega \big(A^{\mathit{eff}} grad(u_\infty), grad(u_\infty)\big)\,dx \tag{6.8}$$

(for H-convergence, see Lemma 10.6). This is a particular example of a notion which was extended to convergence of convex functions by Umberto MOSCO, and later to convergence of general functionals by Ennio DE GIORGI under the name of Γ-convergence.[12]

Sergio SPAGNOLO also studied the corresponding parabolic equations

---

[10] Antonio MARINO and Sergio SPAGNOLO observed in [59] that one can choose $0 < \alpha' \leq \beta' < \infty$ such that any A satisfying (6.2) can be obtained as the G-limit of a scalar sequence $a_n I$ satisfying (6.2) with α and β replaced by α' and β'.

[11] This is where my approach to H-convergence with François MURAT is less precise, since we only proved a similar result if ω is an *open* subset of Ω.

[12] This book being on homogenization, i.e., H-convergence, there will not be many reasons to mention Γ-convergence again, since it is related to questions of minimization and of finding lower semi-continuous envelopes, while homogenization is interested in understanding oscillations (and concentration effects) in elliptic, parabolic or hyperbolic partial differential equations or systems (or any system which appears in a reasonable physical setting).

$$\frac{\partial u_n}{\partial t} - div\big(A^n grad(u_n)\big) = g \text{ in } \Omega \times (0,T), u_n\,|_{t=0} = w \text{ in } \Omega, \tag{6.9}$$

with Dirichlet conditions, and A^n satisfying (6.2); a uniform bound for $u_n \in L^2\big((0,T); H_0^1(\Omega)\big) \bigcap C\big([0,T]; L^2(\Omega)\big)$ being true for $g \in L^2\big((0,T); H^{-1}(\Omega)\big) + L^1\big((0,T); L^2(\Omega)\big)$ and $w \in L^2(\Omega)$.

Sergio SPAGNOLO then studied the corresponding hyperbolic equations

$$\frac{\partial^2 u_n}{\partial t^2} - div\big(A^n grad(u_n)\big) = h \text{ in } \Omega \times (0,T), u_n\,|_{t=0} = w, \frac{\partial u_n}{\partial t}\,|_{t=0} = z \text{ in } \Omega, \tag{6.10}$$

with Dirichlet conditions, and A^n satisfying (6.2), and he found a different situation, because the known existence theorems in a variational framework all required some smoothness of the coefficient A^n with respect to t. I only described in the mid 1990s a program for finding a natural class of A^n, by identifying a class of coefficients including the case of moving bodies, first rigid and then elastic, first not colliding and then colliding, and taking into account the energy of these moving bodies, one might get a realistic class of not so smooth coefficients; for such a class, a homogenization result might hold, analogous to what was observed in spectroscopy (which physicists call absorption and emission of light at specific frequencies), which should appear as a nonlocal effect in the effective equation, and I shall discuss similar questions in a much simpler setting in Chaps. 23 and 24. With A^n independent of t, the result is simple and follows from the elliptic case.[13]

There is an easy abstract elliptic framework, which will show the difference between G-convergence and H-convergence, and the approach that I followed with François MURAT will appear in a natural way.[14] The basic result in this abstract framework is Lemma 6.2, to compare to Theorem 6.5.

Let V be a real (infinite-dimensional) separable Banach space,[15] corresponding to $H_0^1(\Omega)$ in our example; let $||\cdot||$ be the norm on V, $||\cdot||_*$ the norm on V', and $\langle\cdot,\cdot\rangle$ the duality product between V' and V. Let $T_n \in \mathcal{L}(V; V')$ be a bounded sequence of (linear continuous) operators from V into V', corresponding to $T_n u = -div\big(A^n\, grad(u)\big)$ in our example, satisfying a uniform V-ellipticity condition, i.e., there exist $0 < \alpha \le M < \infty$ with

$$\langle T_n u, u\rangle \ge \alpha\,||u||^2 \text{ and } ||T_n u||_* \le M\,||u|| \text{ for all } u \in V, \tag{6.11}$$

---

[13] For A^n independent of t, but with sequences of initial data, there may be resonance effects, as was mentioned to me by Joe KELLER in the spring of 1975.

[14] It was rediscovered independently by Leon SIMON. It was also rediscovered by Olga OLEINIK, although not entirely independently, because she heard talks by Jacques-Louis LIONS in the periodic case, but our former advisor probably forgot to mention that he was following the general theory that we developed.

[15] It is a Hilbert space in disguise, because (6.11) implies that an equivalent norm corresponds to the scalar product $\langle T_n u, v\rangle + \langle T_n v, u\rangle$.

corresponding to $(A(x)\xi,\xi) \geq \alpha\,|\xi|^2$, and $|A(x)\xi| \leq M\,|\xi|$ for all $\xi \in \mathbb{R}^N$, a.e. $x \in \Omega$ in our example. By the Lax–Milgram lemma, T_n is an isomorphism from V onto V'.

Lemma 6.2. *There exists a subsequence T_m, and $T_\infty \in \mathcal{L}(V;V')$, such that for all $f \in V'$, the sequence of solutions u_m of $T_m u_m = f$ converges to u_∞ in V weak, solution of $T_\infty u_\infty = f$, and T_∞ satisfies*

$$\langle T_\infty u, u\rangle \geq \alpha\,||u||^2 \text{ and } ||T_\infty u||_* \leq \frac{M^2}{\alpha}\,||u|| \text{ for all } u \in V. \tag{6.12}$$

In other words $(T_m)^{-1}$ converges weakly to $(T_\infty)^{-1}$ in $\mathcal{L}(V';V)$.

Proof. $T_n u_n = f$ implies $\alpha\,||u_n||^2 \leq \langle T_n u_n, u_n\rangle = \langle f, u_n\rangle \leq ||f||_*||u_n||$, so that $||u_n|| \leq \frac{1}{\alpha}\,||f||_*$, i.e., $||T_n^{-1}||_{\mathcal{L}(V';V)} \leq \frac{1}{\alpha}$, and one can extract a subsequence u_m converging to u_∞ in V weak. Repeating this extraction for f belonging to a countable dense family F of V',[16] and using a Cantor diagonal subsequence,[17] one extracts a subsequence T_m such that for all $f \in F$, the sequence $u_m = (T_m)^{-1} f$ converges to a limit $S(f)$ in V weak. Since the sequence $(T_m)^{-1}$ is uniformly bounded and F is dense, one deduces that $(T_m)^{-1} f$ converges to a limit $S(f)$ in V weak for all $f \in V'$, and S is then automatically a linear continuous operator from V' into V, with $||S\,f|| \leq \frac{1}{\alpha}\,||f||_*$ for all $f \in V'$. One then needs to show that S is invertible, and the fact that the operators $(T_n)^{-1}$ are uniformly bounded is not sufficient for ensuring that, since one can construct a sequence of symmetric surjective isometries converging weakly to 0, in any infinite-dimensional Hilbert space,[18] but the ellipticity condition prevents this kind of problem. Since $\langle S\,f, f\rangle = \lim_m \langle u_m, f\rangle = \lim_m \langle T_m u_m, u_m\rangle \geq \alpha \liminf_m ||u_m||^2$, and $M\,||u_m|| \geq ||T_m u_m||_* = ||f||_*$, one deduces that $\langle S\,f, f\rangle \geq \frac{\alpha}{M^2}\,||f||_*^2$, so that S is invertible by the Lax–Milgram lemma, and its inverse T_∞ has a norm bounded by $\frac{M^2}{\alpha}$. Since u_m converges to $S\,f$ in V weak, one has $\liminf_m ||u_m||^2 \geq ||S\,f||^2$, so that $\langle S\,f, f\rangle \geq \alpha\,||S\,f||^2$ for all $f \in V'$; equivalently, $\langle T_\infty u_\infty, u_\infty\rangle \geq \alpha\,||u_\infty||^2$ for all $u_\infty \in V$, since S is invertible. □

As most results in functional analysis, Lemma 6.2 only gives a general framework, and does not help much for identifying T_∞ in concrete cases, but one often uses the information that T_∞ is invertible and that one can choose $f \in V'$ so that the weak limit u_∞ of the sequence of solutions u_m

[16] The separability of V is not necessary for applying the Lax–Milgram lemma, but it is needed here; this restriction is not really a limitation for applications.

[17] Georg Ferdinand Ludwig Philipp CANTOR, Russian-born mathematician, 1845–1918. He worked in Halle, Germany.

[18] The space contains an isometric copy of l^2, hence isometric to $L^2(0,1)$, and in that case one considers the operator L_n of multiplication by $sign(\sin n\,x)$, so that L_n is its own inverse and converges weakly to 0.

can be any element of V that one wants. As will be seen later, even if all T_n are differential operators, it may happen that T_∞ is not a purely differential operator and that a nonlocal integral correction must be taken into account.

When dealing with a sequence u_n of solutions of $-div\big(A^n\, grad(u_n)\big) = f$, converging to u_∞ in $H^1_0(\Omega)$ weak, we observed that one should not be only interested in the fact that $E^n = grad(u_n)$ converges to $E^\infty = grad(u_\infty)$ in $L^2(\Omega;\mathbb{R}^N)$ weak, but that one should also identify the weak limit D^∞ of $D^n = A^n\, grad(u_n)$ (eventually after extracting another subsequence), which is $A^{\mathit{eff}}\, grad(u_\infty)$. Although we introduced this idea in order to treat the case of scalar coefficients, i.e., $A^n = a_n I$, and then symmetric (positive definite) matrices, it is important to notice that in the nonsymmetric case the point of view of G-convergence is not good enough, because the operators T_n (or T_n^{-1}) cannot see the limit of $A^n\, grad(u_n)$, because adding to A^n a constant skew-symmetric matrix B (small enough in norm for keeping a uniform ellipticity condition), does not change the operator T_n.[19] One could repeat in the nonsymmetric context arguments similar to Lemma 6.2, but we also wanted to avoid a difficulty encountered there, that starting with a bound M for the norm of T_n, one ends with the greater bound $\frac{M^2}{\alpha}$ for the norm of T_∞, and for this purpose we introduced the following definition of $\mathcal{M}(\alpha,\beta;\Omega)$.

Definition 6.3. *For $0 < \alpha \le \beta < \infty$, $\mathcal{M}(\alpha,\beta;\Omega)$ denotes the set of $A \in L^\infty\big(\Omega;\mathcal{L}(\mathbb{R}^N;\mathbb{R}^N)\big)$ satisfying*

$$\forall \xi \in \mathbb{R}^N, (A(x)\xi,\xi) \ge \alpha\,|\xi|^2, (A(x)\xi,\xi) \ge \frac{1}{\beta}\,|A(x)\xi|^2, \;\; a.e.\ x \in \Omega. \tag{6.13}$$

Equivalently $\mathcal{M}(\alpha,\beta;\Omega)$ is the set of $A \in L^\infty\big(\Omega;\mathcal{L}(\mathbb{R}^N;\mathbb{R}^N)\big)$ satisfying

$$\forall \xi \in \mathbb{R}^N, (A(x)\xi,\xi) \ge \alpha\,|\xi|^2, (A^{-1}(x)\xi,\xi) \ge \frac{1}{\beta}\,|\xi|^2, \;\; a.e.\ x \in \Omega. \tag{6.14}$$

If $P \in \mathcal{L}(\mathbb{R}^N;\mathbb{R}^N)$ satisfies $(P\,\xi,\xi) \ge \frac{1}{\beta}\,|P\,\xi|^2$ for all $\xi \in \mathbb{R}^N$, then $|P\,\xi| \le \beta\,|\xi|$ for all $\xi \in \mathbb{R}^N$, but if P is not symmetric and satisfies $(P\,\xi,\xi) \ge \alpha\,|\xi|^2$ and $|P\,\xi| \le M\,|\xi|$ for all $\xi \in \mathbb{R}^N$, then one can only deduce $(P\,\xi,\xi) \ge \frac{\alpha}{M^2}\,|P\,\xi|^2$,[20] while if P is symmetric one deduces that $(P\,\xi,\xi) \ge \frac{1}{M}\,|P\,\xi|^2$, of course.

---

[19] Actually, Sergio SPAGNOLO also introduced the limit of D^n in his proof, and that may be why some do not see much difference between G-convergence and H-convergence. In this book, I use the term G-convergence for describing only the symmetric case, and more important differences with H-convergence will appear later, for questions of correctors, and for the change of form of the effective equation by addition of nonlocal effects.

[20] For $N = 2$, $a > 0$ and $P = \begin{pmatrix} a & b \\ -b & a \end{pmatrix}$ one has $\alpha = a$ and $M = \sqrt{a^2+b^2}$, and for $P^{-1} = \frac{1}{a^2+b^2}\begin{pmatrix} a & -b \\ b & a \end{pmatrix}$ one has $\alpha' = \frac{a}{a^2+b^2} = \frac{\alpha}{M^2}$, and $M' = \frac{1}{M}$.

The reason for using the sets $\mathcal{M}(\alpha,\beta;\Omega)$ lies in the compactness result of Theorem 6.5, for the topology of H-convergence defined now.

Definition 6.4. *For a bounded set $\Omega \subset \mathbb{R}^N$, a sequence $A^n \in \mathcal{M}(\alpha,\beta;\Omega)$ H-converges to $A^{\mathit{eff}} \in \mathcal{M}(\alpha',\beta';\Omega)$ for some $0<\alpha'\leq\beta'<\infty$, if for all $f \in H^{-1}(\Omega)$, the sequence of solutions $u_n \in H_0^1(\Omega)$ of $-div\big(A^n\, grad(u_n)\big) = f$ converges to u_∞ in $H_0^1(\Omega)$ weak, and the sequence $A^n\, grad(u_n)$ converges to $A^{\mathit{eff}}\, grad(u_\infty)$ in $L^2(\Omega;\mathbb{R}^N)$ weak, where u_∞ is the solution of $-div\big(A^{\mathit{eff}}\, grad(u_\infty)\big) = f$ in Ω.*

From Theorem 6.5, one can always take $\alpha'=\alpha$ and $\beta'=\beta$.

H-convergence comes from a topology on $X=\bigcup_{n\geq 1}\mathcal{M}\big(\frac{1}{n},n;\Omega\big)$, union of all $\mathcal{M}(\alpha,\beta;\Omega)$, which is the coarsest topology that makes a list of mappings continuous. For $f\in H^{-1}(\Omega)$ one such mapping is $A\mapsto u$ from X into $H_0^1(\Omega)$ weak, and another one is $A \mapsto A\, grad(u)$ from X into $L^2(\Omega;\mathbb{R}^N)$ weak, where u is the solution of $-div\big(A\, grad(u)\big) = f$. When one restricts that topology to $\mathcal{M}(\alpha,\beta;\Omega)$, it is equivalent to considering only f belonging to a countable bounded set whose combinations are dense in $H^{-1}(\Omega)$; then u and $A\, grad(u)$ belong to bounded sets respectively of $H_0^1(\Omega)$ and $L^2(\Omega;\mathbb{R}^N)$ which are metrizable for the weak topology, so that the restriction of that topology to $\mathcal{M}(\alpha,\beta;\Omega)$ is defined by a countable number of semi-distances and is then defined by a semi-distance. That it is indeed a distance can be seen by showing uniqueness of a limit: if a sequence A^n H-converges to both A^{eff} and to B^{eff}, then one deduces that $A^{\mathit{eff}}\, grad(u_\infty) = B^{\mathit{eff}}\, grad(u_\infty)$ a.e. $x\in\Omega$ for all $f\in H^{-1}(\Omega)$, hence for all $u_\infty\in H_0^1(\Omega)$; then choosing u_∞ to coincide successively with x_j, $j=1,\ldots,N$ on an open subset ω with $\overline{\omega}\subset\Omega$, one must have $A^{\mathit{eff}} = B^{\mathit{eff}}$ a.e. $x\in\omega$.

Of course, one never needs much from this topology, but some arguments use the fact that it exists and that $\mathcal{M}(\alpha,\beta;\Omega)$ is metrizable, and also compact as asserted by Theorem 6.5.

From the point of view of continuum mechanics or physics, homogenization is related to understanding macroscopic properties of mixtures, and when one looks at one mixture there is no sequence of coefficients A^n, but *considering sequences serves in identifying the correct topology* which permits one to say that a fine mixture (corresponding to rapidly varying coefficients in space) resembles a material with slowly varying properties.

Theorem 6.5. *For any sequence $A^n \in \mathcal{M}(\alpha,\beta;\Omega)$ there exists a subsequence A^m and an element $A^{\mathit{eff}}\in\mathcal{M}(\alpha,\beta;\Omega)$ such that A^m H-converges to A^{eff}.*

Proof. Using the same argument as in Lemma 6.2, F being a countable dense set of $H^{-1}(\Omega)$, one can extract a subsequence A^m such that for all $f\in F$ the sequence $u_m\in H_0^1(\Omega)$ of solutions of $-div\big(A^m\, grad(u_m)\big) = f$ converges to $u_\infty = S(f)$ in $H_0^1(\Omega)$ weak, and $A^m\, grad(u_m)$ converges to $R(f)$ in $L^2(\Omega;\mathbb{R}^N)$ weak. The same is then true for all $f\in H^{-1}(\Omega)$, the operator S is invertible, and $R(f) = C\, u_\infty$, where C is a linear continuous operator from $H_0^1(\Omega)$ into $L^2(\Omega;\mathbb{R}^N)$. It remains to show that C is local, of the form $C\,v = A^{\mathit{eff}}\,grad(v)$ for all $v \in H_0^1(\Omega)$, with $A^{\mathit{eff}} \in \mathcal{M}(\alpha,\beta;\Omega)$, and a first step is to show that for all $v \in H_0^1(\Omega)$ one has $\big(C\,v, grad(v)\big) \geq \alpha\,|grad(v)|^2$ and $\big(C\,v, grad(v)\big) \geq \frac{1}{\beta}\,|C\,v|^2$ a.e. $x \in \Omega$.

For $v \in H_0^1(\Omega)$, let $f = -div(C\,v)$, so that $u_\infty = v$, and let φ be a smooth function so that one may use $\varphi\,u_m$ and $\varphi\,v$ as test functions. One gets $\langle f, \varphi\,u_m\rangle = \int_\Omega \big(A^m\,grad(u_m), \varphi\,grad(u_m) + u_m\,grad(\varphi)\big)\,dx$, and u_m converges strongly to v in $L^2(\Omega)$ since $H_0^1(\Omega)$ is compactly embedded into $L^2(\Omega)$, so that

$$\lim_m \langle f, \varphi\,u_m\rangle = \lim_m \int_\Omega \varphi\Big(A^m\,grad(u_m), grad(u_m)\Big)\,dx + \int_\Omega \Big(C\,v, v\,grad(\varphi)\Big)\,dx. \tag{6.15}$$

Since $\langle f, \varphi\,v\rangle = \int_\Omega \big(C\,v, \varphi\,grad(v) + v\,grad(\varphi)\big)\,dx$, and $\langle f, \varphi\,u_m\rangle \to \langle f, \varphi\,v\rangle$, one deduces that for all smooth functions φ one has

$$\int_\Omega \varphi\,\big(A^m\,grad(u_m), grad(u_m)\big)\,dx \to \int_\Omega \varphi\,\big(C\,v, grad(v)\big)\,dx. \tag{6.16}$$

With $\varphi \geq 0$, and the first part of the definition of $\mathcal{M}(\alpha,\beta;\Omega)$, one has

$$\int_\Omega \varphi\,\Big(C\,v, grad(v)\Big)\,dx \geq \alpha\,\liminf_m \int_\Omega \varphi\,|grad(u_m)|^2\,dx \geq \alpha \int_\Omega \varphi\,|grad(v)|^2\,dx, \tag{6.17}$$

the second inequality following from $grad(u_m) \rightharpoonup grad(v)$ in $L^2(\Omega;\mathbb{R}^N)$ weak. Since (6.17) holds for all smooth $\varphi \geq 0$, one deduces that

$$\text{for all } v \in H_0^1(\Omega), \big(C\,v, grad(v)\big) \geq \alpha\,|grad(v)|^2 \text{ a.e. } x \in \Omega. \tag{6.18}$$

Using the second part of the definition of $\mathcal{M}(\alpha,\beta;\Omega)$, one has

$$\int_\Omega \varphi\,\Big(C\,v, grad(v)\Big)\,dx \geq \frac{1}{\beta}\,\liminf_m \int_\Omega \varphi\,|A^m\,grad(u_m)|^2\,dx \geq \frac{1}{\beta} \int_\Omega \varphi\,|C\,v|^2\,dx, \tag{6.19}$$

since $A^m\,grad(u_m) \rightharpoonup C\,v$ in $L^2(\Omega;\mathbb{R}^N)$ weak. Since (6.19) also holds for all smooth $\varphi \geq 0$, one deduces that

$$\text{for all } v \in H_0^1(\Omega), \big(C\,v, grad(v)\big) \geq \frac{1}{\beta}\,|C\,v|^2 \text{ a.e. } x \in \Omega. \tag{6.20}$$

If one shows that $C\,v = A\,grad(v)$ for a measurable A, then (6.18) and (6.20) imply that $A \in \mathcal{M}(\alpha,\beta;\Omega)$ since one can take v to be any affine function in an open set ω with $\overline{\omega} \subset \Omega$. From (6.21) one deduces that

$$\text{for all } v \in H_0^1(\Omega), |C\,v| \leq \beta\,|grad(v)| \text{ a.e. } x \in \Omega, \tag{6.21}$$

and since C is linear, (6.21) implies that

$$\text{if } grad(v) = grad(w) \text{ a.e. in } \omega, \text{ then } C\,v = C\,w \text{ a.e. in } \omega. \tag{6.22}$$

Writing Ω as the union of an increasing sequence ω_n of open sets with $\overline{\omega_n} \subset \Omega$, one defines A in the following way: for $\xi \in \mathbb{R}^N$, one chooses $v_n \in H_0^1(\Omega)$ such that $grad(v_n) = \xi$ on ω_n, and one defines $A\,\xi$ on ω_n as the restriction of $C(v_n)$ to ω_n, and this defines $A\,\xi$ as a measurable function in Ω since $C\,v_n$ and $C\,v_m$ coincide on $\omega_n \cap \omega_m$ by (6.22), and (6.22) also implies that A is linear in ξ. If $w \in H_0^1(\Omega)$ is piecewise affine so that $grad(w)$ is piecewise constant, then (6.22) implies that $C\,w = A\,grad(w)$ a.e. $x \in \Omega$. Since piecewise affine functions are dense in $H_0^1(\Omega)$, for each $v \in H_0^1(\Omega)$ there is a sequence w_k of piecewise affine functions such that $grad(w_k)$ converges to $grad(v)$ strongly in $L^2(\Omega; \mathbb{R}^N)$, and since $|C\,v - A\,grad(w_k)| = |C\,v - C\,w_k| \leq \beta\,|grad(v - w_k)|$ a.e. $x \in \Omega$, one deduces $C\,v = A\,grad(v)$ a.e. $x \in \Omega$. □

In the spring of 1975, I gave a talk in Ann Arbor, MI, and I think it was Bogdan BOJARSKI who mentioned the theory of quasi-conformal mappings,[21] but I never tried to learn that theory; Leon SIMON said that he was not aware of that comment,[22] and Tadeusz IWANIEC said that Bogdan BOJARSKI probably meant that this idea could be used for problems in quasi-conformal mappings,[23] but not that something was already done in that direction. More recently, a connection between homogenization and quasi-conformal mappings appeared in relation with bounds on effective coefficients, and it was investigated by Enzo NESI [76].

For the particular problem of optimal design that François MURAT and myself were interested in, we needed to go beyond Theorem 6.5, because we wanted more precise bounds on the effective coefficients A^{eff}. Our case was $A^n = a_n I$, with $a_n = \alpha\,\chi_n + \beta\,(1 - \chi_n)$ for a sequence of characteristic functions χ_n, converging to θ in $L^\infty(\Omega)$ weak $\star$, so that θ is the local proportion of the material of conductivity α, and we wanted to characterize what A^{eff} could be, depending upon θ.[24] Without imposing as much, we knew that when a_n only depends upon one variable the $L^\infty(\Omega)$ weak $\star$ limits of a_n and $\frac{1}{a_n}$ appear, which we denoted a_+ and $\frac{1}{a_-}$. The sequences $E^n = grad(u_n)$

[21] Bogdan Tadeusz BOJARSKI, Polish mathematician. He works at the Polish Academy of Sciences in Warsaw, Poland, being the director of the Stefan Banach International Mathematical Center.

[22] He asked his student MCCONNELL to extend Sergio SPAGNOLO's approach to (linearized) elasticity, without too much success [61], so that he looked himself at the question and found the approach to H-convergence, and it was the referee of his article who pointed out an article of mine.

[23] Tadeusz IWANIEC, Polish-born mathematician, born in 1947. He works at Syracuse University, Syracuse, NY.

[24] We only found the characterization at the end of 1980.

and $D^n = a_n E^n$ converge in $L^2(\Omega;\mathbb{R}^N)$ weak to E^∞ and D^∞, and by (6.16) (E^n, D^n) converges to (E^∞, D^∞) in $L^1(\Omega)$ weak $\star$,[25] a simple instance of the *div–curl lemma*, so that $a_n|E^n|^2$ converges to (E^∞, D^∞) in the sense of measures. Lemma 4.2 suggested to look at the convex hull of

$$K = \Big\{\Big(E, a\,E, a\,|E|^2, a, \frac{1}{a}\Big) \mid a \in [\alpha,\beta], E \in \mathbb{R}^N\Big\} \tag{6.23}$$

and to investigate what relations between D^∞ and E^∞ follow from

$$\Big(E^\infty, D^\infty, (E^\infty, D^\infty), a_+, \frac{1}{a_-}\Big) \in \overline{conv(K)}. \tag{6.24}$$

For minimizing a linear form on K, one first minimizes in $E \in \mathbb{R}^N$, and for $v, w \in \mathbb{R}^N$ one has

$$a_n|E^n|^2 - 2(E_n, v + a_n w) \ge -\frac{|v + a_n w|^2}{a_n} = -\frac{|v|^2}{a_n} - 2(v,w) - a_n|w|^2, \tag{6.25}$$

which gives at the limit

$$(E^\infty, D^\infty) - 2(E^\infty, v) - 2(D^\infty, w) \ge -\frac{|v|^2}{a_-} - 2(v,w) - a_+|w|^2. \tag{6.26}$$

The best choice for v and w is obtained by solving the system $\frac{v}{a_-} + w = E^\infty$ and $v + a_+ w = D^\infty$. There is a problem if $a_- = a_+$, since one needs to have $D^\infty = a_+ E^\infty$; this is not surprising, since one always has $a_- \le a_+$, and equality only occurs if a_n converges to a_+ in $L^p(\Omega)$ strong for all $p < \infty$ since a_n is bounded in $L^\infty(\Omega)$. If $a_- < a_+$, the best choice for v and w is given by

$$v = \frac{a_-(a_+E^\infty - D^\infty)}{a_+ - a_-}, w = \frac{D^\infty - a_-E^\infty}{a_+ - a_-}, \tag{6.27}$$

and leads to

$$(a_+ - a_-)(E^\infty, D^\infty) - a_-\Big(E^\infty, (a_+E^\infty - D^\infty)\Big) - \Big(D^\infty, (D^\infty - a_-E^\infty)\Big) \ge 0, \tag{6.28}$$

i.e.,

$$(D^\infty - a_-E^\infty, D^\infty - a_+E^\infty) \le 0, \tag{6.29}$$

so that D^∞ belongs to the sphere with diameter $[a_-E^\infty, a_+E^\infty]$.[26]

---

[25] $L^1(\Omega)$ is isometrically embedded in $\mathcal{M}_b(\Omega)$, the space of Radon measures in Ω with finite total mass, which is the dual of $C_0(\Omega)$, the space of continuous functions tending to 0 at the boundary $\partial\Omega$, equipped with the sup norm.

[26] The preceding argument shows that the function defined as $\frac{1}{a-b}(D - a\,E, D - b\,E)$ is convex in $\Big(E, D, (E,D), a, \frac{1}{b}\Big)$ for $b < a$ (it must be taken to be 0 if $b = a$ and $D = a\,E$ and $+\infty$ otherwise or if $b > a$).

When François MURAT and myself started this computation in the early 1970s, we only deduced $|D^\infty| \leq a_+|E^\infty|$, which is more precise than $|D^\infty| \leq \beta\,|E^\infty|$, and this helped us prove that a local relation $D^\infty = A^{eff} E^\infty$ holds. I shall come back later to this computation in a more general setting, but in the early 1970s we found a convexity argument for showing that the eigenvalues of A^{eff} lie between a_- and a_+, i.e., $a_- I \leq A^{eff} \leq a_+ I$, and we immediately extended it to the general symmetric case, as Lemma 6.6.

Lemma 6.6. *If $v_n \rightharpoonup v_\infty$ in $L^2(\Omega; \mathbb{R}^N)$ weak, if $M^n \in \mathcal{M}(\alpha, \beta; \Omega)$ is symmetric a.e. $x \in \Omega$ and $(M^n)^{-1} \rightharpoonup (M^-)^{-1}$ in $L^\infty(\Omega; \mathcal{L}(\mathbb{R}^N; \mathbb{R}^N))$ weak $\star$, then for all nonnegative continuous functions $\varphi \in C_c(\Omega)$, one has*

$$\liminf_n \int_\Omega \varphi\,(M^n v_n, v_n)\,dx \geq \int_\Omega \varphi\,(M^- v_\infty, v_\infty)\,dx, \tag{6.30}$$

i.e., if $(M^n v_n, v_n)$ converges to a Radon measure ν in the sense of distributions, i.e., in $\mathcal{M}_b(\Omega)$ weak $\star$ since $(M^n v_n, v_n)$ is bounded in $L^1(\Omega)$, then $\nu \geq (M^- v_\infty, v_\infty)$ in the sense of measures in Ω.

Proof. If $\mathcal{L}_{sym+}(\mathbb{R}^N; \mathbb{R}^N)$ is the convex cone of symmetric positive definite operators from $\mathbb{R}^N$ into itself, Lemma 6.6 follows from the convexity of $(M, v) \mapsto (M^{-1}v, v)$ on $\mathcal{L}_{sym+}(\mathbb{R}^N; \mathbb{R}^N) \times \mathbb{R}^N$. Indeed, for $M^0 \in \mathcal{L}_{sym+}(\mathbb{R}^N; \mathbb{R}^N)$

$$\begin{aligned} &(M^{-1}v, v) = \big((M^0)^{-1}v_0, v_0\big) + Lin + Rem \\ &Lin = 2\big((M^0)^{-1}v_0, v - v_0\big) - \big((M^0)^{-1}(M - M^0)(M^0)^{-1}v_0, v_0\big) \\ &Rem = \big(M(M^{-1}v - (M^0)^{-1}v_0), (M^{-1}v - (M^0)^{-1}v_0)\big) \geq 0. \end{aligned} \tag{6.31}$$

□

As an application of Lemma 6.6, anticipating the fact that A^{eff} is symmetric in the case when all the A^n are symmetric (Lemma 10.2), one deduces Lemma 6.7 which gives upper bounds as well as lower bounds for A^{eff} in terms of weak $\star$ limits of A^n and of its inverse $(A^n)^{-1}$.

Lemma 6.7. *If $A^n \in \mathcal{M}(\alpha, \beta; \Omega)$ satisfies $A^n(x) \in \mathcal{L}_{sym+}(\mathbb{R}^N; \mathbb{R}^N)$ a.e. $x \in \Omega$ and H-converges to A^{eff}, if $A^n \rightharpoonup A^+$ and $(A^n)^{-1} \rightharpoonup (A^-)^{-1}$ in $L^\infty\big(\Omega; \mathcal{L}(\mathbb{R}^N; \mathbb{R}^N)\big)$ weak $\star$, then one has*

$$A^- \leq A^{eff} \leq A^+ \text{ a.e. } x \in \Omega. \tag{6.32}$$

Proof. In the proof of Theorem 6.5 one constructed a sequence $grad(u_n)$ converging to $grad(u_\infty)$ in $L^2(\Omega; \mathbb{R}^N)$ weak, such that $A^n\,grad(u_n)$ converges to $A^{eff}\,grad(u_\infty)$ in $L^2(\Omega; \mathbb{R}^N)$ weak, and moreover $\big(A^n\,grad(u_n), grad(u_n)\big)$ converges to $\big(A^{eff}\,grad(u_\infty), grad(u_\infty)\big)$ in $L^1(\Omega)$ weak $\star$.

By using Lemma 6.6 with $M^n = (A^n)^{-1}$ and $v_n = A^n\,grad(u_n)$ one obtains $\big(A^{eff}\,grad(u_\infty), grad(u_\infty)\big) \geq \big((A^+)^{-1}A^{eff}\,grad(u_\infty), A^{eff}\,grad(u_\infty)\big)$

in the sense of measures; since both sides of the inequality belong to $L^1(\Omega)$ the inequality is valid a.e. $x \in \Omega$. From the fact that u_∞ can be any element of $H_0^1(\Omega)$, one can choose $grad(u_\infty)$ to be any constant vector on an open subset ω with $\overline{\omega} \subset \Omega$, so that it means $(A^{eff})^{-1} \geq (A^+)^{-1}$, i.e., $A^{eff} \leq A^+$. Choosing $M^n = A^n$ and $v_n = grad(u_n)$ in Lemma 6.6 gives $\big(A^{eff}\, grad(u_\infty), grad(u_\infty)\big) \geq \big(A^-\, grad(u_\infty), grad(u_\infty)\big)$, and similarly, it means $A^{eff} \geq A^-$. □

Additional footnotes: McCONNELL,[27] Olga OLEINIK,[28] RADON,[29] Michael VOGELIUS.[30]

[27] William H. McCONNELL, American mathematician. He worked at IBM (International Business Machines Corporation), San Jose, CA.

[28] Olga Arsen'evna OLEINIK, Ukrainian-born mathematician, 1925–2001. She worked in Moscow, Russia.

[29] Johann RADON, Czech-born mathematician, 1887–1956. He worked in Vienna, Austria.

[30] Michael VOGELIUS, Danish-born mathematician. He worked at University of Maryland, College Park, MD, and he works now at Rutgers University, Piscataway, NJ.

Chapter 7
The Div–Curl Lemma

François MURAT and myself could have discovered the *div–curl lemma* when we first encountered the analogue of (6.16), but we only found it in the spring of 1974, while analyzing cases for which we could calculate A^{eff} explicitly, and besides coefficients depending only upon x_1, or more generally upon (x, e) for a unit vector e, we knew a case where the coefficients are products.

Lemma 7.1. *If $A^n \in \mathcal{M}(\alpha, \beta; \Omega)$, for a bounded open set $\Omega \subset \mathbb{R}^N$, and*

$$\text{for all } i, A^n_{i,i}(x) = f^n_i(x_i) g^n_i(x) \text{ with } 1 \le f^n_i \le M, \frac{\partial g^n_i}{\partial x_i} = 0, \text{ in } \Omega, \tag{7.1}$$
$$\text{for all } i, j, i \ne j, \frac{\partial A^n_{i,j}}{\partial x_j} = 0, \text{ in } \Omega; \tag{7.2}$$

if

$$\begin{aligned} &\frac{1}{f^n_i} \rightharpoonup \frac{1}{f^-_i} \text{ in } L^\infty(\Omega) \text{ weak} \star, \text{ for all } i, \\ &\frac{A^n_{i,j}}{f^n_i} \rightharpoonup \frac{B_{i,j}}{f^-_i} \text{ in } L^\infty(\Omega) \text{ weak} \star, \text{for all } i, j, \end{aligned} \tag{7.3}$$

then A^n H-converges to B.

Proof. For $u_n \rightharpoonup u_\infty$ in $H^1_0(\Omega)$ weak, and $D^n = A^n grad(u_n) \rightharpoonup D^\infty$ in $L^2(\Omega; \mathbb{R}^N)$ weak, one has $\frac{D^n_i}{f^n_i} \rightharpoonup \frac{D^\infty_i}{f^-_i}$ in $L^2(\Omega)$ weak by the same argument used by François MURAT for Lemma 5.1. Then $\frac{D^n_i}{f^n_i} = \sum_j \frac{A^n_{i,j}}{f^n_i} \frac{\partial u_n}{\partial x_j}$, but since $C^n_{i,j} = \frac{A^n_{i,j}}{f^n_i}$ is independent of x_j and $u_n \in H^1_0(\Omega)$, one has $C^n_{i,j} \frac{\partial u_n}{\partial x_j} = \frac{\partial (C^n_{i,j} u_n)}{\partial x_j}$ in the sense of distributions; since $C^n_{i,j} \rightharpoonup \frac{B_{i,j}}{f^-_i}$ in $L^2(\Omega)$ weak, and $u_n \to u_\infty$ in $L^2(\Omega)$ strong, $C^n_{i,j} u_n \rightharpoonup \frac{B_{i,j} u_\infty}{f^-_i}$ in $L^2(\Omega)$ weak, and $\frac{\partial (C^n_{i,j} u_n)}{\partial x_j}$ converges in the sense of distributions to $\frac{\partial (B_{i,j} u_\infty / f^-_i)}{\partial x_j}$, which is $\frac{B_{i,j}}{f^-_i} \frac{\partial u_\infty}{\partial x_j}$, since $u_\infty \in H^1_0(\Omega)$ and $\frac{B_{i,j}}{f^-_i}$ is independent of x_j, and by summing in j one deduces that $D^\infty_i = \sum_j B_{i,j} \frac{\partial u_\infty}{\partial x_j}$. ☐

L. Tartar, *The General Theory of Homogenization*, Lecture Notes of the Unione Matematica Italiana 7, DOI 10.1007/978-3-642-05195-1_7,

François MURAT saw that all examples showed a pattern, a scalar product of a vector field with a good divergence with a gradient vector field, or more generally a vector field with a good curl, so that we conjectured the following first version of the *div–curl lemma*, which I immediately knew how to prove.

Lemma 7.2. *If* $E^n \rightharpoonup E^\infty$, $D^n \rightharpoonup D^\infty$ *in* $L^2(\Omega;\mathbb{R}^N)$ *weak, if*

$$\tfrac{\partial E_i^n}{\partial x_j} - \tfrac{\partial E_j^n}{\partial x_i} \text{ is bounded in } L^2(\Omega) \text{ for } i,j = 1,\ldots,N, \tag{7.4}$$

$$\textstyle\sum_{i=1}^N \tfrac{\partial D_i^n}{\partial x_i} \text{ is bounded in } L^2(\Omega), \tag{7.5}$$

then

$$\int_\Omega \varphi \left(\sum_{i=1}^N E_i^n D_i^n\right) dx \to \int_\Omega \varphi \left(\sum_{i=1}^N E_i^\infty D_i^\infty\right) dx \text{ for all } \varphi \in C_c(\Omega). \tag{7.6}$$

Proof. Considering $\psi_1(E^n - E^\infty)$ and $\psi_2(D^n - D^\infty)$ for $\psi_1, \psi_2 \in C_c^1(\Omega)$, extended by 0 outside Ω, one may assume that E^n and D^n have their support in a fixed compact set, and that $E^\infty = D^\infty = 0$. One wants to show that $\int_{\mathbb{R}^N} (E^n, D^n)\, dx \to 0$, which will prove (7.6) for $\varphi = \psi_1\psi_2$; then, one concludes by an argument of approximation of a function in $C_c(\Omega)$ uniformly by functions in $C_c^1(\Omega)$. Denoting by $\mathcal{F}$ the Fourier transform,[1] one has

$$\int_{\mathbb{R}^N} (E^n, D^n)\, dx = \int_{\mathbb{R}^N} (\mathcal{F}E^n, \mathcal{F}D^n)\, d\xi, \tag{7.7}$$

by the Plancherel formula,[2] where $(\cdot,\cdot)$ is now the Hermitian product on $\mathbb{C}^N$,[3] and one observes that $\mathcal{F}E^n$ and $\mathcal{F}D^n$ converge weakly to 0 in $L^2(\mathbb{R}^N)$, but also strongly in $L^2_{loc}(\mathbb{R}^N)$ by the Lebesgue dominated convergence theorem, since they are bounded in $L^\infty(\mathbb{R}^N)$ and converge pointwise to 0, due to the weak convergence to 0 in $L^2(\mathbb{R}^N)$, and to the supports staying in a fixed compact set. The problem is to bound $|(\mathcal{F}E^n, \mathcal{F}D^n)|$ for large $|\xi|$, but (7.4) implies that $|\xi|$ times the component of $\mathcal{F}E^n$ along ξ is bounded in L^2 and (7.5) implies that $|\xi|$ times the component of $\mathcal{F}D^n$ perpendicular to ξ is bounded in L^2, so that provides a bound for $|\xi|\,|(\mathcal{F}E^n, \mathcal{F}D^n)|$ in L^1 for large $|\xi|$. □

---

[1] I use the notation of Laurent SCHWARTZ, i.e., $\mathcal{F}f(\xi) = \int_{\mathbb{R}^N} f(x) e^{-2i\pi(x,\xi)}\, dx$ for $\xi \in \mathbb{R}^N$ if $f \in L^1(\mathbb{R}^N)$; $\mathcal{F}$ extends into an isometry of $L^2(\mathbb{R}^N)$ with inverse $\overline{\mathcal{F}}$, defined on $L^1(\mathbb{R}^N)$ by $\overline{\mathcal{F}}f(\xi) = \int_{\mathbb{R}^N} f(x) e^{+2i\pi(x,\xi)}\, dx$ for $\xi \in \mathbb{R}^N$.

[2] Michel PLANCHEREL, Swiss mathematician, 1885–1967. He worked at ETH (Eidgenössische Technische Hochschule) in Zürich, Switzerland.

[3] Charles HERMITE, French mathematician, 1822–1901. He worked in Paris, France.

The reason why I found this proof easily was that I knew a proof by Lars HÖRMANDER of the compact embedding of $H_0^1(\Omega)$ into $L^2(\Omega)$ using the Fourier transform, valid for Ω with finite Lebesgue measure. Jacques-Louis LIONS always taught a different approach for the compact embedding of $W_0^{1,p}(\Omega)$ into $L^p(\Omega)$, for Ω bounded, attributed to RELLICH and to KONDRAŠOV,[4,5] following an approach due to FRÉCHET,[6] and to KOLMOGOROV. I find therefore that it is important to teach a few different proofs, or at least mention that other approaches exist, for developing the culture of the students. I heard about the proof of Lars HÖRMANDER in a working group which was meeting on Saturdays at IHP in Paris, when I was a student, so I find that it is a good idea for researchers to go listen to seminars outside their own specialty, or outside the group into which they grew. Unfortunately, one sees opposing schools or mere chapels, developing for nationalistic or political reasons, but often also for personal reasons.

The second version of the *div–curl lemma* consisted in observing that for (7.6) to hold, which corresponds to a convergence in $\mathcal{M}(\Omega)$ weak $\star$, it is enough that the components of E^n and D^n converge in $L^2_{loc}(\Omega)$ weak, and that the convergences in (7.4) and (7.5) be in $H^{-1}_{loc}(\Omega)$ strong.[7]

In the case where $curl(E^n) = 0$, i.e., $E^n = grad(u_n)$ with u_n converging to u_∞ in $H^1(\Omega)$ weak, there is an easier proof by integration by parts, and this is what we did for obtaining (6.16): for $\varphi \in C^1_c(\Omega)$, $\varphi\, u_n \rightharpoonup \varphi\, u_\infty$ in $H^1_0(\Omega)$ weak, and since $div(D^n) \to div(D^\infty)$ in $H^{-1}(\Omega)$ strong, one deduces $\lim_n \langle div(D^n), \varphi\, u_n\rangle = \langle div(D^\infty), \varphi\, u_\infty\rangle$, which means

$$\lim_n \int_\Omega \big(D^n, \varphi\, E^n + u_n\, grad(\varphi)\big)\, dx = \int_\Omega \big(D^\infty, \varphi\, E^\infty + u_\infty\, grad(\varphi)\big)\, dx, \tag{7.8}$$

but $\int_\Omega \big(D^n, u_n\, grad(\varphi)\big)\, dx$ converges to $\int_\Omega \big(D^\infty, u_\infty\, grad(\varphi)\big)\, dx$ since u_n converges to u_∞ in $L^2_{loc}(\Omega)$ strong, so that $\int_\Omega \varphi(D^n, E^n)\, dx$ converges to $\int_\Omega \varphi(D^\infty, E^\infty)\, dx$. Since (D^n, E^n) is bounded in $L^1_{loc}(\Omega)$, it shows that (D^n, E^n) converges to (D^∞, E^∞) in $L^1_{loc}(\Omega)$ weak $\star$.[8]

---

[4] Franz RELLICH, German mathematician, 1906–1955. He worked at Georg-August-Universität, Göttingen, Germany.

[5] Vladimir Iosifovich KONDRAŠOV, Russian mathematician, 1909–1971.

[6] Maurice René FRÉCHET, French mathematician, 1878–1973. He worked in Poitiers, in Strasbourg and in Paris, France.

[7] I introduced the convergence in $H^{-1}_{loc}(\Omega)$ strong for purely mathematical reasons, and after introducing *H-measures* in the late 1980s, it appeared to be the precise condition for an equation to hold for *H-measures*. It is used without any explanation by those who use my ideas without mentioning my name.

[8] It is the topology $\sigma\big(L^1_{loc}(\Omega), C_c(\Omega)\big)$, induced by the weak $\star$ topology on $\mathcal{M}(\Omega)$, which is the dual of $C_c(\Omega)$.

If E^n and D^n converge in $L^2(\Omega)$ weak, (D^n, E^n) is bounded in $L^1(\Omega)$, but if $N \geq 2$ one cannot deduce that the convergence holds in $L^1(\Omega)$ weak,[9] because of the counter-example of Lemma 7.3 which I constructed; of course, for $N = 1$, one has $D^n \to D^\infty$ in $L^2(\Omega)$ strong.

Lemma 7.3. *If $N \geq 2$, if ω is a ball with $\overline{\omega} \subset \Omega$, there exist sequences E^n, D^n converging to 0 in $L^2(\Omega; \mathbb{R}^N)$ weak, with $E^n = grad(u_n)$ and $u_n \in H_0^1(\Omega)$ (converging to 0 in $H_0^1(\Omega)$ weak), $div(D^n) = 0$, but $\int_\omega (E^n, D^n)\, dx \not\to 0$.*

Proof. One solves

$$-\Delta u_n = 0 \text{ in } \omega, u_n = \psi_n \in H^{1/2}(\partial\omega) \text{ on } \partial\omega, \tag{7.9}$$

which has a unique solution $u_n \in H^1(\omega)$ since $H^{1/2}(\partial\omega)$ is the space of traces on $\partial\omega$ of functions in $H^1(\omega)$. One chooses the sequence ψ_n such that

$$\psi_n \rightharpoonup 0 \text{ in } H^{1/2}(\partial\omega) \text{ weak but not in } H^{1/2}(\partial\omega) \text{ strong}, \tag{7.10}$$

which is possible if $H^{1/2}(\partial\omega)$ is infinite dimensional, i.e., for $N \geq 2$. One extends u_n in $\Omega \setminus \overline{\omega}$ by solving

$$-\Delta u_n = 0 \text{ in } \Omega \setminus \omega, u_n = \psi_n \text{ on } \partial\omega, u_n = 0 \text{ on } \partial\Omega, \tag{7.11}$$

so that $u_n \rightharpoonup 0$ in $H_0^1(\Omega)$ weak. One defines $v_n \in H^1(\Omega \setminus \overline{\omega})$ by solving

$$-\Delta v_n = 0 \text{ in } \Omega \setminus \omega, v_n = 0 \text{ on } \partial\Omega, \frac{\partial v_n}{\partial \nu} = \frac{\partial u_n}{\partial \nu} \text{ on } \partial\omega, \tag{7.12}$$

with ν the interior normal to $\partial\omega$, which makes sense since $\frac{\partial u_n}{\partial \nu} \in H^{-1/2}(\partial\omega)$, the dual of $H^{1/2}(\partial\omega)$, and can be computed from the restriction of u_n to ω by the trace theorem of Jacques-Louis LIONS for $H(div; \omega)$. One defines

$$E^n = grad(u_n) \text{ in } \Omega, D^n = grad(u_n) \text{ in } \omega, D^n = grad(v_n) \text{ in } \Omega \setminus \overline{\omega}, \tag{7.13}$$

so that $D^n \rightharpoonup 0$ in $L^2(\Omega; \mathbb{R}^N)$ weak, and $div(D^n) = 0$. Then the *div–curl lemma* asserts that $\int_\Omega \varphi\, (E^n, D^n)\, dx \to 0$ for every $\varphi \in C_c(\Omega)$, but

$$\int_\omega (E^n, D^n)\, dx = \int_\omega |grad(u_n)|^2\, dx \not\to 0, \tag{7.14}$$

since the convergence to 0 would mean that ψ_n converges to 0 in $H^{1/2}(\partial\omega)$ strong. □

---

[9] It is the topology $\sigma\Big(L^1(\Omega), L^\infty(\Omega)\Big)$.

In 1975, Joel ROBBIN showed me a proof of the *div–curl lemma* which I shall describe in Chap. 9, where he used differential forms and the Hodge theorem,[10] and his proof explains some of the variants found later, and it extends to some examples in a more general theory that I shall describe later, which I developed in 1976 with François MURAT under the name *compensated compactness*, a term coined for the *div–curl lemma* by Jacques-Louis LIONS, because it results from a compensation effect, since for $N \geq 2$ one does not usually have $D_i^n E_i^n \rightharpoonup D_i^\infty E_i^\infty$ weak $\star$ for each i, and it looked to him like a compactness argument, since one can pass to the limit in a non-affine quantity for some weakly converging sequences; the name is misleading, since we already noticed in 1974 that (E, D) is the "only" non-affine function for which this happens under information on $curl(E)$ and $div(D)$, a more precise statement being Lemma 7.4, which I shall prove later, in the general framework of *compensated compactness*.

Lemma 7.4. *Let F be real continuous on $\mathbb{R}^N \times \mathbb{R}^N$ such that, whenever*

$$E^n \rightharpoonup E^\infty, D^n \rightharpoonup D^\infty \text{ in } L^\infty(\Omega; \mathbb{R}^N) \text{ weak } \star, \tag{7.15}$$

$$curl(E^n) = 0, div(D^n) = 0 \text{ in } \Omega, \tag{7.16}$$

one can deduce that

$$F(E^n, D^n) \rightharpoonup F(E^\infty, D^\infty) \text{ in } L^\infty(\Omega) \text{ weak } \star, \tag{7.17}$$

then F has the form

$$F(E, D) = c\,(E.D) + \text{affine function}(E, D), \text{ for a constant } c. \tag{7.18}$$

François MURAT extended the *div–curl lemma* to a L^p setting, for $1 < p < \infty$. He assumed that $E^n \rightharpoonup E^\infty$ in $L^p(\Omega; \mathbb{R}^N)$ weak, with the components of $curl(E^n)$ staying bounded in $L^p(\Omega)$, and that $D^n \rightharpoonup D^\infty$ in $L^{p'}(\Omega; \mathbb{R}^N)$ weak, with $div(D^n)$ staying bounded in $L^{p'}(\Omega)$, and he proved that $\int_\Omega \varphi\,(E^n, D^n)\,dx \to \int_\Omega \varphi\,(E^\infty, D^\infty)\,dx$ for every $\varphi \in C_c(\Omega)$.

The proof by integration by parts for the case of gradients is valid for $1 \leq p \leq \infty$.[11] For the general case, François MURAT used Fourier multipliers, and the Hörmander–(Mikhlin) theorem.[12,13] One may use instead the

---

[10] William Vallance Douglas HODGE, Scottish mathematician, 1903–1975. He worked in Cambridge, England.

[11] In that case $div(D^n)$ may stay in a compact of $W^{-1,p'}_{loc}(\Omega)$, but for $p = \infty$ one must recall that $W^{-1,1}(\Omega)$ denotes the distributions of the form $f = \sum_j \frac{\partial g_j}{\partial x_j}$ with $g_j \in L^1(\Omega)$ for $j = 1, \ldots, N$, and not the dual of $W^{1,\infty}_0(\Omega)$.

[12] Solomon Grigorevich MIKHLIN, Russian mathematician, 1908–1990. He worked in Leningrad (now St Petersburg), Russia.

[13] I was told that MIKHLIN's "proof" was incomplete.

Calderón–Zygmund theorem for the (M.) Riesz operators.[14] For $\psi \in C^1_c(\mathbb{R}^N)$, considering $\psi\,(E^n - E^\infty)$ and $\psi\,D^n$, one may assume that the sequences are defined on $\mathbb{R}^N$, and that $E^\infty = 0$, and one projects E^n onto gradients by defining

$$\begin{aligned} &v_n = (\Delta)^{-1} div(E^n), \text{ so that } \tfrac{\partial v_n}{\partial x_j} = -\textstyle\sum_{k=1}^N R_j R_k E^n_k \text{ and} \\ &\tfrac{\partial v_n}{\partial x_j} - E^n_j = -\textstyle\sum_{k=1}^N R_k(R_j E^n_k - R_k E^n_j) \text{ for } j = 1,\ldots,N, \end{aligned} \tag{7.19}$$

where the R_j are the (M.) Riesz operators, defined for $w \in L^2(\mathbb{R}^N)$ by

$$\text{for } j = 1,\ldots,N, \mathcal{F}R_j w(\xi) = i\,\frac{\xi_j}{|\xi|}\,\mathcal{F}w(\xi) \text{ a.e, } \xi \in \mathbb{R}^N, \tag{7.20}$$

so that for $1 < p < \infty$, $||grad(v_n) - E^n||_{L^p(\mathbb{R}^N;\mathbb{R}^N)} \to 0$, since

$$\text{for } j,k = 1,\ldots,N, R_j E^n_k - R_k E^n_j = (-\Delta)^{-1/2}\Big(\frac{\partial E^n_k}{\partial x_j} - \frac{\partial E^n_j}{\partial x_k}\Big). \tag{7.21}$$

Of course, this is the same scenario as the proof of Joel ROBBIN!

There is an improvement of the *div–curl lemma* (at least in its initial form), that Bernard HANOUZET and Jean-Luc JOLY obtained in the early 1980s,[15,16] but when François MURAT showed me their first version I pointed out that it was false, because of my counter-example of Lemma 7.3, and they published a corrected version.[17] Their idea was that (E,D) can be defined in a continuous way in a negative Sobolev space if the components of E and $curl(E)$ belong to a space $H^s_{loc}(\Omega)$ and $div(D)$ and the components of D belong to a space $H^\sigma_{loc}(\Omega)$, for some pairs of negative numbers s,σ, and then to use compactness imbedding theorems of a classical type. This proof also follows the same scenario as the proof of Joel ROBBIN!

Around 1990, a group of four mathematicians wrongly claimed to improve the *div–curl lemma*, and after them a few mathematicians used the term *compensated compactness* for something quite different, which I describe

---

[14] Marcel RIESZ, Hungarian-born mathematician, 1886–1969. He worked in Stockholm and in Lund, Sweden. He was the younger brother of Frederic RIESZ.

[15] Bernard HANOUZET, French mathematician. He works in Bordeaux, France.

[16] Jean-Luc JOLY, French mathematician. He works in Bordeaux, France.

[17] When Jean-Luc JOLY boasted later about avoiding attributing my results to me in order to hurt me, I could not understand why a friend would turn against me in that way, since he would make a fool of himself by publishing something wrong without my help, and I decided not to read anything by him again before it was published. After all, I prefer his honesty in mentioning his position against me, since others behave in much worse way, although for the same apparent reason, to gain power by joining the group of my political opponents, some of whom were expert in falsifying some voting results.

using the names *compensated integrability* and *compensated regularity* [115]. Considering the chaotic situation which already exists in the academic world, I found it quite surprising that some mathematicians with a reasonable stature would try to increase the amount of delusion by misleading students and researchers about some of my ideas!

Chapter 8
Physical Implications of Homogenization

François MURAT and myself were led to Theorem 6.5 and to the *div–curl lemma* 7.2 for purely mathematical reasons, but we learned from the work of Évariste SANCHEZ-PALENCIA that our work was related to an interesting question of continuum mechanics or physics, that of describing the relations between a "microscopic" level and a macroscopic level.

Some people thought that my use of microscopic was misleading, because it suggested the level of atoms, and that I should use the term mesoscopic instead (used to mean any intermediate scale between microscopic and macroscopic), but one should understand the difference between the point of view of a mathematician and that of a specialist of continuum mechanics, or of physics. If a mathematician studies oscillations and concentration effects compatible with a partial differential equation, it may happen that this equation was used for modeling something in continuum mechanics or physics; in that case, there is some physical intuition behind the equation, and this intuition may lead to conjectures about the mathematical properties of the solutions of the equation. The mathematician knows that some conjectures may be false, either because the equation is not a good model for the piece of reality that one thought, or because those who made the conjectures were misled for various reasons: the only way for a mathematician to be sure if the intuitions/conjectures are right is to prove theorems about the equation, and he/she should not be intimidated into confusing conjectures for theorems.[1]

Actually, there were enough wrong guesses made on the side of specialists of continuum mechanics or physics for questions which belong to the mathematical theory of homogenization now that it exists (thanks to the pioneering work of Sergio SPAGNOLO, François MURAT, and myself), that one should be careful not to accept conjectures without explaining if they are compatible with the actual state of the mathematical knowledge on the question.

Theorems in homogenization were showing me that a few things were wrongly guessed in continuum mechanics or physics. Using my approach to

[1] This mistake is common among mathematicians who try to become "applied" by speaking the language of engineers: they often show clearly to anyone who was well trained that they talk about questions that they do not understand.

L. Tartar, *The General Theory of Homogenization*, Lecture Notes of the Unione Matematica Italiana 7, DOI 10.1007/978-3-642-05195-1_8,

different scales based on weak convergence for some quantities and other topologies like G/H-convergence for others, I understood what *internal energy* is, and what the *div–curl lemma* says for a model of *electrostatics*, and then about *equipartition of hidden energy* (which is not about counting degrees of freedom). I understood why the first principle of thermodynamics is obvious, and why thermodynamics is not about dynamics and is nonsense for describing what happens during evolution, because those who developed the theory did not understand what *internal energy* is, and how it could move around, and having called it *heat* and invoked probabilistic games was like having sacrificed to idols. In the late 1970s, looking at the homogenization problem for a first-order scalar hyperbolic equation, which shows nonlocal effects in the effective equation, I understood why quantum mechanics started wrongly, and what *turbulence* is about (which is not about playing probabilistic games).

In 1983, thanks to Robert DAUTRAY who offered me a position at CEA,[2] I read about the *Dirac* equation, and I understood a few things: how quantum mechanics went astray talking about nonexistent "particles" and letting the speed of light c tend to ∞, what *mass* probably is, and what is wrong about the Boltzmann equation, because of completely unadapted ideas concerning "collisions." I actually told Pierre-Louis LIONS that the Boltzmann equation is not a good physical model,[3] and I am surprised that he never mentioned the known defects of the equation.[4] After that, I started thinking about a general programme for giving better mathematical foundations to twentieth century mechanics, *plasticity* and *turbulence*, and twentieth century physics, *atomic physics* and *phase transitions*; later, I called my programme *beyond partial differential equations*, after having already developed *H-measures* and variants, a hint of which I had in 1984, before an illuminating idea of BOSTICK about "electrons" [9].[5]

[2] I met with disaster in Orsay, France, opposing alone an infamous method of falsifying voting results, and unlike others who took the side of my political opponents against me (for reasons which they rarely explained), Robert DAUTRAY gave me the possibility to escape this hell by giving me a job at CEA (Commissariat à l'Énergie Atomique), and he helped me a lot more by telling me what to read for my almost impossible task of understanding physics (in opposition to what physicists say, which is rarely the same thing!).

[3] Pierre-Louis LIONS, French mathematician, born in 1956. He received the Fields Medal in 1994 for his work on partial differential equations. He worked at Université Paris IX-Dauphine, Paris, France, and he holds now a chair (équations aux dérivées partielles et applications, 2002) at Collège de France, Paris.

[4] In 1983, I told Pierre-Louis LIONS in too cryptic a way that the Boltzmann equation is not a good physical model, saying that only two mathematical questions remained, to let the "mean free path" go to 0, and to avoid the angular cut-off introduced by Harold GRAD, and I told him in 1990 an idea for that, based on using restriction theorems on spheres.

[5] I liked what BOSTICK wrote about "electrons" in his article of January 1985 [9], and it confirmed my idea about *mass*, and it gave me a more precise idea about other "particles," but I could not guess what his idea about "photons" means.

For the basic example of a scalar second-order elliptic equation used for Theorem 6.5, I had already used a notation of *electrostatics*, despite the fact that real physical situations lead to symmetric cases. François MURAT and myself had no practical motivation for considering nonsymmetric cases, and we were just following a classical path for mathematicians, to understand more about our subject by finding the limitations of our framework, but in the mid 1980s, Graeme MILTON found a practical situation where nonsymmetric matrices arise, for $N = 2$, in connection with the (classical) Hall effect.[6]

For *electrostatics*, a simplification of the *Maxwell–Heaviside* equation valid for *stationary solutions with no magnetic field and no current*, important quantities are the *electrostatic potential* U, defined up to addition of a constant, the *electric field* $E = -grad(U)$, the *polarization field* D, the *density of electric charge* ϱ, and the *density of electrostatic energy* $e = \frac{1}{2}(E, D)$. Moreover, the balance equation $div(D) = \varrho$, and the constitutive relation $D = A\,E$ hold,[7] where the *dielectric permittivity* A is a symmetric positive definite tensor.

The quantities indexed by n, U_n, E^n, D^n, ϱ_n, and e_n, correspond to physical quantities at a small scale, which it is useful to call mesoscopic for recalling that a different set of equations is valid at the level of "atoms,"[8] and at this mesoscopic level specialists of materials talk about grains with a given crystallographic orientations, arranged along grain boundaries to form a poly-crystal, which may share interfaces with poly-crystals of different materials. Looking at this assembly from a macroscopic level, one would like to define macroscopic quantities, and understand what *effective equations* they satisfy. In my framework, macroscopic quantities are defined as weak limits (or strong limits) in natural spaces, here the weak convergence in $H^1_{loc}(\Omega)$ for U, the weak convergence in $L^2(\Omega; \mathbb{R}^N)$ for E or D, the weak convergence in $\mathcal{M}_b(\Omega)$ for e, but for ϱ the strong convergence in $H^{-1}_{loc}(\Omega)$ appears as a technical constraint.

Actually, using weak convergence for relating different levels was not a new idea, and it was used implicitly each time a *discrete distribution of masses* or a *discrete distribution of charges* was replaced by a *density of mass* or a

[6] Edwin Herbert HALL, American physicist, 1855–1938. He worked at Harvard University, Cambridge, MA.

[7] This is for a linear material, and nonlinear questions will be discussed later.

[8] I conjecture that a good model is to consider the full *Maxwell–Heaviside* equations (in a *vacuum*, so that $D = \varepsilon_0 E$, $B = \mu_0 H$ and $\varepsilon_0 \mu_0 c^2 = 1$), coupled with the *Dirac* equation (without mass term). This is a semi-linear hyperbolic system, too difficult to study at the moment, and I conjecture that a homogenization process should give something like the models which are used, except that in the relation $D = A\,E$, A is supposed to depend upon frequency, i.e., the operator is not local. After all, it is better to start at a mesoscopic level with the equations at hand and, after mentioning that they may not be such good physical models, to state clearly which mathematical questions are of interest!

density of charge, but in my framework, based on my joint work with François MURAT on homogenization and *compensated compactness*, there is something new, related to handling some nonlinear effects.

As Sergio SPAGNOLO found before François MURAT and myself, for $N \geq 2$,[9] the effective dielectric permittivity A^{eff} is not obtained by computing weak limits of functions of A^n, and one needs a new topology of weak type, G-convergence (or H-convergence for the nonsymmetric case); an intuitive reason for this difference is that one *identifies* A from measurements of E and D (by linear methods).[10] The *div–curl lemma* provides another observation of a nonlinear character, that one has $e^\infty = \frac{1}{2}(E^\infty, D^\infty)$, so that the same formula holds at mesoscopic level and at macroscopic level; in other words, in a context of *electrostatics*, it is not necessary to introduce an *internal energy* for keeping track of a part of the energy hidden at an intermediate level.

For fluids showing oscillations in their velocity field u^n, converging weakly to u^∞, the limit of the *density of kinetic energy* $\frac{\varrho|u^n|^2}{2}$ is then $\frac{\varrho|u^\infty|^2}{2} + \varrho\, e$, and the *internal energy* per unit of mass e is ≥ 0, and may be > 0 at some places if the convergence is not strong.

When one understands that effective properties cannot always be computed from a few macroscopic averages/limits, one deduces that apart from the first principle of thermodynamics which is but the *conservation of energy* (when one does not forget about the different ways in which energy may be stored at an intermediate level), *the rules of thermodynamics are rather misleading*, because energy can be stored at various intermediate levels, in different modes, each one following its own rule for moving around, so that expecting a rule for the evolution of the sum is naive, if not rather silly!

Of course, specialists of continuum mechanics or physics dealt with effective quantities for a long time before precise mathematical definitions were proposed, and this is a perfectly normal behavior for engineers and physicists, because not much would be done in continuum mechanics, in physics, and in technology, if one waited for mathematicians to understand what equations to use for the phenomena which engineers and physicists were interested in. As one should expect for any game with unwritten rules, not all players understood the same thing, and a price paid is that a few guesses published in the engineering or physics literature were found to be incorrect when a

---

[9] For $N = 1$, $\frac{1}{A^n} \rightharpoonup \frac{1}{A^{eff}}$ in $L^\infty(\Omega)$ weak $\star$.

[10] One measures values of U_n at a few points (actually averages on small sets), and by interpolation one obtains an approximation U_{app}, which is near U_∞, from which one deduces an approximation $E^{app} = -grad(U_{app})$, which is near E^∞ in a weak topology. For D^n, one measures fluxes through a few surfaces, and by interpolation one deduces an approximation D^{app}, which is near D^∞ in a weak topology.

mathematical theory was finally developed.[11] It also happens that wrong results are published in mathematical journals, since the refereeing process is not perfect, and mathematicians have not found a way to correct this type of mistake.[12]

After François MURAT and myself found that one cannot deduce the effective properties of a mixture from the proportions of materials used (which Sergio SPAGNOLO knew before us), apart from one-dimensional situations like the laminated cases that I shall discuss again in Chap. 12, I was quite puzzled to discover in the spring of 1974 that a book by LANDAU and LIFSHITZ contained a section giving a formula for the conductivity of a mixture. It was easy to guess why they implicitly considered that their mixtures would be isotropic, but they did not seem to know that when mixing two isotropic conductors (with conductivities $\alpha \neq \beta$) using given proportions, one may obtain various isotropic effective conductivities for the mixtures obtained, because they did not mention that their formula could only be an approximation! They should notice an obvious discrepancy, that their formula is not symmetric in α and β if one uses a proportion of 50% for each material!

A year later, in May 1975, at a meeting at UMD, College Park, MD, I heard Ivo BABUŠKA report on his checking the accuracy of the formulas published in the literature for a particular periodic design (where there is an effective value),[13] and the range of answers was quite large. I shall describe in Chap. 21 the question of bounds for the effective conductivity of a mixture in terms of the proportions of its various components (valid for all possible arrangements at a small scale), and quite surprisingly, the formula derived by LANDAU and LIFSHITZ is a good approximation for the effective conductivity of an effectively isotropic mixture in the case of small amplitude oscillations for the conductivities of its isotropic components, a quite puzzling fact considering the complete absence of logical inference in their "derivation," from an explicit computation of a sphere of one isotropic material embedded into an infinite medium made of the other isotropic material. This curious efficiency was my motivation for developing the theory of *H-measures* (Chap. 28), for proving formulas of *small amplitude homogenization* (Chap. 29).

Not much is known mathematically about what concerns realistic mixing processes, and grinding is understood by engineers to mean different things, depending upon grinding cereals or minerals with various degrees of hardness,

---

[11] In such cases, one should be careful to check if the mathematical framework proposed describes in a correct way all the cases which were considered.

[12] It is more important in mathematics to avoid publishing wrong results, but I saw an example of a questionable behavior of rejecting a good paper on homogenization because an incompetent referee (who confused homogenization and Γ-convergence) thought that it contradicted a published paper on Γ-convergence, obviously published without a correct refereeing process.

[13] Strangely enough, one person in the audience disagreed that there should be only one answer for this problem.

but no mathematician knows what it could mean.[14] Having ground materials in fine powders, and poured them in given proportions into a container, other difficulties arise for defining terms like shaking and compressing the mixture, so that it becomes isotropic and traps no air.

The advantage of my framework, using various types of weak convergences, is that, unlike for the "ensemble averages" or the "thermodynamic limits" of the probabilistic methods, the notions that I use are adapted to partial differential equations, and to continuum mechanics or physics, and they help understand an adapted notion of distance between mixtures.

Electricity is another simplification of the *Maxwell–Heaviside* equation also valid for *stationary solutions with no magnetic field*, but now with a *density of current* satisfying the *Ohm law* $j = \sigma E$ for a *conductivity* σ, where σ is a symmetric positive definite tensor (whose inverse σ^{-1} is the *resistivity* tensor). Besides $E = -grad(U)$, the equation of *conservation of charge* becomes $div(j) = 0$, and the *div–curl lemma* tells that one can pass to the limit in (j, E), which begs for a different physical interpretation.

The *Ohm law* comes out of the *Lorentz force*, that a "particle" or an "ion" of charge q_k and velocity v feels a force $f = q_k(E + v \times B)$, which is $q_k E$ because one assumes $B = 0$, but if one admits that this type of particle/ion is slowed down by a drag force $-K_k\, v$, where K_k is a symmetric positive definite tensor, this type of particle/ion is accelerated to attain a limiting velocity v_k,[15] satisfying $q_k\, E = K_k\, v$, i.e., $v_k = q_k\, K_k^{-1} E$,[16] and the *density of current* j is the average of $q_k v_k$, which is then $\sigma\, E$, with σ being the average of $q_k^2 K_k^{-1}$.

Independent of this computation giving an intuition about the *Ohm law*, (j, E) is the average of $(q_k v_k, E)$, which is the power of the *Lorentz force* (per unit volume).

Another application of the *div–curl lemma* concerns *equipartition of hidden energy* for a scalar *wave equation* in an open $\Omega \subset \mathbb{R}^N$

$$\varrho\, \frac{\partial^2 u}{\partial t^2} - div_x\big(A\, grad_x(u)\big) = 0 \text{ in } \Omega \times (0, T), \tag{8.1}$$

where ϱ and A are independent of t, and A is a symmetric tensor, so that any smooth solution of (8.1) satisfies the conservation law

---

[14] Many mathematicians are already quite deluded about what concerns elastic behavior, so that going beyond an elastic range for considering plasticity and cracks is much beyond their grasp. Of course, using probabilistic ideas is just a way to sweep the dirt under the rug, and it cannot result in the cleaning process that mathematicians are responsible for, among models used by engineers or physicists.

[15] As one mentions a limiting velocity, one must exclude high-frequency excitations and only consider either stationary situations or slow variations in time.

[16] That force is related to acceleration is a crucial observation of NEWTON, and it might be because people observed limiting velocities, due to friction effects, that one thought earlier that force was proportional to velocity.

$$\frac{\partial}{\partial t}\Big(\frac{1}{2}\varrho\,\Big|\frac{\partial u}{\partial t}\Big|^2+\frac{1}{2}(A\,grad_x(u),grad_x(u))\Big)-div\Big(A\,grad_x(u)\frac{\partial u}{\partial t}\Big)=0,\quad (8.2)$$

which expresses the *conservation of energy* in the case where ϱ is positive and A is positive definite, so that (8.1) is a *wave equation.* The *density of energy* is the sum of the *density of kinetic energy* $\frac{1}{2}\varrho\,\big|\frac{\partial u}{\partial t}\big|^2$ and of the *density of potential energy* $\frac{1}{2}\big(A\,grad_x(u),grad_x(u)\big)$.[17] If a sequence of solutions u_n of (8.1) converges to 0 in $H^1\big(\Omega\times(0,T)\big)$ weak, then one has

$$\frac{1}{2}\varrho\,\Big|\frac{\partial u_n}{\partial t}\Big|^2-\frac{1}{2}(A\,grad_x(u_n),grad_x(u_n))\rightharpoonup 0 \text{ in } L^1\big(\Omega\times(0,T)\big) \text{ weak } \star, \quad (8.3)$$

i.e., in $L^1\big(\Omega\times(0,T)\big)$ with the weak $\star$ topology of $\mathcal{M}_b\big(\Omega\times(0,T)\big)$, dual of $C_0\big(\Omega\times(0,T)\big)$. One can construct such sequences of solutions of (8.1) by imposing *Dirichlet conditions* or *Neumann conditions* on $\partial\Omega$ for instance, and initial data

$$u_n(\cdot,0)=v_n\rightharpoonup 0 \text{ in } H^1(\Omega) \text{ weak}, \frac{\partial u_n}{\partial t}(\cdot,0)=w_n\rightharpoonup 0 \text{ in } L^2(\Omega) \text{ weak}, \quad (8.4)$$

and (8.3) says that there is a macroscopic *equipartition of hidden energy*, i.e., the weak limits of the *density of kinetic energy* and of the *density of potential energy* are the same. Computing this common limit can be done with *H-measures*, which require more information on u_n and v_n than their weak limits.

In order to obtain (8.3) by applying the *div–curl lemma*, one replaces t by x_0, one defines $E^n=grad_{t,x}u_n$ and one defines D^n by $D^n_0=\varrho\frac{\partial u_n}{\partial t}$, and $D^n_i=-\big(A\,grad_x(u^n)\big)_i$ for $i=1,\ldots,N$.

Of course, the form of *equipartition of hidden energy* that I obtained does not resemble that which I was taught by my physics teachers, where degrees of freedom are counted, but I finally understood what it really meant. Discussing the same question of *equipartition of hidden energy* for the *Maxwell–Heaviside* equation is quite instructive too, but it requires more than the *div–curl lemma*, and I shall describe it later, after the more general form of *compensated compactness* that I also developed with François MURAT, and it will also use the framework of *differential forms* that I shall discuss now in Chap. 9.

Additional footnotes: Harold GRAD.[18]

---

[17] One interpretation in the case $N=2$ is to consider u as the vertical displacement of a membrane, ϱ being its density and A being an elasticity tensor, so that in this model the *density of potential energy* has an elastic origin.

[18] Harold GRAD, American mathematician, 1923–1987. He worked at NYU (New York University), New York, NY.

Chapter 9
A Framework with Differential Forms

During the year 1974–1975 which I spent at UW, Madison, WI, Joel ROBBIN showed me a different proof of the *div–curl lemma*, using differential forms and the Hodge theorem, but it was only a few years later, after obtaining general results of *compensated compactness* with François MURAT, that I fully understood the example of differential forms, and what Joel ROBBIN said.

In the fall of 1975, I heard about the sequential weak continuity of Jacobian determinants proven by Yuri RESHETNYAK,[1] and I saw that it is just the *div–curl lemma* for $N = 2$, and I deduced the case $N = 3$ from the *div–curl lemma* by noticing that $grad(u) \times grad(v)$ is divergence free, but I did not see that the corresponding algebraic manipulations to perform for $N > 3$ are natural in the framework of differential forms, and that Yuri RESHETNYAK's result almost follows from Lemma 9.1, a natural extension of Joel ROBBIN's proof.[2]

Lemma 9.1. *In an open set Ω of $\mathbb{R}^N$, if a^n is a sequence of p-forms with bounded coefficients in $L^2(\Omega)$ which converges weakly to a^∞, if b^n is a sequence of q-forms with bounded coefficients in $L^2(\Omega)$ which converges weakly to b^∞, and if the* exterior derivatives *da^n and db^n have bounded coefficients in $L^2(\Omega)$, then, the* exterior product *$a^n \wedge b^n$ converges to $a^\infty \wedge b^\infty$ in $L^1(\Omega)$ weak $\star$ (i.e., $L^1(\Omega)$ with the weak $\star$ topology of $\mathcal{M}_b(\Omega)$, dual of $C_0(\Omega)$).*

Proof. One uses the Hodge decomposition in order to write locally $a^n = dA^n + \widetilde{a}^n$ with A^n and $\widetilde{a}^n$ having coefficients converging strongly in $L^2_{loc}(\Omega)$ and then one uses the formula $d(A^n \wedge b^n) = dA^n \wedge b^n + (-1)^{p-1} A^n \wedge db^n$, since A^n is a $(p-1)$-form. Since $A^n \wedge b^n$ and $A^n \wedge db^n$ converge weakly, respectively to $A^\infty \wedge b^\infty$ and $A^\infty \wedge db^\infty$, because A^n converges strongly to A^∞, one deduces

[1] Yuriĭ Grigor'evich RESHETNYAK, Russian mathematician, born in 1930. He works at the Sobolev Institute of Mathematics, Novosibirsk, Russia.

[2] The sequential weak continuity of Jacobian determinants also reminded me of something that I noticed earlier, that most of the properties of the Brouwer topological degree follow from the fact that for u a smooth mapping from an open set Ω of $\mathbb{R}^N$ into $\mathbb{R}^N$, and ∇u its Jacobian, $\int_\Omega \Phi(u) det(\nabla u)\, dx$ only depends upon the boundary values of u; I was not surprised then to learn that $det(\nabla u)$ is a robust quantity with respect to oscillations in ∇u.

L. Tartar, *The General Theory of Homogenization*, Lecture Notes of the Unione Matematica Italiana 7, DOI 10.1007/978-3-642-05195-1_9,

that $dA^n \wedge b^n$ converges to $d(A^\infty \wedge b^\infty) + (-1)^p A^\infty \wedge db^\infty = dA^\infty \wedge b^\infty$ in the sense of distributions; since $\widetilde{a}^n \wedge b^n$ converges weakly to $\widetilde{a}^\infty \wedge b^\infty$, because $\widetilde{a}^n$ converges strongly to $\widetilde{a}^\infty$, one deduces that $a^n \wedge b^n$ converges to $a^\infty \wedge b^\infty$ in the sense of distributions. Since each coefficient of $a^n \wedge b^n$ is bounded in $L^1(\Omega)$, it converges to the corresponding coefficient of $a^\infty \wedge b^\infty$ in $L^1(\Omega)$ weak $\star$, i.e., with test functions in $C_0(\Omega)$. □

Of course, the *div–curl lemma* is the case $p = 1$ and $q = N - 1$. This proof obviously extends to the case where the sequences a^n and da^n have bounded coefficients in $L^\alpha(\Omega)$, and the sequences b^n and db^n have bounded coefficients in $L^\beta(\Omega)$, with $\alpha, \beta > 1$ and $\frac{1}{\alpha} + \frac{1}{\beta} \leq 1$, the Hodge decomposition part using then the Calderón–Zygmund theorem, or the Hörmander–(Mikhlin) theorem for Fourier multipliers.

The first version of *compensated compactness* which François MURAT developed followed the same argument, and in order to perform the analogue of the Hodge decomposition, he assumed a *constant rank condition*, which is not needed in the final version of *compensated compactness*, which I proved by following my original proof of the *div–curl lemma*.

Since the results of *compensated compactness* were proven for equations with constant coefficients, the case of differential forms was for some time the only general situation where one uses (smooth) variable coefficients, since the theory is valid on differentiable manifolds. I corrected this defect of *compensated compactness* in the late 1980s, by introducing *H-measures*.

Joel ROBBIN also showed me how to write the Maxwell–Heaviside equation using differential forms, in a better way than what I heard in a talk by Laurent SCHWARTZ, whose point of view was restricted to the equation in a vacuum,[3] and he convinced me that it is because they are coefficients of differential forms that the classical weak topology is natural for some physical quantities, but there is more to understand for other quantities.

In 1970, I heard about the framework used by differential geometers for writing the transport term in fluid dynamics, for the Euler equation or for the Navier–Stokes equation,[4] using affine connections and covariant derivatives. I have wondered since that time if it is of any use, apart from pleasing the people who speak the language of differential geometers, and probably despise those who do not: in order to show that one gains something by this approach, one should at least show that the framework is robust, and tells something useful for limits of sequences of such flows! In other words, is this framework of any use for turbulent flows? My feeling is that nonlocal effects appear for describing the effective equations to use for turbulence,

---

[3] However, as Joel ROBBIN pointed out later, it is not so clear how invariance by action of the Lorentz group is lost in presence of matter. I shall describe later a tentative answer, related to the appearance of nonlocal effects by homogenization.

[4] Leonhard EULER, Swiss-born mathematician, 1707–1783. He worked in Saint Petersburg, Russia.

and I conjecture that the framework of differential geometers will not be so useful. Despite some useless fashions,[5] my feeling is that affine connections and covariant derivatives could be useful, but nothing that I have heard yet shows that.

Electrostatics can be described using the language of differential forms: the *electrostatic potential* U corresponds to the 0-form $\widetilde{[U]}$, which is just the function U, the *electric field* E corresponds to the 1-form

$$\widetilde{[E]} = \sum_{i=1}^{N} E_j\,dx_j, \tag{9.1}$$

the *polarization field* D corresponds to the $(N-1)$-form

$$\widetilde{[D]} = \sum_{i=1}^{N} D_i\,dx_1 \wedge \cdots \wedge dx_{i-1} \wedge dx_{i+1} \wedge \cdots \wedge dx_N, \text{ written } \sum_{i=1}^{N} D_i\,\widehat{dx_i}, \tag{9.2}$$

the *density of charge* ϱ corresponds to the N-form $\widetilde{[\varrho]} = \varrho\,dx$, and the *density of electrostatic energy* e corresponds to the N-form $\widetilde{[e]} = e\,dx$. The equations

$$d\widetilde{[U]} = -\widetilde{[E]}, d\widetilde{[D]} = \widetilde{[\varrho]}, \tag{9.3}$$

correspond to

$$E = -grad(U), div(D) = \varrho. \tag{9.4}$$

For coefficients of differential forms, the weak convergence is natural, since differential forms are integrated on manifolds, and the relation

$$\widetilde{[e]} = \frac{1}{2}\widetilde{[E]} \wedge \widetilde{[D]} \tag{9.5}$$

passes to the limit because of Lemma 7.2, the *div–curl lemma*, a particular case of Lemma 9.1. On the other hand, the *dielectric permittivity* tensor A does not correspond to a differential form, and the weak convergence is not adapted for it; what A does is to transform the 1-form $\widetilde{[E]}$, which has a good exterior derivative, into the $(N-1)$-form $\widetilde{[D]}$, which also has a good exterior derivative, and as a consequence the natural convergence for A is H-convergence!

[5] In the late 1970s, a fashion started about the Yang–Mills equation, which was fake physics since it concerned elliptic situations instead of hyperbolic ones (although the hyperbolic case would be more a problem of physicists than a problem of physics!). Geometers playing with affine connections and covariant derivatives boasted to do something important, since it interested theoretical "physicists," who lost track of physics a long time before, but felt important since their questions interested some geometers, a typical Comte complex!

More generally, the *Maxwell–Heaviside* equation can be described using differential forms, as I learned from Joel ROBBIN. The *density of charge* ϱ and the *density of current* j are coefficients of a 3-form

$$\widetilde{[\varrho\, j]} = \varrho\, dx_1 \wedge dx_2 \wedge dx_3 - (j_1\, dx_2 \wedge dx_3 + j_2\, dx_3 \wedge dx_1 + j_3\, dx_1 \wedge dx_2) \wedge dt, \quad (9.6)$$

which satisfies

$$d(\widetilde{[\varrho\, j]}) = 0, \quad (9.7)$$

corresponding to the *conservation of charge*

$$\frac{\partial \varrho}{\partial t} + div(j) = 0. \quad (9.8)$$

The polarization field D and the magnetic field H are coefficients of a 2-form

$$\begin{aligned}\widetilde{[D\, H]} &= D_1\, dx_2 \wedge dx_3 + D_2\, dx_3 \wedge dx_1 + D_3\, dx_1 \wedge dx_2 \\ &\quad + (H_1\, dx_1 + H_2\, dx_2 + H_3\, dx_3) \wedge dt,\end{aligned} \quad (9.9)$$

which satisfies

$$d(\widetilde{[D\, H]}) = \widetilde{[\varrho\, j]} \quad (9.10)$$

corresponding to

$$div(D) = \varrho, -\frac{\partial D}{\partial t} + curl(H) = j. \quad (9.11)$$

The *induction field* B and the *electric field* E are coefficients of a 2-form

$$\widetilde{[B\, E]} = B_1 dx_2 \wedge dx_3 + B_2 dx_3 \wedge dx_1 + B_3 dx_1 \wedge dx_2 - (E_1 dx_1 + E_2 dx_2 + E_3 dx_3) \wedge dt, \quad (9.12)$$

which satisfies

$$d(\widetilde{[B\, E]}) = 0, \quad (9.13)$$

corresponding to

$$div(B) = 0, \frac{\partial B}{\partial t} + curl(E) = 0. \quad (9.14)$$

The *vector potential* A and the *scalar potential* U are coefficients of a 1-form

$$\widetilde{[A\, U]} = A_1\, dx_1 + A_2\, dx_2 + A_3\, dx_3 + U\, dt, \quad (9.15)$$

which satisfies

$$d(\widetilde{[A\, U]}) = \widetilde{[B\, E]}, \quad (9.16)$$

corresponding to

$$B = -curl(A), E = \frac{\partial A}{\partial t} - grad(U). \quad (9.17)$$

Of course, $\widetilde{[A\,U]}$ is only defined up to addition of an exact form $d\varphi$, so that one does not change the values of the fields by changing A into $A + grad(\varphi)$ and U into $U + \frac{\partial \varphi}{\partial t}$.

In the Dirac equation, there is another unknown $\psi \in \mathbb{C}^4$ for describing *matter* (since the electro-magnetic field is about "light"), and ϱ and j are sesqui-linear quantities in ψ (compatible with *conservation of charge*), and the equation for ψ has a bilinear term,[6] linear in ψ and in the *vector potential* A and the *scalar potential* U; when one changes A into $A + grad(\varphi)$ and U into $U + \frac{\partial \varphi}{\partial t}$, one must also change ψ into $\psi e^{i\,\lambda\,\varphi}$, so that ϱ and j are not changed. Gauge transformations seem then to be related to the fact that some changes at a small scale have no effect on macroscopic values of the physical quantities; however, the *Aharonov–Bohm effect* showed that "electrons," which are waves, are sensitive to A and U.[7,8]

Using the general theory of *compensated compactness* that I shall describe later, I looked for quadratic quantities in B, D, E, H which are sequentially weakly continuous for sequences of solutions of the *Maxwell–Heaviside* equation. I found that they are the linear combinations of (D, H), (B, E) and $(B, H) - (D, E)$; the third one tells about *equipartition of hidden energy.* Then I noticed that these quantities appear in the framework of differentials forms, so that the sequential weak continuity follows from Lemma 9.1:

$$\widetilde{[D\,H]} \wedge \widetilde{[D\,H]} = (D, H)\, dx_1 \wedge dx_2 \wedge dx_3 \wedge dt, \tag{9.18}$$

$$\widetilde{[B\,E]} \wedge \widetilde{[B\,E]} = -(B, E)\, dx_1 \wedge dx_2 \wedge dx_3 \wedge dt, \tag{9.19}$$

$$\widetilde{[D\,H]} \wedge \widetilde{[B\,E]} = \big((B, H) - (D, E)\big)\, dx_1 \wedge dx_2 \wedge dx_3 \wedge dt. \tag{9.20}$$

Since the *density of electromagnetic energy* is $\frac{1}{2}\big((B, H) + (D, E)\big)$, for a sequence of solutions of the *Maxwell–Heaviside* equation such that the fields B, H, D, E converge to 0 in $L^2\big(\Omega \times (0, T)\big)$ weak, one finds that $(B, H) - (D, E)$ converges to 0 in $L^1\big(\Omega \times (0, T)\big)$ weak $\star$ (i.e., $L^1\big(\Omega \times (0, T)\big)$ with the weak $\star$ topology of $\mathcal{M}_b\big(\Omega \times (0, T)\big)$, dual of $C_0\big(\Omega \times (0, T)\big)\big)$, i.e., there is a macroscopic *equipartition of hidden energy* between the *density of magnetostatic*

[6] There is a factor in front of the bilinear term, inversely proportional to the Planck constant h, so that it is natural that h appears in questions of interaction of light with matter. I find it silly that quantum mechanics makes h appear everywhere!

[7] Yakir AHARONOV, Israeli-born physicist, born in 1932. He received the Wolf Prize in Physics in 1998, jointly with Sir Michael V. BERRY, for the discovery of quantum topological and geometrical phases, specifically the Aharonov–Bohm effect, the Berry phase, and their incorporation into many fields of physics. He worked at Yeshiva University, New York, NY, at Tel Aviv University, Tel Aviv, Israel, and at University of South Carolina, Columbia, SC.

[8] David Joseph BOHM, American-born physicist, 1917–1992. He worked at Princeton University, Princeton, NJ, in São Paulo, Brazil, at the Technion, Haifa, Israel, in Bristol, England, and at Birkbeck College, London, England.

energy $\frac{1}{2}(B,H)$ and the *density of electrostatic energy* $\frac{1}{2}(D,E)$. Moreover, (D,H) and (B,E) also converge to 0 in $L^1\big(\Omega\times(0,T)\big)$ weak $\star$, but I do not know a physical interpretation of this property.

Like for the similar property of the *wave equation*, this question becomes clearer when one uses my *H-measures*, since they are able to describe how oscillations (and concentration effects) propagate.

Applying Lemma 9.1 only addresses the sufficient condition for the sequential weak continuity of a function of B,D,E,H, while the theory of *compensated compactness* also addresses the necessary condition, and the result of technical computations is that besides a combination of the three quadratic quantities mentioned, one can only add an affine function in B,D,E,H.

It was an interesting surprise for me to discover that questions which were natural from my point of view led to the same mathematical objects which geometers introduced for other reasons, more natural to them. For a general list of partial differential equations, Joel ROBBIN believed that there would exist some geometrical reason for explaining the list of functions which are sequentially weakly continuous, but it is important to observe that this is not the only question addressed by the theory of *compensated compactness*!

I believe that it was in order to develop a sound mathematical basis for some questions of classical mechanics (which is eighteenth century mechanics) and classical physics (which is nineteenth century physics) that mathematicians like POINCARÉ and É. CARTAN introduced many of the mathematical tools which are often used by geometers now in too abstract a way, maybe because they feel unable to say something useful about nineteenth century continuum mechanics, or twentieth century continuum mechanics and physics, in part since they tend to trust physicists having a Comte complex, so that they work on problems of physicists instead of problems of physics.

The preceding computations made me aware that there were difficulties concerning other quantities, which were introduced either on the geometrical side or on the physical side. For example, one can certainly obtain other sequentially weakly continuous quantities by using the 1-form $[\widetilde{A\,U}]$ which makes A and U play a role, but if one restricts attention to the physical quantities B, E, ϱ, and j, one can consider the *density of the Lorentz force* $\varrho\,E+j\times B$ and its power (j,E) (which I already used in a simple model), but these quantities do not pass to the limit: the exterior product of the 2-form $[\widetilde{B\,E}]$ and the 3-form $[\widetilde{\varrho\,j}]$ is 0, and there are actually no sequentially weakly continuous functions using B, E, ϱ and j which are not affine in B, E, and (B,E). How should one interpret this negative result about the force (per unit volume) $F=\varrho\,E+j\times B$? One way is to say that a force is not a coefficient of a differential form, so that the weak convergence is not adapted for it.

One already needed a different type of weak convergence for the *dielectric permittivity* tensor A, for which the H-convergence is adapted, and I shall show in Chap. 10 how to derive properties of H-convergence by a method very close to the actual way of "identifying" A^{eff}, i.e., imposing a macroscopic field

E^∞ and measuring the corresponding macroscopic field D^∞, giving the information that $D^\infty = A^{\mathit{eff}} E^\infty$, which characterizes A^{eff} if one uses N linearly independent values of E^∞. The adapted topology for a physical quantity seems then to be found once one knows the way to "identify" the effective value of that quantity.

What is a force, or more precisely a force field, and how does one identify it? If one has an oscillating electromagnetic field in a region Ω and one introduces a test particle with charge q_0 at a point x_0 (usually on the boundary $\partial\Omega$) with a velocity v_0, it will experience the Lorentz force $q_0 F^n(x,t;v) = q_0\big(E^n(x,t) + v \times B^n(x,t)\big)$, and its position $x^n(t)$ will satisfy the equation of motion

$$m_0 \frac{d^2 x^n(t)}{dt^2} = q_0 F^n\Big(x^n(t), t; \frac{dx^n(t)}{dt}\Big), \text{ with } x^n(0) = x_0, \frac{dx^n(0)}{dt} = v_0, \tag{9.21}$$

where m_0 is the mass of the particle.[9] More generally, releasing a cloud of test particles having all the same charge to mass ratio $\frac{q_0}{m_0}$, one is led to describe the density $f_n(x,t,v)$ of particles around x at time t having velocity v, which solves the transport equation

$$\frac{\partial f_n}{\partial t} + \sum_{i=1}^{3} v_i \frac{\partial f_n}{\partial x_i} + \frac{q_0}{m_0} \sum_{i=1}^{3} F_i^n(x,t;v) \frac{\partial f_n}{\partial v_i} = 0, \tag{9.22}$$

with an initial condition, and a boundary condition in order to explain what happens to the particles on $\partial\Omega$ which are heading out of Ω. Identifying an effective force field, or some more general concept, then consists in understanding how to pass to the limit in (9.22), i.e., pass to the limit in $\sum_{i=1}^3 F_i^n \frac{\partial f_n}{\partial v_i}$, which is $\sum_{i=1}^3 \frac{\partial (F_i^n f_n)}{\partial v_i}$ in the case where $div_v(F^n) = 0$, like for the Lorentz force.

In the physical problem, passing to the limit in $\varrho_n E^n + j^n \times B^n$ should take into account the fact that ϱ_n and j^n come from repartition of moving charges feeling the Lorentz force, and one must then pass to the limit in (9.22). This leads to the question of *compactness by averaging*, which I first heard from Benoît PERTHAME,[10] and which I described in [119], showing the proof of Patrick GÉRARD, who introduced independently *H-measures* for that purpose.[11]

[9] Of course, there is a corresponding relativistic framework.

[10] Benoît PERTHAME, French mathematician, born in 1959. He worked in Orléans, France, and he now works at UPMC (Université Pierre et Marie Curie), Paris, France.

[11] He called them micro-local defect measures, and developed a more general theory for sequences taking their values in a Hilbert space, but his argument can be followed with my finite-dimensional version of *H-measures*, as I showed in [119].

It is important that (9.21) is a second-order equation in t, since first-order equations in t with oscillating coefficients correspond to equations like

$$\frac{\partial u_n}{\partial t} + \sum_{i=1}^{N} a_i^n(x,t)\frac{\partial u_n}{\partial x_i} = 0, \tag{9.23}$$

for which there are cases with a natural effective equation of a different form, containing some nonlocal terms in x, t. Understanding how to describe this question for a general equation (9.23) seems to be a crucial step for understanding *turbulence*, which, whatever the precise definition of the term is, results from oscillations in the velocity field in a fluid flow. Of course, the appearance of nonlocal terms in some examples suggests that some approaches to turbulence cannot succeed: for example, one should not expect to find an added diffusion term with a "turbulent viscosity,"[12] and a geometrical framework using affine connections should not be useful, since it does not allow for nonlocal effects.

More recently, a different type of equation was introduced by Amit ACHARYA for following a *density of dislocations*,[13] in a way which suggests that there is still something around the framework of differential forms which is not so well understood yet, and further research seems necessary in that direction.

Additional footnotes: Sir Michael BERRY,[14] BROUWER,[15] LEE,[16] MILLS,[17] YANG.[18]

---

[12] Before understanding the importance of nonlocal effects (in 1980), I introduced a different idea, with a model showing an added dissipation in the effective equation, quadratic in u and not in $grad(u)$, which I describe in Chap. 9; it was in working on this idea that I had the first hint about why *H-measures* are needed.

[13] Amit ACHARYA, Indian-born engineer, born in 1965. He works at CMU (*Carnegie Mellon University*), Pittsburgh, PA.

[14] Sir Michael Victor BERRY, British physicist, born in 1941. He received the Wolf Prize in Physics in 1998, jointly with Yakir AHARONOV, for the discovery of quantum topological and geometrical phases, specifically the Aharonov–Bohm effect, the Berry phase, and their incorporation into many fields of physics. He works in Bristol, England.

[15] Luitzen Egbertus Jan BROUWER, Dutch mathematician, 1881–1966. He worked in Amsterdam, The Netherlands.

[16] Tsung-Dao LEE, Chinese-born physicist, born in 1926. He received the Nobel Prize in Physics in 1957, jointly with Chen-Ning YANG, for their penetrating investigation of the so-called parity laws which has led to important discoveries regarding the elementary particles. He worked at Columbia University, New York, NY.

[17] Robert L. MILLS, American physicist, 1927–1999. He worked at OSU (Ohio State University), Columbus, OH.

[18] Chen-Ning YANG, Chinese-born physicist, born in 1922. He received the Nobel Prize in Physics in 1957, jointly with Tsung-Dao LEE, for their penetrating investigation of the so-called parity laws which has led to important discoveries regarding the elementary particles. He worked at IAS (Institute for Advanced Study), Princeton, NJ, and at SUNY (State University of New York) at Stony Brook, NY.

Chapter 10
Properties of H-Convergence

During the year 1974–1975, which I spent at UW, Madison, WI, I simplified the results obtained with François MURAT, by a repeated use of our *div–curl lemma* (Lemma 7.2). Afterward, the same idea was developed independently by Leon SIMON. Although I did not describe it explicitly, it should be considered as a simple case of the way to use the general theory of *compensated compactness*, which I shall describe later.

My new approach made the method easily applicable to all sorts of variational situations, and I first explained it to Jacques-Louis LIONS at a meeting in Marseille, France, in the fall of 1975, and he mentioned it in a footnote of his article for the proceedings. However, although he used my method all the time in his lectures during the following years, he rarely mentioned my name for what he called the energy method,[1] which is a bad name for my method, which I call the *method of oscillating test functions.*

At a conference in Roma (Rome), Italy, in the spring of 1974,[2] I apparently upset Ennio DE GIORGI by my claim that my method, which was actually the joint work with François MURAT that I described in Chap. 6, was more general than the method developed by the Italian school,[3] but I only learned about that from Paolo MARCELLINI at a conference in Évora,[4] Portugal, in June 1996, and I decided that on our next encounter I would apologize to

---

[1] I complained to Jacques-Louis LIONS that he never mentioned my name when using my method in his lectures, and his answer was that *everybody knew that it was my method*! I was quite upset then to find that in his book with Alain BENSOUSSAN and George PAPANICOLAOU "he" wrote that I only did the second-order elliptic case, as if he was unable to recognize that my idea is for general variational equations!

[2] It was the first international conference to which I was invited, and it was organized by Umberto MOSCO.

[3] Since I did not know at the time that Ennio DE GIORGI generously supplied so many insightful ideas to young Italian analysts, my comment was certainly not meant against him: my point was just to say that the Meyers theorem, which Sergio SPAGNOLO used, made generalizations difficult.

[4] Paolo MARCELLINI, Italian mathematician, born in 1947. He works at Università degli Studi di Firenze, Firenze (Florence), Italy.

L. Tartar, *The General Theory of Homogenization*, Lecture Notes of the Unione Matematica Italiana 7, DOI 10.1007/978-3-642-05195-1_10,

Ennio DE GIORGI for what he perceived as arrogance in these early days (while I was actually extremely shy in those days); unfortunately, he died before the end of that year.

I could hardly explain in 1974 how to perform all the computations for higher-order equations, not necessarily symmetric, for systems like linearized elasticity,[5] or for nonlinear elliptic equations, but less than a year after, the extension to general linear elliptic systems in variational form became straightforward, as well as some simple nonlinear equations of monotone type which I shall describe in Chap. 11. The reason was that I could prove most of the important known properties of homogenization of second-order variational elliptic equations in divergence form by repeated applications of the *div–curl lemma*, which I proved with François MURAT just after the conference in Roma (Rome).

There were too many articles written later on, each considering the homogenization of a particular problem, often with an unnecessary restriction to a periodic setting, showing that not everyone understood that all linear variational problems of continuum mechanics could be treated from the point of view of homogenization in the general framework that I devised. It is so upsetting to witness the useless application of a general method to many examples, often uninteresting ones, that I find it worth looking at the reasons for such a behavior.[6]

It was my first mistake that I did not write notes for my Peccot lectures, taught in the beginning of 1977 at Collège de France, in Paris, but my difficulties with writing were genuine. François MURAT wrote notes for similar lectures that he gave shortly after in Alger (Algiers), Algeria, but they did not circulate enough. When there is no obvious written reference to quote for one idea, honest people correctly attribute the idea to its author when they use it, but not so honest people try to take advantage of such situations by claiming as theirs some ideas from others,[7] and they often continue this behavior after learning about a reference to quote.

That my joint work with François MURAT was not quoted correctly was partly due to the curious omissions of Jacques-Louis LIONS, who applied a

---

[5] I only understood much later that it is not wise to study homogenization for linearized elasticity, apart possibly for some engineering applications, because the multiplication of interfaces enhances the defects of linearization (which consists in replacing the group of rotations $S\mathbb{O}(3)$ by its tangent space at I).

[6] Such misplaced efforts of replication must be quite common in every field, of mathematics or other sciences, and it might be difficult to avoid them completely in the future, if not impossible because of the slowing down of academic systems by forcing into them too many who showed no competence for research, a defect which they hide behind this kind of replicating behavior.

[7] One need not be a very good detective in order to find people who talk about things that they do not really understand, and one should suspect that the reason is that they use ideas which are certainly not theirs (although it sometimes happens that bright people have difficulties expressing their own ideas!).

similar policy towards Sergio SPAGNOLO,[8] or towards Évariste SANCHEZ-PALENCIA, preferring to refer to Ivo BABUŠKA for the importance of asymptotic expansions for periodic structures. Our former advisor was a good enough mathematician to realize that without my general method he and his collaborators would only be able to discuss the case of Dirichlet conditions, or equations for which the maximum principle applies. Although there was some formal work done before in the USSR by Nikolai BAKHVALOV,[9] or by Evgeny KHRUSLOV, it seems that Olga OLEINIK was led to work on homogenization because of talks given by Jacques-Louis LIONS in Moscow, where he obviously forgot to mention that he was applying in a periodic setting the general theory developed by François MURAT and myself, and I found it awkward to discover that Olga OLEINIK wrote in some difficult notation what we did earlier in a simpler way.[10]

It was my second mistake that I did not emphasize enough the danger of only studying the periodic case, since it is just a particular case of the general approach of Sergio SPAGNOLO, François MURAT and myself, and because few people have the mathematical ability which Olga OLEINIK showed, to develop a general theory from the knowledge of a few examples. I noticed on many occasions that it is much more difficult for a person who first learned periodic homogenization than for another person who did not hear about it to understand the general setting of homogenization, maybe because the first one mistakenly thinks that he/she already understands what homogenization is about. It is difficult to lose a bad habit when one likes it![11]

Becoming a mathematician requires some ability with abstract concepts, and it is a part of the training to check that one can apply a general method to particular examples. One measures the scientific taste of someone by the subjects that he/she chooses. Then, one measures the scientific stature of someone by the exercises that he/she thinks worth publishing. Finally, one measures the technical ability of someone by the lengths of his/her articles,

---

[8] Jacques-Louis LIONS checked in one of his lectures at Collège de France that he could extend the Meyers theorem to the case of mixing isotropic materials in linearized elasticity, but I did not hear him mention the name of Sergio SPAGNOLO, whose method he was trying to generalize. He did not mention either that one should consider general materials satisfying a *very strong ellipticity condition*, and without such a generalization, he could not mix materials already obtained as mixtures of isotropic materials!

[9] Nikolai Segeevich BAKHVALOV, Russian mathematician, born in 1934. He worked at Moscow State University, Moscow, Russia.

[10] Because of my fight against a method of vote-rigging, organized by a pro-communist group in Orsay, I thought that it was on purpose that Olga OLEINIK avoided mentioning my name when she was using my ideas, but she mentioned once that she did not know about François MURAT's course in Alger (Algiers), Algeria, and she probably rediscovered our general method from the knowledge of its application to periodic media, heard in talks by Jacques-Louis LIONS.

[11] It also seems that most people prefer to be wrong with a crowd than use their brain and agree with the critics who point out that the crowd is obviously heading in the wrong direction, where nothing interesting can be found.

which show the ability or the inability to discover simple proofs. Of course, the fact that one finds so many similar examples published in the literature is not a good sign concerning the health of the academic system, and a lot of improvement would be necessary for improving the scientific culture, and the technical abilities of a majority of its members.

When a method uses only a variational structure, i.e., no special property is used like the maximum principle, it can be extended with minor changes to most of the *linear* partial differential equations of continuum mechanics.

Not much is understood for nonlinear equations, so that it would be useful to see more people work in that direction, but one observes instead a deluded crowd, probably steered by bad "shepherds," who confuse Γ-convergence and homogenization!

My method, which I call the *method of oscillating test functions*, was wrongly called the energy method by Jacques-Louis LIONS, but I heard others call it the duality method, which is not as wrong; it is not really adapted, as it suggests linearity, and I shall use my method for the homogenization of (nonlinear) monotone operators in Chap. 11.

One starts with the same abstract analysis, i.e., Lemma 6.2 and the beginning of the proof of Theorem 6.5, i.e., one extracts a subsequence A^m for which there is a linear continuous operator C from $H_0^1(\Omega)$ into $L^2(\Omega;\mathbb{R}^N)$ such that for all $f\in H^{-1}(\Omega)$ the sequence of solutions $u_m\in H_0^1(\Omega)$ of $-div\big(A^m grad(u_m)\big)=f$ converges to u_∞ in $H_0^1(\Omega)$ weak and $A^m grad(u_m)$ converges to $R(f)=C(u_\infty)$ in $L^2(\Omega;\mathbb{R}^N)$ weak. The *method of oscillating test functions* consists in using the *div–curl lemma* for obtaining a new proof that C is a local operator of the form $C(v)=A^{\mathit{eff}} grad(v)$ with $A^{\mathit{eff}}\in L^\infty\big(\Omega;\mathcal{L}(\mathbb{R}^N;\mathbb{R}^N)\big)$.

One constructs a sequence of oscillating test functions v_m satisfying

$$v_m\rightharpoonup v_\infty \text{ in } H^1(\Omega) \text{ weak,} \tag{10.1}$$
$$-div\big((A^m)^T grad(v_m)\big) \text{ converges in } H^{-1}_{loc}(\Omega) \text{ strong,} \tag{10.2}$$
$$(A^m)^T grad(v_m)\rightharpoonup w_\infty \text{ in } L^2(\Omega;\mathbb{R}^N) \text{ weak,} \tag{10.3}$$

where $(A^m)^T$ is the transpose of A^m. Then one passes to the limit in two different ways in the quantity

$$\big(A^m grad(u_m), grad(v_m)\big)=\big(grad(u_m),(A^m)^T grad(v_m)\big). \tag{10.4}$$

The *div–curl lemma* applies to the left side of (10.4) which then converges to $\big(C(u_\infty),grad(v_\infty)\big)$ in $L^1(\Omega)$ weak $\star$ (i.e., $L^1(\Omega)$ with the weak $\star$ topology of $\mathcal{M}_b(\Omega)$, dual of $C_0(\Omega)$), since $div\big(A^m grad(u_m)\big)$ is a fixed element of $H^{-1}(\Omega)$ and $grad(v_m)$ converges to $grad(v_\infty)$ in $L^2(\Omega;\mathbb{R}^N)$ weak; the *div–curl lemma* also applies to the right side of (10.4) which then converges to $(grad(u_\infty),w_\infty)$ in $L^1(\Omega)$ weak $\star$, because of (10.1)–(10.3). One deduces that

$$\big(C(u_\infty),grad(v_\infty)\big)=(grad(u_\infty),w_\infty) \text{ a.e. in } \Omega. \tag{10.5}$$

One can construct v_m satisfying (10.1)–(10.3) by choosing an open set Ω' with $\overline{\Omega} \subset \Omega'$, by extending A^m in $\Omega' \setminus \Omega$ for example by $A^m(x) = \alpha I$ for $x \in \Omega' \setminus \Omega$, and by choosing $v_m \in H_0^1(\Omega')$ solution of

$$- \, div\big((A^m)^T grad(v_m)\big) = g \text{ in } \Omega', \tag{10.6}$$

for some $g \in H^{-1}(\Omega')$; one obtains a sequence v_m bounded in $H_0^1(\Omega')$, so that its restriction to Ω is bounded in $H^1(\Omega)$, and only a subsequence will satisfy (10.1)–(10.3), but that is enough to obtain (10.5). By Lemma 6.2 one can choose $g \in H^{-1}(\Omega')$ so that v_∞ is any element of $H_0^1(\Omega')$, and in particular for each $j = 1, \ldots, N$, there exists $g_j \in H^{-1}(\Omega')$ such that $v_\infty = x_j$ a.e. in Ω, and using these N choices of g_j, (10.5) means $C\, u_\infty = A^{\textit{eff}} grad(u_\infty)$ for some $A^{\textit{eff}} \in L^2(\Omega; \mathcal{L}(\mathbb{R}^N; \mathbb{R}^N))$.

This method quickly gives an intermediate result, it applies to any linear variational setting, but it is not good for questions of bounds, which unfortunately are not so well understood for general equations or systems; optimal bounds, as I shall define them later, are not even completely understood in the model case. As pointed out by François MURAT, one can easily show that $A^{\textit{eff}} \in L^\infty(\Omega; \mathcal{L}(\mathbb{R}^N; \mathbb{R}^N))$ by using Lemma 10.1, based on the continuity of C, but the information $A^{\textit{eff}} \in \mathcal{M}(\alpha, \beta; \Omega)$ is not natural in this approach.

This method does not require the operators to be elliptic, since it is enough to know how to construct enough sequences v_n, and I shall show such examples when discussing the formulas for laminated materials in Chap. 12.[12]

Lemma 10.1. *If $A^{\textit{eff}} \in L^2(\Omega; \mathcal{L}(\mathbb{R}^N; \mathbb{R}^N))$ and the operator C defined by $C(v) = A^{\textit{eff}} grad(v)$ for all $v \in H_0^1(\Omega)$ is linear continuous from $H_0^1(\Omega)$ into $L^2(\Omega; \mathbb{R}^N)$ of norm $\le \gamma$, then one has $A^{\textit{eff}} \in L^\infty(\Omega; \mathcal{L}(\mathbb{R}^N; \mathbb{R}^N))$ and*

$$|A^{\textit{eff}}(x)|_{\mathcal{L}(\mathbb{R}^N;\mathbb{R}^N)} \le \gamma \text{ a.e. } x \in \Omega. \tag{10.7}$$

Proof. For $\xi \in \mathbb{R}^N \setminus 0$ and $\varphi \in C_c^1(\Omega)$, one defines φ_n by

$$\varphi_n(x) = \varphi(x) \frac{\sin n(\xi, x)}{n} \text{ for } x \in \Omega, \tag{10.8}$$

[12] It is a mathematician's training to know why something is true, and that behavior is not always appreciated by physicists: if a first formal computation made by physicists was given a sound mathematical basis, while a second one did not receive a satisfactory mathematical treatment but seems well corroborated by experiment, mathematicians talk about theorems for the first case and conjectures for the second case, while physicists do not perceive any difference between the two computations. In the case of laminated materials, the limiting behavior is identified without using the general theory, because it does not cover all the cases of laminated materials for which the limit is understood; as will appear in later discussions, this fact was grossly misunderstood by a few mathematicians who like to believe that they understand what homogenization is about.

which is bounded in $H^1_0(\Omega)$, with

$$\lim_n ||grad(\varphi_n)||_{L^2(\Omega;\mathbb{R}^N)} = \frac{1}{\sqrt{2}}\,||\xi\,\varphi||_{L^2(\Omega;\mathbb{R}^N)} = \frac{|\xi|}{\sqrt{2}}\,||\varphi||_{L^2(\Omega)}. \tag{10.9}$$

Since $C(\varphi_n) = A^{\mathit{eff}}\big(grad(\varphi)\frac{\sin n(\xi,\cdot)}{n} + \xi\,\varphi \cos n(\xi,\cdot)\big)$, with $A^{\mathit{eff}} grad(\varphi)$ and $A^{\mathit{eff}}\xi\,\varphi$ belonging to $L^2(\Omega;\mathbb{R}^N)$, one has

$$\lim_n ||C(\varphi_n)||_{L^2(\Omega;\mathbb{R}^N)} = \frac{1}{\sqrt{2}}\,||A^{\mathit{eff}}\xi\,\varphi||_{L^2(\Omega)}. \tag{10.10}$$

One deduces that

$$||A^{\mathit{eff}}\xi\,\varphi||_{L^2(\Omega)} \le \gamma\,|\xi|\,||\varphi||_{L^2(\Omega)}, \tag{10.11}$$

for all $\varphi \in C^1_c(\Omega)$, and for all $\varphi \in L^2(\Omega)$ by density, which means

$$||A^{\mathit{eff}}\xi||_{L^\infty(\Omega;\mathbb{R}^N)} \le \gamma\,|\xi|, \tag{10.12}$$

and since this is valid for all ξ, one obtains (10.7). □

The preceding method is not adapted to nonlinear problems, but there is a variant where the oscillating test functions are asked to satisfy the initial equation, instead of the transposed equation. I shall describe this variant in Chap. 11, for a nonlinear (monotone) setting which extends the linear case.

I now show properties of H-convergence, using the *div–curl lemma.*

Lemma 10.2. *If a sequence $A^n \in \mathcal{M}(\alpha,\beta;\Omega)$ H-converges to A^{eff}, then the transposed sequence $(A^n)^T$ H-converges to $(A^{\mathit{eff}})^T$. In particular, if a sequence A^n H-converges to A^{eff} and $(A_n)^T = A_n$ for all n, a.e. $x \in \Omega$, then $(A^{\mathit{eff}})^T = A^{\mathit{eff}}$ a.e. $x \in \Omega$.*

Proof. $A \in \mathcal{M}(\alpha,\beta;\Omega)$ implies (and is equivalent to) $A^T \in \mathcal{M}(\alpha,\beta;\Omega)$ since $A \in \mathcal{M}(\alpha,\beta;\Omega)$ means $(A(x)\xi,\xi) \ge \alpha\,|\xi|^2$ and $(A^{-1}(x)\xi,\xi) \ge \frac{1}{\beta}\,|\xi|^2$ for all $\xi \in \mathbb{R}^N$, a.e. $x \in \Omega$, and since $(A^{-1})^T = (A^T)^{-1}$, this is the same as $(A^T(x)\xi,\xi) \ge \alpha\,|\xi|^2$ and $((A^T)^{-1}(x)\xi,\xi) \ge \frac{1}{\beta}\,|\xi|^2$ for all $\xi \in \mathbb{R}^N$, a.e. $x \in \Omega$. By Theorem 6.5, a subsequence $(A^m)^T$ H-converges to B^{eff}. One defines $u_m, v_m \in H^1_0(\Omega)$ by

$$-div\big(A^m grad(u_m)\big) = f \text{ in } \Omega, \tag{10.13}$$
$$-div\big((A^m)^T grad(v_m)\big) = g \text{ in } \Omega, \tag{10.14}$$

for $f, g \in H^{-1}(\Omega)$ chosen so that

$$\begin{aligned} &u_m \rightharpoonup u_\infty \text{ in } H^1_0(\Omega) \text{ weak} \\ &A^m grad(u_m) \rightharpoonup A^{\mathit{eff}} grad(u_\infty) \text{ in } L^2(\Omega;\mathbb{R}^N) \text{ weak,} \end{aligned} \tag{10.15}$$

$$\begin{aligned}&v_m \rightharpoonup v_\infty \text{ in } H_0^1(\Omega) \text{ weak}\\&(A^m)^T grad(v_m) \rightharpoonup B^{\mathit{eff}} grad(v_\infty) \text{ in } L^2(\Omega;\mathbb{R}^N) \text{ weak.}\end{aligned}\tag{10.16}$$

Then, as for (10.4), one uses the *div–curl lemma* to pass to the limit in

$$\big(A^m grad(u_m), grad(v_m)\big) = \big(grad(u_m), (A^m)^T grad(v_m)\big),\tag{10.17}$$

by using (10.13)–(10.16), and one deduces that

$$\big(A^{\mathit{eff}} grad(u_\infty), grad(v_\infty)\big) = \big(grad(u_\infty), B^{\mathit{eff}} grad(v_\infty)\big),\tag{10.18}$$

for $u_\infty, v_\infty \in H_0^1(\Omega)$, which implies $(A^{\mathit{eff}})^T = B^{\mathit{eff}}$ a.e. in Ω. The second part of Lemma 10.2 results from uniqueness of H-limits. □

The next result shows that H-convergence inside Ω is not related to any particular boundary condition imposed on $\partial\Omega$.

Lemma 10.3. *If a sequence $A^n \in \mathcal{M}(\alpha,\beta;\Omega)$ H-converges to A^{eff} and $u_n \rightharpoonup u_\infty$ in $H^1_{loc}(\Omega)$ weak with $div\big(A^n grad(u_n)\big)$ belonging to a compact set of $H^{-1}_{loc}(\Omega)$ strong, then $A^n grad(u_n) \rightharpoonup A^{\mathit{eff}} grad(u_\infty)$ in $L^2_{loc}(\Omega;\mathbb{R}^N)$ weak.*

Proof. For $\varphi \in C_c^1(\Omega)$, $\varphi\, u_n$ converges to $\varphi\, u_\infty$ in $H_0^1(\Omega)$ weak, $\varphi\, grad(u_n)$ converges to $\varphi\, grad(u_\infty)$ in $L^2(\Omega;\mathbb{R}^N)$ weak, and $curl\big(\varphi\, grad(u_n)\big)$ has its components bounded in $L^2(\Omega)$, since they are of the form $\frac{\partial\varphi}{\partial x_j}\frac{\partial u_n}{\partial x_k} - \frac{\partial\varphi}{\partial x_k}\frac{\partial u_n}{\partial x_j}$. Since $div\big(A^n\varphi\, grad(u_n)\big) = \varphi\, div\big(A^n grad(u_n)\big) + \big(A^n grad(u_n), grad(\varphi)\big)$, it belongs to a compact set of $H^{-1}(\Omega)$ strong, because multiplication by φ maps $H^{-1}_{loc}(\Omega)$ into $H^{-1}(\Omega)$, and $\big(A^n grad(u_n), grad(\varphi)\big)$ is bounded in $L^2(\Omega)$. One extracts a subsequence such that $A^m\varphi\, grad(u_m)$ converges to w_∞ in $L^2(\Omega;\mathbb{R}^N)$ weak. For $f \in H^{-1}(\Omega)$, one defines $v_n \in H_0^1(\Omega)$ by

$$-div\big((A^n)^T grad(v_n)\big) = f,\tag{10.19}$$

so that v_n converges to v_∞ in $H_0^1(\Omega)$ weak, and by Lemma 10.2 $(A^n)^T grad(v_n)$ converges to $(A^{\mathit{eff}})^T grad(v_\infty)$ in $L^2(\Omega;\mathbb{R}^N)$, and using the *div–curl lemma*, one passes to the limit in

$$\big(A^m\varphi\, grad(u_m), grad(v_m)\big) = \big(\varphi\, grad(u_m), (A^m)^T grad(v_m)\big),\tag{10.20}$$

and one obtains the relation

$$\big(w_\infty, grad(v_\infty)\big) = \big(\varphi\, grad(u_\infty), (A^{\mathit{eff}})^T grad(v_\infty)\big) \text{ a.e. in } \Omega,\tag{10.21}$$

and since v_∞ is arbitrary by Lemma 6.2, $w_\infty = \varphi\, A^{\mathit{eff}} grad(u_\infty)$ a.e. in Ω. Since φ is arbitrary and the limit does not depend upon which subsequence was chosen, one deduces that all the sequence $A^n grad(u_n)$ converges to $A^{\mathit{eff}} grad(u_\infty)$ in $L^2_{loc}(\Omega;\mathbb{R}^N)$ weak. □

In the preceding proof, $div\big(A^n\,grad(\varphi\,u_n)\big)$ may not belong to a compact set of $H^{-1}(\Omega)$ strong, since it is $div\big(A^n\varphi\,grad(u_n)\big) + div\big(u_n A^n\,grad(\varphi)\big)$ and $div\big(A^n\varphi\,grad(u_n)\big)$ belongs to a compact set of $H^{-1}(\Omega)$ strong as was already used, but it is not clear if $div\big(u_n A^n grad(\varphi)\big)$ does, since $u_n A^n grad(\varphi)$ may only converge weakly in $L^2(\Omega;\mathbb{R}^N)$. The complete form of the *div–curl lemma* was used, and not only the special case for gradients.

Lemma 10.3 tells that boundary conditions for u_n are not important, as noticed by Sergio SPAGNOLO for G-convergence. H-convergence was defined with Dirichlet conditions, but the result inside Ω is the same for other boundary conditions, if the Lax–Milgram lemma applies for existence, since one needs to start by using Lemma 6.2. Using Dirichlet conditions has the advantage that no smoothness is necessary for $\partial\Omega$, but for other boundary conditions, non-homogeneous Dirichlet conditions, Neumann conditions, or other variational conditions, it is simpler to assume $\partial\Omega$ locally Lipschitz,[13] and to use γ_0, the trace operator to $\partial\Omega$, which maps $H^1(\Omega)$ onto $H^{1/2}(\partial\Omega)$. What happens on $\partial\Omega$ at the limit can be done at once for many different boundary conditions in the framework of variational inequalities, and one may even allow some nonlinearity in the boundary conditions (the nonlinearity inside Ω is a different matter that I shall describe in Chap. 11).

Lemma 10.4. *Let $A^n \in \mathcal{M}(\alpha,\beta;\Omega)$ be a sequence which H-converges to A^{eff}, and a_n, a_{eff} be the bilinear continuous forms on $H^1(\Omega)$ defined by*

$$a_n(u,v) = \int_\Omega \big(A^n grad(u), grad(v)\big)\,dx \text{ for all } u,v \in H^1(\Omega), \quad (10.22)$$

$$a_{\mathit{eff}}(u,v) = \int_\Omega \big(A^{\mathit{eff}} grad(u), grad(v)\big)\,dx \text{ for all } u,v \in H^1(\Omega). \quad (10.23)$$

For L in the dual of $H^1(\Omega)$, J a proper convex lower semi-continuous function on $H^{1/2}(\partial\Omega)$, $K \subset H^1(\Omega)$ a nonempty closed convex set satisfying

$$K + H^1_0(\Omega) \subset K, \quad (10.24)$$

one considers the variational inequality

$$\begin{aligned} &a_n(u_n, u_n - v) + J\big(\gamma_0(u_n)\big) - J\big(\gamma_0(v)\big)\\ &\quad\le L(u_n - v) \text{ for all } v \in K, u_n \in K. \quad (10.25)\\ &\text{If } u_n \rightharpoonup u_\infty \text{ in } H^1(\Omega) \text{ weak}, \quad (10.26)\end{aligned}$$

(so that (10.25) has a solution u_n for each n), then

$$A^n grad(u_n) \rightharpoonup A^{\mathit{eff}} grad(u_\infty) \text{ in } L^2(\Omega;\mathbb{R}^N) \text{ weak}, \quad (10.27)$$

---

[13] Rudolf Otto Sigismund LIPSCHITZ, German mathematician, 1832–1903. He worked in Bonn, Germany.

and u_∞ is a solution of

$$a_{\mathit{eff}}(u_\infty, u_\infty - v) + J(\gamma_0(u_\infty)) - J(\gamma_0(v)) \le L(u_\infty - v) \text{ for all } v \in K, u_\infty \in K. \tag{10.28}$$

Proof. Choosing $v = u_n \pm \varphi$ in (10.25) with $\varphi \in H_0^1(\Omega)$, which is allowed because of (10.24), shows that (10.25) implies the equation

$$a_n(u_n, \varphi) = L(\varphi) \text{ for all } \varphi \in H_0^1(\Omega), \tag{10.29}$$

i.e.,

$$-div(A^n grad(u_n)) = f \text{ for some } f \in H^{-1}(\Omega). \tag{10.30}$$

Then Lemma 10.2 implies (10.27). By the Hahn–Banach theorem, K is weakly closed in $H^1(\Omega)$ (since it is closed and convex), so that one has $u_\infty \in K$, and inequality (10.28) follows from (10.25) if for all $v \in H^1(\Omega)$ one shows that

$$\liminf_n \big[a_n(u_n, u_n - v) + J(\gamma_0(u_n))\big] \ge a_{\mathit{eff}}(u_\infty, u_\infty - v) + J(\gamma_0(u_\infty)). \tag{10.31}$$

Since $a_n(u_n, v) \to a_{\mathit{eff}}(u_\infty, v)$ by (10.27), and J is lower semi-continuous for the weak topology (Hahn–Banach), one has $\liminf_n J(\gamma_0(u_n)) \ge J(\gamma_0(u_\infty))$ since $\gamma_0(u_n) \rightharpoonup \gamma_0(u_\infty)$ in $H^{1/2}(\partial\Omega)$ weak, and (10.31) follows from

$$\liminf_n a_n(u_n, u_n) \ge a_{\mathit{eff}}(u_\infty, u_\infty). \tag{10.32}$$

The *div–curl lemma* implies that for all $\varphi \in C_c(\Omega)$

$$\int_\Omega \varphi\,(A^n grad(u_n), grad(u_n))\,dx \to \int_\Omega \varphi\,(A^{\mathit{eff}} grad(u_\infty), grad(u_\infty))\,dx. \tag{10.33}$$

Using $(A^n grad(u_n), grad(u_n)) \ge 0$, and $\varphi \in C_c^1(\Omega)$ with $0 \le \varphi \le 1$ in Ω,

$$\liminf_n a_n(u_n, u_n) \ge \int_\Omega \varphi\,(A^{\mathit{eff}} grad(u_\infty), grad(u_\infty))\,dx, \tag{10.34}$$

and then $\int_\Omega (A^{\mathit{eff}} grad(u_\infty), grad(u_\infty))\,dx$ is obtained as supremum of the right side for $\varphi \in C_c^1(\Omega)$ with $0 \le \varphi \le 1$ in Ω. □

Of course, the existence of solutions of (10.25) satisfying uniform $H^1(\Omega)$ bounds requires some compatibility between J, K, and L, and a sufficient condition is that for some $\delta > 0$ and $C \in \mathbb{R}$ one has

$$J(\gamma_0(v)) - L(v) \ge \delta\,||v||_{L^2(\partial\Omega)} - C \text{ for all } v \in K. \tag{10.35}$$

The regularity for $\partial\Omega$ can be weakened to cases where γ_0 can be defined, like for $\partial\Omega$ continuous, if one writes $X = \gamma_0\big(H^1(\Omega)\big)$ and one asks J to be lower semi-continuous on X.

Lemma 10.5. *If a sequence $A^n \in \mathcal{M}(\alpha,\beta;\Omega)$ H-converges to A^{eff}, and ω is an open subset of Ω, then the sequence $A^n|_\omega$ of the restrictions of A^n to ω H-converges to $A^{\mathit{eff}}|_\omega$. Therefore if a sequence $B^n \in \mathcal{M}(\alpha,\beta;\Omega)$ H-converges to B^{eff} and $A^n = B^n$ for all n, a.e. $x \in \omega$, then $A^{\mathit{eff}} = B^{\mathit{eff}}$ a.e. $x \in \omega$.*

Proof. If all A^n belong to $\mathcal{M}(\alpha,\beta;\Omega)$, then all $A^n|_\omega$ belong to $\mathcal{M}(\alpha,\beta;\omega)$ and by Theorem 6.5 a subsequence $A^m|_\omega$ H-converges to some $M^{\mathit{eff}} \in \mathcal{M}(\alpha,\beta;\omega)$. For $f \in H^{-1}(\omega)$ and $g \in H^{-1}(\Omega)$, one solves

$$-div\big(A^m|_\omega grad(u_m)\big) = f \text{ in } \omega, -div\big((A^m)^T grad(v_m)\big) = g \text{ in } \Omega, \quad (10.36)$$

so that

$$\begin{aligned} u_m &\rightharpoonup u_\infty \text{ in } H^1_0(\omega) \text{ weak},\\ A^m|_\omega grad(u_m) &\rightharpoonup M^{\mathit{eff}} grad(u_\infty) \text{ in } L^2(\omega;\mathbb{R}^N) \text{ weak}, \end{aligned} \quad (10.37)$$

$$\begin{aligned} v_m &\rightharpoonup v_\infty \text{ in } H^1_0(\Omega) \text{ weak},\\ (A^m)^T grad(v_m) &\rightharpoonup (A^{\mathit{eff}})^T grad(v_\infty) \text{ in } L^2(\Omega;\mathbb{R}^N) \text{ weak}. \end{aligned} \quad (10.38)$$

Extending u_n, u_∞ by 0 in $\Omega \setminus \omega$, one applies the *div–curl lemma* in ω to

$$\big(A^m|_\omega grad(u_m), grad(v_m)\big) = \big(grad(u_m), (A^m)^T grad(v_m)\big), \quad (10.39)$$

and one obtains

$$\big(M^{\mathit{eff}} grad(u_\infty), grad(v_\infty)\big) = \big(grad(u_\infty), (A^{\mathit{eff}})^T grad(v_\infty)\big) \text{ a.e. in } \omega. \quad (10.40)$$

By Lemma 6.2, u_∞ is arbitrary in $H^1_0(\omega)$ and v_∞ is arbitrary in $H^1_0(\Omega)$, so that $M^{\mathit{eff}} = A^{\mathit{eff}}$ a.e. in ω; since the H-limit is independent of the subsequence used, all the sequence $A^n|_\omega$ H-converges to $A^{\mathit{eff}}|_\omega$. □

Actually, if for a *measurable* subset ω of Ω, one has $A^n = B^n$ for all n, a.e. $x \in \omega$, and $A^n, B^n \in \mathcal{M}(\alpha,\beta;\Omega)$ H-converge in Ω to $A^{\mathit{eff}}, B^{\mathit{eff}}$, then one has $A^{\mathit{eff}} = B^{\mathit{eff}}$ a.e. $x \in \omega$, and it can be proven as Sergio SPAGNOLO did in the symmetric case, using the Meyers theorem.

It is equivalent to prove that $A^{\mathit{eff}} = B^{\mathit{eff}}$ a.e. in $\omega(\varepsilon)$ for each $\varepsilon > 0$, where $\omega(\varepsilon)$ is the set of points of ω at a distance at least ε from $\partial\Omega$. Defining v_n as in (10.36), but choosing $f \in H^{-1}(\Omega)$ and $u_n \in H^1_0(\Omega)$ there, the problem is to use $\chi_{\omega(\varepsilon)}$, the characteristic function of $\omega(\varepsilon)$, as a test function in the *div–curl lemma* in Ω, excluding the type of counter-example of Lemma 7.3. One takes

$f, g \in W^{-1,p}(\Omega)$ with $p > 2$, and by the Meyers theorem $grad(u_n), grad(v_n)$ stay bounded in $L^{q(\varepsilon)}\big(\omega(\varepsilon)\big)$ for some $q(\varepsilon) \in (2, p]$, so that

$$\big(A^n grad(u_n), grad(v_n)\big) = \big(grad(u_n), (B_n)^T grad(v_n)\big) \text{ in } \omega(\varepsilon), \tag{10.41}$$

and stay bounded in $L^{q(\varepsilon)/2}\big(\omega(\varepsilon)\big)$, and converge in $L^{q(\varepsilon)/2}\big(\omega(\varepsilon)\big)$ weak to

$$\big(A^{\text{eff}} grad(u_\infty), grad(v_\infty)\big) = \big(grad(u_\infty), (B^{\text{eff}})^T grad(v_\infty)\big) \text{ in } \omega(\varepsilon). \tag{10.42}$$

As $W^{-1,p}(\Omega)$ is dense in $H^{-1}(\Omega)$, one can pass to the limit in this equality in $\omega(\varepsilon)$ and obtain it for arbitrary $f, g \in H^{-1}(\Omega)$, i.e., for arbitrary $u_\infty, v_\infty \in H^1_0(\Omega)$, and that gives $A^{\text{eff}} = B^{\text{eff}}$ a.e. in $\omega(\varepsilon)$, hence a.e. in ω.

The argument of Lemma 10.5 extends to all variational situations, while the preceding argument requires extending the Meyers theorem, and I do not know in what generality this was done.

The next result is due to Ennio DE GIORGI and Sergio SPAGNOLO, who used characteristic functions of measurable sets for φ, as by using the Meyers theorem, one can extend the preceding result to $\varphi \ge 0$, $\varphi \in L^\infty(\Omega)$.

Lemma 10.6 is more precise than Lemma 6.6, which gives A_- on the right side of (10.42) instead of A^{eff}, where A_-^{-1} is the $L^\infty\big(\Omega; \mathcal{L}(\mathbb{R}^N; \mathbb{R}^N)\big)$ weak $\star$ limit of a subsequence $(A^m)^{-1}$, and $A_- \le A^{\text{eff}}$ by Lemma 6.7.

Lemma 10.6. *If a sequence $A^n \in \mathcal{M}(\alpha, \beta; \Omega)$ satisfies $(A^n)^T = A^n$ a.e. $x \in \Omega$ and H-converges to A^{eff}, if a sequence w_n converges to w_∞ in $H^1_0(\Omega)$ weak and if $\varphi \ge 0$ in Ω with $\varphi \in C_c(\Omega)$, then one has*

$$\liminf_n \int_\Omega \varphi\,\big(A^n grad(w_n), grad(w_n)\big)\,dx \ge \int_\Omega \varphi\,\big(A^{\text{eff}} grad(w_\infty), grad(w_\infty)\big)\,dx. \tag{10.43}$$

For all $w_\infty \in H^1_0(\Omega)$ there exists a sequence u_n converging to u_∞ in $H^1_0(\Omega)$ weak and such that for all $\varphi \in C_c(\Omega)$ one has

$$\lim_n \int_\Omega \varphi\,\big(A^n grad(u_n), grad(u_n)\big)\,dx = \int_\Omega \varphi\,\big(A^{\text{eff}} grad(w_\infty), grad(w_\infty)\big)\,dx. \tag{10.44}$$

Proof. Let $u_n \in H^1_0(\Omega)$ be the solution of

$$-div\big(A^n grad(u_n)\big) = f = -div\big(A^{\text{eff}} grad(w_\infty)\big) \text{ in } \Omega, \tag{10.45}$$

so that u_n converges to some u_∞ in $H^1_0(\Omega)$ weak, and since u_∞ is the solution of $-div\big(A^{\text{eff}} grad(u_\infty)\big) = f$ in Ω, one has $u_\infty = w_\infty$ a.e. in Ω, hence

$$u_n \rightharpoonup w_\infty \text{ in } H^1_0(\Omega) \text{ weak}, A^n grad(u_n) \rightharpoonup A^{\text{eff}} grad(w_\infty) \text{ in } L^2(\Omega; \mathbb{R}^N) \text{ weak}. \tag{10.46}$$

One develops

$$\liminf_n \int_\Omega \varphi \left(A^n \left(grad(w_n) - grad(u_n)\right), grad(w_n) - grad(u_n)\right) dx, \quad (10.47)$$

which is ≥ 0. One term is $\int_\Omega \varphi \left(A^n grad(w_n), grad(w_n)\right) dx$ whose $\liminf_n$ is what one is interested in. Using the symmetry of A^n, the other terms can be written $\int_\Omega \varphi \left(A^n grad(u_n), grad(-2w_n + u_n)\right) dx$, and the *div–curl lemma* applies, so the limit is $-\int_\Omega \varphi \left(A^{\textit{eff}} grad(w_\infty), grad(w_\infty)\right) dx$, and this gives (10.43). Of course, (10.44) is obtained by using the sequence u_n just constructed and applying the *div–curl lemma.* □

The next result deals with the compatibility of H-convergence with the usual preorder relation on $\mathcal{L}(\mathbb{R}^N;\mathbb{R}^N)$, i.e., for $A, B \in \mathcal{L}(\mathbb{R}^N;\mathbb{R}^N)$, $A \leq B$ means $(A\,\xi,\xi) \leq (B\,\xi,\xi)$ for all $\xi \in \mathbb{R}^N$; this preorder is not an order on $\mathcal{L}(\mathbb{R}^N;\mathbb{R}^N)$, but it is a partial order when restricted to symmetric operators.

Lemma 10.7. *If $A^n \in \mathcal{M}(\alpha,\beta;\Omega)$ satisfies $(A^n)^T = A^n$ a.e. $x \in \Omega$ and H-converges to $A^{\textit{eff}}$, if $B^n \in \mathcal{M}(\alpha,\beta;\Omega)$ H-converges to $B^{\textit{eff}}$ and if $B^n \geq A^n$ a.e. in Ω for all n, then one has $B^{\textit{eff}} \geq A^{\textit{eff}}$ a.e. $x \in \Omega$.*

Proof. Let $g \in H^{-1}(\Omega)$ and let $v_n \in H^1_0(\Omega)$ be the solution of

$$-div\left(B^n grad(v_n)\right) = g \text{ in } \Omega, \quad (10.48)$$

$$v_n \rightharpoonup v_\infty \text{ in } H^1_0(\Omega) \text{ weak}, B^n grad(v_n) \rightharpoonup B^{\textit{eff}} grad(v_\infty) \text{ in } L^2(\Omega;\mathbb{R}^N) \text{ weak}. \quad (10.49)$$

Then for $\varphi \geq 0$, $\varphi \in C_c(\Omega)$, one passes to the limit in

$$\int_\Omega \varphi \left(B^n grad(v_n), grad(v_n)\right) dx \geq \int_\Omega \varphi \left(A^n grad(v_n), grad(v_n)\right) dx. \quad (10.50)$$

The left side converges to $\int_\Omega \varphi \left(B^{\textit{eff}} grad(v_\infty), grad(v_\infty)\right) dx$ by the *div–curl lemma*, and one applies Lemma 10.6 for the $\liminf_n$ of the right side, giving

$$\int_\Omega \varphi \left(B^{\textit{eff}} grad(v_\infty), grad(v_\infty)\right) dx \geq \int_\Omega \varphi \left(A^{\textit{eff}} grad(v_\infty), grad(v_\infty)\right) dx. \quad (10.51)$$

Varying $v_\infty \in H^1_0(\Omega)$ gives $\varphi\, B^{\textit{eff}} \geq \varphi A^{\textit{eff}}$ a.e. in Ω, and one varies φ. □

The second part of Lemma 10.6 is valid without symmetry requirement, but the first part is not always true without symmetry. Lemma 10.7 is not true for a general A^n, even if all B^n are symmetric. In Lemma 6.7 the symmetry hypothesis on A^n is also important for comparing $A^{\textit{eff}}$ with A_+, the limit in $L^\infty(\Omega;\mathcal{L}(\mathbb{R}^N;\mathbb{R}^N))$ weak $\star$ of A^n. For constructing counter-examples for $N \geq 2$ (since every operator is symmetric if $N = 1$),[14] one defines A^n by

$$A^n = I + \psi_n(x_1)(e_1 \otimes e_2 - e_2 \otimes e_1), \tag{10.52}$$
$$\text{with } \psi_n \rightharpoonup \Psi_1, (\psi_n)^2 \rightharpoonup \Psi_2 \text{ in } L^\infty(\mathbb{R})\text{weak} \star, \text{ with } \Psi_2 > (\Psi_1)^2. \tag{10.53}$$

Lemma 5.3 for laminated materials (proven in Chap. 12), says that

$$A^n \text{ H-converges to } A^{\mathit{eff}} = I + \Psi_1(e_1 \otimes e_2 - e_2 \otimes e_1) + \big(\Psi_2 - (\Psi_1)^2\big)e_2 \otimes e_2. \tag{10.54}$$

As $A_+ = I + \Psi_1(e_1 \otimes e_2 - e_2 \otimes e_1)$, this gives an example with $A^{\mathit{eff}} > A_+$. One has $A^n \leq I$ for all n, but not $A^{\mathit{eff}} \leq I$ (one has $A^{\mathit{eff}} \geq I$, as it must be from Lemma 10.7, from $A^n \geq I$ for all n). Taking $u_n = u_\infty$ for all n, one has

$$\begin{aligned} &\textstyle\int_\Omega \varphi\,(A^n grad(u_n), grad(u_n))\,dx = \int_\Omega \varphi\,|grad(u_\infty)|^2\,dx = X\\ &\textstyle\int_\Omega \varphi\,(A^{\mathit{eff}} grad(u_\infty), grad(u_\infty))\,dx = X + \int_\Omega \varphi\,(\Psi_2 - \Psi_1^2)\left|\frac{\partial u_\infty}{\partial x_2}\right|^2 dx. \end{aligned} \tag{10.55}$$

I describe now useful estimates for perturbations and continuous dependence of H-limits with respect to parameters.

Lemma 10.8. *Let $A \in \mathcal{M}(\alpha,\beta;\Omega)$ and $B \in L^\infty(\Omega;\mathcal{L}(\mathbb{R}^N;\mathbb{R}^N))$ with*

$$||B||_{L^\infty(\Omega;\mathcal{L}(\mathbb{R}^N;\mathbb{R}^N))} \leq \delta < \alpha, \tag{10.56}$$

then

$$A + B \in M\Big(\alpha - \delta, \frac{\alpha\beta - \delta^2}{\alpha - \delta}; \Omega\Big). \tag{10.57}$$

Proof. Of course $\big((A+B)\xi,\xi\big) \geq \alpha\,|\xi|^2 - |B\,\xi|\,|\xi| \geq (\alpha-\delta)\,|\xi|^2$ for all $\xi \in \mathbb{R}^N$. For the other inequality,[15] one notices that $(A\,\xi,\xi) \geq \frac{1}{\beta}\,|A\,\xi|^2$ means $\big|A\,\xi - \frac{\beta}{2}\xi\big| \leq \frac{\beta}{2}\,|\xi|$, and if $2L = \frac{\alpha\,\beta-\delta^2}{\alpha-\delta}$ one wants to show that for all $\xi \in \mathbb{R}^N$ one has $|(A+B)\xi - L\,\xi| \leq L\,|\xi|$. It is a consequence of $|A\,\xi - L\,\xi| \leq (L-\delta)\,|\xi|$ for all $\xi \in \mathbb{R}^N$, i.e., of $|A\,\xi|^2 - 2L\,(A\,\xi,\xi) \leq (-2\delta\,L + \delta^2)\,|\xi|^2$, and by the definition of L one has $-2\delta\,L + \delta^2 = (\beta - 2L)\,\alpha$; then one notices that $|A\,\xi|^2 - 2L\,(A\,\xi,\xi) \leq (\beta-2L)\,(A\,\xi,\xi)$ which is $\leq (\beta-2L)\,\alpha\,|\xi|^2$ since $\beta - 2L \leq 0$. □

---

[14] François MURAT once showed me a letter which Paolo MARCELLINI sent him, with similar computations, done for the purpose of showing that some bounds in a non-symmetric case are optimal. The following computations were made many years after, and might coincide with those of Paolo MARCELLINI.

[15] For symmetric A, B, the upper bound is $\beta + \delta$, of course.

Lemma 10.9. *If $A^n \in \mathcal{M}(\alpha,\beta;\Omega)$ and $B^n \in \mathcal{M}(\alpha',\beta';\Omega)$ H-converge to A^{eff} and B^{eff}, and $|B^n - A^n|_{\mathcal{L}(\mathbb{R}^N;\mathbb{R}^N)} \le \varepsilon$ for all n, a.e. $x \in \Omega$, then*

$$||B^{\text{eff}} - A^{\text{eff}}||_{L^\infty(\Omega;\mathcal{L}(\mathbb{R}^N;\mathbb{R}^N))} \le \varepsilon \frac{\sqrt{\beta\,\beta'}}{\sqrt{\alpha\,\alpha'}}. \tag{10.58}$$

Proof. For $f,g \in H^{-1}(\Omega)$, one solves

$$-div\big(A^n grad(u_n)\big) = f \text{ in } \Omega, -div\big((B^n)^T grad(v_n)\big) = g \text{ in } \Omega, \tag{10.59}$$

so that

$$u_n \rightharpoonup u_\infty, H^1_0(\Omega) \text{ weak}, A^n grad(u_n) \rightharpoonup A^{\text{eff}} grad(u_\infty), L^2(\Omega;\mathbb{R}^N) \text{ weak}, \tag{10.60}$$

$$v_n \rightharpoonup v_\infty, H^1_0(\Omega) \text{ weak}, (B^n)^T grad(v_n) \rightharpoonup (B^{\text{eff}})^T grad(v_\infty), L^2(\Omega;\mathbb{R}^N) \text{ weak}. \tag{10.61}$$

$\big(A^n grad(u_n), grad(v_n)\big)$ and $\big(grad(u_n), (B^n)^T grad(v_n)\big)$ converge in $L^1(\Omega)$ weak $\star$ to $\big(A^{\text{eff}} grad(u_\infty), grad(v_\infty)\big)$ and $\big(grad(u_\infty), (B^{\text{eff}})^T grad(v_\infty)\big)$ by the *div–curl lemma*, so for $\varphi \in C_c(\Omega)$ one has

$$\begin{aligned} &\lim_n \textstyle\int_\Omega \varphi\,\big((B^n - A^n)\, grad(u_n), grad(v_n)\big)\, dx = X \\ &X = \textstyle\int_\Omega \varphi\,\big((B^{\text{eff}} - A^{\text{eff}})\, grad(u_\infty), grad(v_\infty)\big)\, dx. \end{aligned} \tag{10.62}$$

Choosing $\varphi \ge 0$, one deduces that

$$|X| \le \varepsilon \limsup_n \int_\Omega \varphi\, |grad(u_n)|\, |grad(v_n)|\, dx. \tag{10.63}$$

Using $|grad(u_n)|\, |grad(v_n)| \le a\,\alpha\, |grad(u_n)|^2 + b\,\alpha'\, |grad(v_n)|^2$ if $4a\,b\,\alpha\,\alpha' \ge 1$, $A^n \in \mathcal{M}(\alpha,\beta;\Omega)$, and $B^n \in \mathcal{M}(\alpha',\beta';\Omega)$ one deduces that

$$|X| \le \varepsilon \limsup_n \int_\Omega \varphi\, \big[a\,\big(A^n grad(u_n), grad(u_n)\big) + b\,\big(B^n grad(v_n), grad(v_n)\big)\big]\, dx, \tag{10.64}$$

which gives

$$|X| \le \varepsilon \int_\Omega \varphi\, \big[a\,\big(A^{\text{eff}} grad(u_\infty), grad(u_\infty)\big) + b\,\big(B^{\text{eff}} grad(v_\infty), grad(v_\infty)\big)\big]\, dx, \tag{10.65}$$

hence

$$|X| \le \varepsilon \int_\Omega \varphi\, [a\,\beta\, |grad(u_\infty)|^2 + b\,\beta'\, |grad(v_\infty)|^2]\, dx. \tag{10.66}$$

Varying $\varphi \geq 0$, $\varphi \in C_c(\Omega)$, one deduces that

$$\big|\big((B^{\mathit{eff}}-A^{\mathit{eff}})grad(u_\infty), grad(v_\infty)\big)\big| \leq \varepsilon\left(a\,\beta\,|grad(u_\infty)|^2+b\,\beta'\,|grad(v_\infty)|^2\right) \tag{10.67}$$

a.e. in Ω. Optimizing on $a,b \in \mathbb{Q}$ satisfying $4a\,b\,\alpha\,\alpha' \geq 1$, one obtains

$$\big|\big((B^{\mathit{eff}}-A^{\mathit{eff}})grad(u_\infty), grad(v_\infty)\big)\big| \leq \varepsilon\,\frac{\sqrt{\beta\,\beta'}}{\sqrt{\alpha\,\alpha'}}\,|grad(u_\infty)|\,|grad(v_\infty)|, \text{ in } \Omega, \tag{10.68}$$

and since u_∞ and v_∞ are arbitrary, one deduces (10.58). □

Lemma 10.10. *Let P be an open set of $\mathbb{R}^p$. Let A^n be a sequence defined on $\Omega \times P$, such that $A^n(\cdot,p) \in \mathcal{M}(\alpha,\beta;\Omega)$ for each $p \in P$, and such that the mapping $p \to A^n(\cdot,p)$ is of class C^k (or real analytic) from P into $L^\infty\big(\Omega;\mathcal{L}(\mathbb{R}^N;\mathbb{R}^N)\big)$, with bounds of derivatives up to order k independent of n. Then there exists a subsequence A^m such that for all $p \in P$ the sequence $A^m(\cdot,p)$ H-converges to $A^{\mathit{eff}}(\cdot,p)$ and $p \to A^{\mathit{eff}}(\cdot,p)$ is of class C^k (or real analytic) from P into $L^\infty\big(\Omega;\mathcal{L}(\mathbb{R}^N;\mathbb{R}^N)\big)$.*

Proof. One considers a countable dense set Π of P and, using a diagonal subsequence, one extracts a subsequence A^m such that for all $p \in \Pi$ the sequence $A^m(\cdot,p)$ H-converges to a limit $A^{\mathit{eff}}(\cdot,p)$. Using the fact that A is uniformly continuous on compact subsets of P and Lemma 10.9, one then deduces that $p \to A^{\mathit{eff}}(\cdot,p)$ is continuous from P into $L^\infty\big(\Omega;\mathcal{L}(\mathbb{R}^N;\mathbb{R}^N)\big)$ and that for all $p \in P$ the sequence $A^m(\cdot,p)$ H-converges to $A^{\mathit{eff}}(\cdot,p)$.

Defining the operators $T_m(p)$ from $V = H^1_0(\Omega)$ into $V' = H^{-1}(\Omega)$ by $T_m(p)v = -div\big(A^m(\cdot,p)\,grad(v)\big)$, one finds that the mappings $p \to T_m(p)$ are of class C^k (or real analytic) from P to $\mathcal{L}(V;V')$ and similarly $p \to \big(T_m(p)\big)^{-1}$ are of class C^k (or real analytic) from P to $\mathcal{L}(V';V)$, and finally the operators R_m defined by $R_m v = A^m(\cdot,p)\,grad(v_m)$ with v_m defined by $T_m(v_m) = T_\infty(v)$ are of class C^k (or real analytic) from P into $\mathcal{L}(V;L^2(\Omega;\mathbb{R}^N))$; all the bounds of derivatives up to order k being independent of m, the limit inherits of the same bounds, and since $R_\infty v = A^{\mathit{eff}}(\cdot,p)\,grad(v)$, one deduces that $p \to A^{\mathit{eff}}(\cdot,p)$ is of class C^k (or real analytic) from P into $L^\infty\big(\Omega;\mathcal{L}(\mathbb{R}^N;\mathbb{R}^N)\big)$. □

Chapter 11
Homogenization of Monotone Operators

Although Eduardo ZARANTONELLO first introduced monotone operators for solving a problem in continuum mechanics,[1] the theory of monotone operators quickly became taught as a part of functional analysis. In his course on nonlinear partial differential equations in the late 1960s, Jacques-Louis LIONS taught about a dichotomy, the compactness method, and the monotonicity method. During my stay in Madison in 1974–1975, I found that the *div–curl lemma* gave a natural framework to the monotonicity method for (stationary) diffusion equations, and that it was not so natural to classify the convexity method as being a part of the monotonicity method. A few years later, after developing the theory of *compensated compactness* with François MURAT, I unified all these methods in the *compensated compactness method.* I shall only discuss here the homogenization of monotone operators in the simple framework that I adopted in my Peccot lectures at the beginning of 1977.

At an abstract level, one considers a real separable Hilbert space V, equipped with the norm $||\cdot||$, and a sequence of (nonlinear) operators $\mathcal{A}^n$ from V into V' (equipped with the dual norm $||\cdot||_*$), which are uniformly monotone, i.e., there exists $\alpha > 0$ such that

$$\langle \mathcal{A}^n(u) - \mathcal{A}^n(v), u - v\rangle \geq \alpha\,||u-v||^2 \text{ for all } n \in \mathbb{N}, u, v \in V, \tag{11.1}$$

and globally Lipschitz continuous, i.e., there exists M $(\geq \alpha)$ such that

$$||\mathcal{A}^n(u) - \mathcal{A}^n(v)||_* \leq M\,||u-v|| \text{ for all } n \in \mathbb{N}, u, v \in V. \tag{11.2}$$

For a more concrete homogenization example, one chooses $V = H^1_0(\Omega)$ and

$$\mathcal{A}^n(u) = -div\big(A^n(x, grad(u))\big) \text{ in } \Omega, \tag{11.3}$$

where each A^n belongs to the class $\mathcal{M}on(\alpha,\beta;\Omega)$ of Definition 11.1.

[1] George MINTY also introduced monotone operators, but for a problem of electrical circuits, which do not involve partial differential equations.

L. Tartar, *The General Theory of Homogenization*, Lecture Notes of the Unione Matematica Italiana 7, DOI 10.1007/978-3-642-05195-1_11,

Definition 11.1. *One says that a Carathéodory function A defined on $\Omega \times \mathbb{R}^N$ belongs to $\mathcal{M}on(\alpha, \beta; \Omega)$ if*

$$(A(x,a) - A(x,b), a - b) \geq \frac{1}{\beta}\,|A(x,a) - A(x,b)|^2,$$
$$(A(x,a) - A(x,b), a - b) \geq \alpha\,|a - b|^2 \text{ for all } a, b \in \mathbb{R}^N, \text{ a.e. } x \in \Omega. \quad (11.4)$$

If $A^n \in \mathcal{M}on(\alpha, \beta; \Omega)$, then the operator $\mathcal{A}^n$ defined by (11.3) satisfies (11.1) and (11.2) with $M = \beta$, but the choice in Definition 11.1 is more adapted to homogenization, for proving an analogue of Theorem 6.5.

Theorem 11.2. *Let $A^n \in \mathcal{M}on(\alpha, \beta; \Omega)$ be such that*

$$\mathcal{A}^n(0) \text{ is bounded in } V', \quad (11.5)$$

then there is a subsequence A^m and $A^{\text{eff}} \in \mathcal{M}on(\alpha, \beta; \Omega)$ such that for all $f \in H^{-1}(\Omega)$ the solutions $u_m \in H_0^1(\Omega)$ of

$$-\,div\big(A^m(x, grad(u_m))\big) = f \text{ in } \Omega, \quad (11.6)$$

satisfy

$$u_m \rightharpoonup u_\infty \text{ in } H_0^1(\Omega) \text{ weak},$$
$$A^m(x, grad(u_m)) \rightharpoonup A^{\text{eff}}(x, grad(u_\infty)) \text{ in } L^2(\Omega; \mathbb{R}^N) \text{ weak}, \quad (11.7)$$

so that u_∞ is the solution of

$$-\,div\big(A^{\text{eff}}(x, grad(u_\infty))\big) = f \text{ in } \Omega. \quad (11.8)$$

Proof. From (11.1) and (11.2) each operator $\mathcal{A}^n$ is invertible and its inverse $\mathcal{B}^n = (\mathcal{A}^n)^{-1}$ is Lipschitz continuous from V' into V with constant $\frac{1}{\alpha}$.[2] For $f \in V'$, the sequence $u_n = (\mathcal{A}^n)^{-1}(f) = \mathcal{B}^n(f)$ is bounded in V by (11.5), since $\alpha\,||u_n||^2 \leq \langle \mathcal{A}^n(u_n) - \mathcal{A}^n(0), u_n\rangle = \langle f - \mathcal{A}^n(0), u_n\rangle$ implies

$$||u_n|| \leq \frac{1}{\alpha}\,||f - \mathcal{A}^n(0)||_*. \quad (11.9)$$

[2] If Λ is the (F.) Riesz isometry of the Hilbert space V onto its dual, a proof of the invertibility of $\mathcal{A}^n$ uses a continuation argument in θ for showing that $(1-\theta)\Lambda + \theta\,\mathcal{A}^n$ is invertible for all $\theta \in [0,1]$. If $\mathcal{A}^n$ is of class C^1, (11.1) and (11.2) imply that its derivative satisfies the hypothesis of the Lax–Milgram lemma at all points, and the inverse of its derivative being uniformly bounded, $\mathcal{A}^n$ is a global diffeomorphism from V onto V'.

If X is a countable dense set of V', one extracts a diagonal subsequence indexed by m such that, for all $f \in X$,

$$u_m = \mathcal{B}^m(f) \rightharpoonup u_\infty = \mathcal{B}^\infty(f) \text{ in } V \text{ weak}, \tag{11.10}$$
$$A^m\big(x, grad(u_m)\big) \rightharpoonup R(f) \text{ in } L^2(\Omega; \mathbb{R}^N) \text{ weak}, \tag{11.11}$$

and one writes $R(f) = C(u_\infty)$ after proving that $\mathcal{B}^\infty$ is invertible. Since the Lipschitz constants of $\mathcal{A}^n$ and $\mathcal{B}^n$ are bounded (by β, and $\frac{1}{\alpha}$), the limits exist then for all $f \in V'$. All $\mathcal{B}^m$ are uniformly monotone with constant $\alpha^* = \frac{\alpha}{M^2}$ and Lipschitz continuous with constant $\beta^* = \frac{1}{\alpha}$,[3] and $\mathcal{B}^\infty$ inherits these properties since the norm in V is sequentially lower semi-continuous for the weak topology,[4] so that $\mathcal{B}^\infty$ is invertible.

It remains to show that $C(u_\infty) = A^{eff}\big(x, grad(u_\infty)\big)$ a.e. in Ω for some $A^{eff} \in \mathcal{M}on(\alpha, \beta; \Omega)$. Let ω be an open set with $\overline{\omega} \subset \Omega$ and let $\varphi \in C^1_c(\Omega)$ be equal to 1 on ω. For $p \in \mathbb{R}^N$, one chooses f defining a sequence v^m such that $v^\infty(x) = \varphi(x)(p, x)$ a.e. in ω, and one defines $A^{eff}(x,p)$ in ω by

$$A^{eff}(x,p) = R(f)(x) \text{ a.e. in } \omega, \text{ if } \mathcal{B}^\infty(f) = (p, \cdot) \text{ a.e. in } \omega. \tag{11.12}$$

Of course, one must show that this definition makes sense, that A^{eff} belongs to $\mathcal{M}on(\alpha, \beta; \Omega)$ and that $C(u_\infty) = A^{eff}\big(x, grad(u_\infty)\big)$ a.e. in Ω.

Let ω_1, φ_1, p_1 correspond to a choice f_1 and a sequence v_1^m, and ω_2, φ_2, p_2 correspond to a choice f_2 and a sequence v_2^m. Writing $E_j^m = grad(v_j^m)$ and $D_j^m = A^m\big(grad(v_j^m)\big)$, the *div–curl lemma* implies

$$\int_\Omega \psi\,(D_2^m - D_1^m, E_2^m - E_1^m)\,dx \to L = \int_\Omega \psi\,\Big(A^{eff}(\cdot, p_2) - A^{eff}(\cdot, p_1), p_2 - p_1\Big)\,dx \tag{11.13}$$

for all $\psi \in C_c(\omega_1 \cap \omega_2)$. Assuming also that $\psi \geq 0$, one deduces that

$$\begin{aligned} L &\geq \liminf_m \int_\Omega \frac{\psi}{\beta}\,|D_2^m - D_1^m|^2\,dx \geq \int_\Omega \frac{\psi}{\beta}\,|A^{eff}(\cdot, p_2) - A^{eff}(\cdot, p_1)|^2\,dx \\ L &\geq \liminf_m \int_\Omega \psi\,\alpha\,|E_2^m - E_1^m|^2\,dx \geq \int_\Omega \psi\,\alpha\,|p_2 - p_1|^2\,dx. \end{aligned} \tag{11.14}$$

The integrals may be restricted to $\omega_1 \cap \omega_2$, and varying ψ gives

$$\begin{aligned} \big(A^{eff}(\cdot, p_2) - A^{eff}(\cdot, p_1), p_2 - p_1\big) &\geq \tfrac{\psi}{\beta}\,|A^{eff}(\cdot, p_2) - A^{eff}(\cdot, p_1)|^2 \\ \big(A^{eff}(\cdot, p_2) - A^{eff}(\cdot, p_1), p_2 - p_1\big) &\geq \alpha\,|p_2 - p_1|^2 \text{ a.e. in } \omega_1 \cap \omega_2. \end{aligned} \tag{11.15}$$

---

[3] If $\mathcal{A}^n v_n = g$, then $\alpha\,||u_n - v_n||^2 \leq \langle \mathcal{A}^n u_n - \mathcal{A}^n v_n, u_n - v_n\rangle = \langle f - g, u_n - v_n\rangle \leq ||f - g||_*||u_n - v_n||$, so that $||\mathcal{B}^n f - \mathcal{B}^n g||_* = ||u_n - v_n|| \leq \frac{1}{\alpha}\,||f - g||_*$. Then, $\frac{\alpha}{M^2}\,||f - g||_*^2 \leq \alpha\,||u_n - v_n||^2 \leq \langle \mathcal{A}^n u_n - \mathcal{A}^n v_n, u_n - v_n\rangle = \langle f - g, \mathcal{B}^n f - \mathcal{B}^n g\rangle$.

[4] Since $||\mathcal{B}^\infty f - \mathcal{B}^\infty g|| \leq \liminf_m ||\mathcal{B}^m f - \mathcal{B}^m g|| \leq \beta^*||f - g||_*$, and $\langle f - g, \mathcal{B}^\infty f - \mathcal{B}^\infty g\rangle = \lim_m \langle f - g, \mathcal{B}^m f - \mathcal{B}^m g\rangle \geq \alpha^*||f - g||_*^2$.

By choosing $p_1 = p_2$ one deduces from (11.15) that $A^{eff}(x,p_2) = A^{eff}(x,p_1)$ a.e. in $\omega_1 \cap \omega_2$, so that the definition of A^{eff} makes sense in Ω, and then, by choosing $p_1 \neq p_2$ one deduces that A^{eff} belongs to $\mathcal{M}on(\alpha,\beta;\Omega)$. Then one replaces E_2^m by $grad(u_m)$ and D_2^m by $A^m(\cdot,grad(u_m))$ and one deduces in the same way that for $\psi \in C_c(\omega_1)$ with $\psi \geq 0$, one has

$$L' = \int_\Omega \psi\Big(C(u_\infty) - A^{eff}(\cdot,p_1), grad(u_\infty) - p_1\Big)\,dx \geq \int_\Omega \psi\,\alpha\,|grad(u_\infty) - p_1|^2\,dx$$
$$L' \geq \int_\Omega \frac{\psi}{\beta}\,|C(u_\infty) - A^{eff}(\cdot,p_1)|^2\,dx, \tag{11.16}$$

from which one deduces that

$$|C(u_\infty) - A^{eff}(x,p_1)| \leq \beta\,|grad(u_\infty) - p_1| \text{ a.e. in } \omega_1. \tag{11.17}$$

Varying ω_1 and $p_1 \in \mathbb{R}^N$ implies $C(u_\infty) = A^{eff}(x, grad(u_\infty))$ a.e. in Ω. □

Using the same basic ideas, the analogue of Lemma 10.3 holds: the boundary conditions do not matter as long as one deals with sequences u_n which are bounded in $H^1_{loc}(\Omega)$ with $div\big(A^n(x,grad(u_n))\big)$ staying in a compact of $H^{-1}_{loc}(\Omega)$ strong, assuming that $\mathcal{A}^n(\cdot,0)$ stays bounded in $L^2(\Omega;\mathbb{R}^N)$.

The analogue of Lemma 10.4 holds, i.e., assuming that u_n stays bounded in $H^1(\Omega)$, and satisfies a variational inequality involving A^n in Ω, with $\mathcal{A}^n(\cdot,0)$ bounded in $L^2(\Omega;\mathbb{R}^N)$, then the limit u_∞ of u_m (in $H^1(\Omega)$ weak) satisfies a similar variational inequality involving A^{eff} in Ω.

The analogue of Lemma 10.5 holds in the case of an open set ω. For a measurable set ω, there is a nonlinear analogue of the Meyers theorem, using the Caccioppoli estimates,[5] and the Gehring reverse Hölder inequality.[6]

Since there is no transposition for monotone operators, one way to extend Lemma 10.2 to the monotone case is to observe that $A \in \mathcal{L}(V;V')$ is symmetric if and only if it is the gradient of a functional ($u \mapsto \frac{1}{2}(A\,u,u)$, convex if $A \geq 0$), so that one considers the case where A^n is a gradient in p,

$$A^n(x,p) = \frac{\partial W_n(x,p)}{\partial p} \text{ for all } p \in \mathbb{R}^N, \text{ a.e. in } \Omega, \tag{11.18}$$

with W_n convex in p. Using the *div–curl lemma*, it is natural to extend Theorem 11.2 to k-monotone operators, and then show that A^{eff} is a gradient by the characterization of cyclic monotonicity of Terry ROCKAFELLAR.[7]

---

[5] Renato CACCIOPPOLI, Italian mathematician, 1904–1959. He worked in Napoli (Naples), Italy.

[6] Frederick William GEHRING, American mathematician, born in 1925. He works at University of Michigan, Ann Arbor, MI.

[7] Ralph Tyrrell ROCKAFELLAR, American mathematician, born in 1935. He works at University of Washington, Seattle, WA.

Lemma 11.3. *For $k \geq 3$, if all $A^n \in \mathcal{M}on(\alpha,\beta;\Omega)$ are k-monotone, i.e.,*

$$\sum_{i=1}^{k}(A^n(x,a^i),a^i-a^{i+1}) \geq 0 \text{ for all } a^1,\ldots,a^k \in \mathbb{R}^N, \text{ a.e. in } \Omega, \tag{11.19}$$

where a^{k+1} means a^1, and if A^n H-converges to A^{eff}, then A^{eff} is k-monotone. In particular if all A^n satisfy (11.18), they are cyclically monotone (i.e., k-monotone for all $k \geq 2$), then A^{eff} is cyclically monotone, so that there exists W_{eff} such that

$$A^{\mathit{eff}}(x,p) = \frac{\partial W_{\mathit{eff}}(x,p)}{\partial p} \text{ for all } p \in \mathbb{R}^N, \text{ a.e. in } \Omega. \tag{11.20}$$

Proof. As in the proof of Theorem 11.2, after choosing ω and φ one selects $f_j \in V'$ corresponding to sequences v_j^m with $v_j^\infty(x) = \varphi(x)(x,a^j)$ in Ω, and the *div–curl lemma* shows that for all $\psi \in C_c(\omega)$, $\psi \geq 0$ in ω,

$$\begin{aligned}0 \leq \int_\Omega \psi \sum_{i=1}^{k}(A^m(x,grad(v_i^m)),grad(v_i^m)-grad(v_{i+1}^m))\,dx \\ \to \int_\omega \psi \sum_{i=1}^{k}(A^{\mathit{eff}}(x,a^i),a^i-a^{i+1})\,dx,\end{aligned} \tag{11.21}$$

and varying ψ shows that A^{eff} is k-monotone in p, a.e. $x \in \omega$.[8]

If (11.18) holds then $W_n(x,a^{i+1}) \geq W_n(x,a^i) + (A^n(x,a^i),a^{i+1}-a^i)$ a.e. in Ω since W_n is convex in its second argument, and summing in i gives (11.19). Having then shown that A^{eff} is k-monotone for all $k \geq 2$, one defines $W_{\mathit{eff}}(x,p)$ (normalized by $W_{\mathit{eff}}(x,0)=0$) by the Rockafellar formula

$$W_{\mathit{eff}}(x,p) = \sup\Big\{(A^{\mathit{eff}}(x,0),a^1) + \sum_{i=1}^{n-1}(A^{\mathit{eff}}(x,a^i),a^{i+1}-a^i) + (A^{\mathit{eff}}(x,a^n),p-a^n)\Big\}, \tag{11.22}$$

where the supremum is taken over $n \geq 2$ and $a^1,\ldots,a^n \in \mathbb{R}^N$. ☐

One can then obtain the following generalization of Lemma 10.6.

Lemma 11.4. *Assume moreover that*

$$\mathcal{A}^n(0) \text{ belongs to a compact of } H^{-1}_{loc}(\Omega) \text{ strong, and } W_n(x,0)=0 \text{ a.e. in } \Omega. \tag{11.23}$$

[8] Outside a set of measure 0, k-monotonicity holds for all $a^j \in \mathbb{Q}^N$, and since A^{eff} is Lipschitz continuous in p, k-monotonicity holds for all $a^j \in \mathbb{R}^N$.

If $v_n \rightharpoonup v_\infty$ in $H^1_{loc}(\Omega)$ weak, and $\psi \in C_c(\Omega)$ with $\psi \geq 0$ in Ω, then

$$\liminf_m \int_\Omega \psi\, W_m(x, grad(v_m))\, dx \geq \int_\Omega \psi\, W_{\mathit{eff}}(x, grad(v_\infty))\, dx. \tag{11.24}$$

For $w_\infty \in H^1_0(\Omega)$ there exists $w_m \rightharpoonup w_\infty$ in $H^1_0(\Omega)$ weak, with

$$\lim_m \int_\Omega \chi\, W_m(x, grad(w_m))\, dx = \int_\Omega \chi\, W_{\mathit{eff}}(x, grad(w_\infty))\, dx, \tag{11.25}$$

for all $\chi \in C_c(\Omega)$.

Proof. Let $u_\infty \in H^1_0(\Omega)$ be equal to v_∞ a.e. on the support of ψ, and let $f \in V'$ be such that it generates a sequence u_m converging to u_∞ in $H^1_0(\Omega)$ weak. It will be shown that (11.25) holds for $w_\infty = u_\infty$ and $w_m = u_m$. Then, assuming (11.25), one uses the convexity of W_m, which implies

$$W_m(\cdot, grad(v_m)) \geq W_m(\cdot, grad(u_m)) + (A^m(\cdot, grad(u_m)), grad(v_m) - grad(u_m)) \tag{11.26}$$

in Ω. Multiplying by ψ (chosen ≥ 0) one obtains

$$\begin{aligned}\liminf_m \int_\Omega \psi\, W_m(x, grad(v_m))\, dx &\geq \limsup_m \int_\Omega \psi\, W_m(x, grad(u_m))\, dx\\ = \int_\Omega \psi\, W_{\mathit{eff}}(x, grad(u_\infty))\, dx &= \int_\Omega \psi\, W_{\mathit{eff}}(x, grad(v_\infty))\, dx,\end{aligned} \tag{11.27}$$

since $v_\infty = u_\infty$ on the support of ψ, and the *div–curl lemma* gives

$$\int_\Omega \psi(A^m(x, grad(u_m)), grad(v_m) - grad(u_m))\, dx \to 0. \tag{11.28}$$

Let ω be open containing the support of ψ and $\overline{\omega} \subset \Omega$, and $\varphi \in C^1_c(\Omega)$ equal to 1 on ω. For $a^1, \ldots, a^k \in \mathbb{R}^N$, let $f_j \in V'$ correspond to sequences v_j^m with $v_j^\infty(x) = \varphi(x)(x, a^j)$ in Ω. Using the convexity of W_m in p, and $W_m(\cdot, 0) = 0$,

$$\begin{aligned} W_m(\cdot, grad(u_m)) &\geq (A^m(\cdot, 0), grad(v_1^m))\\ &+ \sum_{i=1}^{k-1} (A^m(\cdot, grad(v_i^m)), grad(v_{i+1}^m) - grad(v_i^m))\\ &+ (A^m(\cdot, grad(v_k^m)), grad(u_m) - grad(v_k^m)) \text{ in } \Omega. \end{aligned} \tag{11.29}$$

By (11.23), a subsequence $\mathcal{A}^{m'}(0) \to g$ in $H^{-1}_{loc}(\Omega)$ strong, so that $g = \mathcal{A}^{\mathit{eff}}(0)$, and

$$\int_\Omega \chi(A^m(x, 0), grad(v_1^m))\, dx \to \int_\Omega \chi(A^{\mathit{eff}}(x, 0), a^1)\, dx, \tag{11.30}$$

for all $\chi \in C^1_c(\omega)$. Multiplying (11.29) by ψ and using (11.30) and the *div–curl lemma* for the other terms on the right, one obtains

$$\liminf_m \int_\Omega \psi\, W_m(x, grad(u_m))\, dx \geq \int_\Omega \psi(A^{\text{eff}}(x,0), a^1)\, dx$$
$$+ \int_\Omega \psi \left[\sum_{i=1}^{k-1} (A^{\text{eff}}(x,a^i), a^{i+1} - a^i) + (A^{\text{eff}}(x,a^k), grad(u_\infty) - a^k) \right] dx. \tag{11.31}$$

If a subsequence of $W_m(\cdot, grad(u_m))$ converges to μ in $\mathcal{M}(\Omega)$ weak $\star$, then $\mu = \mu_s + h\,dx$ with a singular part μ_s which is nonnegative since W_m is convex, and (11.31) means that h satisfies

$$h(\cdot) \geq (A^{\text{eff}}(\cdot,0), a^1) + \sum_{i=1}^{k-1} (A^{\text{eff}}(\cdot,a^i), a^{i+1} - a^i) + (A^{\text{eff}}(\cdot,a^k), grad(u_\infty) - a^k), \tag{11.32}$$

a.e. in Ω. An opposite inequality follows in the same way from

$$0 = W_m(\cdot,0) \geq W_m\Big(\cdot, grad(u_m)\Big) + \Big(A^m\Big(\cdot, grad(u_m)\Big), grad(v^m_k) - grad(u_m)\Big)$$
$$+ \sum_{i=1}^{k} \Big(A^m\Big(\cdot, grad(v^m_i)\Big), grad(v^m_{i-1}) - grad(v^m_i)\Big) \text{ in } \Omega, \tag{11.33}$$

where $v^m_0 = 0$. Multiplying by $\psi \in C_c(\omega)$ with $\psi \geq 0$ and using the *div–curl lemma* for taking the limit shows that $\mu_s \leq 0$ (hence $\mu_s = 0$), and that h satisfies

$$0 \geq h(\cdot) + \Big(A^{\text{eff}}\Big(\cdot, grad(u_\infty)\Big), a^k - grad(u_\infty)\Big) + \sum_{i=1}^{k} \Big(A^{\text{eff}}(\cdot,a^i), a^{i-1} - a^i\Big) \text{ in } \Omega. \tag{11.34}$$

For x outside a set of arbitrarily small Lebesgue measure, one can take $a^i = \frac{i}{k} grad(u_\infty)(x)$ and (11.32) and (11.34) become two Riemann sums and letting k tend to ∞, one obtains $h(x) = W_{\text{eff}}\big(x, grad(u_\infty)(x)\big)$. □

One can then generalize Lemma 6.7.

Lemma 11.5. *Under the hypotheses of Lemma 11.4, if*

$$W_n(x,p) \rightharpoonup W_+(x,p) \text{ in } L^\infty(\Omega) \text{ weak } \star \text{ for all } p \in \mathbb{R}^N, \text{ a.e. } x \in \Omega$$
$$(W_n)^*(x,q) \rightharpoonup (W_-)^*(x,q) \text{ in } L^\infty(\Omega) \text{ weak } \star \text{ for all } q \in \mathbb{R}^N, \text{ a.e. } x \in \Omega \tag{11.35}$$

where W_* denotes the conjugate (convex) function of W,[9] *then one has*

$$W_-(x,p) \le W_{\mathit{eff}}(x,p) \le W_+(x,p) \text{ for all } p \in \mathbb{R}^N, \text{ a.e. } x \in \Omega. \tag{11.36}$$

Proof. For $p \in \mathbb{R}^N$ one has

$$W_n(\cdot,p) \ge W_n\big(\cdot, grad(u_n)\big) + \big(A^n\big(\cdot, grad(u_n)\big), p - grad(u_n)\big) \text{ in } \Omega, \tag{11.37}$$

and, using Lemma 11.3 and the *div–curl lemma*, one deduces

$$W_+(x,p) \ge W_{\mathit{eff}}\big(x, grad(u_\infty)\big) + \big(A^{\mathit{eff}}\big(x, grad(u_\infty)\big), p - grad(u_\infty)\big), \tag{11.38}$$

a.e. in Ω and varying p gives

$$W_+\big(x, grad(u_\infty)\big) \ge W_{\mathit{eff}}\big(x, grad(u_\infty)\big), \text{ a.e. } x \in \Omega, \tag{11.39}$$

which is the right inequality in (11.36). For $q \in \mathbb{R}^N$ one has

$$(W_n)^*(x,q) + W_n\big(x, grad(u_n)\big) \ge \big(q, grad(u_n)\big), \text{ a.e. } x \in \Omega, \tag{11.40}$$

and, using Lemma 11.3, one deduces

$$(W_-)^*(x,q) + W_{\mathit{eff}}\big(x, grad(u_\infty)\big) \ge \big(q, grad(u_\infty)\big), \text{ a.e. } x \in \Omega, \tag{11.41}$$

and varying q gives

$$(W_-)^*\big(x, grad(u_\infty)\big) \ge (W_{\mathit{eff}})^*\big(x, grad(u_\infty)\big), \text{ a.e. } x \in \Omega, \tag{11.42}$$

which is equivalent to the left inequality in (11.36). □

Additional footnotes: George MINTY.[10]

---

[9] If a proper function f (i.e., taking its values in $(-\infty, +\infty]$ but not identical to $+\infty$) is bounded below by an affine continuous function on a locally convex space E, the conjugate function f^* is defined on the dual E' by $f^*(\xi) = \sup_{e \in E}\big((\xi, e) - f(e)\big)$.

[10] George James MINTY Jr., American mathematician, 1930–1986. He worked at Indiana University, Bloomington, IN.

Chapter 12
Homogenization of Laminated Materials

I have already shown at Lemma 4.1 a one-dimensional homogenization result observed with François MURAT around 1970, but we also noticed afterward a natural generalization (a little academic too), which is a step towards the N-dimensional case of laminated material, i.e., where the coefficients only depend upon x_1, or more generally upon (x, e) for a unit vector $e \in \mathbb{R}^N$.

Lemma 12.1. *Let $\Omega = (x_-, x_+) \subset \mathbb{R}$ be a bounded open interval. For $a_n \in \mathcal{M}(\alpha, \beta; \Omega)$, b_n, c_n bounded in $L^2(\Omega)$, and d_n bounded in $L^1(\Omega)$, one assumes that u_n converges to u_∞ in $H^1_{loc}(\Omega)$ weak and satisfies*

$$-\frac{d}{dx}\Big(a_n \frac{du_n}{dx} + b_n u_n\Big) + c_n \frac{du_n}{dx} + d_n u_n \to f \text{ in } H^{-1}_{loc}(\Omega) \text{ strong.} \tag{12.1}$$

Assume that for a subsequence one has

$$\begin{aligned} &\frac{1}{a_m} \rightharpoonup \frac{1}{a_{\mathit{eff}}} \text{ in } L^\infty(\Omega) \text{ weak } \star,\\ &\frac{b_m}{a_m} \rightharpoonup \frac{b_{\mathit{eff}}}{a_{\mathit{eff}}} \text{ in } L^2(\Omega) \text{ weak},\\ &\frac{c_m}{a_m} \rightharpoonup \frac{c_{\mathit{eff}}}{a_{\mathit{eff}}} \text{ in } L^2(\Omega) \text{ weak},\\ &d_m - \frac{b_m c_m}{a_m} \rightharpoonup d_{\mathit{eff}} - \frac{b_{\mathit{eff}} c_{\mathit{eff}}}{a_{\mathit{eff}}} \text{ in } \mathcal{M}(\Omega) \text{ weak } \star. \end{aligned} \tag{12.2}$$

Then one has

$$a_n \frac{du_n}{dx} + b_n u_n \to a_{\mathit{eff}} \frac{du_\infty}{dx} + b_{\mathit{eff}} u_\infty \text{ in } L^2_{loc}(\Omega) \text{ strong}, \tag{12.3}$$

$$c_n \frac{du_n}{dx} + d_n u_n \rightharpoonup c_{\mathit{eff}} \frac{du_\infty}{dx} + d_{\mathit{eff}} u_\infty \text{ in } L^1(\Omega) \text{ weak } \star \tag{12.4}$$

$$-\frac{d}{dx}\Big(a_{\mathit{eff}} \frac{du_\infty}{dx} + b_{\mathit{eff}} u_\infty\Big) + c_{\mathit{eff}} \frac{du_\infty}{dx} + d_{\mathit{eff}} u_\infty = f \text{ in } \Omega. \tag{12.5}$$

L. Tartar, *The General Theory of Homogenization*, Lecture Notes of the Unione Matematica Italiana 7, DOI 10.1007/978-3-642-05195-1_12,

Proof. Because u_n converges to u_∞ in $H^1_{loc}(\Omega)$ weak, $\frac{du_n}{dx}$ converges to $\frac{du_\infty}{dx}$ in $L^2_{loc}(\Omega)$ weak, and u_n converges to u_∞ in $C(\Omega)$ strong,[1] since $\Omega \subset \mathbb{R}$. Because $c_n\frac{du_n}{dx} + d_n u_n$ is bounded in $L^1_{loc}(\Omega)$, it stays in a compact of $H^{-1}_{loc}(\Omega)$ strong, so $\frac{d}{dx}\big(a_n\frac{du_n}{dx} + b_n u_n\big)$ stays in a compact of $H^{-1}_{loc}(\Omega)$ strong, and $a_n\frac{du_n}{dx} + b_n u_n$ stays in a compact of $L^2_{loc}(\Omega)$ strong, since $\Omega \subset \mathbb{R}$. From u_m one extracts a subsequence u_p such that

$$a_p\frac{du_p}{dx} + b_p u_p \to g \text{ in } L^2_{loc}(\Omega) \text{ strong,} \tag{12.6}$$

for some $g \in L^2_{loc}(\Omega)$. Multiplying (12.6) by $\frac{1}{a_p}$ and using (12.2) one obtains

$$\frac{du_{\mathit{eff}}}{dx} + \frac{b_{\mathit{eff}}}{a_{\mathit{eff}}}u_\infty = \frac{1}{a_{\mathit{eff}}}g \text{ in } \Omega, \tag{12.7}$$

so that g is independent of the subsequence, and (12.3) holds. Then,

$$c_p\frac{du_p}{dx} + d_p u_p = \frac{c_p}{a_p}\left(a_p\frac{du_p}{dx} + b_p u_p\right) + \left(d_p - \frac{b_p c_p}{a_p}\right)u_p, \tag{12.8}$$

so that using (12.2) one obtains

$$c_p\frac{du_p}{dx} + d_p u_p \rightharpoonup \frac{c_{\mathit{eff}}}{a_{\mathit{eff}}}g + \left(d_{\mathit{eff}} - \frac{b_{\mathit{eff}}c_{\mathit{eff}}}{a_{\mathit{eff}}}\right)u_\infty \text{ in } \mathcal{M}(\Omega) \text{ weak } \star. \tag{12.9}$$

and with the value of g it means (12.4), which with (12.3) implies (12.5). □

I have already shown at Lemma 5.1 a two-dimensional result of François MURAT, which he generalized as Lemma 5.2, whose proof I postponed, in order to explain the general principle following Lemma 12.2, which gives Lemma 5.3 for a second-order scalar equation.

Lemma 5.1 has a simple physical interpretation for the model of *electricity* described in Chap. 8: it is the rule that resistances in series are added, but for resistances in parallel one adds their inverses, the conductivities. Indeed, if one imposes a macroscopic electric field E^∞ parallel to e_1 the current flows everywhere in the direction e_1, and creates a situation of resistances in series, while if E^∞ is parallel to e_2 the current flows everywhere in the direction e_2, and creates a situation of resistances in parallel. This example shows that for a general mixture one cannot assert easily how the current will flow, creating resistances in series along the current lines, and putting them in parallel after; it is useful then to know bounds on effective properties, like Lemma 6.7.

---

[1] $C(\Omega)$ is a Fréchet space, and the strong convergence means the uniform convergence on every compact of Ω.

One should notice that the model of *electrostatics* does not provide the same physical intuition, and that talking about primal and dual minimization problems is about mathematics but not about physics, since nature creates the solution by an (hyperbolic) evolution process which has nothing to do with minimization.

An equation may then have properties which are more or less intuitive depending upon one's training,[2] and using physical examples in partial differential equations is a pedagogical approach similar to using drawings in geometry, since drawings are not proofs but are often sufficient hints for trained people to understand how to write a proof,[3] if they need to. It does not give the physical interpretations more value than hints, but it assumes that the student heard about physics, and is eager to understand it in a more mathematical way. Unfortunately, mathematics and physics are often taught now in too ideological a manner, and few students plan to acquire a vast knowledge, so most of them do not understand these examples which could be useful if they learned more. One rarely hears the motto of Hugo of Saint Victor "Learn everything, and you will see afterward that nothing is useless," but before learning something which does not seem related to what one already knows, one should take the time to understand fully what one learned.

When putting resistances in series or in parallel, physicists do not consider the case of anisotropic materials, but they might guess the formulas of Lemma 5.2, or give the effective properties of laminated materials in other physical theories. The job of a mathematician goes further than proving a few particular results, and it is to discover the simplifying and unifying structures behind the results. All effective properties of laminated materials, for linear systems of partial differential equations, having a physical interpretation or not, can be obtained by repeated application of Lemma 12.2.

Lemma 12.2. *If* $D^n \rightharpoonup D^\infty$ *in* $L^2(\Omega; \mathbb{R}^N)$ *weak, with* $div(D^n)$ *staying in a compact of* $H^{-1}_{loc}(\Omega)$ *strong, if* f_n only depends upon x_1 *and* $f_n \rightharpoonup f_\infty$ *in* $L^2(\Omega)$ *weak, then*

$$D_1^n(x) f_n(x_1) \rightharpoonup D_1^\infty(x) f_\infty(x_1) \text{ in } L^1(\Omega) \text{ weak } \star. \tag{12.10}$$

---

[2] The equation $u_t - (u^m)_{xx} = 0$ is used for $m > 1$ as a model of flow in a porous medium, $u \geq 0$ denoting a density of mass; the total mass $m = \int_\mathbb{R} u(x,t)\,dx$ is independent of t, as well as $m\,x_* = \int_\mathbb{R} x\,u(x,t)\,dx$, where x_* is the centre of mass of the distribution of mass. However, $m = 1$ gives the heat equation, and in the interpretation that u is a temperature (near equilibrium), no physical meaning is attached to $\int_\mathbb{R} u(x,t)\,dx$ and $\int_\mathbb{R} x\,u(x,t)\,dx$, which are independent of t.

[3] I was told that Jean DIEUDONNÉ once lectured on an abstract question of geometry, and at some point he did not remember the next step in the proof. As a member of Bourbaki, he opposed using drawings, but he went near the blackboard, hiding from many what he was doing (since he was tall and strong), and he drew a picture; after figuring out how to proceed, he erased it and finished the proof. This behavior should be avoided for pedagogical reasons: if drawings are useful hints for teachers, why not explain how to use them to the students.

It follows from the *div–curl lemma*, but since $E^n = (f_n, 0, \ldots, 0)$ is a gradient, a simple integration by parts is needed, and there is a corresponding statement in L^p for $1 \leq p \leq \infty$.

Definition 12.3. *If* $\varphi_n \in L^2(\Omega)$ *converges to* φ_∞ *in* $L^2(\Omega)$ *weak, one says that* φ_n does not oscillate in (x, e), *for a nonzero vector* $e \in \mathbb{R}^N$, *if for all* f_n depending only upon (x, e) *converging to* f_∞ *in* $L^2(\Omega)$ *weak, one has*

$$\varphi_n(x) f_n\big((x,e)\big) \rightharpoonup \varphi_\infty(x) f_\infty\big((x,e)\big) \text{ in } L^1(\Omega) \text{ weak } \star. \tag{12.11}$$

If $D^n \rightharpoonup D^\infty$ in $L^2(\Omega; \mathbb{R}^N)$ weak, and $div(D^n)$ stays in a compact of $H^{-1}_{loc}(\Omega)$ strong, the *div–curl lemma* implies that D^n_1 does not oscillate in x_1, as stated in Lemma 12.2, but more generally, for every nonzero vector $e \in \mathbb{R}^N$,

$$\sum_{i=1}^N e_j D^n_j \text{ does not oscillate in } (x, e). \tag{12.12}$$

Another sufficient condition uses *H-measures*: if $\varphi_n \rightharpoonup \varphi_\infty$ in $L^2(\Omega)$ weak and, for any subsequence φ_m defining an H-measure $\mu \in \mathcal{M}(\Omega \times \mathbb{S}^{N-1})$,[4] and μ does not charge $\Omega \times \frac{e}{|e|}$ and $\Omega \times \frac{-e}{|e|}$, then φ_n *does not oscillate in* (x, e).[5]

Proof of Lemma 5.2 and Lemma 5.3. One applies Lemma 12.2 not only to D^n but also to $E^n = grad(u_n)$, using

$$\frac{\partial E^n_j}{\partial x_1} - \frac{\partial E^n_1}{\partial x_j} \in \text{compact of } H^{-1}_{loc}(\Omega) \text{ strong}, j = 2, \ldots, N, \tag{12.13}$$

so that E^n_j does not oscillate in x_1 for $j = 2, \ldots, N$. Instead of using E^n and deducing $D^n = A^n E^n$ (or using D^n and deducing $E^n = (A^n)^{-1} D^n$), one uses a vector made of those components of E^n and D^n which do not oscillate in x_1, and deduce from it the other components of E^n and D^n: one defines

$$\text{a } \textit{good} \text{ vector } G^n : G^n_1 = D^n_1, G^n_j = E^n_j \text{ for } j = 2, \ldots, N, \tag{12.14}$$
$$\text{an } \textit{oscillating} \text{ vector } O^n : O^n_1 = E^n_1, G^n_j = D^n_j \text{ for } j = 2, \ldots, N, \tag{12.15}$$

and one notices that

$$D^n = A^n E^n \text{ is equivalent to } O^n = B^n G^n, \text{ with } B^n = \Phi_{e_1}(A^n), \tag{12.16}$$

---

[4] In my initial definition of H-measures, I associated the H-measure to $\varphi_m - \varphi_\infty$, but it is useful to talk about the H-measure associated to φ_m, to mean that the weak limit φ_∞ exists, and that one considers the H-measure associated to $\varphi_m - \varphi_\infty$.

[5] Using another subsequence for the H-measure π of (φ_p, f_p) to exist, one has $\pi_{1,1} = \mu$, and $\pi_{2,2}$ is supported in $\Omega \times \left\{\frac{-e}{|e|}, \frac{+e}{|e|}\right\}$, so that $\pi_{1,2} = 0$, implying (12.11), because π is Hermitian nonnegative.

and, an easy computation using only $A^n_{1,1} \geq \alpha > 0$ gives

$$\begin{aligned} B^n_{1,1} &= \frac{1}{A^n_{1,1}}, \text{ so that } B^n_{1,1} \geq \alpha' > 0 \\ B^n_{1,j} &= -\frac{A^n_{1,j}}{A^n_{1,1}} \text{ for } j = 2, \ldots, N, \\ B^n_{i,1} &= \frac{A^n_{i,1}}{A^n_{1,1}} \text{ for } i = 2, \ldots, N, \\ B^n_{i,j} &= A^n_{i,j} - \frac{A^n_{i,1} A^n_{1,j}}{A^n_{1,1}} \text{ for } i, j = 2, \ldots, N, \end{aligned} \tag{12.17}$$

and Φ_{e_1} is involutive (i.e., is its own inverse). Because $O^n_i = \sum_j B^n_{i,j} G^n_j$, each $B^n_{i,j}$ only depends upon x_1, each G^n_j does not oscillate in x_1, the weak limit of each term $B^n_{i,j} G^n_j$ is the product of weak limits, and one deduces that

$$O^\infty = B^\infty G^\infty, \text{ where } B^n \rightharpoonup B^\infty \text{ in } L^\infty\big(\Omega; \mathcal{L}(\mathbb{R}^N; \mathbb{R}^N)\big) \text{ weak } \star. \tag{12.18}$$

Because $O^\infty = B^\infty G^\infty$ is equivalent to $D^\infty = \Phi_{e_1}(B^\infty) E^\infty$, one deduces that

$$B^\infty = \Phi_{e_1}(A^{\mathit{eff}}), \text{ i.e., } A^{\mathit{eff}} = \Phi_{e_1}(B^\infty), \text{ a.e. in } \Omega, \tag{12.19}$$

and (12.17) for B^∞ corresponds to the formulas (5.6)–(5.9) for A^{eff}. □

My approach using Lemma 12.2 can be easily followed for any elliptic system, and a much weaker notion than ellipticity is necessary, and it explains which nonlinear operations one must perform before taking weak limits. As Lemma 5.3 is valid under the condition $(A^n e_1, e_1) \geq \alpha > 0$ a.e. in Ω for all n (or $(A^n e_1, e_1) \leq -\alpha < 0$ by changing all signs), it even applies to the one-dimensional *wave equation*, which is hyperbolic if $a_n, \varrho_n \geq \alpha > 0$

$$\frac{\partial}{\partial t}\left(\varrho_n(x) \frac{\partial u_n}{\partial t}\right) - \frac{\partial}{\partial x}\left(a_n(x) \frac{\partial u_n}{\partial x}\right) = f. \tag{12.20}$$

If $\varrho_n \rightharpoonup \varrho_\infty$ and $\frac{1}{a_n} \rightharpoonup \frac{1}{a_{\mathit{eff}}}$ in $L^\infty(\Omega)$ weak $\star$, then the effective equation has the form (12.20) with ϱ_n and a_n replaced by ϱ_∞ and a_{eff}.[6]

The main reason for using the Lax–Milgram lemma in Lemma 6.2 is to construct enough sequences $E^n = grad(u_n)$ converging weakly, to enough vectors E^∞, in ω with $\overline{\omega} \subset \Omega$, for example constant vectors, and with $D^n = (A^n)^T E^n$ such that $div(D^n)$ stays in a compact of $H^{-1}_{loc}(\Omega)$. One does not need the Lax–Milgram lemma in the laminated case since there are explicit sequences to use, by taking G a constant vector, and writing $O^n = B^n(x_1) G$, which gives a vector E^n which is a gradient, and a vector D^n which is divergence free.

[6] This was observed in the early 1970s by Alain BAMBERGER.

In the framework of differential forms of Chap. 9, one can restrict differential forms to manifolds, and here it is natural to consider the family of hyperplanes $x_1 = constant$: this selects precisely the components of G^n. As

$$(E^n, D^n) = (O^n, G^n) = (B^n G^n, G^n), \tag{12.21}$$

the symmetric part of the tensor B^n also appears.

In the spring of 1975, I saw a preprint by McConnell, who computed the formulas for laminated materials in (linearized) elasticity, and the algebraic computations were more intricate than those for Lemma 5.2.

Instead of using the usual (linearized) strain–stress relation,

$$\sigma^n_{i,j} = \sum_{k,\ell=1}^{N} C^n_{i,j,k,\ell}(x_1)\varepsilon^n_{k,\ell} \text{ for } i,j = 1,\ldots,N, \tag{12.22}$$

expressing the symmetric Cauchy stress tensor σ^n in terms of the (linearized) strain tensor ε^n,[7] one should use the list of components which do not oscillate in x_1. That $\sigma^n_{i,1}$ does not oscillate in x_1 follows from the equilibrium equations

$$\sum_{j=1}^{N} \frac{\partial \sigma^n_{i,j}}{\partial x_j} = f_i \text{ in } \Omega \text{ for } i = 1,\ldots,N, \tag{12.23}$$

and $\sigma^n_{1,i}$ does not oscillate in x_1 either, by symmetry of σ^n. Then,[8] $\frac{\partial u^n_i}{\partial x_j}$ does not oscillate in x_1 for $j \geq 2$, so that $\varepsilon^n_{i,j}$ does not oscillate in x_1 for $i,j \geq 2$ One defines the *good* tensor G^n, and the *oscillating* tensor O^n by

$$G^n_{i,j} = \sigma^n_{i,j} \text{ if } i \text{ or } j = 1, G^n_{i,j} = \varepsilon^n_{i,j} \text{ if } i \text{ and } j \geq 2, \tag{12.24}$$

$$O^n_{i,j} = \varepsilon^n_{i,j} \text{ if } i \text{ or } j = 1, O^n_{i,j} = \sigma^n_{i,j} \text{ if } i \text{ and } j \geq 2, \tag{12.25}$$

and one replaces the relation (12.22) by the equivalent relation

$$O^n = K^n G^n, \text{ i.e., } O^n_{i,j} = \sum_{k,\ell=1}^{N} K^n_{i,j,k,\ell}(x_1) G^n_{k,\ell}, \tag{12.26}$$

defining a nonlinear mapping Ψ_{e_1} such that

$$K^n = \Psi_{e_1}(C^n). \tag{12.27}$$

[7] Defined from the displacement u^n by $\varepsilon^n_{i,j} = \frac{1}{2}\left(\frac{\partial u^n_i}{\partial x_j} + \frac{\partial u^n_j}{\partial x_i}\right)$ for $i,j = 1,\ldots,N$.

[8] By the Korn inequality, all $\varepsilon^n_{i,j}$ in $L^2(\Omega)$ imply all $\frac{\partial u^n_i}{\partial x_j}$ in $L^2_{loc}(\Omega)$.

Writing $X_{sym} = \mathcal{L}_{sym}(\mathbb{R}^N;\mathbb{R}^N)$, Lemma 12.2 implies then

$$O^\infty = K^\infty G^\infty, \text{ where } K^n \rightharpoonup K^\infty \text{ in } L^\infty\big(\Omega;\mathcal{L}(X_{sym};X_{sym})\big) \text{ weak } \star, \tag{12.28}$$

and because $O^\infty = K^\infty G^\infty$ is equivalent to $\sigma^\infty = \Psi_{e_1}(K^\infty)\varepsilon^\infty$, one deduces

$$K^\infty = \Psi_{e_1}(C^{eff}), \text{ i.e., } C^{eff} = \Psi_{e_1}(K^\infty), \text{ a.e. in } \Omega, \tag{12.29}$$

which explains the computations of McConnell, if $K^\infty \in Range(\Psi_{e_1})$.

Applying the homogenization techniques of Chap. 6 to linearized elasticity requires using the Lax–Milgram lemma, i.e., assuming that C^n satisfies a uniform *very strong ellipticity condition*, that there exists $\alpha > 0$ such that

$$\sum_{i,j,k,l=1}^N C^n_{i,j,k,\ell} M_{i,j} M_{k,\ell} \ge \alpha \sum_{i,j=1}^N M^2_{i,j} \text{ for all } M \in X_{sym}. \tag{12.30}$$

In the laminated case, assuming that C^n only depends upon x_1, a first condition to impose on C^n is to be able to invert the relation (12.22) and write it as (12.26) with $K^n = \Psi_{e_1}(C^n)$ bounded, so that a subsequence converges in L^∞ weak $\star$ to K^∞; a second condition is to be able to solve $K^\infty = \Psi_{e_1}(C^{eff})$.

For $\xi \in \mathbb{S}^{N-1}$, the *acoustic tensor* $A^n(\xi)$ is defined by

$$A^n(\xi)_{i,k} = \sum_{j,\ell=1}^N C^n_{i,j,k,\ell}\xi_j\xi_\ell, \text{ for } \xi \in \mathbb{S}^{N-1}, i,k = 1,\ldots,N, \tag{12.31}$$

and one finds that for defining Ψ_{e_1}, one must invert $A^n(e_1)$, because

$$\sigma^n_{i,1} = \sum_{k,\ell=1}^N C^n_{i,1,k,\ell}\varepsilon^n_{k,\ell} = \sum_k A^n(e_1)_{i,k}\varepsilon^n_{k,1} + \tau^n_i, \tag{12.32}$$

where τ^n_i only uses the $\varepsilon^n_{k,\ell}$ for $k,\ell \ge 2$. The *ellipticity condition* for C^n is that $A^n(\xi)$ has an inverse for all $\xi \in \mathbb{S}^{N-1}$, but even adding that the inverse of $A^n(e_1)$ is bounded is not enough, since it is the L^∞ weak $\star$ limit of $\big(A^n(e_1)\big)^{-1}$ which appears, and one needs the weak $\star$ limit to have an inverse; a sufficient condition for that is that there exists $\alpha > 0$ such that for all n

$$(A^n(e_1)\lambda,\lambda) \ge \alpha\,|\lambda|^2 \text{ for all } \lambda \in \mathbb{R}^N, \text{ a.e. in } \Omega. \tag{12.33}$$

The *strong ellipticity condition*, or strict Legendre–Hadamard condition, is precisely that $A^n(\xi) \ge \alpha\, I$ for all $\xi \in \mathbb{S}^{N-1}$.[9]

[9] In the range where the linearization of elasticity is valid, i.e., if all $\frac{\partial u_i}{\partial x_j}$ are small in $L^\infty(\Omega)$, the strict Legendre–Hadamard condition ensures that the evolution problem

I conclude with Theorem 12.4, about correctors for laminated materials, the general question of correctors being the subject of Chaps. 13 and 14.

Theorem 12.4. *If $A^n \in \mathcal{M}(\alpha,\beta;\Omega)$ H-converges to A^{eff}, and only depends upon x_1, if u_n converges to u_∞ in $H^1_{loc}(\Omega)$ weak, with $div(A^n\,grad(u_n))$ staying in a compact of $H^{-1}_{loc}(\Omega)$ strong, then*

$$\big(A^n\,grad(u_n)\big)_1 \rightarrow \big(A^{\mathit{eff}}\,grad(u_\infty)\big)_1 \text{ in } L^2_{loc}(\Omega) \text{ strong,} \tag{12.34}$$

$$\frac{\partial u_n}{\partial x_i} \rightarrow \frac{\partial u_\infty}{\partial x_i} \text{ in } L^2_{loc}(\Omega) \text{ strong, for } i=2,\ldots,N. \tag{12.35}$$

$$grad(u_n) - P^n\,grad(u_\infty) \rightarrow 0 \text{ in } L^2_{loc}(\Omega;\mathbb{R}^N) \text{ strong.} \tag{12.36}$$

with $P^n \in L^\infty\big(\Omega;\mathcal{L}(\mathbb{R}^N;\mathbb{R}^N)\big)$ defined by

$$\begin{aligned} P^n_{1,1} &= \frac{A^{\mathit{eff}}_{1,1}}{A^n_{1,1}},\\ P^n_{1,j} &= \frac{A^{\mathit{eff}}_{1,j}}{A^{\mathit{eff}}_{1,1}} - \frac{A^n_{1,j}}{A^n_{1,1}} \text{ for } j=2,\ldots,N,\\ P^n_{i,j} &= \delta_{i,j} \text{ for } i=2,\ldots,N, \text{ and } j=1,\ldots,N, \end{aligned} \tag{12.37}$$

Proof. Writing $E^n = grad(u_n)$ and $D^n = A^n\,grad(u_n)$, one uses the vector G^n of (12.14), then (12.34) and (12.35) mean that G^n converges to G^∞ in $L^2_{loc}(\Omega;\mathbb{R}^N)$ strong. In proving this statement, one notices that

$$\big(B^n(G^n-G^\infty), G^n-G^\infty\big) \text{ converges to 0 in } \mathcal{M}(\Omega) \text{ weak } \star. \tag{12.38}$$

Indeed, (12.14) implies $(B^nG^n,G^n) = (D^n,E^n)$ which converges in $L^1(\Omega)$ weak $\star$ to $(D^\infty,E^\infty) = (B^\infty G^\infty, G^\infty)$ by the *div–curl lemma*, and since G^n does not oscillate in x_1 and B^n only depends upon x_1 and converges to B^∞ in $L^\infty\big(\Omega;\mathcal{L}(\mathbb{R}^N;\mathbb{R}^N)\big)$ weak $\star$, both (B^nG^n,G^∞) and (B^nG^∞,G^n) converge to $(B^\infty G^\infty,G^\infty)$ in $L^1_{loc}(\Omega)$ weak. Then, there exists $\gamma>0$ such that

$$(B^n\lambda,\lambda) \geq \gamma|\lambda|^2 \text{ for all } \lambda \in \mathbb{R}^N, \tag{12.39}$$

and one may take $\gamma = \frac{\alpha}{\beta^2+1}$, since $(B^nG^n,G^n) = (A^nE^n,E^n) \geq \alpha|E^n|^2$ and $|G^n|^2 \leq |D^n|^2 + |E^n|^2 \leq (\beta^2+1)|E^n|^2$. Finally, (12.37) follows from writing

$$\frac{\partial u_n}{\partial x_1} = \frac{1}{A^n_{1,1}}\Big(\big(A^n\,grad(u_n)\big)_1 - \sum_{j=2}^N A^n_{1,j}\frac{\partial u_n}{\partial x_j}\Big), \tag{12.40}$$

---

is hyperbolic with finite and nonzero phase velocities, but it is not clear if the finite propagation speed property holds.

and using the fact that $\frac{1}{A^n_{1,1}}$ is uniformly bounded by $\frac{1}{\alpha}$. □

The strong convergence result of Theorem 12.4 is not true without the ellipticity condition: let A be constant with $A_{1,1} > 0$ and $(A\,\xi,\xi) = 0$ for some $\xi \neq 0$, $u_n = \frac{1}{n}\sin\big(n\,(\xi,x)\big)$ satisfies $div\big(A\,grad(u_n)\big) = 0$ and G^n converges to 0 in $L^2(\Omega;\mathbb{R}^N)$ weak, but not in $L^2_{loc}(\Omega;\mathbb{R}^N)$ strong.

Additional footnotes: BOURBAKI,[10] Bourbaki,[11] Jean DIEUDONNÉ,[12] KORN.[13]

[10] Charles Denis Sauter BOURBAKI, French general, 1816–1897; of Greek ancestry, he declined an offer of the throne of Greece in 1862.

[11] Nicolas Bourbaki is the pseudonym of a group of mathematicians, mostly French; those who chose the name certainly knew about a French general named BOURBAKI.

[12] Jean Alexandre Eugène DIEUDONNÉ, French mathematician, 1906–1992. He worked in Paris and Nice, France. There is a Laboratoire Jean Alexandre Dieudonné at Université de Nice–Sophia Antipolis, Nice, France.

[13] Arthur KORN, German-born mathematician, 1870–1945. He worked in München (Munich), and in Berlin, Germany, and at the Stevens Institute of Technology, Hoboken, NJ.

Chapter 13
Correctors in Linear Homogenization

I understood the necessity of defining *correctors* from a remark of Ivo BABUŠKA, when I first met him in May 1975, at a conference that he organized at UMD, College Park, MD. He told me that in elasticity,[1] it is not the average stress which is important but the maximum stress, since plastic behaviour or cracks may start at some points where the boundary of the elastic domain is reached, while the average stress is still much below the critical level where non-elastic behaviour occurs.[2] One then needs to study amplifying factors, for computing local stresses from an average stress.

Ivo BABUŠKA thought of *correctors* in the periodic case, in relation with the asymptotic expansions which Évariste SANCHEZ-PALENCIA used before, but I considered the periodic setting too special, and in the fall of 1975, I then developed a theory of *correctors* for the general framework of homogenization developed with François MURAT (not yet called H-convergence).

I proved Theorem 12.4 for laminated materials later, and it is for pedagogical reasons that I described it first; in the general framework, one cannot prove a result as strong, and the basic result is the following Theorem 13.1.

Theorem 13.1. *If $A^n \in \mathcal{M}(\alpha,\beta;\Omega)$ H-converges to A^{eff}, then there is a subsequence A^m and an associated sequence P^m of* correctors *such that*

[1] Ivo BABUŠKA probably thought about linearized elasticity, since he did not warn me about something that I found later, that it is not wise to use linearized elasticity for homogenization, since the multiplication of interfaces is often incompatible with the small strains necessary for using a linearization. This point actually suggests that alloys may not behave elastically, but a reason why Ivo BABUŠKA did not mention this point may be that he thought about periodic engineering designs, where ε is not small, having faith that good engineers only use designs which do not produce large displacements or large deformations.

[2] In the fall of 1975, I heard that for cyclic loadings in temperature, which induce large stresses, it happens that after a crack appears the stresses generated are not as high, because of the crack opening and closing periodically. Of course, there is then the danger that a crack may propagate, and become a threat to security.

L. Tartar, *The General Theory of Homogenization*, Lecture Notes of the Unione Matematica Italiana 7, DOI 10.1007/978-3-642-05195-1_13,

$$\begin{aligned} &P^m \rightharpoonup I \textit{ in } L^2\big(\Omega; \mathcal{L}(\mathbb{R}^N; \mathbb{R}^N)\big) \textit{ weak},\\ &A^m P^m \rightharpoonup A^{\mathit{eff}} \textit{ in } L^2\big(\Omega; \mathcal{L}(\mathbb{R}^N; \mathbb{R}^N)\big) \textit{ weak}, \end{aligned} \tag{13.1}$$

$$\begin{aligned} &curl(P^m\lambda) = 0 \textit{ in } \Omega \textit{ for all } \lambda \in \mathbb{R}^N,\\ &div(A^m P^m\lambda) \textit{ stays in a compact of } H^{-1}_{loc}(\Omega) \textit{ strong, for all } \lambda \in \mathbb{R}^N. \end{aligned} \tag{13.2}$$

For any sequence $u_m \in H^1_{loc}(\Omega)$ *satisfying*

$$\begin{aligned} &u_m \rightharpoonup u_\infty \textit{ in } H^1_{loc}(\Omega) \textit{ weak},\\ &div\big(A^m\, grad(u_m)\big) \in \textit{ compact of } H^{-1}_{loc}(\Omega) \textit{ strong}, \end{aligned} \tag{13.3}$$

one has

$$grad(u_m) - P^m grad(u_\infty) \to 0 \textit{ in } L^1_{loc}(\Omega; \mathbb{R}^N) \textit{ strong}. \tag{13.4}$$

Proof. For an open set Ω' of $\mathbb{R}^N$ containing $\overline{\Omega}$, one extends A^n by αI in $\Omega' \setminus \overline{\Omega}$ and one extracts a subsequence A^m which H-converges to a limit on Ω', extending A^{eff} (but still called A^{eff}), and then one chooses functions

$$\varphi_i \in H^1_0(\Omega'), grad(\varphi_i) = e_i \text{ on } \Omega, \text{ for } i = 1, \ldots, N, \tag{13.5}$$

and one defines the sequence P^m by

$$P^m e_i = grad(v_i^m) \text{ in } \Omega, \text{ for } i = 1, \ldots, N, \tag{13.6}$$

where the sequences v_i^m are defined for $i = 1, \ldots, N$ by

$$v_i^m \in H^1_0(\Omega'), div\big(A^m grad(v_i^m) - A^{\mathit{eff}} grad(\varphi_i)\big) = 0 \text{ in } \Omega'. \tag{13.7}$$

By this construction $v_i^m \rightharpoonup \varphi_i$ in $H^1_0(\Omega')$ weak, so that $grad(v_i^m) \rightharpoonup grad(\varphi_i)$ and $A^m grad(v_i^m) \rightharpoonup A^{\mathit{eff}} grad(\varphi_i)$ in $L^2(\Omega'; \mathbb{R}^N)$ weak; by restriction to Ω, $P^m e_i \rightharpoonup e_i$ and $A^m P^m e_i \rightharpoonup A^{\mathit{eff}} e_i$ in $L^2(\Omega; \mathbb{R}^N)$ weak, i.e., (13.1) holds.[3] Then, $curl(P^m e_i) = 0$ in Ω and $div(A^m P^m e_i)$ is a fixed element of $H^{-1}(\Omega)$, for $i = 1, \ldots, N$, so that (13.2) holds.

Since P^n is bounded in $L^2\big(\Omega; \mathcal{L}(\mathbb{R}^N; \mathbb{R}^N)\big)$, $P^n grad(u_\infty)$ is bounded in $L^1(\Omega; \mathbb{R}^N)$, and this can be improved by using the Meyers theorem, or by assuming a better integrability property for $grad(u_\infty)$. Choosing $g \in C(\Omega; \mathbb{R}^N)$ and $\varphi \in C_c(\Omega)$, one computes the limit of

[3] The construction has P^m satisfying a more precise condition than (13.1), but it is useful to impose only (13.1) since one may prefer different definitions for P^n: in the laminated case, it is more natural to take P^n as in Theorem 12.4, depending only upon x_1, and in the periodic case, it is more natural to take P^n periodic.

$$X_m = \int_\Omega \varphi \big(A^m(grad(u_m) - P^m g), grad(u_m) - P^m g\big)\,dx. \tag{13.8}$$

Writing $g = \sum_{k=1}^N g_k e_k$, one expands the integrand in (13.8) and by (13.2) the *div–curl lemma* applies to each term, so that $\big(A^m grad(u_m), grad(u_m)\big)$ converges to $\big(A^{\mathit{eff}} grad(u_\infty), grad(u_\infty)\big)$, $(A^m grad(u_m), P^m e_\ell)$ converges to $(A^{\mathit{eff}} grad(u_\infty), e_\ell)$, $\big(A^m P^m e_k, grad(u_m)\big)$ converges to $\big(A^{\mathit{eff}} e_k, grad(u_\infty)\big)$ and $(A^m P^m e_k, P^m e_\ell)$ converges to $(A^{\mathit{eff}} e_k, e_\ell)$, all these convergences being in $\mathcal{M}(\Omega)$ weak $\star$. Since φ and each $\varphi\, g_k$ belong to $C_c(\Omega)$, one deduces that

$$X_m \to X_\infty = \int_\Omega \varphi \big(A^{\mathit{eff}}(grad(u_\infty) - g), grad(u_\infty) - g\big)\,dx. \tag{13.9}$$

If $u_\infty \in C^1(\Omega)$, one can take $g = grad(u_\infty)$, so that $X_m \to 0$; by taking $0 \le \varphi \le 1$ and $\varphi = 1$ on a compact K of Ω, one deduces that $grad(u_m) - P^m grad(u_\infty) \to 0$ in $L^2(K;\mathbb{R}^N)$ strong for every compact of Ω.

If $u_\infty \in H^1(\Omega)$, one approaches $grad(u_\infty)$ by $g \in C(\Omega;\mathbb{R}^N)$, so that

$$||grad(u_\infty) - g||_{L^2(\Omega;\mathbb{R}^N)} \le \varepsilon \tag{13.10}$$

and $X_\infty \le \beta \int_\Omega |grad(u_\infty) - g|^2\,dx \le \beta\,\varepsilon^2$. By (13.8) and (13.9) one has

$$\limsup_m \int_K \alpha\,|grad(u_m) - P^m g|^2\,dx \le \beta\,\varepsilon^2, \tag{13.11}$$

from which one deduces that

$$\limsup_m \int_K |grad(u_m) - P^m g|\,dx \le \varepsilon\,\frac{\sqrt{\beta\, meas(K)}}{\sqrt{\alpha}}. \tag{13.12}$$

If C is a bound for the norm of P^m in $L^2\big(\Omega;\mathcal{L}(\mathbb{R}^N;\mathbb{R}^N)\big)$, (13.10) implies

$$\limsup_m \int_K |grad(u_m) - P^m grad(u_\infty)|\,dx \le \varepsilon\,\frac{\sqrt{\beta\, meas(K)}}{\sqrt{\alpha}} + C\,\varepsilon, \tag{13.13}$$

so that $grad(u_m) - P^m grad(u_\infty) \to 0$ in $L^1(K;\mathbb{R}^N)$ strong, i.e., (13.4). □

One can prove that $grad(u_m) - P^m grad(u_\infty)$ converges to 0 in $L^p_{loc}(\Omega;\mathbb{R}^N)$ strong for some $p > 1$ if one uses $grad(u_\infty) \in L^r(\Omega;\mathbb{R}^N)$ for some $r \ge 2$, and if P^m is bounded in $L^q_{loc}\big(\Omega;\mathcal{L}(\mathbb{R}^N,\mathbb{R}^N)\big)$ for some $q > 2$, using the Meyers theorem, or directly as in the laminated case (where $p = \infty$). One may take $g \in L^s(\Omega;\mathbb{R}^N)$ in (13.8) and (13.9) with $s = \frac{2q}{q-2}$, and if g is near $grad(u_\infty)$ in $L^r(\Omega;\mathbb{R}^N)$ or equal to $grad(u_\infty)$ if $r \ge s$, then $P^m(grad(u_\infty) - g)$ is small in $L^t_{loc}(\Omega;\mathbb{R}^N)$ with $t = \frac{qr}{q+r}$, and one may take $p = \min\{s,t\}$.

Using *correctors* and Theorem 13.1, one can discuss the effect of lower-order terms. As a simplification, I shall not use the Meyers theorem.

Lemma 13.2. *If* $A^n \in \mathcal{M}(\alpha,\beta;\Omega)$ *H-converges to* $A^{\textit{eff}}$, c^n *is bounded in* $L^p(\Omega;\mathbb{R}^N)$, $p > N$ *if* $N \geq 2$, $p = 2$ *if* $N = 1$, $u_n \rightharpoonup u_\infty$ *in* $H^1_{loc}(\Omega)$ *weak and*

$$- \,div\big(A^n grad(u_n)\big) + \big(c^n, grad(u_n)\big) \to f \text{ in } H^{-1}_{loc}(\Omega) \text{ strong,} \tag{13.14}$$

then u_∞ *satisfies an equation with an effective coefficient* $c^{\textit{eff}}$,

$$- \,div\big(A^{\textit{eff}} grad(u_\infty)\big) + \big(c^{\textit{eff}}, grad(u_\infty)\big) = f \text{ in } \Omega, \tag{13.15}$$

where, for a subsequence of correctors P^m *associated to* A^m,

$$\begin{aligned}(P^m)^T c^m \rightharpoonup c^{\textit{eff}} &\text{ in } L^{2p/(p+2)}(\Omega) \text{ weak if } N \geq 2,\\ &\text{in } L^1(\Omega) \text{ weak} \star \text{ if } N = 1.\end{aligned} \tag{13.16}$$

Proof. One extracts a subsequence with (13.16).[4] By the Sobolev embedding theorem, $\big(c^m, grad(u_m)\big)$ stays in a compact of $H^{-1}_{loc}(\Omega)$, so by Theorem 13.1 $grad(u_m) - P^m grad(u_\infty)$ converges to 0 in $L^1_{loc}(\Omega;\mathbb{R}^N)$ strong, and one has

$$\begin{aligned}\big(c^m, grad(u_m)\big) \rightharpoonup \big(c^{\textit{eff}}, grad(u_\infty)\big) &\text{ in } L^{\frac{2p}{p+2}}(\Omega) \text{ weak if } N \geq 2,\\ \text{in } L^1(\Omega) \text{ weak} \star \text{ if } N = 1.&\end{aligned} \tag{13.17}$$

Indeed, if $g \in C(\Omega;\mathbb{R}^N)$ satisfies (13.10), and by (13.16) $(c^m, P^m g)$ converges to $(c^{\textit{eff}}, g)$ in $L^1(\Omega)$ weak $\star$, i.e., $L^1(\Omega)$ equipped with the weak $\star$ topology of $\mathcal{M}_b(\Omega)$, dual of $C_0(\Omega)$. Then, by (13.11), both $(c^m, grad(u_m) - P^m g)$ and $(c^{\textit{eff}}, grad(u_\infty) - g)$ have small norms in $L^1(\Omega)$, and one deduces that $\big(c^m, grad(u_m)\big)$ converges to $\big(c^{\textit{eff}}, grad(u_\infty)\big)$ in $L^1(\Omega)$ weak $\star$, from which one deduces (13.17). □

If one adds a term $d_n u_n$ in the equation, with d_n bounded in $L^q(\Omega)$ with $q > \frac{N}{2}$ for $N \geq 2$, $q = 1$ for $N = 1$, $d_n u_n$ stays in a compact of $H^{-1}_{loc}(\Omega)$ strong, and this term can be put into the right side converging in $H^{-1}_{loc}(\Omega)$ strong; if one extracts a subsequence with d_m converging to d_∞ in $L^q(\Omega)$ weak for $N \geq 2$, or in $L^1(\Omega)$ weak $\star$ if $N = 1$, then $d_m u_m$ converges to $d_\infty u_\infty$.

Then, François MURAT and myself proved an interesting variant.

Lemma 13.3. *If* $A^n \in \mathcal{M}(\alpha,\beta;\Omega)$ *H-converges to* $A^{\textit{eff}}$, b^n *is bounded in* $L^2(\Omega;\mathbb{R}^N)$, u_n *converges to* u_∞ *in* $H^1_{loc}(\Omega)$ *weak and*

$$- \,div\big(A^n grad(u_n) + b^n\big) \to f \text{ in } H^{-1}_{loc}(\Omega) \text{ strong.} \tag{13.18}$$

[4] There could be different subsequences $(P^m)^T c^m$ converging to different limits, but Lemma 13.2 tells us that all these limits give the same value for $(c^{\textit{eff}}, grad(u_\infty))$.

If for a subsequence of correctors Π^m *associated to* $(A^m)^T$,

$$(\Pi^m)^T b^m \rightharpoonup b^{\mathit{eff}} \textit{ in } \mathcal{M}(\Omega; \mathbb{R}^N) \textit{ weak } \star, \tag{13.19}$$

then, b^{eff} *belongs to* $L^2(\Omega; \mathbb{R}^N)$*, and*

$$A^n grad(u_n) + b^n \rightharpoonup A^{\mathit{eff}} grad(u_\infty) + b^{\mathit{eff}} \textit{ in } L^2(\Omega; \mathbb{R}^N) \textit{ weak }, \tag{13.20}$$
$$-div\big(A^{\mathit{eff}} grad(u_\infty) + b^{\mathit{eff}}\big) = f \textit{ in } \Omega. \tag{13.21}$$

Proof. One extracts a subsequence such that (13.19) holds and such that

$$A^m grad(u_m) + b^m \text{ converges to } \xi \text{ in } L^2(\Omega; \mathbb{R}^N) \text{ weak.} \tag{13.22}$$

For $v_\infty \in C_c^1(\Omega)$, let $v_n \in H_0^1(\Omega)$ be the solution of

$$div\big((A^m)^T grad(v_m) - (A^{\mathit{eff}})^T grad(v_\infty)\big) = 0 \text{ in } \Omega, \tag{13.23}$$

so that

$$\begin{aligned} &v_m \rightharpoonup v_\infty \text{ in } H_0^1(\Omega) \text{ weak},\\ &(A^m)^T grad(v_m) \rightharpoonup (A^{\mathit{eff}})^T grad(v_\infty) \text{ in } L^2(\Omega; \mathbb{R}^N) \text{ weak},\\ &grad(v_m) - \Pi^m grad(v_\infty) \to 0 \text{ in } L^2_{loc}(\Omega; \mathbb{R}^N) \text{ strong.} \end{aligned} \tag{13.24}$$

By the *div–curl lemma*

$$\big(A^m grad(u_m) + b^m, grad(v_m)\big) \rightharpoonup \big(\xi, grad(v_\infty)\big) \text{ in } L^1(\Omega) \text{ weak } \star, \tag{13.25}$$
$$\big(grad(u_m), (A^m)^T grad(v_m)\big) \rightharpoonup \big(grad(u_\infty), (A^{\mathit{eff}})^T grad(v_\infty)\big) \tag{13.26}$$
$$\text{in } L^1(\Omega) \text{ weak } \star.$$

By (13.24), $\big(b^m, grad(v_m)\big)$ has the same limit as $\big(b^m, \Pi^m grad(v_\infty)\big)$, which is $\big(b^{\mathit{eff}}, grad(v_\infty)\big)$, and one deduces that

$$\big(\xi - A^{\mathit{eff}} grad(u_\infty) - b^{\mathit{eff}}, grad(v_\infty)\big) = 0 \text{ in } \Omega. \tag{13.27}$$

By choosing v_∞ affine on an open set ω with $\overline{\omega} \subset \Omega$, one deduces that

$$b^{\mathit{eff}} = \xi - A^{\mathit{eff}} grad(u_\infty) \in L^2(\Omega; \mathbb{R}^N), \tag{13.28}$$

which with (13.22) gives (13.20), and (13.21). □

The formula of Theorem 12.4, for P^n in the laminated case, shows that $\Pi^m \neq (P^m)^T$ in general. If $(A^n)^T = A^n$, one may choose $\Pi^n = P^n$.

Despite the convergence in (13.19) being in $\mathcal{M}(\Omega)$ weak $\star$, the limit belongs to $L^2(\Omega; \mathbb{R}^N)$, and François MURAT noticed a more general property.

Lemma 13.4. *If $A^n \in \mathcal{M}(\alpha,\beta;\Omega)$ H-converges to A^{eff}, if γ_n is bounded in $L^q(\Omega)$, for $q \in [2,\infty]$, and if*

$$P^m_{i,j}\gamma_m \rightharpoonup \ell_{i,j} \text{ in } \mathcal{M}(\Omega) \text{ weak} \star \text{ if } q=2, \\ \text{in } L^{2q/(q+2)}(\Omega) \text{ weak if } q>2, \tag{13.29}$$

for a subsequence of correctors P^m, *and some* $i,j \in \{1,\ldots,N\}$, *then*

$$\ell_{i,j} \in L^q(\Omega). \tag{13.30}$$

Proof. The case $q=2$ is obtained by solving

$$-\,div\big((A^n)^T grad(u_n) + \gamma_n e_j\big) = 0 \text{ in } \Omega, \tag{13.31}$$

and applying Lemma 13.3. For $q>2$, one may assume that one also has

$$\gamma_m^2 \rightharpoonup \delta_\infty = \gamma_\infty^2 \text{ in } L^{q/2}(\Omega) \text{ weak}, \tag{13.32}$$
$$(P^m_{i,j})^2 \rightharpoonup Q_{i,j} \text{ in } \mathcal{M}(\Omega) \text{ weak} \star. \tag{13.33}$$

Then, by (13.2) and the *div–curl lemma*, one has

$$\text{for } \lambda \in \mathbb{R}^N, (A^m P^m\lambda, P^m\lambda) \rightharpoonup (A^{\mathrm{eff}}\lambda,\lambda) \text{ in } L^1(\Omega) \text{ weak} \star, \tag{13.34}$$

and, using $\alpha\,(P^m_{i,j})^2 \le (A^m P^m e_j, P^m e_j)$ and $(A^{\mathrm{eff}} e_j, e_j) \le \beta$, one has

$$Q_{i,j} \le \frac{\beta}{\alpha} \text{ a.e. in } \Omega. \tag{13.35}$$

Then, using

$$\pm\, P^m_{i,j}\gamma_m \le \frac{\varepsilon}{2}\,(P^m_{i,j})^2 + \frac{1}{2\varepsilon}\,\gamma_m^2 \text{ a.e. in } \Omega, \tag{13.36}$$

for every $\varepsilon > 0$, one deduces

$$\pm\ell_{i,j} \le \frac{\varepsilon}{2}\frac{\beta}{\alpha} + \frac{1}{2\varepsilon}\,\gamma_\infty^2 \text{ a.e. in } \Omega, \tag{13.37}$$
$$\text{i.e., } |\ell_{i,j}| \le \frac{\sqrt{\beta}}{\sqrt{\alpha}}\gamma_\infty \text{ a.e. in } \Omega, \tag{13.38}$$

by minimizing for $\varepsilon \in \mathbb{Q}_+$. □

One may give an analogous proof for $q=2$, but in this case δ_∞ in (13.32) is a nonnegative Radon measure, and for $\psi \in C_c(\Omega)$ (13.37) is replaced by

$$\int_\Omega \ell_{i,j}\psi\,dx \le \frac{\varepsilon}{2}\frac{\beta}{\alpha}\int_\Omega |\psi|\,dx + \frac{1}{2\varepsilon}\,\langle \delta_\infty, |\psi|\rangle. \tag{13.39}$$

One uses the Radon–Nikodym decomposition of δ_∞,[5]

$$\delta_\infty = f\,dx+\nu, \text{ with } f \in L^1(\Omega), \text{ and } \nu \text{ singular with respect to } dx, \quad (13.40)$$

and since ν lives on a set of Lebesgue measure 0, one deduces from (13.39) that

$$\ell_{i,j} \in L^2(\Omega) \text{ and } |\ell_{i,j}| \le \frac{\sqrt{\beta}}{\sqrt{\alpha}}\sqrt{f} \text{ a.e. in } \Omega. \quad (13.41)$$

François MURAT also studied correctors in the situation of Lemma 13.3.

Lemma 13.5. *Under the hypotheses of Lemma 13.3, one has*

$$grad(u_m) - P^m grad(u_\infty) - r^m \to 0 \text{ in } L^1_{loc}(\Omega) \text{ strong}, \quad (13.42)$$

for some r^m (constructed explicitly) which satisfies

$$r^m \rightharpoonup 0 \text{ in } L^1(\Omega;\mathbb{R}^N) \text{ weak}, \quad (13.43)$$
$$A^m r^m + b^m \rightharpoonup b^{eff} \text{ in } L^1(\Omega;\mathbb{R}^N) \text{ weak}. \quad (13.44)$$

Proof. Let $\rho_n \in H^1_0(\Omega)$ be the solution of

$$div\big(A^n grad(\rho_n) + b^n\big) = 0 \text{ in } \Omega, \quad (13.45)$$

so that by Lemma 13.3 a subsequence $\rho_m \rightharpoonup \rho_\infty$ in $H^1_0(\Omega)$ weak, with

$$div\big(A^{eff} grad(\rho_\infty) + b^{eff}\big) = 0 \text{ in } \Omega. \quad (13.46)$$

Then one notices that

$$div\big(A^m grad(u_m) - A^m grad(\rho_m)\big) \to f \text{ in } H^{-1}_{loc}(\Omega) \text{ strong}, \quad (13.47)$$

implying

$$grad(u_m) - grad(\rho_m) - P^m\big(grad(u_\infty) - grad(\rho_\infty)\big) \to 0 \text{ in } L^1_{loc}(\Omega;\mathbb{R}^N) \text{ strong}, \quad (13.48)$$

and one chooses

$$r^m = grad(\rho_m) - P^m grad(\rho_\infty), \quad (13.49)$$

which implies (13.42), (13.43) and (13.44). □

Corollary 13.6. *If $A^n \in \mathcal{M}(\alpha,\beta;\Omega)$ H-converges to A^{eff}, if b^n is bounded in $L^2(\Omega;\mathbb{R}^N)$, if c^n is bounded in $L^p(\Omega;\mathbb{R}^N)$ with $p > N$ if $N \ge 2$, $p = 2$ if $N = 1$, if $u_n \rightharpoonup u_\infty$ in $H^1_{loc}(\Omega)$ weak and*

[5] Otton Marcin NIKODÝM, Polish-born mathematician, 1887–1974. He worked at Kenyon College, Gambier, OH.

$$-div\big(A^n grad(u_n) + b^n\big) + \big(c^n, grad(u_n)\big) \to f \text{ in } H^{-1}_{loc}(\Omega) \text{ strong}, \quad (13.50)$$

$$(c^n, r^n) \rightharpoonup e^{\mathit{eff}}, \quad (13.51)$$

then, one has

$$-\,div\big(A^{\mathit{eff}} grad(u_\infty) + b^{\mathit{eff}}\big) + \big(c^{\mathit{eff}}, grad(u_\infty)\big) + e^{\mathit{eff}} = f \text{ in } \Omega, \quad (13.52)$$

using b^{eff} *and* c^{eff} *given by (13.16) and (13.19).*

I want to conclude with the case of periodic coefficients, which is just a special case of the general setting that François MURAT and myself developed. Although it was the asymptotic expansions used in a periodic framework by Évariste SANCHEZ-PALENCIA in the early 1970s which motivated my renewed interest in questions of continuum mechanics, the assumption of periodicity is an unnecessary restriction for questions of continuum mechanics or physics, but one may accept it as a first step in situations where one is not sure about the correct class of partial differential equations (or a generalization *beyond partial differential equations*) to consider. However, periodicity is useful for some technological applications, since engineers choose to create some periodic patterns,[6] like in the nuclear engineering application which motivated Ivo BABUŠKA, the correctors helping for computing variations of temperature or stress on a period cell.

Nature uses periodicity in crystals, but it is at atomic scale, much smaller than the scale which I called microscopic in early days, but I used the basic partial differential equations of continuum mechanics at this level, so that it is the scale which specialists of material science call mesoscopic, and they observe poly-crystals at this level, with grain boundaries between crystals of different orientations; however, real crystals have defects due to dislocations, so that it is only in some approximation that one may say that nature creates periodic patterns.[7]

N linearly independent vectors $y^1, \ldots, y^N \in \mathbb{R}^N$ generate the period cell

$$Y = \Big\{y \in \mathbb{R}^N \mid y = \sum_{i=1}^{N} \xi_i y^i, 0 \le \xi_i \le 1 \text{ for } i = 1, \ldots, N\Big\}, \quad (13.53)$$

[6] The same vicious circle exists concerning fractals. Rough objects are created by nature, but no one has shown a natural process which creates a self-similar fractal structure: it is just that there are people who use self-similar fractal sets as models for rough objects!

[7] Independently of the defects, which affect the movement of grain boundaries, the laws of movement of grain boundaries are not so well understood, since it is not a local question (despite naive "specialists" of material science playing with models where only the orientation of the normals to the interfaces plays a role).

and a function g defined on $\mathbb{R}^N$ is said to be *Y-periodic* if

$$g(y+y^i) = g(y) \text{ a.e. } y \in \mathbb{R}^N, \text{ for } i = 1, \ldots, N. \tag{13.54}$$

If $A \in \mathcal{M}(\alpha, \beta; \mathbb{R}^N)$ is Y-periodic, and $\varepsilon_n \to 0$, one defines

$$A^n(x) = A\Big(\frac{x}{\varepsilon_n}\Big) \text{ a.e. } x \in \Omega. \tag{13.55}$$

Lemma 13.7. *Under (13.55), A^n H-converges to a constant A^{eff}. For $\lambda \in \mathbb{R}^N$, w_λ is the Y-periodic solution (defined up to addition of a constant) of*

$$div\big(A(grad(w_\lambda) + \lambda)\big) = 0 \textit{ in } \mathbb{R}^N, w_\lambda \in H^1_{loc}(\mathbb{R}^N) \tag{13.56}$$

and $P \in H^1_{loc}\big(\mathbb{R}^N; \mathcal{L}(\mathbb{R}^N; \mathbb{R}^N)\big)$ is Y-periodic, defined by

$$P\lambda = grad(w_\lambda) + \lambda \textit{ a.e. in } \mathbb{R}^N. \tag{13.57}$$

Then A^{eff} and a sequence of correctors P^n are defined by

$$A^{\mathit{eff}}\lambda = \frac{1}{meas(Y)} \int_Y A(grad(w_\lambda) + \lambda)\, dy \textit{ for all } \lambda \in \mathbb{R}^N, \tag{13.58}$$

$$P^n(x) = P\left(\frac{x}{\varepsilon_n}\right) \textit{ a.e. } x \in \Omega. \tag{13.59}$$

Proof. Using (13.60) for defining $u_n \in H^1(\Omega)$, which satisfies (13.61),

$$u_n(x) = (\lambda, x) + \varepsilon_n w_\lambda\Big(\frac{x}{\varepsilon_n}\Big), \text{ a.e. } x \in \Omega, \tag{13.60}$$

$$u_n \in H^1_{loc}(\Omega), div\big(A^n grad(u_n)\big) = 0 \text{ in } \Omega, \tag{13.61}$$

one observes that the restriction to Ω of a Y-periodic function in $L^2_{loc}(\mathbb{R}^N)$ converges in $L^2(\Omega)$ weak to a constant, its average on Y. One finds that u_n converges to u_∞ in $H^1(\Omega)$ weak, with $u_\infty(x) = (\lambda, x)$, and then $grad(u_n)$ converges in $L^2(\Omega; \mathbb{R}^N)$ weak to $grad(u_\infty) = \lambda$, and $A^n grad(u_n)$ converges in $L^2(\Omega; \mathbb{R}^N)$ weak to $A^{\mathit{eff}}\lambda$, as defined by (13.58). Using N linearly independent $\lambda \in \mathbb{R}^N$ shows that A^n H-converges to A^{eff}, and P^n are *correctors*, since $grad(u_n) = P^n grad(u_\infty)$, and P^n satisfies (13.1) and (13.2). □

Additional footnotes: KENYON.[8]

[8] George KENYON, 2nd Baron KENYON, British statesman, 1776–1855. Kenyon College, Gambier, OH, is named after him.

Chapter 14
Correctors in Nonlinear Homogenization

Apart from the case of monotone operators, not much is understood for homogenization of nonlinear partial differential equations. At the beginning of 1977, in my Peccot lectures at Collège de France, in Paris, I described my result for monotone operators, which I discussed in Chap. 11, but in the summer of 1977, at a conference in Rio de Janeiro, Brazil, I reported about my difficulties in developing a similar homogenization theory for finite (i.e., nonlinear) elasticity: I did not find a reasonable class of strain–stress relations which is stable by homogenization (and monotonicity is not a reasonable assumption). However, I thought for many years that such a class existed, and later I agreed that in this case the Γ-convergence approach would give information on the stored energy functional of the effective material.

In 1987, Owen RICHMOND suggested that the homogenization of perforated plates requires higher-order gradients, i.e., that the effective equation is not that of an elastic material, but I was not sure why, and I thought that his comment was special to thin plates, and due to the holes; in the mid 1990s, I discussed with Gilles FRANCFORT the possible reason that the edges of the holes would go through large rotations and go out of plane.

There are other reasons which led me to think that there is no possible theory of homogenization for (nonlinear) elasticity, since the effective equations of reasonable materials must include nonelastic effects.[1] If this is the case, the Γ-convergence approach leads nowhere, since Γ-convergence is not homogenization, and it cannot say anything relevant if one does not understand which topology to use, and this is precisely about understanding how to describe a class of materials which is stable by homogenization, and it seems to include the possibility of nonelastic behaviour.

In the case of a sequence of monotone operators $A^n \in \mathcal{M}on(\alpha, \beta; \Omega)$ which only depend upon x_1 (and $p \in \mathbb{R}^N$), one can give an implicit description of what the homogenized operator A^{eff} is, but the explicit computations are not easy (since they require solving nonlinear equations), although the method is

[1] One of the reasons why the ideas which were used in (nonlinear) elasticity are not reasonable is the fact that they were for science-fiction materials, which can sustain infinite strains and stresses, and never break.

L. Tartar, *The General Theory of Homogenization*, Lecture Notes of the Unione Matematica Italiana 7, DOI 10.1007/978-3-642-05195-1_14,

quite similar to that used in the linear case, and consists in using the same sequences (12.9) of vectors $G^n, O^n \in L^2(\Omega;\mathbb{R}^N)$, respectively made from the *good* (non-oscillating) components and from the *oscillating* components of $E^n = grad(u_n)$ and $D^n = A^n\big(x_1, grad(u_n)\big)$: one writes the nonlinear constitutive relation between E^n and D^n in the equivalent form

$$O^n = B^n(x_1, G^n), \tag{14.1}$$

and one obtains the following analogue of Lemma 12.3.

Lemma 14.1. *Let $A^n \in \mathcal{M}on(\alpha,\beta;\Omega)$ depend upon x_1 (and $p \in \mathbb{R}^N$), and define B^n on $\Omega \times \mathbb{R}^N$ so that (14.1) holds (with notation (12.14) and (12.15)). If*

$$B^n(\cdot,p) \rightharpoonup B^\infty(\cdot,p) \text{ in } L^\infty(\Omega) \text{ weak } \star, \text{ for all } p \in \mathbb{R}^N, \tag{14.2}$$

then

$$D^\infty = A^{\text{eff}}(x_1, E^\infty) \text{ a.e. in } \Omega \text{ means } O^\infty = B^\infty(x_1, G^\infty) \text{ a.e. in } \Omega. \tag{14.3}$$

Proof. For $A \in \mathcal{M}on(\alpha,\beta;\Omega)$, one writes the relation $D = A(x,E)$ in the equivalent form $O = B(x,G)$.[2] $A \in \mathcal{M}on(\alpha,\beta;\Omega)$ implies $B \in \mathcal{M}on(\alpha',\beta';\Omega)$, with $\alpha' = \frac{\alpha}{\beta^2+1}, \beta' = \frac{\beta\,(\alpha^2+1)}{\alpha^2}$ for example, since

$$(A(\cdot,E') - A(\cdot,E), E' - E) = (B(\cdot,G') - B(\cdot,G), G' - G), \tag{14.4}$$
$$|G' - G|^2 + |O' - O|^2 = |D' - D|^2 + |E' - E|^2, \tag{14.5}$$

and

$$|D' - D|^2 + |E' - E|^2 \le \min\Big\{(\beta^2+1)\,|E' - E|^2, \frac{1+\alpha^2}{\alpha^2}\,|D' - D|^2\Big\}. \tag{14.6}$$

For $p \in \mathbb{R}^N$, one constructs a sequence of solutions $u'_n \in H^1(\Omega)$ of

$$-\,div\big(A^n(x_1, grad(u'_n))\big) = 0 \text{ in } \Omega, \text{ by } G'^n = p, O'^n = B^n(x_1,p), \tag{14.7}$$

so that the limits in $L^2(\Omega;\mathbb{R}^N)$ weak (for a subsequence) satisfy

$$grad(u'_m) \rightharpoonup E'^\infty, A^m\Big(x_1, grad(u'_m)\Big) \rightharpoonup D'^\infty, \text{ with } G'^\infty = p, O'^\infty = B^\infty(x_1,p). \tag{14.8}$$

---

[2] One only needs to know that $t \mapsto (A(\cdot, E + t\,e_1).e_1)$ is invertible, so that knowing $E_2,\ldots,E_N$, there is only one value of E_1 corresponding to a given value of D_1.

Then one uses the *div–curl lemma* for taking the limit of

$$(D'^m - D^m, E'^m - E^m) \geq 0 \text{ a.e. in } \Omega, \tag{14.9}$$

and one obtains

$$(B^\infty(x_1,p)-O^\infty, p-G^\infty) = (D'^\infty - D^\infty, E'^\infty - E^\infty) \geq 0 \text{ a.e. in } \Omega. \tag{14.10}$$

(14.10) is true for all $p \in \mathbb{Q}^N$ outside a set of measure 0, and one deduces that

$$O^\infty = B^\infty(x_1, G^\infty) \text{ a.e. in } \Omega, \tag{14.11}$$

by using $p \in \mathbb{R}^N$ by continuity, and then p near $G^\infty(x)$. □

The analogue of Lemma 12.4 for correctors in the laminated monotone case is then the following lemma.

Lemma 14.2. *Let $A^n \in \mathcal{M}on(\alpha, \beta; \Omega)$ depend only upon x_1 (and $p \in \mathbb{R}^N$), and let A^{eff} be defined as in Lemma 14.1. If $u_n \rightharpoonup u_\infty$ in $H^1_{loc}(\Omega)$ weak, and $div\big(A^n(\cdot, grad(u_n))\big)$ stays in a compact of $H^{-1}_{loc}(\Omega)$ strong, then*

$$\begin{aligned} &\big(A^n(\cdot, grad(u_n))\big)_1 \to \big(A^{\mathit{eff}}(\cdot, grad(u_\infty))\big)_1 \text{ in } L^2_{loc}(\Omega) \text{ strong,}\\ &\frac{\partial u_n}{\partial x_i} \to \frac{\partial u_\infty}{\partial x_i} \text{ in } L^2_{loc}(\Omega) \text{ strong, for } i = 2, \ldots, N. \end{aligned} \tag{14.12}$$

If one defines the sequence P^n from $\Omega \times \mathbb{R}^N$ into $\mathbb{R}^N$ by

$$\begin{aligned} &q_1(x_1,\lambda) = \big(A^{\mathit{eff}}(x_1,\lambda)\big)_1 \text{ a.e. in } \Omega, q_j(x_1,\lambda) = \lambda_j, j \geq 2,\\ &P^n_1(x,\lambda) = \big(B^n(x_1, q(x_1,\lambda))\big)_1, P^n_j(x,\lambda) = \lambda_j, j \geq 2, \text{ a.e. in } \Omega, \end{aligned} \tag{14.13}$$

then one has

$$grad(u_n) - P^n(\cdot, grad(u_\infty)) \to 0 \text{ in } L^2_{loc}(\Omega; \mathbb{R}^N) \text{ strong.} \tag{14.14}$$

Proof. If $E^n = grad(u_n)$ and $D^n = A^n\big(\cdot, grad(u_n)\big)$ and G^n is as in (12.14), the statement (14.12) means that G^n converges to G^∞ in $L^2_{loc}(\Omega; \mathbb{R}^N)$ strong. In order to prove this statement, one first notices that

$$\big(B^n(\cdot, G^n) - B^n(\cdot, G^\infty), G^n - G^\infty\big) \text{ converges to 0 in } \mathcal{M}(\Omega) \text{ weak } \star. \tag{14.15}$$

One has $(B^n(\cdot, G^n), G^n) = (D^n, E^n)$, which by using the *div–curl lemma* converges to $(D^\infty, E^\infty) = (O^\infty, G^\infty)$, which is also the limit of $(B^n(\cdot, G^n), G^\infty)$; $(B^n(\cdot, \lambda), G^n)$ converges to $(B^\infty(\cdot, \lambda), G^\infty)$ in $L^2_{loc}(\Omega)$ weak for fixed $\lambda \in \mathbb{R}^N$ since G^n does not oscillate in x_1, and then, since B^n is uniformly Lipschitz continuous, one deduces that $(B^n(\cdot, G^\infty), G^\infty)$ converges to $(B^\infty(\cdot, G^\infty), G^\infty) = (O^\infty, G^\infty)$ in $L^1_{loc}(\Omega)$ weak. Then one uses the fact that $B^n \in \mathcal{M}on(\alpha', \beta'; \Omega)$. Similarly $B^n(\cdot, G^n) - B^n(\cdot, G^\infty)$ converges to 0 in $L^2_{loc}(\Omega; \mathbb{R}^N)$ strong, and that expresses how E^n_1 oscillates, proving (14.14). □

Unlike in the linear case, the formula for laminated materials does not extend easily to nonlinear situations which are not elliptic: even without dependence in x, passing to the limit for weakly convergent sequences of a nonlinear wave equation is not a simple matter, and one needs to use "entropy" conditions, and my *compensated compactness method*; Ron DiPerna was the first to apply it with success for some systems.

In trying to define a general framework for correctors in the monotone case, one stumbles on the question whether or not some naturally defined correctors $P^m(x,\lambda)$ are Carathéodory functions. This difficulty was avoided by François Murat who emphasized an intermediate step in my proof of Theorem 13.1, but replacing smooth functions by piecewise constant functions, which are more natural for nonlinear settings. Instead of trying to prove that $grad(u_m) - P^m(\cdot, grad(u_\infty))$ tends to 0 in L^1, he showed that $grad(u_m) - P^m(\cdot, g)$ is uniformly small in L^2 when g is a piecewise constant function such that $grad(u_\infty) - g$ is small in L^2, noticing that one can still deduce from this statement a few interesting consequences.

Lemma 14.3. *If a sequence $A^n \in \mathcal{M}on(\alpha,\beta;\Omega)$ satisfies $A^n(\cdot,0) = 0$ and defines at the limit $A^{\rm eff} \in \mathcal{M}on(\alpha,\beta;\Omega)$, then there is a subsequence A^m and an associated sequence P^m of* "correctors," *from $\Omega \times \mathbb{R}^N$ into $\mathbb{R}^N$, satisfying*

$$\begin{aligned} &P^m(\cdot,0) = 0,\ P^m(\cdot,\lambda) \rightharpoonup \lambda \text{ in } L^2(\Omega) \text{ weak} \\ &A^m\big(\cdot, P^m(\cdot,\lambda)\big) \rightharpoonup A^{\rm eff}(\cdot,\lambda) \text{ in } L^2(\Omega) \text{ weak}, \\ &curl\big(P^m(\cdot,\lambda)\big) = 0 \text{ in } \Omega, \\ &div\big(A^m\big(\cdot, P^m(\cdot,\lambda)\big)\big) \text{ stays in a compact of } H^{-1}_{loc}(\Omega) \text{ strong}, \end{aligned} \tag{14.16}$$

for all $\lambda \in \mathbb{R}^N$. If $u_m \rightharpoonup u_\infty$ in $H^1_{loc}(\Omega)$ weak, if $div\big(A^m(\cdot, grad(u_m))\big)$ stays in a compact of $H^{-1}_{loc}(\Omega)$ strong, if g is piecewise constant in Ω, with level sets which are compact modulo sets of measure 0, one has

$$\limsup_m \int_\Omega \psi\, |grad(u_m) - P^m(\cdot,g)|^2\,dx \le \frac{\beta}{\alpha} \int_\Omega \psi\, |grad(u_\infty) - g|^2\,dx, \tag{14.17}$$

for all $\psi \in C_c(\Omega), \psi \ge 0$. The constructed correctors satisfy

$$||P^m(\cdot,\lambda) - P^m(\cdot,\lambda')||_{L^2(\Omega)} \le C\,|\lambda - \lambda'|, \tag{14.18}$$

for all $\lambda, \lambda' \in \mathbb{R}^N$, and all m, and

$$\limsup_m \int_\Omega \psi\, |P^m(\cdot,\lambda) - P^m(\cdot,\lambda')|^2\,dx \le \frac{\beta \int_\Omega \psi\,dx}{\alpha}\,|\lambda - \lambda'|^2, \tag{14.19}$$

for all $\lambda, \lambda' \in \mathbb{R}^N$, and all $\psi \in C_c(\Omega), \psi \ge 0$.

Proof. For a bounded open set Ω' of $\mathbb{R}^N$ containing $\overline{\Omega}$, one extends A^n in $\Omega' \setminus \overline{\Omega}$ so that $A^n \in \mathcal{M}on(\alpha, \beta; \Omega')$ and $A^n(\cdot, 0) = 0$, and one extracts a subsequence A^m for which a limiting operator exists on Ω', extending A^{eff} (by the analogue of Lemma 10.5 mentioned after Theorem 11.2). One then chooses $\varphi \in C_c^1(\Omega')$ equal to 1 on Ω, and for every $\lambda \in \mathbb{R}^N$, one defines $v^\lambda, v_m^\lambda \in H_0^1(\Omega')$ by

$$v^\lambda(x) = (\lambda, x)\,\varphi(x) \text{ in } \Omega', \text{ i.e., } grad(v^\lambda) = \varphi\,\lambda + (\lambda, \cdot)\,grad(\varphi) \text{ in } \Omega'$$
$$div\big(A^m(\cdot, grad(v_m^\lambda)) - A^{\text{eff}}(\cdot, grad(v^\lambda))\big) = 0 \text{ in } \Omega', \tag{14.20}$$

so that v_m^λ converges to v^λ in $H_0^1(\Omega')$ weak. One defines P^m on $\Omega \times \mathbb{R}^N$ by

$$P^m(\cdot, \lambda) = grad(v_m^\lambda) \text{ a.e. in } \Omega, \tag{14.21}$$

so that (14.16) is satisfied on Ω. Using the hypothesis on u_m and (14.16), the *div–curl lemma* implies that, for $\psi \in C_c(\Omega)$ and $\lambda \in \mathbb{R}^N$, one has

$$\int_\Omega \psi\,\big(A^m(\cdot, grad(u_m)) - A^m(\cdot, grad(v_m^\lambda)), grad(u_m) - grad(v_m^\lambda)\big)\,dx$$
$$\to \quad \int_\Omega \psi\,\big(A^{\text{eff}}(\cdot, grad(u_\infty)) - A^{\text{eff}}(\cdot, \lambda), grad(u_\infty) - \lambda\big)\,dx, \tag{14.22}$$

so that if $\psi \ge 0$ one obtains

$$\limsup_m \int_\Omega \psi\,|grad(u_m) - P^m(\cdot, \lambda)|^2\,dx \le \frac{\beta}{\alpha} \int_\Omega \psi\,|grad(u_\infty) - \lambda|^2\,dx. \tag{14.23}$$

For a partition of Ω into measurable subsets $\omega_i, i = 1, \ldots, k$, let

$$g = \sum_{i=1}^k \chi_{\omega_i} \lambda^i, \tag{14.24}$$

where χ_{ω_i} is the characteristic functions of ω_i, so that

$$P^m(\cdot, g) = \sum_{i=1}^k \chi_{\omega_i} P^m(\cdot, \lambda^i), \tag{14.25}$$

and for deducing (14.17) from (14.23), one assumes that each ω_i is compact (modulo a set of measure 0): using (14.23) for a decreasing sequence of functions ψ shows that (14.23) is still true when ψ is replaced by $\psi\,\chi_\omega$ with ω a compact subset of Ω, and summing in i the inequalities for ω_i gives (14.17). Since $A^m, A^{\text{eff}} \in \mathcal{M}on(\alpha, \beta; \Omega')$, one deduces from (14.20) that

$$\int_{\Omega'} |grad(v_m^\lambda) - grad(v_m^{\lambda'})|^2\,dx \le \frac{\beta}{\alpha} \int_{\Omega'} |grad(v^\lambda) - grad(v^{\lambda'})|^2\,dx, \tag{14.26}$$

from which (14.18) follows (with a constant C depending upon the choice of φ). On the other hand, the *div–curl lemma* implies

$$\begin{aligned}\int_\Omega \psi\,(A^m(\cdot,grad(v_m^\lambda)) - A^m(\cdot,grad(v_m^{\lambda'})), grad(v_m^\lambda) - grad(v_m^{\lambda'}))\,dx \\ \to \int_\Omega \psi\,(A^{\mathit{eff}}(\cdot,grad(v^\lambda)) - A^{eff}(\cdot,grad(v^{\lambda'})), grad(v^\lambda) - grad(v^{\lambda'}))\,dx \\ = \int_\Omega \psi\,(A^{\mathit{eff}}(\cdot,\lambda) - A^{\mathit{eff}}(\cdot,\lambda'), \lambda-\lambda'))\,dx \text{ for } \psi\in C_c(\Omega), \quad (14.27)\end{aligned}$$

and if $\psi\geq 0$, one deduces (14.19) by using $A^m, A^{\mathit{eff}}\in \mathcal{M}on(\alpha,\beta;\Omega)$. □

The restriction on the subsets ω_i can be removed by using the Meyers theorem, since for each $\lambda\in\mathbb{R}^N$, the sequence $P^m(\cdot,\lambda)$ stays bounded in $L^p_{loc}(\Omega;\mathbb{R}^N)$ for some $p>2$.

There are other ways to define correctors (like choosing them to be periodic when one works with periodically modulated problems), hence it is useful to base applications of correctors on (14.17) and (14.19), and not on their explicit description. One does not write $P^m(x,grad(u_\infty))$ since P^m may not be Carathéodory functions, but this is formally what $grad(u_m)$ looks like, and a precise way to handle this statement is to use (14.17) and (14.19), and then to let g converge to $grad(u_\infty)$. A crucial step in this argument of François MURAT is the following result which shows that some limits are more regular than one might expect.

Lemma 14.4. *If P^n is any sequence of correctors with (14.17) and (14.19), then after extracting a subsequence, there exists a function H on $\Omega\times\mathbb{R}^N$, measurable in $x\in\Omega$ and locally Lipschitz continuous in $\lambda\in\mathbb{R}^N$, such that*

$$\begin{aligned}&|H(x,\lambda)-H(x,\lambda')|\leq\frac{\beta}{\alpha}|\lambda-\lambda'|(|\lambda|+|\lambda'|) \text{ for all } \lambda,\lambda'\in\mathbb{R}^N, \text{ a.e. } x\in\Omega\\ &H(x,0)=0 \text{ a.e. } x\in\Omega\\ &|P^m(\cdot,\lambda)|^2\rightharpoonup H(\cdot,\lambda) \text{ in } L^1(\Omega) \text{ weak}\,\star, \text{ for all } \lambda\in\mathbb{R}^N, \qquad (14.28)\end{aligned}$$

i.e., $\int_\Omega |P^m(\cdot,\lambda)|^2\varphi\,dx\to\int_\Omega H(\cdot,\lambda)\varphi\,dx$ for all $\varphi\in C_0(\Omega)$. If $u_m\rightharpoonup u_\infty$ in $H^1_{loc}(\Omega)$ weak and (14.17) holds, one has

$$|grad(u_m)|^2\rightharpoonup H(\cdot,grad(u_\infty)) \text{ in } L^1(\Omega) \text{ weak}\,\star. \qquad (14.29)$$

Proof. If for a given $\lambda\in\mathbb{R}^N$, $|P^m(\cdot,\lambda)|^2$ converges in $\mathcal{M}(\Omega)$ weak $\star$ to a Radon measure μ_λ, then using (14.19) with $\lambda'=0$ gives

$$\langle\mu_\lambda,\psi\rangle\leq\frac{\beta\,|\lambda|^2}{\alpha}\int_\Omega\psi\,dx, \qquad (14.30)$$

so that μ_λ actually belongs to $L^\infty(\Omega)$, and is denoted $H(\cdot,\lambda)$. By a diagonal argument, one extracts a subsequence with $|P^m(\cdot,\lambda)|^2$ converging to $H(\cdot,\lambda)$ in $\mathcal{M}(\Omega)$ weak $\star$ for all $\lambda \in \mathbb{Q}^N$, and it is then true for all $\lambda \in \mathbb{R}^N$ by the locally Lipschitz character stated in (14.28), which follows from

$$\begin{aligned}\Big|\int_\Omega \psi\,\big(H(\cdot,\lambda)-H(\cdot,\lambda')\big)\,dx\Big| &= \Big|\lim_m \int_\Omega \psi\,\big(|P^m(\cdot,\lambda)|^2-|P^m(\cdot,\lambda')|^2\big)\,dx\Big|\\ &\le \limsup_m \int_\Omega |\psi|\,\big(|P^m(\cdot,\lambda)-P^m(\cdot,\lambda')|\big)\big(|P^m(\cdot,\lambda)|+|P^m(\cdot,\lambda')|\big)\,dx\\ &\le \frac{\beta\,|\lambda-\lambda'|(|\lambda|+|\lambda'|)}{\alpha}\int_\Omega |\psi|\,dx \text{ for all } \psi\in C_c(\Omega), \quad (14.31)\end{aligned}$$

by application of (14.19). For any $\varepsilon > 0$, one chooses a piecewise constant function g such that $||grad(u_\infty) - g||_{L^2(\Omega)} \le \varepsilon$, and one can use (14.17) if the level sets of g are compact modulo sets of measure 0, giving $||grad(u_m) - P^m(\cdot,g)||_{L^2_{loc}(\Omega)} \le C\varepsilon$, hence $||\,|grad(u_m)|^2 - |P^m(\cdot,g)|^2||_{L^1_{loc}(\Omega)} \le C'\varepsilon$. In order to be sure that $|P^m(\cdot,g)|^2$ converges in $L^1(\Omega)$ weak $\star$ to $H(\cdot,g)$, one also chooses g such that each level set differs from its interior by a set of measure 0. Then one deduces (14.29) by noticing that the local Lipschitz character of H implies $||H\big(\cdot,grad(u_\infty)\big) - H(\cdot,g)||_{L^1(\Omega)} \le C''\varepsilon$. □

One can describe in a similar way the limit of $\big(b^m, grad(u_m)\big)$ for a sequence b^m bounded in $L^2(\Omega)$, if one already extracted a subsequence such that $|b^m|^2$ converges in $\mathcal{M}(\Omega)$ weak $\star$ to a Radon measure $\pi = \pi_0 dx + \pi_s$, with $\pi_0 \in L^1(\Omega)$ and π_s singular with respect to the Lebesgue measure, and if for λ in a countable dense set of $\mathbb{R}^N$, $\big(b^m, P^m(\cdot,\lambda)\big)$ converges in $\mathcal{M}(\Omega)$ weak $\star$ to a Radon measure ν^λ. By using Lemma 14.4 and the (Bunyakovsky–) Cauchy–Schwarz inequality,[3,4] one deduces that

$$|\langle \nu^\lambda,\psi\rangle| = \lim_m\Big|\int_\Omega \psi\,\big(b^m, P^m(\cdot,\lambda)\big)\,dx\Big| \le \langle \pi,|\psi|\rangle^{1/2}\Big(\int_\Omega |\psi|\,H(x,\lambda)\,dx\Big)^{1/2}, \quad (14.32)$$

for every $\psi \in C_c(\Omega)$. Letting ψ converge to the characteristic function of an arbitrary Borel set E,[5] one deduces that ν^λ has no singular part, and

$$\nu^\lambda = B(\cdot,\lambda)\,dx \text{ with } |B(x,\lambda)| \le \Big(\frac{\beta\,\pi_0(x)}{\alpha}\Big)^{1/2}|\lambda| \text{ a.e. } x\in\Omega, \quad (14.33)$$

[3] Viktor Yakovlevich BUNYAKOVSKY, Ukrainian-born mathematician, 1804–1889. He worked in St Petersburg, Russia. He studied with CAUCHY in Paris (1825), and he proved the Cauchy–Schwarz inequality in 1859, 25 years before SCHWARZ.

[4] Karl Herman Amandus SCHWARZ, German mathematician, 1843–1921. He worked in Berlin, Germany.

[5] Félix Edouard Justin Emile BOREL, French mathematician, 1871–1956. He worked in Paris, France.

so that $B(\cdot,\lambda) \in L^2(\Omega)$. The same analysis applied to $\big(b^m, P^m(\cdot,\lambda) - P^m(\cdot,\lambda')\big)$ gives

$$|B(x,\lambda) - B(x\lambda')| \le \left(\frac{\beta\pi_0(x)}{\alpha}\right)^{1/2} |\lambda - \lambda'| \text{ for all } \lambda,\lambda' \in \mathbb{R}^N \text{ a.e. } x \in \Omega, \tag{14.34}$$

from which one deduces that

$$\big(b^m, grad(u_m)\big) \rightharpoonup B\big(x, grad(u_\infty)\big) \text{ in } L^1(\Omega) \text{ weak } \star\,. \tag{14.35}$$

In the spring of 1979, Georges DUVAUT and myself worked as consultants for INRIA,[6] and we were asked a question concerning the effective behaviour of a laminated material made of rubber and steel, and I then explained to an engineer of the industrial group interested (and to Georges DUVAUT by the same occasion), that there was no theory of homogenization for (nonlinear) elasticity, but if a formula for the effective elastic behaviour of their laminate must be used, it had to be the nonlinear analogue of the formulas (12.24)–(12.29) in the linear case (obtained by MCCONNELL in 1975). Of course, one should not use the linearized strain ε but the gradient of the deformation $F = \nabla\, u$, and one should not use the symmetric Cauchy stress tensor, adapted to the Eulerian point of view, but the Piola stress tensor,[7] usually called the Piola–Kirchhoff stress tensor,[8] adapted to the Lagrangian point of view.[9] One defines then the good tensor G^n, and the oscillating tensor O^n, by

$$\begin{aligned} &(G^n)_{i,j} = (\sigma^n)_{i,j} \text{ if } j = 1, (G^n)_{i,j} = (F^n)_{i,j} = \tfrac{\partial u_i^n}{\partial x_j} \text{ if } j \ge 2,\\ &(O^n)_{i,j} = (F^n)_{i,j} = \tfrac{\partial u_i^n}{\partial x_j} \text{ if } j = 1, (O^n)_{i,j} = (\sigma^n)_{i,j} \text{ if } j \ge 2, \end{aligned} \tag{14.36}$$

and one replaces the nonlinear constitutive relation

$$\sigma^n = \Sigma^n(x_1, F^n), \tag{14.37}$$

by the equivalent relation

$$O^n = K^n(x_1, G^n), \tag{14.38}$$

[6] Georges Jean DUVAUT, French mathematician, born in 1934. He worked at UPMC (Université Pierre et Marie Curie), Paris, France.

[7] Gabrio PIOLA, Italian physicist, 1794–1850. He worked in Milano (Milan), Italy.

[8] Gustav Robert KIRCHHOFF, German physicist, 1824–1887. He worked in Berlin, Germany.

[9] I was told that EULER introduced both the Eulerian point of view and the Lagangian point of view.

and one defines the limiting constitutive relation by

$$\sigma = \sum^{\mathit{eff}}(x_1, F) \text{ a.e. in } \Omega \text{ means } O = K^\infty(x_1, G) \text{ a.e. in } \Omega, \qquad (14.39)$$

where

$$K^n(\cdot, M) \rightharpoonup K^\infty(\cdot, M) \text{ in } L^\infty(\Omega) \text{ weak } \star \text{ for all } M \in \mathcal{L}(\mathbb{R}^N, \mathbb{R}^N). \quad (14.40)$$

In the case of hyper-elasticity, where Σ is the gradient of a stored energy function $W(F)$, a uniform rank-one convexity condition is what one needs in order to transform (14.37) into (14.38), and that condition is formally related to finite propagation speed.

Chapter 15
Holes with Dirichlet Conditions

In the fall of 1975, after studying the homogenization of monotone operators, described in Chap. 11, and the question of correctors, described in Chap. 13, I thought about degenerate elliptic problems, corresponding to holes in the domain, imposing either Dirichlet boundary conditions, described in this Chap. 15, or Neumann boundary conditions, described in Chap. 16.

I started with the case of a homogeneous Dirichlet condition on the boundaries of the holes, which I discarded after proving easily that the solutions converge to 0, by an argument about the constant in the Poincaré inequality.

Soon after, I saw what Évariste SANCHEZ-PALENCIA did,[1] which showed that there was more to investigate, which, in the periodic case, is to identify the weak limit of $\frac{u_\varepsilon}{\varepsilon^k}$ for a correct value of k.

Although my initial work was set in a periodic framework, I then looked at obtaining similar results in a more general situation, of course!

Lemma 15.1. *Assume that for a bounded open set Ω and a closed set T_n of $\mathbb{R}^N$, the sequence of characteristic functions χ_{Ω_n} of $\Omega_n = \Omega \setminus T_n$ satisfies*

$$\chi_{\Omega_n} \rightharpoonup \theta \text{ in } L^\infty(\Omega) \text{ weak } \star. \tag{15.1}$$

For $A^n \in \mathcal{M}(\alpha, \beta; \Omega_n)$ and f_n bounded in $L^2(\Omega)$, one solves[2]

$$-\,div\big(A^n grad(u_n)\big) = f_n \text{ in } \Omega_n, \quad u_n \in H^1_0(\Omega_n), \tag{15.2}$$

and let $\widetilde{u_n}$ be extension of u_n by 0 outside Ω_n. Then

$$\theta(x) < 1 \text{ a.e. } x \in \Omega, \text{ implies } \widetilde{u_n} \to 0 \text{ in } H^1_0(\Omega) \text{ strong}. \tag{15.3}$$

[1] Jacques-Louis LIONS proved similar results in his course at Collège de France.

[2] $H^1_0(\Omega_n)$ is defined as the closure of $C^\infty_c(\Omega_n)$ in $H^1(\Omega_n)$ in order to avoid unnecessary assumptions about the regularity of the boundary $\partial\Omega_n$.

L. Tartar, *The General Theory of Homogenization*, Lecture Notes of the Unione Matematica Italiana 7, DOI 10.1007/978-3-642-05195-1_15,

Proof. Of course, (15.2) corresponds to the variational formulation

$$\int_{\Omega_n} (A^n grad(u_n), grad(v))\,dx = \int_{\Omega_n} f_n v\,dx \text{ for every } v \in H_0^1(\Omega_n), \quad (15.4)$$

and since Ω is bounded, one has the Poincaré inequality

$$\int_{\Omega_n} |v|^2\,dx \leq \gamma(\Omega_n) \int_{\Omega_n} |grad(v)|^2\,dx \text{ for all } v \in H_0^1(\Omega_n), \quad (15.5)$$

where $\gamma(\Omega_n)$ is defined as the best constant in the inequality. Since $\gamma(\Omega_n) \leq \gamma(\Omega) < \infty$, the solution u_n exists and is unique by the Lax–Milgram lemma. Taking $v = u_n$ in (15.4) and using $A^n \in \mathcal{M}(\alpha, \beta; \Omega_n)$, one obtains

$$\alpha \int_{\Omega_n} |grad(u_n)|^2\,dx \leq ||f_n||_{L^2(\Omega_n)} \Big(\gamma(\Omega_n) \int_{\Omega_n} |grad(u_n)|^2\,dx\Big)^{1/2}, \quad (15.6)$$

so that

$$||grad(u_n)||_{L^2(\Omega_n;\mathbb{R}^N)} \leq \frac{\sqrt{\gamma(\Omega_n)}}{\alpha} ||f_n||_{L^2(\Omega_n)}, \quad (15.7)$$

$$||u_n||_{L^2(\Omega_n)} \leq \frac{\gamma(\Omega_n)}{\alpha} ||f_n||_{L^2(\Omega_n)}, \quad (15.8)$$

showing that u_n is bounded in $H_0^1(\Omega_n)$ since f_n is bounded in $L^2(\Omega)$.[3] The conclusion follows from $\gamma(\Omega_n) \to 0$, which will be deduced like Corollary 15.2, so here one concludes by using (15.2): one extracts a subsequence such that

$$\widetilde{u_m} \rightharpoonup u_\infty \text{ in } H_0^1(\Omega) \text{ weak and } L^2(\Omega) \text{ strong}, \quad (15.9)$$

so that

$$\widetilde{u_m} = \chi_{\Omega_n} \widetilde{u_m} \rightharpoonup \theta u_\infty \text{ in } L^2(\Omega) \text{ weak}, \quad (15.10)$$

implying $\theta\, u_\infty = u_\infty$ a.e. $x \in \Omega$, and since $\theta \neq 1$ a.e. $x \in \Omega$, one deduces that $u_\infty = 0$; the uniqueness of the limit implies that the whole sequence $\widetilde{u_n}$ converges to 0 in $H_0^1(\Omega)$ weak and $L^2(\Omega)$ strong. Then, one uses (15.4) again with $v = u_n$, and since $\int_{\Omega_n} f_n u_n\,dx = \int_\Omega f_n \widetilde{u_n}\,dx \to 0$ one deduces that $\widetilde{grad(u_n)} \to 0$ in $L^2(\Omega; \mathbb{R}^N)$ strong. □

Corollary 15.2. *Lemma 15.1 implies*

$$\gamma(\Omega_n) \to 0. \quad (15.11)$$

---

[3] One could define f_n only in Ω_n, and assume that $\widetilde{f_n}$ is bounded in $L^2(\Omega)$.

Proof. Taking $A^n = I$ for all n, one looks at the first eigenvalue of $-\Delta$ in Ω_n

$$-\Delta\varphi_n = \lambda_{1,n}\varphi_n, \quad \varphi_n \in H_0^1(\Omega_n), \varphi_n \neq 0, \tag{15.12}$$

and one has

$$\gamma(\Omega_n) = \frac{1}{\lambda_{1,n}}. \tag{15.13}$$

One normalizes φ_n by $||\widetilde{\varphi_n}||_{L^2(\Omega)} = 1$, and one uses $f_n = \widetilde{\varphi_n}$, so that Lemma 15.1 implies that $\frac{\varphi_n}{\lambda_{1,n}}$ converges to 0 in $L^2(\Omega)$ strong, i.e. $\lambda_{1,n}$ tends to ∞, or $\gamma(\Omega_n)$ tends to 0. □

One has the same result if Ω is unbounded but with $meas(\Omega) < \infty$, since $H_0^1(\Omega)$ is compactly embedded into $L^2(\Omega)$, and the Poincaré inequality holds.[4]

Corollary 15.3. *If (15.1) holds, $A^n \in \mathcal{M}(\alpha, \beta; \Omega_n)$ and f_n is bounded in $H^{-1}(\Omega)$,[5] and one writes $r_{\Omega_n} f_n$ for restriction of f_n to Ω_n, one solves*

$$-\,div\big(A^n grad(u_n)\big) = r_{\Omega_n} f_n \text{ in } \Omega_n, \quad u_n \in H_0^1(\Omega_n), \tag{15.14}$$

and then one has

$$\theta(x) < 1 \text{ a.e. } x \in \Omega, \text{ implies } \widetilde{u_n} \to 0 \text{ in } H_0^1(\Omega) \text{ weak and } L^2(\Omega) \text{ strong.} \tag{15.15}$$

Proof. Multiplying (15.14) by u_n gives

$$\int_{\Omega_n} \big(A^n grad(u_n), grad(u_n)\big)\,dx = \langle r_{\Omega_n} f_n, u_n\rangle = \langle f_n, \widetilde{u_n}\rangle, \tag{15.16}$$

from which one deduces the bounds

$$||grad(\widetilde{u_n})||_{L^2(\Omega_n)} \le \frac{1}{\alpha}\,||f_n||_{H^{-1}(\Omega_n)}, \tag{15.17}$$

$$||\widetilde{u_n}||_{L^2(\Omega)} \le \frac{\sqrt{\gamma(\Omega_n)}}{\alpha}\,||f_n||_{H^{-1}(\Omega)}, \tag{15.18}$$

[4] The Poincaré inequality is a consequence of the compact embedding, as I proved by an argument which I call the *equivalence lemma*, which generalizes a result of Jaak PEETRE. However, the Poincaré inequality also holds for a domain sandwiched between two parallel hyperplanes, although the compactness embedding may fail.

[5] $H^{-1}(\Omega)$ is the dual of $H_0^1(\Omega)$, and since the Poincaré inequality holds, it is the space of distributions in Ω of the form $\sum_{i=1}^N \frac{\partial f_i}{\partial x_i}$, with $f_1, \ldots, f_N \in L^2(\Omega)$.

so that $\widetilde{u_n} \to 0$ in $L^2(\Omega)$ strong by Corollary 15.2, and then $\widetilde{u_n} \rightharpoonup 0$ in $H_0^1(\Omega)$ weak since $\widetilde{u_n}$ is bounded in $H_0^1(\Omega)$. □

For a more precise estimate about $\gamma(\Omega_n)$, Lemma 15.4 is useful.[6]

Lemma 15.4. *Let $Y \subset \mathbb{R}^N$ be a parallelepiped generated by independent vectors $y^i, i = 1, \ldots, N$. Then for every δ with $0 < \delta < 1$, there is a constant $K(Y, \delta) > 0$ such that for all $a \in \mathbb{R}^N$ and all $\varepsilon > 0$ one has*

$$\int_{a+\varepsilon Y} |v|^2\,dx \leq K(Y,\delta)\,\varepsilon^2 \int_{a+\varepsilon Y} |grad(v)|^2\,dx \text{ for all } v \in H^1(a+\varepsilon Y)$$
$$\text{satisfying } meas\{y \in \varepsilon Y \mid v(a+y) = 0\} \geq \delta\,\varepsilon^N meas(Y). \tag{15.19}$$

Proof. By translation one may assume that $a = 0$, and by rescaling one may assume that $\varepsilon = 1$. If the inequality was not true there would exist a sequence $v_n \in H^1(Y)$ satisfying $meas\{y \in Y \mid v_n(y) = 0\} \geq \delta\,meas(Y)$, and

$$1 = \int_Y |v_n|^2\,dx > n \int_Y |grad(v_n)|^2\,dx, \tag{15.20}$$

so that $grad(v_n)$ would converge to 0 in $L^2(Y;\mathbb{R}^N)$ strong, and a subsequence v_m would converge to v_∞ in $H^1(Y)$ weak and $L^2(Y)$ strong, implying

$$grad(v_\infty) = 0 \text{ a.e. in } Y, \tag{15.21}$$
$$1 = \int_Y |v_\infty|^2\,dx, \tag{15.22}$$

but also

$$meas\{y \in Y \mid v_\infty(y) = 0\} \geq \liminf_n meas\{y \in Y \mid v_n(y) = 0\} \geq \delta\,meas(Y), \tag{15.23}$$

contradicting (15.21)–(15.22), which imply that v_∞ is a nonzero constant. □

In the periodic case where χ is the characteristic function of a Y-periodic open set (i.e. χ is lower semicontinuous), with

$$0 < \frac{1}{meas(Y)} \int_Y \chi(y)\,dy = \theta < 1, \tag{15.24}$$

and one defines]

$$\Omega_n = \Big\{x \in \Omega \mid \chi\Big(\frac{x}{\varepsilon_n}\Big) = 1\Big\}, \tag{15.25}$$

[6] I first heard about a similar property from Alain DAMLAMIAN.

with ε_n tending to 0, or more generally, if Ω_n can be covered by a translation of $\varepsilon_n Y$ overlapping only on a set of Lebesgue measure 0, and such that on each of these translated $a_i + \varepsilon_n Y$ (with $i \in I_n$) one has

$$meas\{x \in a_i + \varepsilon_n Y \mid x \notin \Omega_n\} \geq \delta\, \varepsilon_n^N\, meas(Y), i \in I_n, \tag{15.26}$$

for some $\delta > 0$ (equal to $1 - \theta$ in the periodic case), then

$$\gamma(\Omega_n) \leq K(Y, \delta)\, \varepsilon_n^2, \tag{15.27}$$

as the same constant appears on each of these translated parallelepipeds, by Lemma 15.5. In the periodic case one actually has

$$\lim_n \frac{\gamma(\Omega_n)}{\varepsilon_n^2} = K_1, \tag{15.28}$$

where K_1 is the best constant in the Poincaré inequality

$$\int_Y |v|^2\, dx \leq K_1 \int_Y |grad(v)|^2\, dx \text{ for all } v \in H^1_{0,per}(Y), \tag{15.29}$$

where $H^1_{0,per}(Y)$ is the subspace of $H^1_{loc}(\mathbb{R}^N)$ of functions which are Y-periodic and are 0 on the set $\{x \in \mathbb{R}^N, \chi(x) = 0\}$.

After proving the bound (15.8) for the norm in $L^2(\Omega_n)$ of the solution u_n of (15.2), and the upper bound (15.27) for $\gamma(\Omega_n)$ in the case where (15.26) holds, it is natural to ask about a possible limit in $L^2(\Omega)$ weak of $\frac{\widetilde{u_n}}{\varepsilon_n^2}$. In the periodic case (15.25), one has the following result.

Lemma 15.5. *Let Ω_n be given by (15.25), let A^n be given by*

$$A^n(x) = A\Big(\frac{x}{\varepsilon_n}\Big), x \in \Omega \tag{15.30}$$

where A is Y-periodic with $A \in \mathcal{M}(\alpha, \beta; \mathbb{R}^N)$, and assume that

$$f_n \to f_\infty \text{ in } L^2(\Omega) \text{ strong.} \tag{15.31}$$

Then the solutions u_n of (15.2) satisfy

$$\frac{\widetilde{u_n}}{\varepsilon_n^2} \rightharpoonup v_\infty = K_A\, f_\infty \text{ in } L^2(\Omega) \text{ weak,} \tag{15.32}$$

where

$$K_A = \frac{1}{meas(Y)} \int_Y w(y)\, dy = \frac{1}{meas(Y)} \int_Y z(y)\, dy, \tag{15.33}$$

and w, z are the solutions of

$$\begin{aligned} -div\big(A^T grad(w)\big) &= 1 \text{ in } \{x \in \mathbb{R}^N \mid \chi(x) = 1\},\ w \in H^1_{0,per}(Y),\\ -div\big(A\, grad(z)\big) &= 1 \text{ in } \{x \in \mathbb{R}^N \mid \chi(x) = 1\},\ z \in H^1_{0,per}(Y). \end{aligned} \tag{15.34}$$

More generally, if the sequence f_n does not converge strongly, one has

$$\{f_n \widetilde{w_n} \rightharpoonup g \text{ in } L^2(\Omega) \text{ weak}\} \text{ implies } \frac{\widetilde{u_n}}{\varepsilon_n^2} \rightharpoonup g \text{ in } L^2(\Omega) \text{ weak.} \tag{15.35}$$

Proof. Using (15.29), the solutions w, z exist and are unique by application of the Lax–Milgram lemma to $H^1_{0,per}(Y)$, and one has $K_A > 0$ since

$$K_A = \frac{1}{meas(Y)} \int_Y \big(A^T grad(w), grad(w)\big)\, dx \tag{15.36}$$

for example, and w is not 0 since $\theta < 1$.[7] The fact that $\int_Y w(y)\, dy = \int_Y z(y)\, dy$ follows easily from multiplying the equations in (15.34) by z or w since they both give $\int_Y \big(A^T grad(w), grad(z)\big)\, dy$. One then defines

$$w_n(x) = \varepsilon_n^2 w\Big(\frac{x}{\varepsilon_n}\Big), \tag{15.37}$$

which satisfies

$$-\, div\big((A^n)^T grad(w_n)\big) = 1 \text{ in } \Omega_n, \tag{15.38}$$

and since w_n need not be 0 on $\partial\Omega$, one uses

$$\varphi\, w_n \in H^1_0(\Omega_n) \text{ for all } \varphi \in C^1_c(\Omega). \tag{15.39}$$

One extracts a subsequence such that $\frac{\widetilde{u_m}}{\varepsilon_m^2}$ converges to v_∞ in $L^2(\Omega)$ weak, and if one identifies v_∞ to be $K_A f_\infty$, then all the sequence must converge and (15.32) holds. After multiplying (15.2) by φw_n and (15.38) by φu_n, and after subtraction, one obtains

$$\begin{aligned} \int_\Omega \varphi\, \big(f_n \widetilde{w_n} - \widetilde{u_n}\big)\, dx &= \int_\Omega \big(A^n grad(u_n), grad(\varphi)\big) \widetilde{w_n}\, dx \\ &- \int_\Omega \big((A^n)^T grad(w_n), grad(\varphi)\big) \widetilde{u_n}\, dx. \end{aligned} \tag{15.40}$$

[7] This proof, that $K_A > 0$, is preferable to the other one based on the maximum principle which states that $w, z > 0$ in $\{x \in \mathbb{R}^N \mid \chi(x) = 1\}$, since it is better to avoid using the maximum principle if one wants to derive methods which are valid for most problems in continuum mechanics or physics.

One divides by ε_n^2 and one takes the limit, for the subsequence $m \to \infty$. The left side then converges to $\int_\Omega \varphi\,(f_\infty\,K_A - v_\infty)\,dx$ (or $\int_\Omega \varphi\,(g - v_\infty)\,dx$ in the case (15.35)), while the right side tends to 0, since $\widetilde{u_n}, \widetilde{w_n}$ have $L^2(\Omega)$ bounds of order ε_n^2 and $\widetilde{grad(u_n)}, \widetilde{grad(w_n)}$ have $L^2(\Omega)$ bounds of order ε_n, by (15.7)–(15.8) and (15.27). This implies $f_\infty\,K_A - v_\infty = 0$ a.e. on Ω. □

The next result, which I only checked in the early 1990s, describes what the correctors are.

Lemma 15.6. *Assuming (15.30)–(15.31), and $f_\infty \in L^p_{loc}(\Omega)$, $2 \le p \le \infty$, one has*

$$\frac{1}{\varepsilon_n}\left(\widetilde{grad(u_n)} - f_\infty \widetilde{grad(z_n)}\right) \to 0 \text{ in } L^q_{loc}(\Omega;\mathbb{R}^N) \text{ strong}, \tag{15.41}$$

$$\frac{1}{\varepsilon_n^2}\left(\widetilde{u_n} - f_\infty \widetilde{z_n}\right) \to 0 \text{ in } L^q_{loc}(\Omega;\mathbb{R}^N) \text{ strong}, \tag{15.42}$$

where z_n is defined by

$$z_n = \varepsilon_n^2 z\Big(\frac{x}{\varepsilon_n}\Big), \tag{15.43}$$

and z is defined at (15.34), and $\frac{1}{q} = \frac{1}{p} + \frac{1}{2}$.

Proof. If one multiplies (15.2) by $\varphi\,u_n$ with $\varphi \in C^1_c(\Omega)$, one obtains

$$\int_\Omega \varphi\,\big(A^n grad(u_n), grad(u_n)\big)\,dx + \int_\Omega u_n\big(A^n grad(u_n), grad(\varphi)\big)\,dx = \int_\Omega \varphi\,f_n u_n\,dx, \tag{15.44}$$

so that, dividing by ε_n^2 and using (15.7)–(15.8) and (15.27), one obtains

$$\lim_n \frac{1}{\varepsilon_n^2} \int_\Omega \varphi\,\big(A^n grad(u_n), grad(u_n)\big)\,dx = \int_\Omega \varphi\,f_\infty K_A f_\infty\,dx, \tag{15.45}$$

which is equivalent to

$$\frac{1}{\varepsilon_n^2}\,\big(A^n grad(u_n), grad(u_n)\big) \rightharpoonup K_A f_\infty^2 \text{ in } \mathcal{M}(\Omega) \text{ weak } \star. \tag{15.46}$$

Similarly, multiplying (15.2) by $\varphi\,z_n$

$$\frac{1}{\varepsilon_n^2}\big(A^n grad(u_n), grad(z_n)\big) \rightharpoonup K_A f_\infty \text{ in } \mathcal{M}(\Omega) \text{ weak } \star, \tag{15.47}$$

or multiplying the rescaled version of (15.34) for z_n by either $\varphi\,u_n$ or $\varphi\,z_n$, one obtains

$$\frac{1}{\varepsilon_n^2}(A^n grad(z_n), grad(u_n)) \rightharpoonup K_A f_\infty \text{ in } \mathcal{M}(\Omega) \text{ weak } \star, \tag{15.48}$$

$$\frac{1}{\varepsilon_n^2}(A^n grad(z_n), grad(z_n)) \rightharpoonup K_A \text{ in } \mathcal{M}(\Omega) \text{ weak } \star. \tag{15.49}$$

For $h \in C_c(\Omega)$, one deduces from (15.46)–(15.49) that

$$\frac{1}{\varepsilon_n^2}(A^n[grad(u_n) - h\, grad(z_n)], [grad(u_n) - h\, grad(z_n)]) \rightharpoonup K_A|f_\infty - h|^2, \tag{15.50}$$

in $\mathcal{M}(\Omega)$ weak $\star$, implying

$$\limsup_n \frac{\int_\Omega \psi\, |grad(u_n) - h\, grad(z_n)|^2\, dx}{\varepsilon_n^2} \leq \frac{K_A}{\alpha} \int_\Omega \psi\, |f_\infty - h|^2\, dx, \tag{15.51}$$

for all $\psi \geq 0, \psi \in C_c(\Omega)$. One deduces (15.41) by using (15.51) and the following consequence of the Hölder inequality:

$$\limsup_n \frac{\int_\Omega \psi\, |(h - f_\infty)\, grad(z_n)|^q\, dx}{\varepsilon_n^q} \leq C(\psi)\Big(\int_K |f_\infty - h|^p\, dx\Big)^{q/p}, \tag{15.52}$$

where K is a compact containing the support of ψ.

In order to prove (15.42), one chooses $h \in C_c^1(\Omega)$ and one considers the function $u_n - h\, z_n$, which belongs to $H_0^1(\Omega_n)$ and whose gradient is $grad(u_n) - h\, grad(z_n) - z_n\, grad(h)$, so that for every $\psi \in C_c(\Omega), \psi \geq 0$, one has

$$\limsup_n \frac{\int_\Omega \psi\, |grad(u_n - h\, z_n)|^2\, dx}{\varepsilon_n^2} \leq \frac{K_A}{\alpha} \int_\Omega \psi\, |f_\infty - h|^2\, dx; \tag{15.53}$$

using Lemma 15.4, one deduces that

$$\limsup_n \frac{\int_\Omega \psi\, |u_n - h\, z_n|^q\, dx}{\varepsilon_n^{2q}} \leq C(\psi)\left(\int_K |f_\infty - h|^2\, dx\right)^{q/2}, \tag{15.54}$$

and one concludes by approaching f_∞ by functions in $C_c^1(\Omega)$. □

The preceding results cannot be obtained by a simple application of the *div–curl lemma*. The sequence $E^n = \frac{1}{\varepsilon_n} grad(u_n)$ certainly has a good curl and converges to 0 in $L^2(\Omega; \mathbb{R}^N)$ weak, and if one could extend the sequence $D^n = \frac{1}{\varepsilon_n} A^n grad(u_n)$ inside the holes (i.e. $\Omega \setminus \Omega_n$) in such a way that its divergence would stay in a compact set of $H^{-1}_{loc}(\Omega)$, one would have 0 for the weak $\star$ limit of (E^n, D^n). That limit does not depend upon how one extends D^n inside the holes since E^n is 0 inside these holes, and that limit is not usually 0, since it is $K_A f_\infty^2$.

In the case of Dirichlet boundary conditions that I just discussed, E^n is naturally extended by 0 inside the holes, but it is not necessary to extend D^n. However, in the example of the stationary Stokes equation in a perforated periodic medium, a formal asymptotic expansion done by Horia ENÉ and Évariste SANCHEZ-PALENCIA makes the Darcy law appear for the limit of a rescaled velocity field;[8] the velocity u^n is naturally extended by 0 inside the solid, but my method requires an extension of the "pressure" inside the solid, and this is not straightforward,[9] but the extension which I constructed used an unrealistic assumption for three-dimensional situations, and Grégoire ALLAIRE later extended it to a general periodic setting. My idea for the extension was to construct the transposed operator, which is a restriction, but I shall not describe it here, partly because Évariste SANCHEZ-PALENCIA described my construction as an appendix of his book [83], partly because after Robert DAUTRAY put me in contact in the mid 1980s with specialists at IFP, Rueil-Malmaison, France, I learned that real porous media are much more complex than the periodic idealization may suggest.
Additional footnotes: Alain DAMLAMIAN.[10]

[8] When I met him in the mid 1980s, Georges MATHERON mentioned that he derived the Darcy law in the 1960s by probabilistic arguments, but since probabilists rarely bother to check that their assertions are compatible with the partial differential equations of continuum mechanics, I suppose that he only conjectured the result, and I do not know if a mathematical proof along these lines was written.

[9] It was because Jacques-Louis LIONS told me that he was unable to construct the extension that I looked at the question.

[10] Alain DAMLAMIAN, French mathematician, born in 1946. He works at Université Paris XII, Créteil, France.

Chapter 16
Holes with Neumann Conditions

In the fall of 1975, I was not motivated by applications when I thought about degenerate elliptic problems, corresponding to holes in the domain. Although I understood that using a Neumann condition on the boundary of a hole is natural if one fills the hole with a perfect insulator, I lacked intuition about questions of continuum mechanics at the time; as a result of a discussion with Ivo BABUŠKA, at a conference in December 1975, in Versailles, France, I decided to work at developing my intuition.[1] I told him what I proved for a homogeneous Neumann condition on the boundary of the holes, and I said that I expected a similar result for a non-homogeneous Neumann condition, and Ivo BABUŠKA asked me what scaling I would consider (in the periodic case, which was the framework that he used), and he showed me why the scaling that I thought about is wrong.[2]

For the case of homogeneous Neumann conditions, D^n is naturally extended by 0 inside the holes, but an adequate extension of E^n inside the holes is needed in my proof.

For Ω a connected bounded open set of $\mathbb{R}^N$, and $T_n \subset \Omega$ a closed set (like holes cut out of Ω), one writes $\Omega_n = \Omega \setminus T_n$, and one assumes that $\partial\Omega$ and ∂T_n do not intersect, so that $\partial\Omega_n = \partial\Omega \cup \partial T_n$. For $f_n \in L^2(\Omega_n)$, one looks for solutions of

---

[1] Intuition seems to be like experience, in that one cannot transmit it easily from teacher to student, and one must work at acquiring it. Learning the language of engineers does not give much intuition, and I advocate instead that mathematicians continue to behave as mathematicians, but learn more about the physical meaning of the equations that they study. It was my intuitive understanding of continuum mechanics and physics which permitted me to perceive what is wrong with some of the models which are used, but it was my understanding of partial differential equations and other parts of mathematics which permitted me to imagine a plan for correcting the defects that I found.

[2] I wanted to use a condition $\frac{\partial u}{\partial \nu} = g$, instead of $\varepsilon\, g$ in a periodic setting, which Ivo BABUŠKA pointed out to be the correct scaling: the area of the boundary of a hole is $O(\varepsilon^{N-1})$, and one wants a total flux $O(\varepsilon^N)$ through this boundary, like if there is no hole, but a conductor and a uniform source of heat in its place.

L. Tartar, *The General Theory of Homogenization*, Lecture Notes of the Unione Matematica Italiana 7, DOI 10.1007/978-3-642-05195-1_16,

$$\begin{aligned}&-div\big(A^n grad(u_n)\big) = f_n \text{ in } \Omega_n\\&\big(A^n grad(u_n),\nu\big) = 0 \text{ on } \partial T_n\\&u_n = 0 \text{ on } \partial\Omega,\end{aligned}\tag{16.1}$$

where ν is the exterior normal to Ω_n. Writing (16.1) requires some smoothness of ∂T_n, and the condition $\big(A^n grad(u_n),\nu\big) = 0$ is understood in the sense given by Jacques-Louis LIONS for normal traces of functions in $H(div;\Omega_n)$,[3] which belong to the dual of the space of traces of functions in $H^1(\Omega_n)$.

There is a variational formulation, equivalent to (16.1) if ∂T_n is smooth, which makes sense without any smoothness assumption for ∂T_n:

$$\begin{aligned}&u_n \in V_n = \{u \in H^1(\Omega_n) \mid u = 0 \text{ on } \partial\Omega\},\\&\int_{\Omega_n} \big(A^n grad(u_n), grad(v)\big)\,dx = \int_{\Omega_n} f_n v\,dx \text{ for all } v \in V_n,\end{aligned}\tag{16.2}$$

and the condition $u = 0$ on $\partial\Omega$ is understood without regularity of $\partial\Omega$, by considering the closure in $H^1(\Omega_n)$ of the functions vanishing in a variable neighbourhood of the boundary, and this assumes that $dist(T_n,\partial\Omega) > 0$.

Existence and uniqueness of solutions of (16.2) follow from the Lax–Milgram lemma, if the Poincaré inequality holds for Ω_n, and this requires Ω_n to be connected, for example, but in my analysis (first performed in a periodic framework), I assumed that ∂T_n is locally Lipschitz in order to construct an extension operator P_n, i.e. such that for every $v \in V_n$ one has $P_n v \in H^1_0(\Omega)$ and the restriction of $P_n v$ to Ω_n is v, satisfying a uniform bound

$$\begin{aligned}&P_n \in \mathcal{L}(V_n;V) \text{ with } V = H^1_0(\Omega)\\&\int_\Omega |grad(P_n v)|^2\,dx \le C_* \int_{\Omega_n} |grad(v)|^2\,dx \text{ for all } v \in V_n,\end{aligned}\tag{16.3}$$

and a consequence of (16.3) is that the Poincaré inequality holds for Ω_n with

$$\gamma(\Omega_n) \le C_*\gamma(\Omega).\tag{16.4}$$

Lemma 16.1. *If (16.3) holds, if*

$$\chi_{\Omega_n} \rightharpoonup \theta \textit{ in } L^\infty(\Omega) \textit{ weak} \star,\tag{16.5}$$

[3] There is a natural framework with differential forms, which Jacques-Louis LIONS did not notice (as he told me after I asked him about it), that $H(div;\omega) = \{v \in L^2(\omega;\mathbb{R}^N) \mid div(v) \in L^2(\omega)\}$ corresponds to $(N-1)$-forms $w = \sum_i v_i \widehat{dx_i}$ with information on the exterior derivative dw, and that the normal trace (v,ν) is simply the coefficient of the restriction of w to $\partial\omega$. The classical trace theorem for the Sobolev space $H^1(\omega)$ corresponds to traces of 0-forms with a good exterior derivative, and more generally, for $0 \le k \le N-1$ a k-form with a good exterior derivative has a restriction on the boundary $\partial\omega$.

and if for a sequence $u_n \in V_n$ one has

$$P_n u_n \rightharpoonup u_\infty \text{ in } H_0^1(\Omega) \text{ weak,} \tag{16.6}$$

then

$$\widetilde{u_n} \rightharpoonup v_\infty = \theta\, u_\infty \text{ in } L^2(\Omega) \text{ weak.} \tag{16.7}$$

Equivalently, (16.3), (16.5) and

$$\begin{aligned} &u_n \in V_n, \int_{\Omega_n} |grad(u_n)|^2\,dx \le C \text{ for all } n \\ &\widetilde{u_n} \rightharpoonup v_\infty \text{ in } L^2(\Omega) \text{ weak,} \end{aligned} \tag{16.8}$$

imply

$$\frac{v_\infty}{\theta} \in H_0^1(\Omega). \tag{16.9}$$

Proof. By the compact embedding of $H_0^1(\Omega)$ into $L^2(\Omega)$, (16.6) implies $P_n u_n \to u_\infty$ in $L^2(\Omega)$ strong, and (16.7) follows from (16.5) and $\widetilde{u_n} = \chi_{\Omega_n} P_n u_n$. If (16.8) holds, $P_n u_n$ is bounded in $H_0^1(\Omega)$, and for a subsequence $P_m u_m \rightharpoonup u_\infty$ in $H_0^1(\Omega)$ weak, and (16.7) gives $v_\infty = \theta\, u_\infty$, i.e. (16.9). □

Corollary 16.2. *If (16.3), (16.5), and (16.8) hold, and if one assumes that*

$$\begin{aligned} &D^n \rightharpoonup D^\infty \text{ in } L^2(\Omega;\mathbb{R}^N) \text{ weak} \\ &D^n = 0 \text{ a.e. on } T_n, \text{ for all } n \\ &div(D^n) \in \text{ compact set of } H^{-1}_{loc}(\Omega), \end{aligned} \tag{16.10}$$

then one has

$$\int_{\Omega_n} \varphi\,(grad(u_n), D^n)\,dx \to \int_\Omega \varphi\left(grad\Big(\frac{v_\infty}{\theta}\Big), D^\infty\right) dx \text{ for all } \varphi \in C_c(\Omega), \tag{16.11}$$

Proof. By applying the *div–curl lemma* to $E^n = grad(P_n u_n)$ and D^n. □

The precise way in which u_n is extended in T_n does not matter, as long as a uniform bound like (16.3) holds, and the statements can be made independently of what the extensions are. Actually, as pointed out by Grégoire ALLAIRE and François MURAT, one can obtain the same results with fewer regularity assumptions on the holes T_n than what I assumed.

Theorem 16.3. *Assume that (16.3) holds and*

$$\theta(x) \ge \delta > 0 \text{ a.e. } x \in \Omega. \tag{16.12}$$

Then for any sequence $A^n \in \mathcal{M}(\alpha,\beta;\Omega_n)$ there exists a subsequence A^m and $A^{\text{eff}} \in \mathcal{M}(\frac{\alpha}{C_*},\beta;\Omega)$ such that for every $f \in L^2(\Omega)$ the solutions u_m of (16.1)–(16.2) with $f_m = f\chi_{\Omega_m}$ satisfy

$$\begin{aligned}&\widetilde{u_m} \rightharpoonup \theta\, u_\infty \textit{ in } L^2(\Omega) \textit{ weak, with } u_\infty \in H^1_0(\Omega),\\ &A^m \widetilde{grad(u_m)} \rightharpoonup A^{\mathit{eff}} grad(u_\infty) \textit{ in } L^2(\Omega;\mathbb{R}^N) \textit{ weak,}\\ &-div\big(A^{\mathit{eff}} grad(u_\infty)\big) = \theta\, f \textit{ in } \Omega.\end{aligned} \tag{16.13}$$

Proof. For each $f \in L^2(\Omega)$, the solution u_n exists in V_n and satisfies

$$\alpha \int_{\Omega_n} |grad(u_n)|^2\, dx \le \int_{\Omega_n} |f|\, |u_n|\, dx, \tag{16.14}$$

which by the Cauchy–Schwarz inequality and (16.4) gives

$$\int_{\Omega_n} |grad(u_n)|^2\, dx \le \frac{C_* \gamma(\Omega)}{\alpha^2} \int_{\Omega_n} |f|^2\, dx, \tag{16.15}$$

so that one can extract a subsequence u_m, and use Lemma 16.1 to obtain

$$\begin{aligned}&\widetilde{u_m} \rightharpoonup S(f) = \theta\, u_\infty \text{ in } L^2(\Omega) \text{ weak, with } u_\infty \in H^1_0(\Omega),\\ &A^m \widetilde{grad(u_m)} \rightharpoonup R(f) \text{ in } L^2(\Omega;\mathbb{R}^N) \text{ weak.}\end{aligned} \tag{16.16}$$

Since the homogeneous Neumann condition is used on ∂T_n, (16.1) implies

$$-\,div\big(A^n \widetilde{grad(u_n)}\big) = \chi_{\Omega_n} f \text{ in } \Omega, \tag{16.17}$$

which with (16.16) gives

$$-\,div\big(R(f)\big) = \theta\, f \text{ in } \Omega. \tag{16.18}$$

Mutiplying (16.1) by $\psi\, u_n$ with $\psi \in C^1(\overline{\Omega})$ gives

$$\begin{aligned}&\int_{\Omega_n} \psi\,(A^n grad(u_n), grad(u_n))\, dx + \int_\Omega P_n u_n (A^n \widetilde{grad(u_n)}, grad(\psi))\, dx\\ &\quad = \int_{\Omega_n} f\, \psi\, u_n\, dx \to \int_\Omega \theta\, f\, \psi\, u_\infty\, dx, \text{ which, by (16.18) is}\\ &\quad = \int_\Omega \big(R(f), \psi\, grad(u_\infty) + u_\infty\, grad(\psi)\big)\, dx,\end{aligned} \tag{16.19}$$

and one deduces that

$$\int_{\Omega_n} \psi\,(A^n grad(u_n), grad(u_n))\, dx \to \int_\Omega \psi\,(R(f), grad(u_\infty))\, dx, \tag{16.20}$$

for all $\psi \in C^1(\overline{\Omega})$, and since $\big(A^n grad(u_n), grad(u_n)\big)$ is bounded in $L^1(\Omega_n)$ it is valid for all $\psi \in C^0(\overline{\Omega})$ by density.[4] Using $A^n \in \mathcal{M}(\alpha, \beta; \Omega_n)$, one has

$$\frac{1}{\beta}\,|A^n grad(u_n)|^2 \le \big(A^n grad(u_n), grad(u_n)\big), \text{ a.e. } x \in \Omega_n, \tag{16.21}$$

so that (16.16) and (16.20) imply

$$\frac{1}{\beta}\,|R(f)|^2 \le \big(R(f), grad(u_\infty)\big), \text{ a.e. } x \in \Omega, \tag{16.22}$$

which gives

$$|R(f)| \le \beta\,|grad(u_\infty)|, \text{ a.e. } x \in \Omega. \tag{16.23}$$

Then, using (16.20) with $\psi = 1$, and (16.3), one has

$$\begin{aligned}\frac{\alpha}{C_*}\int_\Omega |grad(u_\infty)|^2\,dx &\le \liminf_n \int_{\Omega_n} \big(A^n grad(u_n), grad(u_n)\big)\,dx\\ &= \int_\Omega \big(R(f), grad(u_\infty)\big)\,dx,\end{aligned} \tag{16.24}$$

which by (16.18) implies

$$\frac{\alpha}{C_*}\,||u_\infty||^2_{H^1_0(\Omega)} \le \int_\Omega \theta\, f\, u_\infty\,dx \le ||\theta\, f||_{H^{-1}(\Omega)}||u_\infty||_{H^1_0(\Omega)}, \tag{16.25}$$

and then, using (16.18) and (16.23) gives

$$||\theta\, f||_{H^{-1}(\Omega)} \le ||R(f)||_{L^2(\Omega;\mathbb{R}^N)} \le \beta||u_\infty||_{H^1_0(\Omega)}. \tag{16.26}$$

If one defines the linear mapping T by

$$T(\theta\, f) = u_\infty \text{ for } f \in L^2(\Omega), \tag{16.27}$$

then T is defined on $L^2(\Omega)$ because of (16.12), and can be extended in a unique way as a linear continuous mapping from $H^{-1}(\Omega)$ to $H^1_0(\Omega)$ by (16.26). Like in the proof of Lemma 6.2, T is invertible, by an application of the Lax–Milgram lemma, since (16.25)–(16.26) imply

$$\frac{\alpha}{C_*\beta^2}\,||\theta\, f||^2_{H^{-1}(\Omega)} \le \langle T(\theta\, f), \theta\, f\rangle. \tag{16.28}$$

[4] The case $\psi \in C_c(\Omega)$ can also be deduced by using $D^n = \widetilde{A^n grad(u_n)}$ in (16.10).

Then, varying f in $L^2(\Omega)$ makes u_∞ vary in a dense set of $H^1_0(\Omega)$, hence

$$R(f) = A^{\mathit{eff}} grad(u_\infty) \text{ with } A^{\mathit{eff}} \in M\Big(\frac{\alpha}{C_*}, \beta; \Omega\Big), \tag{16.29}$$

by (16.23) and (16.24). □

It remains to show that the existence of an extension satisfying (16.3) is a reasonable hypothesis.[5] I proved it first for a periodic framework, and since the constructions are local, I easily extended the results to more general cases, but in order to see the limitations of (16.3) it is useful to observe that a physical interpretation of (16.1) consists in having the holes T_n filled with a perfect insulator, and that the geometry of the holes is important in order for the current to have the possibility to flow in every macroscopic direction.

Let us assume that one deals with a Y-periodic situation, using the characteristic length ε_n, and that on a unit period there are a finite number of disjoint connected closed holes with Lipschitz boundary. Around a hole τ_i there is a region of width at least $2d_i$ without any other hole, and one defines

$$O_i = \tau_i \cup \{x \in \mathbb{R}^N \setminus \tau_i \mid dist(x, \tau_i) < d_i\}, \tag{16.30}$$

and one can use classical ideas for extension of Sobolev spaces to construct a linear continuous mapping Q_i from $H^1(O_i \setminus \tau_i)$ into $H^1(O_i)$ such that

$$\begin{aligned}&(Q_i u)(x) = u(x) \text{ a.e. } x \in O_i \setminus \tau_i \text{ for all } u \in H^1(O_i \setminus \tau_i),\\ &\int_{O_i} |grad(Q_i u)|^2\,dx \le C_i \int_{O_i\setminus\tau_i} (|u|^2 + |grad(u)|^2)\,dx \text{ for all } u \in H^1(O_i \setminus \tau_i).\end{aligned} \tag{16.31}$$

Actually, one cannot use any such extension for a rescaled situation, since $|u|$ and $|grad(u)|$ appear in (16.31), and they rescale in a different way: one must construct a better extension $\widetilde{Q_i}$ from $H^1(O_i \setminus \tau_i)$ into $H^1(O_i)$, satisfying

$$\int_{O_i} |grad(\widetilde{Q_i} u)|^2\,dx \le C_i^* \int_{O_i\setminus\tau_i} |grad(u)|^2\,dx \text{ for all } u \in H^1(O_i \setminus \tau_i). \tag{16.32}$$

[5] Most mathematicians do not bother too much about that, but it is better to avoid developing general theories which only have trivial examples, or no example at all! My goal being to develop new mathematical tools for giving sounder foundations to twentieth century mechanics/physics, I must check from time to time that I am heading in a reasonable direction.

This is easily done by choosing[6]

$$\widetilde{Q_i}1 = 1,\ \widetilde{Q_i}u = Q_i u \text{ if } \int_{O_i\setminus\tau_i} u\,dx = 0. \tag{16.33}$$

Since the open sets $O_i \setminus \tau_i$ do not intersect, one can glue all these extensions together and obtain an extension with $C^* = \max_i C_i^*$, and the same constant is kept for all the rescaled domains.

One sees from the preceding construction that one can deal with quite a variety of shapes of holes, if they satisfy some uniform geometrical condition, and Denise CHENAIS carried out precise estimates for the norm of such extensions in the early 1970s [13].[7]

[6] In the fall of 1975, I mentioned that I solved the problem with holes and Neumann conditions, but I did not write it down: I had difficulties with writing and I did not know yet that many have the tendency to publish what others do. A few months after, Georges DUVAUT told me that it was easy, and that the problem was only a question of extension, which was indeed what I had done, but in the fall of 1976, on the eve of leaving for a Franco-Japanese meeting in Tokyo and Kyoto, Japan, he called me because he realized that he had unwanted terms in $\frac{1}{\varepsilon}$ (he was working in a periodic framework, of course!); he missed the difference between (16.31) and (16.32), and a choice like (16.33), which I explained to him.

[7] Denise CHENAIS, French mathematician, born in 1944. She worked at Université de Nice–Sophia Antipolis, Nice, France.

Chapter 17
Compensated Compactness

The *div–curl lemma*, discussed in Chapter 7, was obtained with François MURAT in 1974 while we were trying to unify all the cases where explicit H-limits could be computed.[1] In 1976, Jacques-Louis LIONS asked François MURAT to generalize our *div–curl lemma*, and gave him some articles which he thought related, by SCHULENBERGER and WILCOX,[2,3] and I think that it was at that time that he proposed the name *compensated compactness*, which after all is not so good, and I refer to [115] for the different ideas of *compensated integrability* and *compensated regularity*, which some people still want to confuse with *compensated compactness*.[4]

The first generalization of François MURAT was to study for which bilinear forms B one can compute the weak limit of $B(U^n, V^n)$,[5] under weak convergence hypotheses on U^n and V^n, together with some linear partial differential constraints.[6] Once he obtained the characterization, I thought that the bilinear setting was too restrictive, and that he should study for which

---

[1] I do not call explicit the formula for the case of periodic coefficients, since it only describes an algorithm which cannot be carried out in a simple way, because one must solve partial differential equations on a period cell: it can be used for approaching the coefficients by using numerical codes.

[2] John R. SCHULENBERGER, American mathematician, 2007. He worked in Denver, CO, at University of Utah, Salt Lake City, UT, and at Texas Tech University, Lubbock, TX.

[3] Calvin Hayden WILCOX, American mathematician, 1924–2001. He worked at UW (University of Wisconsin), Madison, WI, and at University of Utah, Salt Lake City, UT.

[4] Could it be that those who proved results of *compensated regularity* and wrongly pretended that they improved *compensated compactness* only thought about the first generalization by François MURAT, and not about the general theory that I developed immediately after, and that they did not understand about my *compensated compactness method*, which unified what Jacques-Louis LIONS taught as a dichotomy for nonlinear partial differential equations in the late 1960s, the compactness method or the monotonicity/convexity method?

[5] And usually not all nonlinear quantities as in classical compactness arguments.

[6] It is a part of what should be generalized, that other adapted topologies of weak type should also be considered.

L. Tartar, *The General Theory of Homogenization*, Lecture Notes of the Unione Matematica Italiana 7, DOI 10.1007/978-3-642-05195-1_17,

quadratic forms Q one can compute the weak limit of $Q(U^n)$ under weak convergence hypotheses on U^n and some linear partial differential constraints, and François MURAT obtained this second generalization by adapting his proof to that more general question. The procedure that he followed is similar to the scenario of Joel ROBBIN's proof based on the Hodge theorem that I described in Lemma 9.1, and it needs a constant rank hypothesis.

I obtained the third generalization by also considering general quadratic forms Q, and using a sequence U^n converging weakly to U^∞ and satisfying linear differential constraints, but I studied when the weak $\star$ limit of $Q(U^n)$ is $\geq Q(U^\infty)$ in the sense of Radon measures. My proof used the same ideas from my original proof of the *div–curl lemma*, following an argument of Lars HÖRMANDER for proving the compactness of the injection of $H^1_0(\Omega)$ into $L^2(\Omega)$ (when Ω has finite Lebesgue measure), using the Fourier transform, and no rank condition is used in this proof!

In the general framework of Theorem 17.3, the coefficients $A_{i,j,k}$ are real constants, and one may use complex constants, but the theory cannot handle variable coefficients; I corrected this defect of *compensated compactness* by the introduction of *H-measures* in the late 1980s. Similarly, Corollary 17.4 contains the example of differential forms of Lemma 9.1, but only for an open set of $\mathbb{R}^N$, while the theory of differential forms extends to any differentiable manifold, and one adds a Riemannian structure for the Hodge theorem.

Lemma 17.1 is a necessary condition for sequential weak "lower semi-continuity," the sufficient condition of Theorem 17.3 being valid for quadratic functions.

Lemma 17.1. *Let Ω be an open set of $\mathbb{R}^N$ and let F be a real function defined on $\mathbb{R}^p$. If for every sequence U^n satisfying*

$$U^n \rightharpoonup U^\infty \text{ in } L^\infty(\Omega;\mathbb{R}^p) \text{ weak } \star, \tag{17.1}$$

$$F(U^n) \rightharpoonup V^\infty \text{ in } L^\infty(\Omega) \text{ weak } \star, \tag{17.2}$$

$$\sum_{j=1}^{p}\sum_{k=1}^{N} A_{i,j,k}\frac{\partial U^n_j}{\partial x_k} = 0, \text{ for } i = 1,\ldots,q, \tag{17.3}$$

one can deduce that

$$V^\infty(x) \geq F\big(U^\infty(x)\big) \text{ a.e. in } \Omega, \tag{17.4}$$

then

$$t \mapsto F(a + t\,\lambda) \text{ is convex for every } a \in \mathbb{R}^p \text{ and every } \lambda \in \Lambda, \tag{17.5}$$

where Λ is the characteristic set

$$\Lambda = \Big\{\lambda \in \mathbb{R}^p \mid \exists \xi \in \mathbb{R}^N \setminus 0, \sum_{j=1}^{p}\sum_{k=1}^{N} A_{i,j,k}\lambda_j\xi_k = 0, \text{ for } i = 1,\ldots,q\Big\}. \tag{17.6}$$

Proof. Let $a \in \mathbb{R}^N$, $\lambda \in \Lambda$, and $\xi \neq 0$ associated to λ in (17.6). Let φ_n be a sequence of smooth real functions of one variable and define U^n by

$$U^n(x) = a + \varphi_n\big((\xi, x)\big)\,\lambda, \tag{17.7}$$

which satisfies (17.3). For $\theta \in (0,1)$, one defines ψ of period 1 by

$$\psi(z) = 1 \text{ for } 0 < z < \theta, \psi(z) = 0 \text{ for } \theta < z < 1, \tag{17.8}$$

and one chooses

$$\varphi_n(z) = \psi(n\,z). \tag{17.9}$$

Of course, φ_n is not smooth, but since (17.7) implies (17.3) for all smooth φ_n, the same is true for limits (interpreting (17.3) in the sense of distributions), and the choice (17.9) still implies (17.3). Letting n tend to ∞ gives

$$U^n \rightharpoonup \theta(a+\lambda) + (1-\theta)a \text{ in } L^\infty(\Omega;\mathbb{R}^p) \text{ weak } \star, \tag{17.10}$$
$$F(U^n) \rightharpoonup \theta\,F(a+\lambda) + (1-\theta)F(a) \text{ in } L^\infty(\Omega) \text{ weak } \star, \tag{17.11}$$

so that one should have

$$\theta\,F(a+\lambda) + (1-\theta)F(a) \geq F(\theta(a+\lambda) + (1-\theta)a), \tag{17.12}$$

and varying $a \in \mathbb{R}^p$, $\lambda \in \Lambda$ and $\theta \in (0,1)$ gives (17.5). □

Corollary 17.2. *If a function F is such that (17.1)–(17.3) imply*

$$V^\infty(x) = F\big(U^\infty(x)\big) \text{ a.e. in } \Omega, \tag{17.13}$$

then

$$t \mapsto F(a + t\,\lambda) \text{ is affine for every } a \in \mathbb{R}^p \text{ and every } \lambda \in \Lambda. \tag{17.14}$$

Proof. One applies Lemma 17.1 to both F and $-F$. □

If X is the subspace spanned by Λ, and F satisfies (17.14), then the restriction of F to any subspace parallel to X is a polynomial of degree $\leq dim(X)$; in particular if Λ spans $\mathbb{R}^p$, every function F satisfying (17.14) is a polynomial of degree $\leq p$. Actually, Lemma 17.5 gives other necessary conditions, which imply that on any subspace parallel to X, the degree of F is also $\leq N$.

The basic results of *compensated compactness*, Theorem 17.3 and Corollary 17.4, show that the conditions of Lemma 17.1 and Corollary 17.2 are sufficient in the case of quadratic functions; the conditions are even strengthened by replacing L^∞ by larger spaces, more natural in a quadratic setting.

Theorem 17.3. *Let $\Omega \subset \mathbb{R}^N$ be open, and let Q be a real quadratic form on $\mathbb{R}^p$ satisfying*

$$Q(\lambda) \geq 0 \text{ for all } \lambda \in \Lambda, \tag{17.15}$$

let U^n be a sequence satisfying

$$U^n \rightharpoonup U^\infty \text{ in } L^2_{loc}(\Omega; \mathbb{R}^p) \text{ weak}, \tag{17.16}$$

and

$$\sum_{j=1}^{p} \sum_{k=1}^{N} A_{i,j,k} \frac{\partial U_j^n}{\partial x_k} \text{ belongs to a compact of } H^{-1}_{loc}(\Omega) \text{ strong, for } i = 1, \ldots, q. \tag{17.17}$$

Then, if a subsequence satisfies

$$Q(U^m) \rightharpoonup \mu \text{ in } \mathcal{M}(\Omega) \text{ weak} \star, \tag{17.18}$$

for a Radon measure μ, then one has

$$\mu \geq Q(U^\infty) \text{ in } \Omega \text{ (in the sense of measures)}. \tag{17.19}$$

Proof. For $\varphi \in C^1_c(\Omega)$ and V^m defined by

$$V^m = \varphi\, U^m - \varphi\, U^\infty, \tag{17.20}$$

one proves that

$$\liminf_{m\to\infty} \int_{\mathbb{R}^N} Q\big(V^m(x)\big)\, dx \geq 0, \tag{17.21}$$

which implies Theorem 17.3. Indeed, if B denotes the symmetric bilinear form on $\mathbb{R}^p$ associated to Q, one has

$$\begin{aligned} \textstyle\int_{\mathbb{R}^N} Q(V^m)\, dx &= \textstyle\int_{\mathbb{R}^N} \varphi^2 \big(Q(U^m) - 2B(U^m, U^\infty) + Q(U^\infty)\big)\, dx \\ \to &\quad \langle \mu, \varphi^2 \rangle - \textstyle\int_{\mathbb{R}^N} \varphi^2 Q(U^\infty)\, dx, \end{aligned} \tag{17.22}$$

by (17.16) and (17.18), so that (17.21) implies

$$\langle \mu, \varphi^2 \rangle - \int_{\mathbb{R}^N} \varphi^2 Q(U^\infty)\, dx \geq 0, \tag{17.23}$$

and since this is true for all $\varphi \in C^1_c(\Omega)$, it is (17.19).

The functions V^m are defined in all $\mathbb{R}^N$ and satisfy

$$support(V^m) \subset K = support(\varphi), \tag{17.24}$$

$$V^m \rightharpoonup 0 \text{ in } L^2(\mathbb{R}^N; \mathbb{R}^p) \text{ weak}, \tag{17.25}$$

$$\sum_{j=1}^{p} \sum_{k=1}^{N} A_{i,j,k} \frac{\partial V_j^m}{\partial x_k} \to 0 \text{ in } H^{-1}(\mathbb{R}^N) \text{ strong, for } i = 1, \ldots, q. \tag{17.26}$$

One uses the Fourier transform $\mathcal{F}V^m$ of V^m, and (17.25) is equivalent to

$$\mathcal{F}V^m \rightharpoonup 0 \text{ in } L^2(\mathbb{R}^N;\mathbb{C}^p) \text{ weak,} \tag{17.27}$$

(17.26) is equivalent to

$$\frac{1}{1+|\xi|}\sum_{j=1}^{p}\sum_{k=1}^{N} A_{i,j,k}\mathcal{F}V_j^m(\xi)\,\xi_k \to 0 \text{ in } L^2(\mathbb{R}^N) \text{ strong, for } i=1,\ldots,q, \tag{17.28}$$

and, by the Plancherel theorem, (17.21) is equivalent to

$$\liminf_{m\to\infty}\int_{\mathbb{R}^N} Q\big(\mathcal{F}V^m(\xi)\big)\,d\xi \ge 0, \tag{17.29}$$

where Q is extended to become a Hermitian form on $\mathbb{C}^p$. The conditions (17.24) and (17.25) imply that $\mathcal{F}V^m$ is bounded in $L^\infty(\mathbb{R}^N,\mathbb{C}^p)$ and

$$\mathcal{F}V^m(\xi) \to 0 \text{ for all } \xi \in \mathbb{R}^N, \tag{17.30}$$

so that by the Lebesgue dominated convergence theorem one has

$$\mathcal{F}V^m \to 0 \text{ in } L^2_{loc}(\mathbb{R}^N,\mathbb{C}^p) \text{ strong.} \tag{17.31}$$

For proving (17.29), the integral on $|\xi|\le 1$ tends to 0 by (17.31), and one uses the following inequality on $|\xi|\ge 1$: for every $\varepsilon>0$, there exists C_ε with

$$\Re Q(W) \ge -\varepsilon\,|W|^2 - C_\varepsilon\sum_{i=1}^{q}\Big|\sum_{j=1}^{p}\sum_{k=1}^{N} A_{i,j,k}W_j\frac{\eta_k}{|\eta|}\Big|^2 \text{ for all } W\in\mathbb{C}^p, \eta\in\mathbb{R}^N\setminus 0. \tag{17.32}$$

One applies (17.32) to $W=\mathcal{F}V^m(\xi)$ and $\eta=\xi$ for $|\xi|\ge 1$ and using (17.28) since $\frac{|\xi|}{1+|\xi|}$ is bounded away from 0, one deduces that

$$\liminf_{m\to\infty}\int_{|\xi|\ge 1}\Re Q\big(\mathcal{F}V^m(\xi)\big)\,d\xi \ge -\varepsilon\,M^2, \tag{17.33}$$

with M an upper bound for the norm of V^m in $L^2(\mathbb{R}^N,\mathbb{C}^p)$; letting ε tend to 0 gives (17.29). Finally, one proves (17.32) by contradiction: if it was not true, there would exist $\varepsilon_0>0$ and for all n a pair $W^n,\eta^n\in\mathbb{C}^p\times\mathbb{R}^N\setminus 0$ with

$$\Re Q(W^n) < -\varepsilon_0|W^n|^2 - n\sum_{i=1}^{q}\Big|\sum_{j=1}^{p}\sum_{k=1}^{N} A_{i,j,k}W_j^n\frac{\eta_k^n}{|\eta|}\Big|^2. \tag{17.34}$$

By homogeneity, one may assume that $|W^n| = 1$ for all n, so that

$$\sum_{j=1}^{p}\sum_{k=1}^{N} A_{i,j,k} W_j^n \frac{\eta_k^n}{|\eta|} \to 0 \text{ for } i = 1, \ldots, q. \tag{17.35}$$

Extracting a subsequence with $W^m \to W^\infty$ and $\eta^m \to \eta^\infty$, one obtains

$$\sum_{j=1}^{p}\sum_{k=1}^{N} A_{i,j,k} W_j^\infty \frac{\eta_k^\infty}{|\eta|} = 0 \text{ for } i = 1, \ldots, q, \tag{17.36}$$

and

$$\Re Q(W^\infty) \leq -\varepsilon_0, \tag{17.37}$$

but (17.36) implies $W^\infty \in \Lambda + i\,\Lambda$, and the hypothesis (17.15) implies

$$\Re Q(\lambda) \geq 0 \text{ for all } \lambda \in \Lambda + i\,\Lambda, \tag{17.38}$$

in contradiction with (17.37). □

Corollary 17.4. *If Q is a real quadratic form on $\mathbb{R}^p$ satisfying*

$$Q(\lambda) = 0 \textit{ for all } \lambda \in \Lambda \tag{17.39}$$

then (17.16) and (17.17) imply

$$Q(U^n) \rightharpoonup Q(U^\infty) \textit{ in } \mathcal{M}(\Omega) \textit{ weak } \star. \tag{17.40}$$

Proof. Applying Theorem 17.3 to $-Q$ and $+Q$, one finds that for any subsequence such that (17.20) holds one has $\mu = Q(U^\infty)$, so that all the sequence $Q(U^n)$ converges to $Q(U^\infty)$ in $\mathcal{M}(\Omega)$ weak $\star$. □

If F is a polynomial of degree 2, then it satisfies (17.5) if and only if its homogeneous quadratic part Q satisfies (17.15).

In the cases where $\Lambda = \mathbb{R}^p$, Theorem 17.3 is a classical *convexity argument* saying that the functional Φ_φ defined on $L^2(\Omega, \mathbb{R}^p)$ by

$$\Phi_\varphi(U) = \int_\Omega \varphi(x) Q\big(U(x)\big)\,dx \text{ with } \varphi \geq 0, \varphi \in C_c(\Omega), \tag{17.41}$$

is (sequentially) weakly lower semicontinuous when restricted to sequences satisfying (17.17) if Q is convex (i.e., convex in all directions), which is but a classical consequence of the Hahn–Banach theorem.

In the case where $\Lambda = \{0\}$, Theorem 17.3 is a classical *compactness argument* saying that the functional Φ_φ is (sequentially) weakly continuous when restricted to sequences satisfying (17.17) if Q is any quadratic form, since sequences satisfying (17.16) and (17.17) converge in $L^2_{loc}(\Omega,\mathbb{R}^p)$ strong, by Theorem 17.3 applied to $U^m - U^\infty$ for a negative definite Q.

In the general case where $0 \neq \Lambda \neq \mathbb{R}^p$, Theorem 17.3 is a "new" argument of *compensated compactness*, saying that the functional Φ_φ is (sequentially) weakly lower semicontinuous when restricted to sequences satisfying (17.17) if Q is a quadratic form which is convex in some directions, given by Λ, which are the directions that cannot be controlled by the differential information (17.17).

When $\Lambda \neq \mathbb{R}^p$, there are nonconvex quadratic forms which are nonnegative on Λ: in an Euclidean space E,[7] if $\varepsilon > 0$ is small and $a \in E$ has norm 1, the quadratic form Q_ε defined by $Q_\varepsilon(U) = |U|^2 - (1+\varepsilon)(U,a)^2$ is positive except in a small conic neighbourhood of the direction of a, and one chooses $a \notin \Lambda$.

The *div–curl lemma* corresponds to the case $p = 2N$, with $U_i = E_i$ and $U_{N+i} = D_i$ for $i = 1,\ldots,N$, the list (17.17) containing the information on $curl(E^n)$ and $div(D^n)$, and in that case one has

$$\Lambda = \Big\{U = (E,D) \mid \exists \xi \neq 0, E_i\xi_j - E_j\xi_i = 0, i,j = 1,\ldots,N, \sum_{i=1}^N D_i\xi_i = 0\Big\}, \tag{17.42}$$

i.e.,

$$\Lambda = \{U = (E,D) \in \mathbb{R}^N \times \mathbb{R}^N \mid (E,D) = 0\}, \tag{17.43}$$

so the quadratic form Q defined by $Q(U) = (E,D)$ satisfies the hypothesis (17.39) of Corollary 17.4, and the conclusion (17.40) is the *div–curl lemma*.

In the case $0 \neq \Lambda \neq \mathbb{R}^p$, and for functions F which are not quadratic (plus affine), there are other conditions than (17.14) for (17.1)–(17.3) to imply (17.13), and cases where these conditions are not automatically satisfied. One important difference is that these conditions do not use the characteristic set Λ, which lost some useful information, but the more complete characteristic set $\mathcal{V}$ defined by

$$\mathcal{V} = \Big\{(\lambda,\xi) \in \mathbb{R}^p \times (\mathbb{R}^N \setminus 0) \mid \sum_{j=1}^p \sum_{k=1}^N A_{i,j,k}\xi_k\lambda_j = 0, i = 1,\ldots,q\Big\}. \tag{17.44}$$

Lemma 17.5. *If F is a function such that (17.1)–(17.3) imply (17.13), if*

$$(\lambda^m,\xi^m) \in \mathcal{V} \text{ for } m = 1,\ldots,r, \text{ and } rank(\xi^1,\ldots,\xi^r) < r, \tag{17.45}$$

then one has

$$\nabla^r F(a).(\lambda^1,\ldots,\lambda^r) = 0 \text{ for every } a \in \mathbb{R}^p. \tag{17.46}$$

---

[7] EUCLID of Alexandria, "Greek/Egyptian" mathematician, about 325 BCE–265 BCE. It is not known where he was born, but he worked in Alexandria, Egypt, shortly after it was founded by Alexander the Great, in 331 BCE.

Proof. Notice that (17.45) is impossible for $r = 1$. For $r = 2$ it means that there exists $\xi \neq 0$, such that $\lambda^1, \lambda^2 \in \Lambda_\xi$ where Λ_ξ is the subspace defined by

$$\Lambda_\xi = \{\lambda \in \mathbb{R}^p \mid (\lambda, \xi) \in \mathcal{V}\}, \tag{17.47}$$

and the condition (17.46) in that case corresponds to (17.14) since the union of all Λ_ξ is precisely the set Λ defined by (17.6), so F is affine in every direction parallel to Λ, so that, if X is the subspace spanned by Λ, the restriction of F to any subspace parallel to X is a polynomial of degree $\leq dim(X) \leq p$, and all the derivatives written make sense.

The result, being true for $r = 2$, is then extended by induction to $r > 2$. By the induction hypothesis, if $rank(\xi^1, \ldots, \xi^r) < r - 1$, then $\nabla^{r-1}F(a).(\lambda^1, \ldots, \lambda^{r-1}) = 0$ for every $a \in \mathbb{R}^p$, so that by deriving this identity in the direction λ^r, one obtains $\nabla^r F(a).(\lambda^1, \ldots, \lambda^r) = 0$ for every $a \in \mathbb{R}^p$. If $rank(\xi^1, \ldots, \xi^r) = r - 1$, one may assume that after a permutation $\xi^1, \xi^2 \ldots, \xi^{r-1}$ are linearly independent and by multiplying the ξ^i by nonzero constants, one may also assume that

$$\xi^r = \sum_{m=1}^{r-1} \xi^m. \tag{17.48}$$

Let U^n be defined by

$$U^n(x) = a + t \sum_{m=1}^{r} \cos\big(n\,(\xi^m, x)\big)\,\lambda^m, \tag{17.49}$$

so that (17.1) is satisfied with $U^\infty = a$, (17.3) is satisfied since $(\lambda^m, \xi^m) \in \mathcal{V}$ for $1 \leq m \leq r$, and (17.2) is true for some V^∞ as shown now. Indeed, since U^n takes its values in $a + X$ and the restriction of F to $a + X$ is a polynomial of degree at most $dim(X)$, $F\big(U^n(x)\big)$ is a combination of terms of the form $\prod_{i=1}^s \cos\big(n\,(\eta^i, x)\big)$, with each η_i being one of the ξ^m, and using (17.48) and trigonometric formulas, $F\big(U^n(x)\big)$ is a combination of terms of the form $\prod_{m=1}^{r-1} G_m\big(n\,(\xi^m, x)\big)$, with each G_m being of the form $\sin^a \cos^b$, and such a term converges in $L^\infty(\mathbb{R}^N)$ weak $\star$ to a constant. Hence (17.2) is true with V^∞ being a constant function and, a being fixed, this constant value depends upon t (and λ_m for $m = 1, \ldots, r-1$) in a polynomial way.

The constant term in the expansion in powers of t is a, and the coefficient of t is 0, since for any $\xi \neq 0$ the sequence $\cos\big(n\,(\xi, x)\big)$ converges to 0 in $L^\infty(\mathbb{R}^N)$ weak $\star$, but more generally, the induction hypothesis implies that the coefficient of t^k is automatically 0 for $1 \leq k \leq r-1$. Indeed, for $1 \leq k \leq r-1$ the coefficient of t^k in $F(U^n)$ is

$$\frac{1}{k!} \sum_\alpha \Big[\nabla^k F(a).(\lambda^{\alpha_1}, \lambda^{\alpha_2}, \ldots, \lambda^{\alpha_k}) \prod_{i=1}^{k} \cos\big(n\,(\xi^{\alpha_i}, x)\big)\Big], \tag{17.50}$$

where the sum is extended over all multi-indices α such that $1 \leq \alpha_i \leq r$ for $i = 1, \ldots, k$. Either all the α_i are distinct and in that case the ξ^{α_i} are linearly independent and the term $\prod_{i=1}^k \cos\big(n\,(\xi^{\alpha_i}, x)\big)$ converges to 0 in $L^\infty(\mathbb{R}^N)$ weak $\star$, or the α_i are not all distinct so $rank(\xi^{\alpha_1}, \xi^{\alpha_2}, \ldots, \xi^{\alpha_k}) < k$ and by the induction hypothesis the coefficient $\nabla^k F(a).(\lambda^{\alpha_1}, \lambda^{\alpha_2}, \ldots, \lambda^{\alpha_k})$ is 0, so at the limit the coefficient of t^k in V^∞ is 0. For $k = r$, there will appear $r!$ identical terms by symmetry of $\nabla^r F(a)$, with the coefficient $\nabla^r F(a).(\lambda^1, \lambda^2, \ldots, \lambda^r)$, which one is trying to identify, multiplied by the weak $\star$ limit of $\prod_{m=1}^r \cos\big(n\,(\xi^m, x)\big)$. Using (17.48), $\prod_{m=1}^r \cos\big(n\,(\xi^m, x)\big)$ has the same weak $\star$ limit than $\prod_{m=1}^{r-1} \cos^2\big(n\,(\xi^m, x)\big)$, which is a positive constant, so that in order to have $V^\infty = 0$, one must have $\nabla^r F(a).(\lambda^1, \lambda^2, \ldots, \lambda^r) = 0$. □

For $r \geq N + 1$, condition (17.45) is automatically satisfied, so that $\nabla^r F(a).(\lambda^1, \ldots, \lambda^r) = 0$ for all $\lambda_i \in \Lambda$, $i = 1, \ldots, r$, showing that on any subspace parallel to X, F is a polynomial of degree $\leq N$; by a preceding observation, its degree is then $\leq \min\{p, N\}$.

For what concerns whether there are other conditions than (17.5) for ensuring that (17.1)–(17.3) imply (17.4), I did not think that my construction of Lemma 17.5 could be useful, but in the 1990s Vladimír ŠVERÁK[8] settled negatively a similar question, that rank-one convexity does not imply quasi-convexity if $N = 3$, and the first part of his argument was similar to mine, while the second part of his argument was of a different nature. I just checked then the following simple generalization of Lemma 17.5.

Lemma 17.6. *Let F be a function of class C^3 such that (17.1)–(17.3) imply (17.4). Then if (17.45) holds with $r = 3$, and if for some $a_0 \in \mathbb{R}^p$ one has*

$$\nabla^2 F(a_0).(\lambda^m, \lambda^m) = 0 \text{ for } m = 1, 2, 3, \tag{17.51}$$

then one must have

$$\nabla^3 F(a_0).(\lambda^1, \lambda^2, \lambda^3) = 0. \tag{17.52}$$

Proof. One uses the same construction as in Lemma 17.5, with a replaced by a_0, so that (17.1) is true with $U^\infty = a_0$, (17.3) is true by construction, and since U^n is obtained by rescaling a fixed periodic function, (17.2) is true for some constant function V^∞, which is now a function of t of class C^3, and one looks at the Taylor expansion around a_0 in powers of t.

The coefficient of t is 0 since cos has average 0, and one shows that the coefficient of t^2 is also zero. If $rank(\xi^i, \xi^j) = 2$, then $\cos\big(n\,(\xi^i, x)\big)\cos\big(n\,(\xi^j, x)\big)$ converges to 0 in $L^\infty(\Omega)$ weak $\star$, while if $rank(\xi^i, \xi^j) = 1$ one deduces that $\lambda^i, \lambda^j \in \Lambda_\xi$ for some unit vector ξ, so that $\lambda^i \pm \lambda^j \in \Lambda_\xi \subset \Lambda$, implying $\nabla^2 F(a_0).(\lambda^i \pm \lambda^j, \lambda^i \pm \lambda^j) \geq 0$, and then (17.51) implies $\nabla^2 F(a_0).(\lambda^i, \lambda^j) = 0$, and all contributions to the coefficient of t^2 vanish.

---

[8] Vladimír ŠVERÁK, Czech-born mathematician. He works at UMN (University of Minnesota), Minneapolis, MN.

Since $\nabla^2 F(a_0 + s\,v)(\lambda^i, \lambda^i) \geq 0$ and since this quantity is 0 for $s = 0$, its derivative at $s = 0$ must also be 0, i.e., one has $\nabla^3 F(a_0).(\lambda^i, \lambda^i, v) = 0$ for $i = 1, 2, 3$, and all $v \in \mathbb{R}^p$. More generally, if $rank(\xi_i, \xi_j) = 1$ one also has $\nabla^3 F(a_0).(\lambda^i, \lambda^j, v) = 0$ for all $v \in \mathbb{R}^p$. Indeed, one has $\nabla^2 F(a_0 + s\,v).(\lambda^i + \lambda^j, \lambda^i + \lambda^j) \geq 0$ and since this quantity is 0 for $s = 0$, its derivative at $s = 0$ must be 0, i.e., $\nabla^3 F(a_0).(\lambda^i + \lambda^j, \lambda^i + \lambda^j, v) = 0$, but since one already knows that $\nabla^3 F(a_0).(\lambda^k, \lambda^k, v) = 0$ one deduces that $\nabla^3 F(a_0).(\lambda^i, \lambda^j, v) = 0$ for all $v \in \mathbb{R}^p$. The only possibly nonzero contributions to the coefficient of t^3 then come from the terms in $\nabla^3 F(a_0).(\lambda^1, \lambda^2, \lambda^3)$ and because of (17.48) the weak $\star$ limit of $\cos\big(n\,(\xi^1, x)\big)\cos\big(n\,(\xi^2, x)\big)\cos\big(n\,(\xi^3\,x)\big)$ is $\neq 0$, so that $\nabla^3 F(a_0).(\lambda^1, \lambda^2, \lambda^3)$ must be 0. □

I was led in a quite natural way to discover the necessary conditions of Lemma 17.5, in relation with the study of oscillations in a discrete velocity model from kinetic theory, the Broadwell model.[9] In order to understand what happens for weakly converging sequences of initial data, I had a situation corresponding to $N = 2$, $p = 3$, and differential information on $u^1_t + u^1_x$, $u^2_t - u^2_x$, and u^3_t being bounded in L^∞, and since Λ is the union of the three axes of coordinates, the quadratic functions $u^1 u^2$, $u^1 u^3$, and $u^2 u^3$ are sequentially weakly $\star$ continuous, and it remained to decide about the polynomial $u^1 u^2 u^3$.

In the situation of Lemma 17.5, François MURAT found the necessary conditions to be almost sufficient under mild restrictions, including a constant rank condition, which is that the dimension of Λ_ξ must be independent of ξ.

Additional footnotes: Alexander the Great.[10]

[9] James E. BROADWELL, American engineer. He worked at Caltech (California Institute of Technology), Pasadena, CA.

[10] Alexandros Philippou Makedonon, king of Macedon as Alexander III, 356–323 BCE. He is referred to as Alexander the Great, in relation with the large empire that he conquered.

Chapter 18
A Lemma for Studying Boundary Layers

In the fall of 1976, on the plane which was taking us to a Franco-Japanese meeting in Tokyo and Kyoto, Japan, Jacques-Louis LIONS asked me a question related to correctors near the boundary in a simple situation of periodic homogenization, when the boundary is a hyperplane containing all but one of the y^i, say $i = 1, \ldots, N-1$. As is usual in the study of boundary layers, it is natural in this case to rescale in the direction of y^N, and to look at a scalar second-order elliptic equation in a semi-infinite strip, and the usual question of matching the boundary layer with the internal behavior of the solution leads one to wonder if the solution in the semi-infinite strip tends to a constant as x_N tends to ∞.

Jacques-Louis LIONS said that there is convergence at an exponential rate, which he proved with probabilistic methods,[1] and he asked me to find an alternative approach, and it did not take me too long to provide him with the purely variational proof of Lemma 18.2, for which I invented an interesting variant of the Lax–Milgram lemma, Lemma 18.1. In one of his books, Jacques-Louis LIONS wrote a sentence which suggests that I only proved Lemma 18.1, but all the constructions of Lemma 18.2 are mine, and he only asked the question to find a variational approach.

Sometimes, asking the right question is an important step in proving an interesting mathematical result, and I usually emphasize who asked the questions that I answer, since it is a part of the discovery process, and it should not be neglected.[2]

One is still far from a good understanding about boundary layers in homogenization, and the question is important, since most of the error made by replacing an oscillating A^n by a non-oscillating A^{eff} seems to come from

---

[1] I thought then that Jacques-Louis LIONS did that with his usual collaborators of the moment, Alain BENSOUSSAN and George PAPANICOLAOU, but he did not mention them, and he probably worked alone on this question.

[2] I do not find honest that those for whom I answer a question would present my work as theirs, or that they would mention it once in an article with only their name, and then give the reference of this article without ever mentioning my name again, since it only seems to me a disguised way of stealing my ideas!

L. Tartar, *The General Theory of Homogenization*, Lecture Notes of the Unione Matematica Italiana 7, DOI 10.1007/978-3-642-05195-1_18,

what happens near the boundary. Apart from describing ideas in an approximately chronological order, I find another reason in presenting my result, that a similar point of view could be useful for various questions in continuum mechanics, but also for general partial differential equations.

The problem is set in an open cylinder $\Omega = \omega \times (0,\infty)$ of $\mathbb{R}^N$, with $\omega \subset \mathbb{R}^{N-1}$ open, and one wants to solve an elliptic equation

$$-div\big(A\,grad(u)\big) = f \text{ in } \Omega, \tag{18.1}$$

with Dirichlet condition on $\omega \times \{0\}$, and Neumann condition on $\partial\omega \times (0,\infty)$, or periodicity condition in x_i for $i = 1,\ldots,N-1$, if ω is a parallelepiped in $\mathbb{R}^{N-1}$, as was the situation considered by Jacques-Louis LIONS,[3] his question being the existence of u converging rapidly to a constant as $x_N \to \infty$.

For $s \in \mathbb{R}$, I considered the Hilbert spaces V^s and $W^s \subset V^s$ defined by

$$V^s = \big\{v \in H^1_{loc}(\Omega) \mid e^{s\,x_N} grad(v) \in L^2(\Omega;\mathbb{R}^N), u = 0 \text{ at } x_N = 0\big\}. \tag{18.2}$$

$$W^s = \big\{w \mid w \in V^s \text{ and } e^{s\,x_N} w \in L^2(\Omega)\big\}. \tag{18.3}$$

Assuming that $A \in L^\infty\big(\Omega;\mathcal{L}(\mathbb{R}^N;\mathbb{R}^N)\big)$ is uniformly elliptic, a classical application of the Lax–Milgram lemma would involve V^0 and would not tell one much about the asymptotic behavior as $x_N \to \infty$, so I looked at proving that $u \in V^\eta$ for some $\eta > 0$, and for controlling the norm of the solution I found it natural to consider the bilinear form b, continuous on $V^\eta \times W^\eta$, defined by

$$b(v,w) = \int_\Omega \big(A\,grad(v), grad(e^{2\eta\,x_N} w)\big)\,dx \text{ for all } v \in V^\eta, w \in W^\eta, \tag{18.4}$$

the linear form L, defined by

$$L(w) = \int_\Omega \big[g\,e^{2\eta\,x_N} w + \big(h, grad(e^{2\eta\,x_N} w)\big)\big]\,dx, \text{ for all } w \in W^\eta, \tag{18.5}$$

which is continuous on W^η if one has

$$e^{\eta\,x_N} g \in L^2(\Omega), e^{\eta\,x_N} h \in L^2(\Omega;\mathbb{R}^N), \tag{18.6}$$

and the problem

$$\text{find } u \in V^\eta \text{ such that } b(u,w) = L(w) \text{ for all } w \in W^\eta. \tag{18.7}$$

[3] Besides periodicity in $x_1,\ldots,x_{N-1}$, Jacques-Louis LIONS also assumed periodicity in x_N. In the summer of 1983, during a CEA–EDF–INRIA summer school at Bréau sans Nappe, near Rambouillet, France, he asked me a similar question for the case where each period contains a hole, and I told him how to create a variant of my method, using spaces ℓ^2 for taking care of the direction x_N.

The formulation (18.7) is equivalent to (18.1) with Neumann condition on the lateral boundary $\partial\omega \times (0,\infty)$, and Dirichlet condition at $x_N = 0$, with

$$f = g - div(h) \text{ in } \Omega, \tag{18.8}$$

but the classical Lax–Milgram lemma does not apply! The lack of symmetry in b made me introduce the space W^η, but I could not expect to find u there, since this implies that u converges to 0 as $x_N \to \infty$. One cannot take $w = u$ in (18.7), but (18.16) permits one to control $u - \overline{u}$ by $grad(u)$, where $\overline{u}$ is the average on u in x' defined by (18.15), which then depends only upon x_N; trying to make $u - \overline{u}$ appear led me to discover which w to use in the proof of Lemma 18.2, and then I looked at the variant of the Lax–Milgram lemma which is needed, and I proved Lemma 18.1. Notice that a simple choice for f is to assume

$$f \in L^2_{loc}(\Omega), e^{\eta\, x_N} f \in L^2(\Omega). \tag{18.9}$$

Lemma 18.1. *If V and W are Banach spaces,*[4] *if b is a continuous bilinear form on $V \times W$, if $M \in \mathcal{L}(V;W)$ is* surjective *and*

$$\textit{there exists } \delta > 0 \textit{ such that } b(v, M\,v) \geq \delta\,||v||_V^2 \textit{ for all } v \in V, \tag{18.10}$$

then for all $L \in W'$ there exists a unique solution satisfying

$$u \in V, b(u,w) = L(w) \textit{ for all } w \in W. \tag{18.11}$$

Proof. One defines $B \in \mathcal{L}(V;W')$ by $\langle B\,v, w\rangle = b(v,w)$ for all $v \in V$, $w \in W$.

$$\delta\,||v||_V^2 \leq b(v, M\,v) = \langle B\,v, M\,v\rangle \leq ||B\,v||_{W'}\,||M\,v||_W \text{ for all } v \in V \tag{18.12}$$

implies

$$||B\,v||_{W'} \geq \delta' ||v||_V \text{ for all } v \in V, \delta' = \frac{\delta}{||M||_{\mathcal{L}(V,W)}} > 0, \tag{18.13}$$

so that B is injective and $R(B)$ is closed. If $w_0 \in W$ is orthogonal to $R(B)$

$$b(v, w_0) = \langle B\,v, w_0\rangle = 0 \text{ for all } v \in V, \tag{18.14}$$

then by surjectivity of M one has $w_0 = M\,v_0$ for some $v_0 \in V$, and taking $v = v_0$ in (18.14) gives $v_0 = 0$ by (18.10), so that $w_0 = 0$, which by the Hahn–Banach theorem proves that $R(B)$ is dense in W', and B is an isomorphism from V onto W'. □

[4] They are actually Hilbert spaces in disguise, since one may use on V the scalar product $((v_1, v_2)) = b(v_1, M\,v_2) + b(v_2, M\,v_1)$, and V is isomorphic to W'.

One assumes that ω is bounded and for $\psi \in H^1(\omega)$ one uses

$$\overline{\psi} = \frac{1}{meas(\omega)} \int_\omega \psi \, dx', \tag{18.15}$$

where $dx' = dx_1 \ldots dx_{N-1}$. For $\eta \geq 0$ and $u \in V^\eta$, $\overline{u}$ is the function defined on $(0, \infty)$ by averaging in $x' = (x_1, \ldots, x_{N-1})$. One assumes that $\partial\omega$ is smooth enough for the Poincaré–Wirtinger inequality to hold,[5]

$$\int_\omega \left|\psi(x') - \overline{\psi}\right|^2 dx' \leq C(\omega)^2 \int_\omega |grad(\psi)|^2 \, dx' \text{ for all } \psi \in H^1(\omega). \tag{18.16}$$

To apply Lemma 18.1 one needs a mapping M, and the guess $M\,u = u - \overline{u}$ does not work, but the proof of Lemma 18.2 naturally leads one to define M by

$$M\,u = u - \tilde{u} \tag{18.17}$$

where $\tilde{u}$ is obtained from $\overline{u}$ by solving the differential equation

$$\frac{d\tilde{u}}{dx_N} + 2\eta\,\tilde{u} = 2\eta\,\overline{u} \text{ for } x_N \in (0, \infty), \text{ and } \tilde{u}(0) = 0. \tag{18.18}$$

Lemma 18.2. *If* $A \in L^\infty\big(\Omega; \mathcal{L}(\mathbb{R}^N; \mathbb{R}^N)\big)$ *satisfies*

$$(A\,\xi, \xi) \geq \alpha\,|\xi|^2 \text{ for all } \xi \in \mathbb{R}^N, \text{ a.e. in } \Omega, \tag{18.19}$$

$$|(A\,\xi, e_N)| \leq \gamma\,|\xi| \text{ for all } \xi \in \mathbb{R}^N, \text{ a.e. in } \Omega, \tag{18.20}$$

and if one chooses $\eta > 0$ *such that*

$$2\eta < \frac{\alpha}{C(\omega)\,\gamma}, \tag{18.21}$$

then the hypotheses of Lemma 18.1 are satisfied for b defined in (18.4), and M defined in (18.17) and (18.18), with notation (18.15).

Proof. Using (18.4) for b and (18.17) for M, one has

$$b(u, M\,u) = \int_\Omega e^{2\eta x_N} \Big[\big(A\,grad(u), grad(u)\big) + 2\eta\,(A\,grad(u), e_N)\Big(u - \tilde{u} - \frac{d\tilde{u}}{dx_N}\Big)\Big]\,dx, \tag{18.22}$$

[5] Wilhelm WIRTINGER, Austrian mathematician, 1865–1945. He worked in Innsbruck and Vienna, Austria.

and using (18.18) for $\tilde{u}$, one has

$$b(u, M\,u) = \int_\Omega e^{2\eta x_N} \big[(A\,grad(u), grad(u)) + 2\eta\,(A\,grad(u), e_N)(u - \overline{u})\big]\,dx, \tag{18.23}$$

and one chooses $\delta = \alpha - 2\eta\,\gamma\,C(\omega) > 0$, since one has

$$\begin{aligned} \Big|\int_\omega (A\,grad(u), e_N)(u - \overline{u})\,dx'\Big| &\le \gamma \int_\omega |grad(u)|\,|u - \overline{u}|\,dx' \\ &\le \gamma\,||grad(u)||_{L^2(\omega)}||u - \overline{u}||_{L^2(\omega)} \le \gamma\,C(\omega) \int_\omega |grad(u)|^2\,dx', \end{aligned} \tag{18.24}$$

a.e. $x_N \in (0, \infty)$ by (18.20), by the Cauchy–Schwarz inequality, and by (18.16).

One must check that $M\,u$ belongs to W^η. Since

$$\frac{\partial(M\,u)}{\partial x_i} = \frac{\partial u}{\partial x_i} - \delta_{i,N}\frac{d\tilde{u}}{dx_N}, i = 1, \ldots, N, \tag{18.25}$$

one has $M\,u \in V^\eta$ if (and only if) $e^{\eta\,x_N}\frac{d\tilde{u}}{dx_N} \in L^2(0, \infty)$. By (18.16)

$$\int_\Omega e^{2\eta\,x_N}|u - \overline{u}|^2\,dx \le C^2(\omega)\sum_{i=1}^{N-1}\int_\Omega e^{2\eta\,x_N}\Big|\frac{\partial u}{\partial x_i}\Big|^2\,dx, \tag{18.26}$$

and since $\psi \mapsto \overline{\psi}$ is a contraction in $L^2(\omega)$, one has

$$meas(\omega)\int_0^\infty e^{2\eta\,x_N}\Big|\frac{d\overline{u}}{dx_N}\Big|^2\,dx_N \le \int_\Omega e^{2\eta\,x_N}\Big|\frac{\partial u}{\partial x_N}\Big|^2\,dx, \tag{18.27}$$

so that one has $e^{\eta\,x_N}M\,u \in L^2(\Omega)$ if (and only if) $e^{\eta\,x_N}(\tilde{u} - \overline{u}) \in L^2(0, \infty)$. The desired properties follow from (18.18), because

$$\frac{d[e^{\eta\,x_N}(\tilde{u} - \overline{u})]}{dx_N} + \eta\,[e^{\eta\,x_N}(\tilde{u} - \overline{u})] = -\,e^{\eta\,x_N}\frac{d\overline{u}}{dx_N} \in L^2(0, \infty), \tag{18.28}$$

and

$$\frac{d\varphi}{dx_N} + \eta\,\varphi \in L^2(0, \infty), \varphi(0) = 0 \text{ imply } \varphi, \frac{d\varphi}{dx_N} \in L^2(0, \infty). \tag{18.29}$$

One must also check that M is surjective, and in order to solve $M\,u = v$ with $v \in V_\eta^0$ given, one seeks a solution of the form $u(x) = v(x) + h(x_N)$, which implies $\overline{u} = \overline{v} + h$ and in order to have $\tilde{u} = h$ one must have

$$\frac{dh}{dx_N} + 2\eta\, h = 2\eta\, \overline{u} = 2\eta\, (\overline{v} + h), h(0) = 0, \tag{18.30}$$

so that $e^{\eta\, x_N} \frac{dh}{dx_N} = 2\eta\, e^{\eta\, x_N} \overline{v}$, which belongs to $L^2(0,\infty)$ since $v \in W^\eta$. $\square$

For $u \in V^\eta$, the limit of $\overline{u}(x_N)$ exists as x_N tends to ∞, and by (18.16) $e^{\eta\, x_N}(u - \overline{u}) \in L^2(\Omega)$, so that u converges to a constant at infinity.

If one uses $\alpha(x_N)$ in (18.19) and $\gamma(x_N)$ in (18.20), then the constraint (18.21) on η involves $\inf_{x_N} \alpha(x_N)$ and $\sup_{x_N} \gamma(x_N)$, although only the behavior of $\alpha(x_N)$ and $\gamma(x_N)$ for x_N very large should play a role for determining the asymptotic behavior of the solution as $x_N \to \infty$.

In the late 1970s, I described my method in a seminar in Nice, France, and Pierre GRISVARD was surprised that one needed a variant of the Lax–Milgram lemma,[6] but he could not see a way to avoid it. He would certainly appreciate my second method, but I only found it in the mid 1990s.

In the early 1980s, Jacques-Louis LIONS told me that Olga OLEINIK found my result connected to the Saint-Venant principle.[7] At a conference in the summer of 1986 in Durham, England, she discussed similar questions without mentioning my result; after her talk, I asked her about that, and she said that she was not aware of my result, which was puzzling, since Jacques-Louis LIONS having no interest in continuum mechanics could hardly think by himself about the Saint-Venant principle, and there was no reason to attribute this remark to Olga OLEINIK if he heard it from someone else.[8]

In 1987, Klaus KIRCHGÄSSNER told me that questions about the Saint-Venant principle are different in nature,[9] since they are posed in doubly infinite strips, and he preferred to talk about results of the Lindelöf type.[10]

In the mid 1990s, I developed a second method for my student Gregor WEISKE,[11] who was going to look at generalizations, and he showed as part of his PhD thesis that it can be used for nonlinearities of monotone type.

My second method uses the Lax–Milgram lemma on V^0, and uses in a better way the asymptotic behavior of $\alpha(x_N)$ and $\gamma(x_N)$ in constraining η.

---

[6] Pierre GRISVARD, French mathematician, 1940–1994. He worked in Nice, France.

[7] Adhémar Jean Claude BARRÉ DE SAINT-VENANT, French mathematician, 1797–1886. He worked in Paris, France.

[8] As for my joint work with François MURAT, Jacques-Louis LIONS probably used my result in one of his talks in Moscow, Russia, but forgot to mention my name, and Olga OLEINIK probably directed to him her comment concerning the Saint-Venant principle, thinking that he described one of his ideas!

[9] Klaus KIRCHGÄSSNER, German mathematician, born in 1931. He worked in Stuttgart, Germany.

[10] Ernst Leonard LINDELÖF, Finnish mathematician, 1870–1946. He worked in Helsinki, Finland.

[11] Gregor Christian WEISKE, German-born mathematician. He studied under my supervision for his PhD (1997), at CMU (*Carnegie Mellon University*), Pittsburgh, PA.

Lemma 18.3. *If* $A \in L^\infty\big(\Omega; \mathcal{L}(\mathbb{R}^N; \mathbb{R}^N)\big)$ *satisfies*

$$(A(x)\xi, \xi) \geq \alpha(x_N)\,|\xi|^2 \text{ for all } \xi \in \mathbb{R}^N \text{ a.e. } x \in \Omega, \tag{18.31}$$
$$|(A(x)\xi, e_N)| \leq \gamma(x_N)\,|\xi| \text{ for all } \xi \in \mathbb{R}^N, \text{ a.e. } x \in \Omega, \tag{18.32}$$

with $\alpha(x_N) \geq \alpha_- > 0$ *a.e. on* $(0,\infty)$, *and* $\eta > 0$ *satisfies*

$$2\eta < \liminf_{s\to\infty} \frac{\alpha(s)}{C(\omega)\,\gamma(s)}, \tag{18.33}$$

then if

$$L(v) = \int_\Omega \big(h, grad(v)\big)\,dx, \text{ with } e^{\eta\,x_N} h \in L^2(\Omega; \mathbb{R}^N), \tag{18.34}$$

the solution of

$$u \in V^0, \langle \mathcal{A}u, v\rangle = L(v) \text{ for all } v \in V^0, \tag{18.35}$$

satisfies $u \in V^\eta$.

Proof. Using $w = \varphi\,u - \psi$, with φ and ψ depending only on x_N gives

$$\langle \mathcal{A}u, \varphi\,u - \psi\rangle = \int_\Omega \big[\varphi\,\big(A\,grad(u), grad(u)\big) + (A\,grad(u), e_N)(\varphi' u - \psi')\big]\,dx, \tag{18.36}$$

where $'$ means $\frac{d}{dx_N}$, and for having $w \in V^0$ one assumes that

$$\varphi, \varphi' \text{ bounded}, \varphi > 0, \psi' = \varphi'\overline{u} \text{ on } (0,\infty), \text{ and } \psi(0) = 0, \tag{18.37}$$

noticing that $\varphi' u - \psi' = \varphi'(u - \overline{u}) \in L^2(\Omega)$, and one deduces that

$$\langle \mathcal{A}u, \varphi\,u - \psi\rangle \geq \int_\Omega |grad(u)|^2 \big(\varphi(x_N)\alpha(x_N) - |\varphi'(x_N)|\,C(\omega)\,\gamma(x_N)\big)\,dx. \tag{18.38}$$

By (18.33), there exists $\varepsilon > 0$ and $S \geq 0$ such that

$$2\eta + \varepsilon \leq \frac{\alpha(s)}{C(\omega)\,\gamma(s)} \text{ for all } s \geq S, \tag{18.39}$$

and for $T > S$ one chooses

$$\varphi_T(s) = \begin{cases} 1 \text{ for } 0 \leq s \leq S \\ e^{2\eta\,(s-S)} \text{ for } S \leq s \leq T \\ e^{2\eta\,(T-S)} \text{ for } s \geq T, \end{cases} \tag{18.40}$$

so that, using $\gamma(s) \geq \alpha(s) \geq \alpha_- > 0$ a.e. on $(0,\infty)$, one has

$$\varphi_T(s)\alpha(s) - |\varphi_T'(s)|\,C(\omega)\,\gamma(s) \geq \delta\,\varphi_T(s) \text{ a.e. on } (0,\infty), \tag{18.41}$$

with $\delta = \min\{1, \varepsilon\,C(\omega)\}\,\alpha_- > 0$. Using (18.37) and then (18.40) gives

$$L(\varphi\,u - \psi) = \int_\Omega \big[\varphi\,(h, grad(u)) + \varphi' h_N(u - \overline{u})\big]\,dx \tag{18.42}$$

$$|L(\varphi_T\,u - \psi_T)| \leq \kappa\left(\int_\Omega \varphi_T|grad(u)|^2\,dx\right)^{1/2}\left(\int_\Omega \varphi_T|h|^2\,dx\right)^{1/2}, \tag{18.43}$$

with $\kappa = 1 + 2\eta\,C(\omega)$. Using (18.38) and (18.41), one deduces from (18.43)

$$\int_\Omega \varphi_T|grad(u)|^2\,dx \leq \frac{\kappa^2}{\delta^2}\int_\Omega \varphi_T|h|^2\,dx. \tag{18.44}$$

Since $\varphi_T(s) \leq e^{2\eta\,s}$ for all $s > 0$, the right side of (18.44) is bounded by $\int_\Omega e^{2\eta\,x_N}|h|^2\,dx < \infty$ by (18.34), and letting T tend to ∞ gives

$$\int_\Omega \max\{1, e^{2\eta\,(x_N - S)}\}|grad(u)|^2\,dx \leq \frac{\kappa^2}{\delta^2}\int_\Omega e^{2\eta\,x_N}|h|^2\,dx, \tag{18.45}$$

so that $u \in V^\eta$. □

I once discussed with Gilles FRANCFORT using my second method with $\varphi(s) = e^{-2\eta\,|s-z|}$, and changing the position of z, for approaching a doubly infinite strip as a limit of long cylinders; it reminded me of a result by Seva SOLONNIKOV,[12] which I heard someone else explain in a seminar in Paris, France, for a flow in an infinite pipe, where the difficulty was to found uniform bounds for the kinetic energy, the energy dissipated by viscosity, and the drop of pressure on intervals of fixed length.

[12] Vsevolod Alekseevich SOLONNIKOV, Russian mathematician, born in 1933. He worked in Leningrad/St Petersburg, Russia, but also at Università di Ferrara, Ferrara, Italy.

Chapter 19
A Model in Hydrodynamics

In December 1976, at a conference at École Centrale in Lyon, France, I presented a result in periodic homogenization for a model which I devised, by playing with the equations of hydrodynamics with the idea of understanding something concerning turbulence. After my talk, Jacques-Louis LIONS gave me pedagogical advice, to never put more than one idea in a talk!

In my continuum mechanics course at École Polytechnique (in Paris, France, 1966–1967), Jean MANDEL did not say much about twentieth century mechanics, plasticity (which was his own research specialty) and turbulence, apart from mentioning the importance of Reynolds numbers, and explaining that wires whistle when the wind is strong because of von Kármán vortices detaching alternatively from one side and the other, at an audible frequency.[1]

I later heard about KOLMOGOROV's ideas concerning "developed isotropic turbulence," and it seemed that he made a similar mistake in guessing a question of homogenization than LANDAU and LIFSHITZ did later, looking for a nonexisting formula for the effective conductivity of a mixture.[2]

Not knowing all the defects of the Navier–Stokes equation, I thought about "turbulence" as letting the kinematic viscosity $\nu = \frac{\mu}{\varrho}$ tend to 0,[3] in which case one loses the bound on $grad(u)$, and one cannot expect to use the compactness method for passing to the limit. It was a few months before I developed the *compensated compactness method*,[4] and a few years before I understood

---

[1] Todor (Theodore) VON SKÖLLÖSKISLAKI KÁRMÁN, Hungarian-born mathematician, 1881–1963. He worked in Aachen, Germany, and at Caltech (California Institute of Technology), Pasadena, CA.

[2] The term "developed" shows that KOLMOGOROV did not want to describe how turbulence builds up, but only what happens once it settles, but did he know of many real examples where the result is "isotropic"?

[3] Olga LADYZHENSKAYA seemed to be the one who first thought about the game of letting time tend to ∞ and considering the Hausdorff dimension of an attractor, but this question has no relation to turbulence, and it is precisely why it became so popular among those who advocate fake mechanics, and rarely mention the name of Olga LADYZHENSKAYA, of course!

[4] It must certainly be improved for solving questions about turbulence.

L. Tartar, *The General Theory of Homogenization*, Lecture Notes of the Unione Matematica Italiana 7, DOI 10.1007/978-3-642-05195-1_19,

the importance of studying a different question of homogenization, for first-order partial differential equations, which I shall describe with the question of appearance of nonlocal effects, in Chaps. 23 and 24. I considered the equality

$$\frac{\partial u}{\partial t}+\left(\sum_{j=1}^{3}u_j\frac{\partial}{\partial x_j}\right)u=\frac{\partial u}{\partial t}+u\times curl(-u)+grad\left(\frac{|u|^2}{2}\right), \tag{19.1}$$

certainly known to D. BERNOULLI,[5] who studied stationary irrotational (i.e., with $curl(u)=0$) solutions of the Euler equation,[6] and found them to satisfy the (D.) Bernoulli law $p+\frac{\varrho|u|^2}{2}=constant$, and my observation was that for small viscosity the vorticity $curl(u)$ could be large, and the force $u\times curl(-u)$ would look like a part of the Lorentz force $E+u\times B$ in electromagnetism,[7] and would make the fluid spin; I then decided to replace $curl(-u)$ by a given oscillating vector field and study its effect on the solution. Not knowing what to expect, I began with the stationary case, and I first used formal asymptotic expansions in a periodic setting, so that I considered

$$-\nu\,\Delta\,u^n+u^n\times\frac{1}{\varepsilon_n}b\left(\frac{x}{\varepsilon_n}\right)+grad\,p_n=f, div(u^n)=0 \text{ in } \Omega, u^n\in H_0^1(\Omega;\mathbb{R}^3), \tag{19.2}$$

for a periodic vector field b. As Michel FORTIN was sharing my office, we did the computations together, and we first noticed that the average of b must be 0, or the whole fluid would spin too fast, and then we derived an equation for the first term of the formal asymptotic expansion. Using my *method of oscillating test functions*, I then proved the result to hold.

Although the force $u^n\times b^n$ is orthogonal to the velocity u^n, and does no work, it induces oscillations in $grad(u^n)$ which dissipate more energy by viscosity (per unit of time, since one is looking at a stationary problem). An interesting feature is that the added dissipation which appears in the limiting

---

[5] Daniel BERNOULLI, Swiss mathematician, 1700–1782. He worked in St Petersburg, Russia, and in Basel, Switzerland.

[6] EULER introduced the Euler equation for an inviscid fluid, i.e., with $\mu=0$, long before NAVIER wrote his equation, which one now calls the Navier–Stokes equation, although STOKES first wrote the linear Stokes equation, and only much later considered the nonlinear case.

[7] I do not remember when I read about the analogy between the Euler equation and electromagnetism, in an article by Keith MOFFATT, where he discussed the conservation of helicity, observed by Jean-Jacques MOREAU.

equation is not quadratic in $grad(u^\infty)$, but quadratic in u^∞, contrary to a general belief about "turbulent viscosity."[8]

I did not check the non-periodic case at the time, but I did it for the occasion of a conference at IMA, Minneapolis, MN, in October 1984. I used a term $u^n \times curl(v^n)$ with v^n converging weakly, but since there is a quadratic effect in the strength of the oscillations, I wrote $v^n = v^0 + \lambda\, w^n$ with w^n converging weakly to 0, for showing that the effective equation has a term in $\lambda^2 M^{\mathit{eff}}\, u^\infty$. Also, I did not impose boundary conditions.[9]

Lemma 19.1. *In an open set $\Omega \subset \mathbb{R}^3$, one assumes that*

$$-\nu\,\Delta\, u^n + \big(u^n \times curl(v^0 + \lambda\, w^n)\big) + grad(p_n) = f^n, div(u^n) = 0 \text{ in } \Omega, \quad (19.3)$$

with $v^0 \in L^3_{loc}(\Omega; \mathbb{R}^3)$ and

$$\begin{aligned} &u^n \rightharpoonup u^\infty \text{ in } H^1_{loc}(\Omega; \mathbb{R}^3) \text{ weak}, w^n \rightharpoonup 0 \text{ in } L^3_{loc}(\Omega; \mathbb{R}^3) \text{ weak},\\ &p_n \rightharpoonup p_\infty \text{ in } L^2_{loc}(\Omega) \text{ weak}, f^n \to f^\infty \in H^{-1}_{loc}(\Omega; \mathbb{R}^3) \text{ strong}. \end{aligned} \quad (19.4)$$

Then, there is a subsequence indexed by m, such that

$$u^m \times curl(w^m) \rightharpoonup \lambda\, M^{\mathit{eff}} u^\infty \text{ in } H^{-1}_{loc}(\Omega; \mathbb{R}^3) \text{ weak}, \quad (19.5)$$

$$\nu\,|grad(u^m)|^2 \rightharpoonup \nu\,|grad(u^\infty)|^2 + \lambda^2 \,(M^{\mathit{eff}} u^\infty, u^\infty) \text{ in } \mathcal{M}(\Omega) \text{ weak} \star, \quad (19.6)$$

$$\begin{aligned} &-\nu\,\Delta\, u^\infty + \big(u^\infty \times curl(v^0)\big) + \lambda^2 M^{\mathit{eff}} u^\infty + grad(p_\infty)\\ &\quad = f^\infty, div(u^\infty) = 0 \text{ in } \Omega, \end{aligned} \quad (19.7)$$

for a symmetric nonnegative tensor $M^{\mathit{eff}} \in L^{3/2}_{loc}\big(\Omega; \mathcal{L}_{sym+}(\mathbb{R}^3; \mathbb{R}^3)\big)$,[10] depending only upon ν and the sequence w^m.

Proof. As all statements are local, one works on a bounded open set ω with smooth boundary such that $\overline{\omega} \subset \Omega$, and after writing Ω as a countable increasing union of such open sets, one uses a diagonal subsequence. For three independent vectors $k \in \mathbb{R}^3$, one defines $z^n \in H^1_0(\omega; \mathbb{R}^3)$ by

$$\text{for } k \in \mathbb{R}^3, -\nu\,\Delta\, z^n + \big(k \times curl(w^n)\big) + grad(q_n) = 0, div(z^n) = 0 \text{ in } \omega, \quad (19.8)$$

---

[8] However, one must be careful in interpreting the mathematical results, since my model has some non-physical aspects.

[9] If one wants to speak about the effective properties of a mixture, one should obtain a result independent of the boundary conditions used, like Lemma 10.3; if one fails to do this, one may be talking about global properties of the "mixture together with its container."

[10] By the Sobolev embedding theorem, $u^\infty \in H^1_{loc}(\Omega; \mathbb{R}^3) \subset L^6_{loc}(\Omega; \mathbb{R}^3)$, so that one has $M^{\mathit{eff}} u^\infty \in L^{6/5}_{loc}(\Omega; \mathbb{R}^3) \subset H^{-1}_{loc}(\Omega; \mathbb{R}^3)$.

and $q_n \in L^2(\omega)$ is normalized to have integral 0, so that one has

$$z^n \rightharpoonup 0 \text{ in } H_0^1(\omega;\mathbb{R}^3) \text{ weak}, q_n \rightharpoonup 0 \text{ in } L^2(\omega) \text{ weak}, \tag{19.9}$$

and one extracts a subsequence such that[11]

$$z^m \times curl(w^m) \rightharpoonup M^{eff} k \text{ in } H^{-1}(\omega;\mathbb{R}^3) \text{ weak, for all } k \in \mathbb{R}^3. \tag{19.10}$$

By elliptic regularity theory (and the Calderón–Zygmund theorem), $grad(z^n)$ is bounded in $L^3(\omega;\mathbb{R}^{3\times 3})$ and $z^n \to 0$ in $L^r(\omega;\mathbb{R}^3)$ strong for all $r < \infty$, by the Sobolev embedding theorem and the Rellich–Kondrašov compactness embedding theorem. One has $z_i^m \frac{\partial w_j^m}{\partial x_\kappa} = \frac{\partial [z_i^m w_j^m]}{\partial x_\kappa} - \frac{\partial z_i^m}{\partial x_\kappa} w_j^m$, and $z_i^m w_j^m$ converges strongly to 0 in $L^2(\omega)$, and $\frac{\partial z_i^m}{\partial x_\kappa} w_j^m$ is bounded in $L^{3/2}(\omega)$, so that it belongs to a compact of $H^{-1}(\omega)$ strong. One deduces that, for all $k \in \mathbb{R}^3$, one has[12]

$$z^m \times curl(w^m) \rightharpoonup M^{eff} k \text{ in } H^{-1}(\omega;\mathbb{R}^3) \text{ strong}, M^{eff} k \in L^{3/2}(\omega;\mathbb{R}^3). \tag{19.11}$$

One assumes that

$$u^m \times curl(w^m) \rightharpoonup g \text{ in } H^{-1}(\omega;\mathbb{R}^3) \text{ weak}, \tag{19.12}$$

and one wants to show that $g = \lambda M^{eff} u^\infty$, which is (19.5), and implies (19.7). For $\varphi \in C_c^1(\omega)$, one multiplies the equation for u^m by $\varphi\, z^m$, and since $div(\varphi\, z^m) = (grad(\varphi), z^m)$ may be $\neq 0$, one uses the fact that p_n is bounded in $L^2(\omega)$; since $z^m \to 0$ in $L^2(\omega;\mathbb{R}^3)$ strong, one has

$$\begin{aligned} &\langle grad(p_m), \varphi\, z^m\rangle = -\textstyle\int_\omega p_m (grad(\varphi), z^m)\, dx \to 0,\\ &\nu \textstyle\int_\omega (grad(u^m), grad(\varphi) \otimes z^m)\, dx \to 0, \end{aligned} \tag{19.13}$$

and also, since $\varphi\, u_i^m z_j^m \rightharpoonup 0$ in $W_0^{1,3/2}(\omega)$ weak for all i, j, one has

$$\langle u^m \times curl(v^0), \varphi\, z^m\rangle \to 0, \tag{19.14}$$

so one deduces that

$$\nu \int_\omega \varphi\, (grad(u^m), grad(z^m))\, dx + \lambda\, \langle u^m \times curl(w^m), \varphi\, z^m\rangle \to 0. \tag{19.15}$$

---

[11] By the Sobolev embedding theorem, $H_0^1(\omega) \subset L^6(\omega)$, and for $v \in L^3(\omega;\mathbb{R}^3)$ the mapping $u \mapsto u \times curl(v)$ is continuous from $H_0^1(\omega;\mathbb{R}^3)$ into $H^{-1}(\omega;\mathbb{R}^3)$, since for $\varphi \in H_0^1(\omega)$, one has $u\,\varphi \in W_0^{1,3/2}(\omega;\mathbb{R}^3)$.

[12] The convergence in (19.11) holds in $W^{-1,s}(\omega;\mathbb{R}^3)$ strong for $2 \le s < 3$.

Since $(a \times b, c) = -(c \times b, a)$ for all $a, b, c \in \mathbb{R}^3$, one deduces that

$$\lambda \langle u^m \times curl(w^m), \varphi\, z^m \rangle = -\lambda \langle z^m \times curl(w^m), \varphi\, u^m \rangle \to -\lambda \langle M^{\mathit{eff}} k, \varphi\, u^\infty \rangle, \tag{19.16}$$

although $curl(w^m)$ is a distribution, and using (19.11). One multiplies the equation for z^m by $\varphi\, u^m$, and using $q_m \rightharpoonup 0$ in $L^2(\omega)$ weak, $grad(z^m) \rightharpoonup 0$ in $L^2(\omega; \mathbb{R}^3)$ weak, and $u^m \to u^\infty$ in $L^2(\omega; \mathbb{R}^3)$ strong, one deduces that

$$\begin{aligned} &\langle grad(q_m), \varphi\, u^m \rangle = -\int_\omega q_m (grad(\varphi), u^m)\, dx \to 0,\\ &\nu \int_\omega (grad(z^m), grad(\varphi) \otimes u^m)\, dx \to 0, \end{aligned} \tag{19.17}$$

and

$$\nu \int_\omega \varphi\, \big(grad(z^m), grad(u^m)\big)\, dx + \langle k \times curl(w^m), \varphi\, u^m \rangle \to 0, \tag{19.18}$$

so that one has

$$\langle k \times curl(w^m), \varphi\, u^m \rangle = -\langle u^m \times curl(w^m), \varphi\, k \rangle \to -\langle g, \varphi\, k \rangle. \tag{19.19}$$

Subtracting (19.15) and (19.18), and then using (19.16) and (19.19), one has

$$\lambda \langle M^{\mathit{eff}} k, \varphi\, u^\infty \rangle = \langle g, \varphi\, k \rangle \text{ for all } \varphi \in C_c^1(\omega), k \in \mathbb{R}^3, \text{ i.e., } g = \lambda\, (M^{\mathit{eff}})^T u^\infty, \tag{19.20}$$

and it remains to show that M^{eff} is symmetric a.e. in ω. This follows by the same method: z^{*n} being the solution for k^*, multiplying the equation for z^{*n} by $\varphi\, z^n$, the equation for z^n by $\varphi\, z^{*n}$, and noticing that

$$\begin{aligned} &\int_\omega (grad(z^m), grad(\varphi) \otimes z^{*m})\, dx \to 0,\\ &\int_\omega (grad(z^{*m}), grad(\varphi) \otimes z^m)\, dx \to 0\\ &\int_\omega q_m (grad(\varphi), z^{*m})\, dx \to 0,\\ &\int_\omega q_{*m} (grad(\varphi), z^m)\, dx \to 0, \end{aligned} \tag{19.21}$$

one obtains

$$\begin{aligned} &\langle k \times curl(w^m), \varphi\, z^{*m} \rangle - \langle k^* \times curl(w^{*m}), \varphi\, z^m \rangle \to 0\\ &\text{i.e.,} \quad \langle M^{\mathit{eff}} k^*, \varphi\, k \rangle = \langle M^{\mathit{eff}} k, \varphi\, k^* \rangle \text{ for all } \varphi \in C_c^1(\omega), \end{aligned} \tag{19.22}$$

which is

$$(M^{eff} k^*, k) = (M^{eff} k, k^*) \text{ a.e. in } \omega. \tag{19.23}$$

By multiplying the equation for u^m by $\varphi\, u^m$, one has

$$\lim_m \int_\omega \varphi\, \nu\, (grad(u^m), grad(u^m))\, dx + \int_\omega \varphi\, \nu\, (grad(u^\infty), grad(\varphi) \otimes u^\infty)\, dx$$
$$- \int_\omega p_\infty(grad(\varphi), u^\infty)\, dx \to \langle f^\infty, \varphi\, u^\infty \rangle \tag{19.24}$$

and by multiplying the equation for u^∞ by $\varphi\, u^\infty$, one has

$$\int_\omega [\varphi\, \nu\, (grad(u^\infty), grad(u^\infty)) + \varphi\, \nu\, (grad(u^\infty), grad(\varphi) \otimes u^\infty)]\, dx$$
$$+ \int_\omega [\lambda^2 (M^{eff} u^\infty, u^\infty) - p_\infty(grad(\varphi), u^\infty)]\, dx = \langle f^\infty, \varphi\, u^\infty \rangle, \tag{19.25}$$

which implies

$$\int_\omega \varphi\, \nu\, (grad(u^m), grad(u^m))\, dx \to \int_\omega [\varphi\, \nu\, (grad(u^\infty), grad(u^\infty))$$
$$+ \lambda^2 (M^{eff} u^\infty, u^\infty)]\, dx, \text{ for all } \varphi \in C^1_c(\omega), \tag{19.26}$$

and since $\nu\, \big(grad(u^m), grad(u^m)\big)$ is bounded in $L^1(\omega)$, (19.26) is then true for all $\varphi \in C_c(\omega)$, so that the convergence is in $\mathcal{M}(\omega)$ weak $\star$, showing (19.6), which implies that $M^{eff} \geq 0$. □

If $div(w^n) = 0$, which is the case for fluid dynamics, one takes[13]

$$z^n = \sum_{j=1}^3 k_j \frac{\partial Z^n}{\partial x_j}, \tag{19.27}$$

with Z^n solving

$$-\nu\, \Delta\, Z^n = w^n, \tag{19.28}$$

so that $div(z^n) = 0$. Since second derivatives of Z^n_ℓ converge to 0 in $L^2_{loc}(\Omega)$ weak, and first derivatives converge to 0 in $L^2_{loc}(\Omega)$ strong, one deduces that

$$\nu \sum_{\kappa,\ell=1}^3 \frac{\partial^2 Z^m_\ell}{\partial x_i \partial x_\kappa} \frac{\partial^2 Z^m_\ell}{\partial x_j \partial x_\kappa} \rightharpoonup M^{eff}_{i,j} \text{ in } \mathcal{M}(\Omega) \text{ weak } \star. \tag{19.29}$$

[13] One has $[k \times curl(w)]_i = \sum_j k_j [\partial_i w_j - \partial_j w_i]$.

In October 1984, I found the quadratic effect of the oscillations quite intriguing, and I saw a similarity with how effective corrections at a macroscopic level are often computed in quantum mechanics, but I was frustrated that my physicist friends at the conference, David BERGMAN and Graeme MILTON, did not show much interest in discussing this remark with me. Formula (19.29) was my first hint about the usefulness of defining *H-measures*,[14] which I only introduced about 2 years after for the different question of "small-amplitude" homogenization, which I describe in Chap. 29. I shall describe then the formula with *H-measures* giving M^{eff}, and what changes must be made for the evolution case, which I also studied at the end of 1984.

I do not remember when George PAPANICOLAOU told me that terms like $M\,u$ are called Brinkman forces in the literature. Such forces appear for flows around obstacles, which induce a drag, and despite some non-physical aspects of my model, I want to interpret the result as saying that vortices oppose a resistance proportional to the difference in velocities, and it suggests describing a turbulent flow by adding a variable like M, which serves in computing an added dissipation, which cannot be accounted for by a "turbulent viscosity," and one would need an evolution equation for M; however, in the spirit of kinetic theory, there could be different modes behaving each in its own way, so that one must be aware of some defects of kinetic theory, like those that I described in [119], and one should be aware of new tools like *H-measures*, which must be improved.

Additional footnotes: HAUSDORFF,[15] Olga LADYZHENSKAYA,[16] Keith MOFFATT,[17] Jean-Jacques MOREAU.[18]

---

[14] With (M.) Riesz operators R_k, (19.29) uses the limits of $R_i R_\kappa w_\ell^m R_j R_\kappa w_\ell^m$.

[15] Felix HAUSDORFF, German mathematician, 1869–1942. He worked in Leipzig, in Greifswalf and in Bonn, Germany. He wrote literary and philosophical work under the pseudonym of Paul MONGRÉ.

[16] Olga Aleksandrovna LADYZHENSKAYA, Russian mathematician, 1922–2004. She worked at the Steklov Mathematical Institute, in St Petersburg, Russia (named Leningrad, USSR, for many years).

[17] Henry Keith MOFFATT, Scottish-born mathematician, born in 1935. He worked in Bristol and Cambridge, England.

[18] Jean-Jacques MOREAU, French mathematician, born in 1923. He worked at Université des Sciences et Techniques de Languedoc (Montpellier II), Montpellier, France.

Chapter 20
Problems in Dimension $N = 2$

In the early 1970s, when I started working with François MURAT, we considered mixtures of two isotropic conductors, corresponding to

$$A^n = \big(\chi_n \alpha + (1 - \chi_n)\, \beta\big)\, I \text{ H-converges to } A^{\mathit{eff}}, \chi_n \rightharpoonup \theta \text{ in } L^\infty(\Omega) \text{ weak } \star, \tag{20.1}$$

and, for discovering a relaxed problem,[1] we looked for a characterization

$$A^{\mathit{eff}} \in \mathcal{K}(\theta) \text{ a.e. in } \Omega. \tag{20.2}$$

With Lemma 6.7, we had a first necessary condition

$$\begin{gathered} \lambda_-(\theta)\, I \le A^{\mathit{eff}} \le \lambda_+(\theta)\, I \text{ a.e. in } \Omega, \\ \frac{1}{\lambda_-(\theta)} = \frac{\theta}{\alpha} + \frac{1-\theta}{\beta}, \lambda_+(\theta) = \theta\, \alpha + (1-\theta)\, \beta, \end{gathered} \tag{20.3}$$

and we did not know that such formulas were conjectured, but not proven.

I want to expand on this important point, since those who do not perceive the boundary between what is understood and what is not are often lured by pseudo-logic arguments: a game A implies a result B, which looks like something observed, and one postulates that nature plays game A. Besides showing a total ignorance of the scientific method in general and logic in particular, it shows a curious lack of imagination.[2] It reminds me of others,

---

[1] We found later that this characterization is not needed for our initial question: our inequality (20.3) permits us to conclude, with the use of laminated materials.

[2] A case related to homogenization is that of quasi-crystals. Experimental physicists heated a metallic ribbon over its Curie temperature, favoring a particular magnetic orientation by imposing an exterior magnetic field; then, by fast tempering, they hoped to freeze the magnetic orientations in the ribbon, forcing the material to change its microstructure, to adapt to the questions of evacuating heat, and releasing elastic tensions. They checked the result by X-ray diffraction, and they saw an unexpected five-fold symmetry! Since a 0.1 millimetre thickness for a ribbon corresponds to about a million atomic distances, what kind of "physicist" must one be to find this related to a tiling of the plane invented by Roger PENROSE?

L. Tartar, *The General Theory of Homogenization*, Lecture Notes of the Unione Matematica Italiana 7, DOI 10.1007/978-3-642-05195-1_20,

who put in their hypotheses what is observed, and are quite naive then to be enthusiastic about the efficiency of their game to predict observations!

I find it instructive to recall a previous interaction between mathematics and physics, when Laurent SCHWARTZ explained (by his theory of distributions) some formulas used by DIRAC, as well as a formal method due to HEAVISIDE, whose name should be better known, since he found the form that one uses of the Maxwell equation, which I prefer to call the Maxwell–Heaviside equation. Physicists still use the notation $\delta(x)$ for Dirac "function," and it is not the fact that it is not a function but a Radon measure which is important,[3] since many formal results simply correspond to the intuition of a point mass, a concept which did not create many philosophical problems to a mathematician like POISSON, almost two centuries ago. A formula like $\lambda^N \delta(\lambda x) = \delta(x)$ in $\mathbb{R}^N$ for all $\lambda > 0$ is already a good test for intuition.[4] DIRAC was more daring, and he used the partial derivatives $\frac{\partial \delta}{\partial x_j}$. In the 1930s, Sergei SOBOLEV gave a meaning to the property for a function in $L^2(\Omega)$ to have partial derivatives in $L^2(\Omega)$, and Jean LERAY used this notion of a "weak derivative" in his work on Navier–Stokes equation, but the theory of distributions of Laurent SCHWARTZ went further, permitting one to take as many derivatives as one wants of these new objects called distributions, extending Radon measures by restricting test functions to be in $C_c^\infty(\Omega)$,[5] with precise bounds. For a distribution T, one defines $\frac{\partial T}{\partial x_j}$ by $\langle \frac{\partial T}{\partial x_j}, \varphi \rangle = -\langle T, \frac{\partial \varphi}{\partial x_j} \rangle$ for all functions $\varphi \in C_c^\infty(\Omega)$, so that the derivative $\frac{\partial \delta_0}{\partial x_j}$ is simply the mapping $\varphi \mapsto -\frac{\partial \varphi(0)}{\partial x_j}$. DIRAC also understood the importance of distinguishing an element of a Hilbert space H,[6] a "ket" $|b\rangle$, from an element of its dual H', a "bra" $\langle a|$,[7] but DIRAC dared to use the functions $e^{\pm 2 i \pi (\cdot, \xi)}$ like an "orthonormal basis" of $L^2(\mathbb{R}^N)$, although these functions do not belong to

---

[3] For $a \in \mathbb{R}^N$, the Dirac mass at a, written δ_a, is the mapping $\varphi \mapsto \langle \delta_a, \varphi \rangle = \varphi(a)$ for all $\varphi \in C_c(\mathbb{R}^N)$, which extends to the Fréchet space $C(\mathbb{R}^N)$ ($\mathcal{E}_0(\mathbb{R}^N)$ in Laurent SCHWARTZ's notation); physicists write $\delta(x - a)$ instead of δ_a.

[4] When one "identifies" a function $f \in L^1_{loc}(\mathbb{R}^N)$ to a Radon measure $\mu_f \in \mathcal{M}(\mathbb{R}^N)$, one uses the definition $\langle \mu_f, \varphi \rangle = \int_{\mathbb{R}^N} f \varphi \, dx$ for all $\varphi \in C_c(\mathbb{R}^N)$, and one often writes f instead of μ_f, which is a bad idea, since one should write $\mu_f = f \, dx$ for showing the crucial role of the Lebesgue measure dx, and the formula for scaling. If for $\lambda > 0$ and $\varphi \in C(\mathbb{R}^N)$ one defines $\psi = T_\lambda \varphi$ by $\psi(x) = \varphi(\lambda x)$ for all $x \in \mathbb{R}^N$, then it is natural for a Radon measure $\nu \in \mathcal{M}(\mathbb{R}^N)$ to define $T_\lambda \nu$ by the formula $\lambda^N \langle T_\lambda \nu, T_\lambda \varphi \rangle = \langle \nu, \varphi \rangle$ for all $\varphi \in C_c(\mathbb{R}^N)$, and to say that $\nu \in \mathcal{M}(\mathbb{R}^N)$ is homogeneous of degree k if $T_\lambda \nu = \lambda^k \nu$ for all $\lambda > 0$; then, for $f \in C(\mathbb{R}^N)$ one has $T_\lambda(\mu_f) = \mu_{(T_\lambda f)}$. For $k > -N$ one has $g = |x|^k \in L^1_{loc}(\mathbb{R}^N)$, and μ_g is homogeneous of degree k, but there is no nonzero function $h \in L^1_{loc}(\mathbb{R}^N)$ for which μ_h is homogeneous of degree $-N$; however, δ_0 is homogeneous of degree $-N$.

[5] $C_c^\infty(\Omega)$ is the space of C^∞ functions with compact support in Ω, $\mathcal{D}(\Omega)$ in Laurent SCHWARTZ's notation.

[6] This is precisely distinguishing the function f from the Radon measure $f \, dx$.

[7] Physicists use a scalar product $\langle a|b \rangle$, linear in b, anti-linear in a, and an operator $|b\rangle\langle a|$; mathematicians use (u, v), linear in u, anti-linear in v, and $b \otimes a$.

$L^2(\mathbb{R}^N)$, and he wrote his famous formula $\int_{\mathbb{R}^N} e^{\pm 2i\,\pi\,(x,\xi)}\,dx = \delta(\xi)$. Laurent SCHWARTZ gave it a meaning, by extending the Fourier transform $\mathcal{F}$ (and $\overline{\mathcal{F}}$) to the space of tempered distributions $\mathcal{S}'(\mathbb{R}^N)$,[8] so that the Dirac formula is simply $\mathcal{F}1 = \overline{\mathcal{F}}1 = \delta_0$, but the extended Fourier transform is not given by an integral, of course!

Some mathematicians understood that one needs Sobolev spaces for partial differential equations from continuum mechanics or physics, with the theory of distributions in the background, since these equations use irregular coefficients and nonlinearities, and they learned to recognize what can be easily transformed into a correct statement among all that physicists say, and what uses "arguments" whose mathematical meaning is not yet clear.

Physicists have a different notion of "knowledge" than mathematicians, and they seem to believe that DIRAC already did what Laurent SCHWARTZ explained. I think that everyone trained in mathematics understood DIRAC's work as conjectures, some of them being settled by the work of Laurent SCHWARTZ.

Until the end of the 1960s, the training in mathematics in Paris was one of the best, and the mathematicians who had not the level for studying there did not dare criticize the much better mathematicians who could study there. Probably because of an important drop in the level of training, some young people trained in mathematics write statements without scientific value, attributing theorems on G-convergence or H-convergence to people who worked tens of years before these notions were defined: with a better training, they would be precise, talk about conjectures, and mention under what hypotheses the results were conjectured. Until recently, many did not think about anisotropic properties, so how could they prove a result in G-convergence or H-convergence? Actually, many of the early writers used a single number for a mixture, like its total energy in a domain, for particular boundary conditions, so that it cannot be confused with the G-convergence of Sergio SPAGNOLO, or the H-convergence of François MURAT and myself, which are local properties of a matrix-valued function, independent of boundary conditions!

More than ever, it is important to tell younger generations how to do mathematics. Being a mathematician interested in continuum mechanics does not mean speaking the words used by practitioners, and putting aside the critical mind of a scientist for discussing the models, and the precision of mathematical reasoning for proving what is right. FEYNMAN described a problem which he encountered in Rio de Janeiro,[9] Brazil, of graduate students learning

---

[8] It is the dual of the Fréchet space $\mathcal{S}(\mathbb{R}^N)$, of all functions $\psi \in C^\infty(\mathbb{R}^N)$ such that $P\,D^\gamma \psi \in L^\infty(\mathbb{R}^N)$ for all polynomials P and all multi-indices γ.

[9] Richard Phillips FEYNMAN, American physicist, 1918–1988. He received the Nobel Prize in Physics in 1965, jointly with Sin-Itiro TOMONAGA and Julian SCHWINGER, for their fundamental work in quantum electrodynamics, with deep-ploughing consequences for the physics of elementary particles. He worked at Cornell University, Ithaca, NY, and at Caltech (California Institute of Technology), Pasadena, CA.

physics as if it was a foreign language, without perceiving its relation to the real world [29]. This defect is becoming classical among "mathematicians" who pretend to be interested in applications: their knowledge in continuum mechanics or physics is limited, so they turn to fashionable areas like biology, where they know even less, hoping to create illusion by using some key words that they read!

Computing the effective conductivity of a particular periodic pattern will show the difficulty in interpreting an old formula without reading its derivation. A checkerboard with squares of conductivity 1 and z has an isotropic effective conductivity $\sqrt{z}$; for conductivities a and b it is $\sqrt{a\,b}$, but I use the notation z for the case $z \in \mathbb{C}$ but not a real ≤ 0, which I shall describe. This result follows from an observation of Joe KELLER,

$$\text{if } N = 2 \text{ and if } b(x) = \frac{1}{a(x)}, \text{ "then" } b_{\mathit{eff}} = \frac{1}{a_{\mathit{eff}}}, \tag{20.4}$$

which is based on a property only valid for $N = 2$,

$$div(R_{\pi/2}u) = -curl(u), curl(R_{\pi/2}u) = +div(u) \text{ in } \mathbb{R}^2 \tag{20.5}$$

with R_θ denoting the rotation of angle θ in the plane, and $curl(u) = \partial_1 u_2 - \partial_2 u_1$. I do not remember what Joe KELLER wrote in 1964 [44], so for the theory of homogenization (20.4) may be interpreted in two ways: either

$$\begin{aligned} &\text{if } N = 2, \text{ if } a_n I \text{ G-converges to } a_{\mathit{eff}} I, \text{ and if } \frac{1}{a_n} I \text{ G-converges to } b_{\mathit{eff}} I \\ &\text{then } b_{\mathit{eff}} = \frac{1}{a_{\mathit{eff}}} \text{ a.e. in } \Omega, \end{aligned} \tag{20.6}$$

or

$$\begin{aligned} &\text{if } N = 2, \text{ and if } a_n I \text{ G-converges to } a_{\mathit{eff}} I \\ &\text{then } \frac{1}{a_n} I \text{ G-converges to } \frac{1}{a_{\mathit{eff}}} I, \end{aligned} \tag{20.7}$$

and (20.7) implies (20.6), and it is true, and follows from Lemma 20.1.

Lemma 20.1. *If $\Omega \subset \mathbb{R}^2$ and $A^n \in \mathcal{M}(\alpha, \beta; \Omega)$, then*

$$A^n \text{ H-converges to } A^{\mathit{eff}} \text{ implies } \frac{(A^n)^T}{det(A^n)} \text{ H-converges to } \frac{(A^{\mathit{eff}})^T}{det(A^{\mathit{eff}})}. \tag{20.8}$$

Of course, one may omit transposition by Lemma 10.2. The case $A^n = a_n I$ implies (20.7), but Lemma 20.1 does not restrict A^{eff} to be isotropic.

Corollary 20.2. *If $\Omega \subset \mathbb{R}^2$ and $A^n \in \mathcal{M}(\alpha,\beta;\Omega)$ H-converges to A^{eff}, then*

$$\textit{if } det(A^n) = \kappa \textit{ for all } n \textit{ a.e. in } \Omega, \textit{ then } det(A^{\mathit{eff}}) = \kappa \textit{ a.e. in } \Omega. \tag{20.9}$$

In the early 1970s, François MURAT and myself noticed that laminating in x_1 a material with conductivity $\begin{pmatrix} a & 0 \\ 0 & b \end{pmatrix}$ in proportion η with the same material rotated by $\frac{\pi}{2}$ in proportion $1-\eta$, gives an effective conductivity $\begin{pmatrix} d_1(\eta) & 0 \\ 0 & d_2(\eta) \end{pmatrix}$ with $\frac{1}{d_1(\eta)} = \frac{\eta}{a} + \frac{1-\eta}{b}$ and $d_2(\eta) = \eta\, b + (1-\eta)\, a$, by Lemma 5.2, proved in (12.13)–(12.19), so that $d_1(\eta) d_2(\eta) = a\, b$. Since our computation was part of our search for (20.1) and (20.2), we did not think of (20.9), and I heard it from Alain BAMBERGER, probably restricted to G-convergence.

Since the effective conductivity of the checkerboard geometry is isotropic, (20.6) permits one to compute it: since $\frac{z}{a}$ corresponds to the same checkerboard with conductivities exchanged, one has $\frac{z}{a_{\mathit{eff}}} = a_{\mathit{eff}}$, so that $a_{\mathit{eff}} = \sqrt{z}$.[10]

In the early 1980s, I found a group of mappings which commute with H-convergence if $N = 2$, and it contains the involutive mapping $A \mapsto \frac{A^T}{det(A)}$; of course there is $A \mapsto A^T$ for which there is no restriction on N. Graeme MILTON found it independently, and he was persuaded to use nonsymmetric matrices for a question related to the classical Hall effect (which occurs in metallic ribbons, reasonably described by two-dimensional domains).[11]

Lemma 20.3. *For $P = \begin{pmatrix} a & b \\ c & d \end{pmatrix}$ with $a\,d - b\,c > 0$, and $M \in \mathcal{L}(\mathbb{R}^2;\mathbb{R}^2)$ positive definite, $a\,I + b\,R_{\pi/2}M$ is invertible, and one writes*

$$\mathcal{T}_P(M) = (-c\,R_{\pi/2} + d\,M)(a\,I + b\,R_{\pi/2}M)^{-1}. \tag{20.10}$$

Then, $\mathcal{T}_P(M)$ is positive definite, and

$$P \mapsto \mathcal{T}_P \textit{ is a group homomorphism.} \tag{20.11}$$

Proof. If $\xi \in \mathbb{R}^2$ and $(a\,I + b\,R_{\pi/2}M)\xi = 0$, then taking the scalar product by $M\,\xi$ gives $a\,(M\,\xi,\xi) = 0$, and taking the scalar product by $R_{\pi/2}\xi$ gives $b\,(M\,\xi,\xi) = 0$; since a or b is $\neq 0$, one deduces that $(M\,\xi,\xi) = 0$, i.e., $\xi = 0$.

For $E \in \mathbb{R}^2$, $E \neq 0$, and $D = M\,E$, one writes $E^* = a\,E + b\,R_{\pi/2}D$ and $D^* = -c\,R_{\pi/2}E + d\,D$, so that $D^* = \mathcal{T}_P(M)E^*$, and $(D^*,E^*) = (a\,d - b\,c)(D,E)$ is > 0, since $a\,d - b\,c > 0$ and $(D,E) = (M\,E,E) > 0$.

---

[10] For $z \in \mathbb{C}$ not a real ≤ 0, there is a natural choice for the square root.

[11] He proposed to add me as a co-author of his article, but I thought that it was enough to mention that I found this group independently.

If $P' = \begin{pmatrix} a' & b' \\ c' & d' \end{pmatrix}$ with $a'd' - b'c' > 0$, and $E^{**} = a'E^* + b'R_{\pi/2}D^*$ and $D^{**} = -c'R_{\pi/2}E^* + d'D^*$, then one has $E^{**} = a'(a\,E + b\,R_{\pi/2}D) + b'R_{\pi/2}(-c\,R_{\pi/2}E + d\,D) = (a'a + b'c)E + (a'b + b'd)R_{\pi/2}D$, and $D^{**} = -c'R_{\pi/2}(a\,E + b\,R_{\pi/2}D) + d'(-c\,R_{\pi/2}E + d\,D) = -(c'a + d'c)R_{\pi/2}E + (c'b + d'd)D$, so that the transformation from E, D to E^{**}, D^{**} corresponds to the matrix $P'P$. □

If $det(P) > 0$, $\mathcal{T}_{\lambda\,P} = \mathcal{T}_P$ for all $\lambda \neq 0$. If $P = \begin{pmatrix} 0 & 1 \\ -1 & 0 \end{pmatrix}$, $\mathcal{T}_P(A) = R_{\pi/2}A^{-1}R_{-\pi/2} = \frac{A^T}{det(A)}$, and Lemma 20.1 follows from Lemma 20.4

Lemma 20.4. *If $\Omega \subset \mathbb{R}^2$ and $A^n \in \mathcal{M}(\alpha, \beta; \Omega)$ H-converges to A^{eff}, then for all $P = \begin{pmatrix} a & b \\ c & d \end{pmatrix}$ with $a\,d - b\,c > 0$, one has*

$$\mathcal{T}_P(A^n) \text{ H-converges to } \mathcal{T}_P(A^{\mathit{eff}}). \tag{20.12}$$

Proof. That A^n H-converges to A^{eff} means that

$$\begin{aligned}
&\text{if } E^n \rightharpoonup E^\infty, D^n \rightharpoonup D^\infty \text{ in } L^2_{loc}(\Omega; \mathbb{R}^2) \text{ weak,}\\
&\text{if } curl(E^n), div(D^n) \text{ stay in a compact of } H^{-1}_{loc}(\Omega) \text{ strong,}\\
&\text{if } D^n = A^n E^n \text{ a.e. in } \Omega,\\
&\text{then } D^\infty = A^{\mathit{eff}} E^\infty \text{ a.e. in } \Omega,
\end{aligned} \tag{20.13}$$

and one constructs sequences (E^n, D^n) by solving $-div\big(A^n grad(u_n)\big) + u_n = f \in H^{-1}(\Omega)$, $u_n \in H^1_0(\Omega)$, for enough fs. One then defines

$$\widetilde{E}^n = a\,E^n + b\,R_{\pi/2}D^n, \widetilde{D}^n = -c\,R_{\pi/2}E^n + d\,D^n, \tag{20.14}$$

so that

$$curl(\widetilde{E}^n) = a\,curl(E^n) + b\,div(D^n), div(\widetilde{D}^n) = c\,curl(E^n) + d\,div(D^n), \tag{20.15}$$

$$\widetilde{D}^n = \mathcal{T}_P(A^n)\widetilde{E}^n. \tag{20.16}$$

From (20.13) one deduces that

$$\begin{aligned}
&\widetilde{E}^n \rightharpoonup \widetilde{E}^\infty = a\,E^\infty + b\,R_{\pi/2}D^\infty \text{ in } L^2_{loc}(\Omega; \mathbb{R}^2) \text{ weak}\\
&\widetilde{D}^n \rightharpoonup \widetilde{D}^\infty = -c\,R_{\pi/2}E^\infty + d\,D^\infty \text{ in } L^2_{loc}(\Omega; \mathbb{R}^2) \text{ weak}\\
&curl(\widetilde{E}^n), div(\widetilde{D}^n) \text{ stay in a compact of } H^{-1}_{loc}(\Omega)\\
&\widetilde{D}^\infty = \mathcal{T}_P(A^{\mathit{eff}})\widetilde{E}^\infty, \text{ a.e. in } \Omega,
\end{aligned} \tag{20.17}$$

and since $\widetilde{E}^\infty = grad(u_\infty)$, and u_∞ can be arbitrary in $H^1_0(\Omega)$, one will deduce (20.12) if one checks that $\mathcal{T}_P(A^n)$ is uniformly bounded and elliptic, for extracting a subsequence which H-converges. This follows from

$$\begin{aligned}
&|\widetilde{E}^n| \le (|a| + \beta\,|b|)|E^n|, |\widetilde{D}^n| \le (|c| + \beta\,|d|)|E^n|,\\
&(\widetilde{D}^n, \widetilde{E}^n) = (a\,d - b\,c)\,(D^n, E^n) \ge \alpha\,(a\,d - b\,c)\,|E^n|^2,\\
&\alpha\,(|a| + |b|)|E^n|^2 \le (|a| + |b|)(D^n, E^n)\\
&\quad = \quad (\widetilde{E}^n, sign(b)R_{\pi/2}E^n + sign(a)D^n) \le (1+\beta)|E^n|\,|\widetilde{E}^n|,
\end{aligned} \tag{20.18}$$

which shows that $|E^n|$ and $|\widetilde{E}^n|$ are equivalent. □

In the mid 1990s, I was in Minneapolis, MN, for a conference at IMA, and Vladimír SVERÁK asked me a question about the Beltrami equation,[12]

$$\overline{\partial} f = \mu\,\partial f \text{ with } |\mu| < 1, \text{ in } \Omega \subset \mathbb{C}, \tag{20.19}$$

saying that it conserves the same form by homogenization;[13] he asked me about precise bounds on effective coefficients. Writing (20.19) as a real system, I found it related to Corollary 20.2 in the symmetric case.

Lemma 20.5. *If $f = P + i\,Q$, $\mu = a + i\,b$ with $a^2 + b^2 < 1$, (20.19) means*

$$\begin{pmatrix} 1-a & -b \\ -b & 1+a \end{pmatrix} grad(P) = \begin{pmatrix} -b & 1+a \\ -1+a & b \end{pmatrix} grad(Q) \text{ in } \Omega \subset \mathbb{R}^2, \tag{20.20}$$

corresponding to

$$div\big(A\,grad(P)\big) = 0, A = \frac{1}{1-a^2-b^2}\begin{pmatrix} (1+a)^2+b^2 & -2b \\ -2b & (1-a)^2+b^2 \end{pmatrix}, \tag{20.21}$$

[12] Eugenio BELTRAMI, Italian mathematician, 1835–1900. He worked in Bologna, Pisa, Roma (Rome), and Pavia, Italy.

[13] I thought that it was the explanation of something that Yves MEYER told me in August 1990 at the International Congress of Mathematicians in Kyoto, Japan, that he worked with Raphaël COIFMAN on a problem of homogenization, but since he was puzzled that I talked about anisotropic materials, I wondered what problem they considered for not finding that the class of isotropic materials is not stable by homogenization.

or to

$$div\big(B\,grad(Q)\big) = 0, B = \frac{1}{1 - a^2 - b^2}\begin{pmatrix} (1-a)^2 + b^2 & -2b \\ -2b & (1+a)^2 + b^2 \end{pmatrix}. \tag{20.22}$$

Proof. For $z = x + i\,y$ and $\overline{z} = x - i\,y$, the notation is $\partial = \frac{\partial}{\partial z}, \overline{\partial} = \frac{\partial}{\partial \overline{z}}$, so that $\partial_x = \partial + \overline{\partial}$, $\partial_y = i\,\partial - i\,\overline{\partial}$, or $2\partial = \partial_x - i\,\partial_y$, $2\overline{\partial} = \partial_x + i\,\partial_y$, and (20.19) is

$$(\partial_x + i\,\partial_y)(P + i\,Q) = \mu\,(\partial_x - i\,\partial_y)(P + i\,Q), \tag{20.23}$$

or by separating the derivatives of P and of Q

$$(1-\mu)\partial_x P + i\,(1+\mu)\partial_y P = i\,(-1+\mu)\partial_x Q + (1+\mu)\partial_y Q. \tag{20.24}$$

Writing the real and the imaginary parts of (20.24) gives (20.20), which is $M_1 grad(P) = M_2 grad(Q)$, and both M_1 and M_2 have determinant $1 - a^2 - b^2 > 0$. The compatibility condition that $M_2^{-1} M_1 grad(P)$ be a gradient gives (20.21) with $A = -R_{\pi/2} M_2^{-1} M_1$, symmetric with determinant $+1$; if Ω is simply connected (and connected), (20.21) permits one to compute Q up to addition of a constant. Similarly for (20.22) with $B = R_{\pi/2} M_1^{-1} M_2$. □

Lemma 20.6. *If M is symmetric, with $det(M) = 1$, and $M_{11}, M_{22} > 0$, so that $Trace(M) \geq 2$, there is a unique way to write*

$$M = \frac{1}{1 - a^2 - b^2}\begin{pmatrix} (1+a)^2 + b^2 & -2b \\ -2b & (1-a)^2 + b^2 \end{pmatrix}, \tag{20.25}$$

with $a^2 + b^2 < 1$.

Proof. Since $Trace(M) = \frac{2(1+a^2+b^2)}{1-a^2-b^2} \in [2, \infty)$, there is only one possible value of $a^2 + b^2$, which is $\frac{Trace(M)-2}{Trace(M)+2}$. Then one must have

$$a = \frac{M_{11} - M_{22}}{Trace(M) + 2}, b = \frac{-2M_{12}}{Trace(M) + 2}, \tag{20.26}$$

and it gives the correct value of $a^2 + b^2$ since $det(M) = +1$. □

For answering Vladimír SVERÁK's question about mixing values of μ from the open unit disc $D \subset \mathbb{C}$, I checked the formula for laminated materials.

Lemma 20.7. *If μ^n in (20.19) only depends upon $x\cos\theta + y\sin\theta$, one finds μ^{eff} by using the inversion of centre $-(\cos 2\theta - i\,\sin 2\theta)$ and power 1, taking a $L^\infty(\Omega)$ weak $\star$ limit, and using the inversion again.*

Proof. By Lemma 5.2, proved in (12.13)–(12.19), identifying A^{eff} in the case of a symmetric (positive definite) A^n depending only upon x_1 requires computing the $L^\infty(\Omega)$ weak $\star$ limit of $\frac{1}{A^n_{1,1}}$, of $\frac{A^n_{1,2}}{A^n_{1,1}}$, and of $A^n_{2,2} - \frac{(A^n_{1,2})^2}{A^n_{1,1}}$, but due to $det(A^n) = 1$ the last quantity is $\frac{1}{A^n_{1,1}}$. Similarly, if ξ is a unit vector, and η is another unit vector orthogonal to ξ, and A^n only depends upon (x, ξ), one needs the $L^\infty(\Omega)$ weak $\star$ limit of $\frac{1}{(A^n\xi,\xi)}$ and of $\frac{(A^n\xi,\eta)}{(A^n\xi,\xi)}$. For A given by (20.21), $\xi = \begin{pmatrix}\cos\theta\\ \sin\theta\end{pmatrix}$, $\eta = R_{\pi/2}\xi = \begin{pmatrix}-\sin\theta\\ \cos\theta\end{pmatrix}$, one finds

$$(A\,\xi,\xi) = \frac{1+a^2+b^2+2a\cos 2\theta - 2b\sin 2\theta}{1-a^2-b^2} = \frac{(a+\cos 2\theta)^2 + (b-\sin 2\theta)^2}{1-a^2-b^2} \tag{20.27}$$

$$\frac{1}{(A\,\xi,\xi)} = -1 + \frac{2+2a\cos 2\theta - 2b\sin 2\theta}{(a+\cos 2\theta)^2+(b-\sin 2\theta)^2}, \tag{20.28}$$

$$(A\,\xi,\eta) = \frac{-2a\sin 2\theta - 2b\cos 2\theta}{1-a^2-b^2}, \tag{20.29}$$

$$\frac{(A\,\xi,\eta)}{(A\,\xi,\xi)} = \frac{-2a\sin 2\theta - 2b\cos 2\theta}{(a+\cos 2\theta)^2+(b-\sin 2\theta)^2}. \tag{20.30}$$

It becomes simpler in complex notation, since

$$\frac{1}{\mu+\cos 2\theta - i\sin 2\theta} = \frac{(a+\cos 2\theta) - i\,(b-\sin 2\theta)}{(a+\cos 2\theta)^2+(b-\sin 2\theta)^2}, \tag{20.31}$$

$$\frac{2(\cos 2\theta - i\sin 2\theta)}{\mu+\cos 2\theta - i\sin 2\theta} = 1 + \frac{1}{(A\,\xi,\xi)} + i\,\frac{(A\,\xi,\eta)}{(A\,\xi,\xi)}, \tag{20.32}$$

so that it is linearly equivalent to work in the plane with coordinates $\frac{1}{(A\,\xi,\xi)}$ and $\frac{(A\,\xi,\eta)}{(A\,\xi,\xi)}$, or to perform an inversion of centre $-(\cos 2\theta - i\sin 2\theta)$ before taking weak limits. □

For the case of mixing two values μ_1, μ_2 from D, if a circle goes through μ_1 and μ_2 and intersects the unit circle ∂D at z_0, one uses an inversion of centre z_0 and one deduces that the effective values of μ lie on the arc of this circle between μ_1 and μ_2; the extreme circles to consider are those which are also tangent to the unit circle,[14] and this is optimal by Lemma 20.8.

Lemma 20.8. *If for a closed disc $D_* \subset \overline{D}$ one has $\mu^n \in D_*$ a.e. in Ω, then $\mu^{eff} \in D_*$ a.e. in Ω.*

Proof. If the centre of the disc is $\begin{pmatrix}X\\ Y\end{pmatrix}$ and its radius is ρ, then one has $X^2+Y^2 = \theta^2$ with $0 \le \theta < 1$ and $0 < \rho < 1-\theta$, and D_* has equation

[14] In 1963–1965, in the classes of "mathématiques supérieures" and "mathématiques spéciales" at Lycée Charlemagne, in Paris, France, I received a good mathematical training which included algebra, analysis, and geometry, and I learned to use inversions for problems in plane geometry involving lines and circles.

$$a^2 + b^2 - 2a\,X - 2b\,Y + \theta^2 - \rho^2 \le 0 \tag{20.33}$$

and using (20.25) and (20.26), it is

$$Trace(M) - 2 - 2X(M_{1,1} - M_{2,2}) + 4Y\,M_{1,2} + (\theta^2 - \rho^2)(Trace(M) + 2) \le 0. \tag{20.34}$$

By Lemma 6.7, if M^n is symmetric, converges weakly $\star$ to M^+ and G-converges to M^{eff}, and K is symmetric nonnegative, then

$$Trace(M^n K) \le \kappa \text{ a.e. for all } n \text{ implies } Trace(M^{\mathit{eff}} K) \le \kappa \text{ a.e.}, \tag{20.35}$$

since it is true for M^+ and $Trace\big((M^+ - M^{\mathit{eff}})K\big) \ge 0$. For (20.34),

$$K = \begin{pmatrix} 1 - 2X + \theta^2 - \rho^2 & 2Y \\ 2Y & 1 + 2X + \theta^2 - \rho^2 \end{pmatrix}, \tag{20.36}$$

and one has $1 \pm 2X + \theta^2 \ge 1 - 2\theta + \theta^2 \ge \rho^2$, and the determinant of K is $(1 + \theta^2 - \rho^2)^2 - 4X^2 - 4Y^2 \ge 0$, since $1 + \theta^2 - \rho^2 \ge 2\theta$. □

If the period $(0,1) \times (0,1)$ is cut into four small squares of side $\frac{1}{2}$ and materials with conductivities α, β, γ, and δ are used for each of the four small squares, then there are symmetries with respect to the parallels to the axes of coordinates passing through the centre of a small square, and the matrix of conductivity is then diagonal.[15] Stefano MORTOLA and Sergio STEFFÈ conjectured a formula [68],[16,17] which was checked numerically by Donatella MARINI,[18] and in the case where $\alpha = \beta = \gamma = 1$ they also checked that the limiting case $\delta \to \infty$ is correct, i.e., when one of the four small squares is filled with a perfect conductor, and their value $\sqrt{3}$ for the effective isotropic conductivity in this case corresponds to what an argument using a conformal transformation gives. The formula was proven to be true, by Graeme MILTON, and independently by CRASTER and OBNOSOV, who treated a more general case where the period is cut into four rectangles.[19,20]

---

[15] In the checkerboard case, there is another symmetry with respect to the diagonal of each small square, and the effective conductivity is then isotropic.

[16] Stefano MORTOLA, Italian mathematician, born in 1951. He works at Politecnico di Milano, Milano (Milan), Italy.

[17] Sergio STEFFÉ, Italian mathematician, born in 1948. He works in Pisa, Italy.

[18] Donatella MARINI, Italian mathematician. She works in Pavia, Italy.

[19] Richard Vaughan CRASTER, British mathematician. He works in London, England.

[20] Yurii Viktorovich OBNOSOV, Russian mathematician. He works in Kazan, Russia.

Additional footnotes: Charlemagne,[21] Raphaël COIFMAN,[22] Yves MEYER,[23] SCHWINGER,[24] TOMONAGA.[25]

[21] Charlemagne, Frankish king, 742/43–814. He ruled over France, Germany, and Italy, and was crowned emperor in Roma (Rome), Italy, in 800, the beginning of the Holy Roman Empire.

[22] Ronald Raphaël COIFMAN, Israeli-born mathematician, born in 1941. He works at Yale University, New Haven, CT.

[23] Yves François MEYER, French mathematician, born in 1939. He worked at Université Paris Sud XI, Orsay (where he was my colleague from 1975 to 1979), at École Polytechnique, Palaiseau, at Université Paris IX-Dauphine, Paris, and at ENS-Cachan (Ecole Normale Supérieure de Cachan), Cachan, France.

[24] Julian Seymour SCHWINGER, American physicist, 1918–1994. He received the Nobel Prize in Physics in 1965, jointly with Sin-Itiro TOMONAGA and Richard Phillips FEYNMAN, for their fundamental work in quantum electrodynamics, with deep-ploughing consequences for the physics of elementary particles. He worked at UCB (University of California at Berkeley), Berkeley, CA, at Purdue University, West Lafayette, IN, and at Harvard University, Cambridge, MA.

[25] Sin-Itiro TOMONAGA, Japanese-born physicist, 1906–1979. He received the Nobel Prize in Physics in 1965, jointly with Richard FEYNMAN and Julian SCHWINGER, for their fundamental work in quantum electrodynamics, with deep-ploughing consequences for the physics of elementary particles. He worked in Tokyo, Japan, in Leipzig, Germany, in Tsukuba, Japan, and at IAS (Institute for Advanced Study), Princeton, NJ.

Chapter 21
Bounds on Effective Coefficients

I mentioned the use of symmetries for showing that the effective conductivity of a checkerboard is isotropic, or simply diagonal in the Mortola–Steffé conjecture; it follows from using mirror symmetries in Lemma 21.1.

Lemma 21.1. *If $A^n \in \mathcal{M}(\alpha,\beta;\Omega)$ H-converges to $A^{\rm eff}$, if φ is a diffeomorphism from Ω onto $\varphi(\Omega)$, and B^n is defined in $\varphi(\Omega)$ by*

$$B^n\big(\varphi(x)\big) = \frac{1}{det\big(\nabla\varphi(x)\big)}\nabla\varphi(x)A^n(x)\nabla\varphi^T(x) \text{ a.e. } x \in \Omega, \tag{21.1}$$

then $B^n \in \mathcal{M}\big(\alpha',\beta';\varphi(\Omega)\big)$ H-converges to $B^{\rm eff}$, defined in $\varphi(\Omega)$ by

$$B^{\rm eff}\big(\varphi(x)\big) = \frac{1}{det\big(\nabla\varphi(x)\big)}\nabla\varphi(x)A^{\rm eff}(x)\nabla\varphi^T(x) \text{ a.e. } x \in \Omega. \tag{21.2}$$

Proof. If $-div\big(A^n\,grad(u_n)\big) = f$ in Ω, one defines v_n in $\varphi(\Omega)$ by

$$v_n = u_n \circ \varphi^{-1} \text{ in } \varphi(\Omega), \text{ i.e., } u_n = v_n \circ \varphi \text{ in } \Omega, \tag{21.3}$$
$$grad(u_n) = \nabla\varphi^T\,grad(v_n)\circ\varphi \text{ in } \Omega. \tag{21.4}$$

In the case $f \in L^2(\Omega)$,[1] one writes the equation in variational form

$$\int_\Omega \big(A^n\,grad(u_n), grad(w)\big)\,dx = \int_\Omega f\,w\,dx \text{ for all } w \in C^1_c(\Omega), \tag{21.5}$$

and, after making the change of variables $x = \varphi(y)$, one deduces that

$$-\,div\big(B^n\,grad(v_n)\big) = g \text{ in } \varphi(\Omega), g\big(\varphi(x)\big) = \frac{1}{det\big(\nabla\varphi(x)\big)}f(x) \text{ a.e. } x \in \Omega. \tag{21.6}$$

[1] The generalization to the case $f \in H^{-1}(\Omega)$ is straightforward.

L. Tartar, *The General Theory of Homogenization*, Lecture Notes of the Unione Matematica Italiana 7, DOI 10.1007/978-3-642-05195-1_21,

The $L^2(\varphi(\Omega);\mathbb{R}^N)$ weak limits of $grad(v_n)$ and $B^n\,grad(v_n)$ are deduced from $grad(u_\infty)$ and $A^{eff}\,grad(v_\infty)$, and it gives (21.2) for B^{eff}. □

One often uses Lemma 21.1 when φ is a rotation (modulo a translation), for example in proving that sets like $\mathcal{K}(\theta)$ in (20.2) are defined by constraints on the eigenvalues of A^{eff}. My initial method with François MURAT for finding bounds, and the generalization that I made, are not restricted to symmetric A^n, so they apply to H-convergence, but I lack physical intuition for the non-symmetric case,[2] for which I do not know what question to ask concerning rotations. For a similar reason, I chose not to work on questions of homogenization in linearized elasticity, a theory which is not frame-indifferent, so that unrealistic effects deprive the mathematical results of much of their value!

Sergio SPAGNOLO proved a generalization of Lemma 21.1, by using a sequence φ^n, with uniform bounds for the partial derivatives of the components of φ^n and those of its inverse ψ^n, and using the Reshetnyak theorem for passing to the limit in the Jacobian determinant appearing in (21.6) for g_n.

Definition 21.2. *For $\theta \in (0,1)$, $0 < \alpha \le \beta < \infty$,*

$$\lambda_-(\theta) = \Big(\frac{\theta}{\alpha} + \frac{1-\theta}{\beta}\Big)^{-1}, \lambda_+(\theta) = \theta\,\alpha + (1-\theta)\,\beta, \tag{21.7}$$

$$\mathcal{B}(\theta) = \{A \in \mathcal{L}_{sym}(\mathbb{R}^N;\mathbb{R}^N) \mid \lambda_-(\theta)I \le A \le \lambda_+(\theta)I\}, \tag{21.8}$$

$$\mathcal{H}(\theta) = \{A \in \mathcal{B}(\theta) \mid det(A) = \lambda_-(\theta)\lambda_+^{N-1}(\theta)\}, \tag{21.9}$$

$$\mathcal{K}(\theta) = \{A^{eff} \mid \textit{for mixtures using proportions } \theta, 1-\theta\}. \tag{21.10}$$

The "definition" (21.10) is only an intuitive idea, and must be explained. In the early 1970s, François MURAT and myself used the intuition that if one mixes materials which were obtained as mixtures of some initial materials, then the result can be obtained by mixing directly the initial materials in an adapted way, and from the mathematical point of view, it is here that the metrizability property of H-convergence is important: one is looking at the closure of a set containing the tensors of the form $\big(\alpha\,\chi + \beta\,(1-\chi)\big)I$ with χ being the characteristic function of an arbitrary measurable set (or an open set); one identifies some first-generation sets contained in the sequential closure of the initial set, then one identifies some second-generation sets contained in the sequential closure of some first-generation sets, and one repeats the process finitely many times, and since the topology is metrizable every set constructed is included in the sequential closure of the initial set.

[2] The only instance that I have heard of non-symmetric tensors occurring in a realistic situation is the Hall effect, which Graeme MILTON studied, but it concerns an electrical current in a thin ribbon, so that the macroscopic direction of the current is imposed, and the situation is not subject to a complete frame indifference!

Using Lemma 21.1 and the local property of H-convergence, Lemma 10.3, one can show that there exist sets $\mathcal{K}(\theta) \subset \mathcal{B}(\theta)$ such that the admissible pairs (θ, A^{eff}) with $\theta \in L^\infty(\Omega)$ and $A^{eff} \in \mathcal{L}_{sym+}\big(\Omega; \mathcal{L}(\mathbb{R}^N;\mathbb{R}^N)\big)$ are characterized by $A^{eff}(x) \in \mathcal{K}\big(\theta(x)\big)$ a.e. $x \in \Omega$, but it involves too much of measure theory, which is not an interesting part of my subject.[3]

Recently, I wrote an article [118] for the case of mixing m anisotropic materials, defining $\mathcal{K}_-(\theta)$ for the constant A^{eff} which one can construct with θ constant on a cube, and $\mathcal{K}_+(\theta)$ for the family of inequalities ψ_i such that every admissible pair satisfies $\psi_i(\theta, A^{eff}) \le 0$. This is how it happens from a practical point of view: the sufficient conditions correspond to constructing precise mixtures and computing their effective properties, the necessary conditions correspond to proving inequalities that all effective properties must satisfy. What is important is to improve the existing methods for doing that.

Lemma 21.3. *For $\Omega \subset \mathbb{R}^N$, $N \ge 2$, and $0 < \theta < 1$ one has*

$$\mathcal{H}(\theta) \subset \mathcal{K}(\theta) \subset \mathcal{B}(\theta), \tag{21.11}$$

and for $N = 2$ one has

$$\bigcup_{0<\theta<1} \mathcal{H}(\theta) = \bigcup_{0<\theta<1} \mathcal{K}(\theta) = \bigcup_{0<\theta<1} \mathcal{B}(\theta), \tag{21.12}$$

which is the set of symmetric A whose eigenvalues $\lambda_1(A) \le \lambda_2(A)$ satisfy

$$\frac{\alpha\,\beta}{\alpha+\beta-\lambda_2(A)} \le \lambda_1(A) \le \lambda_2(A) \le \alpha+\beta-\frac{\alpha\,\beta}{\lambda_1(A)}. \tag{21.13}$$

Proof. Lemma 6.7 gives $\mathcal{K}(\theta) \subset \mathcal{B}(\theta)$.

Laminating $\alpha\,I$ and $\beta\,I$ produces a tensor with one eigenvalue $\lambda_-(\theta)$ and $(N-1)$ eigenvalues $\lambda_+(\theta)$; with such a material one can construct $\mathcal{H}(\theta)$. After Corollary 20.2, I proved that for $N = 2$, laminating a material of conductivity $\begin{pmatrix}\gamma & 0\\ 0 & \delta\end{pmatrix}$ with the same material rotated by $\frac{\pi}{2}$ gives all the diagonal matrices with eigenvalues between γ and δ, and determinant $\gamma\,\delta$. To obtain a material with eigenvalues $\mu_1 \le \mu_2 \le \ldots \le \mu_N$ with determinant $\lambda_-\lambda_+^{N-1}$, one laminates one with eigenvalues $\lambda_-, \lambda_+, \lambda_+, \ldots$ with its rotated variant with eigenvalues $\lambda_+, \lambda_-, \lambda_+, \ldots$, and one obtains one with $\mu_1, \nu_1, \lambda_+, \ldots$, and $\mu_1\nu_1 = \lambda_-\lambda_+$; one laminates it with its rotated variant $\mu_1, \lambda_+, \nu_1, \ldots$, and one obtains one with $\mu_1, \mu_2, \nu_2, \lambda_+, \ldots$, and $\mu_1\mu_2\nu_2 = \lambda_-\lambda_+^2$; and so on.

[3] In the late 1960s, I witnessed a group of mathematicians pretending to work in control theory and spending most of their time on technical questions of measurability; nowadays, others do that and pretend to be interested in mechanics!

The curves parametrized by $\big(\lambda_-(\theta),\lambda_+(\theta)\big)$ or by $\big(\lambda_+(\theta),\lambda_-(\theta)\big)$ are increasing, so the union of the squares $\mathcal{B}(\theta)$ is the region between them, of equation (21.13), as the union of the pieces of hyperbolas $\mathcal{H}(\theta)$. □

Although François MURAT and myself followed the same constructions as Antonio MARINO and Sergio SPAGNOLO, the necessary condition of Lemma 6.7 helped us go further.[4]

We noticed that $\mathcal{H}(\theta)\neq\mathcal{K}(\theta)$,[5] since mixing in proportion $\frac{1}{2}$ a material in $\mathcal{H}(\theta_1)$ and one in $\mathcal{H}(\theta_2)$ does not always give one in $\mathcal{H}(\frac{\theta_1+\theta_2}{2})$.

Finding the characterization of $\mathcal{K}(\theta)$ took us a few more years, and the first step was a generalization of the method that François MURAT and myself used, which I developed in the fall of 1977, during a visit at MRC in Madison, WI: one idea was to replace the *div–curl lemma* by our more general *compensated compactness* theory, and another idea was to consider more than one problem, and for that I used the correctors, which I described in Chap. 13.

Lemma 21.4. *Assume that F is a continuous function on $\mathcal{L}(\mathbb{R}^N;\mathbb{R}^N)\times\mathcal{L}(\mathbb{R}^N;\mathbb{R}^N)$ which has the property that*

$$\begin{aligned}&\widetilde{P}^m\rightharpoonup\widetilde{P}^\infty \text{ in } L^2\big(\Omega;\mathcal{L}(\mathbb{R}^N;\mathbb{R}^N)\big) \text{ weak,}\\ &\widetilde{Q}^m\rightharpoonup\widetilde{Q}^\infty \text{ in } L^2\big(\Omega;\mathcal{L}(\mathbb{R}^N;\mathbb{R}^N)\big) \text{ weak,}\\ &curl(\widetilde{P}^m\,e) \text{ stays in a compact of } H^{-1}_{loc}\big(\Omega;\mathcal{L}_{skew}(\mathbb{R}^N;\mathbb{R}^N)\big) \text{ strong,}\\ &div(\widetilde{Q}^m\,e) \text{ stays in a compact of } H^{-1}_{loc}(\Omega) \text{ strong for all } e\in\mathbb{R}^N,\end{aligned}\tag{21.14}$$

imply

$$\liminf_{m\to\infty}\int_\Omega F(\widetilde{P}^m,\widetilde{Q}^m)\,\varphi\,dx\geq\int_\Omega F(\widetilde{P}^\infty,\widetilde{Q}^\infty)\,\varphi\,dx \text{ for all } \varphi\in C_c(\Omega),\varphi\geq 0.\tag{21.15}$$

One defines the function g on $\mathcal{L}(\mathbb{R}^N;\mathbb{R}^N)$, possibly taking the value $+\infty$, by

$$g(A)=\sup_{P\in\mathcal{L}(\mathbb{R}^N;\mathbb{R}^N)}F(P,A\,P).\tag{21.16}$$

Then, if $A^n\in M(\alpha,\beta;\Omega)$ H-converges to A^{eff}, one has

$$\liminf_{n\to\infty}\int_\Omega g(A^n)\,\varphi\,dx\geq\int_\Omega g(A^{\mathit{eff}})\,\varphi\,dx \text{ for all } \varphi\in C_c(\Omega),\varphi\geq 0.\tag{21.17}$$

[4] I used (21.12) and (21.13) in 1974 for computing necessary conditions of optimality for classical solutions of our initial problem.

[5] This was how I knew that the formula in LANDAU and LIFSHITZ is wrong.

Proof. One assumes that the left side of (21.17) is $< +\infty$, one extracts a subsequence A^m for which correctors P^m exist, and $\liminf_m$ is a limit. For $X \in C^1\big(\Omega; \mathcal{L}(\mathbb{R}^N; \mathbb{R}^N)\big)$, $\widetilde{P}^m = P^m X$ and $\widetilde{Q}^m = Q^m X$ satisfy (21.14) with $\widetilde{P}^\infty = X$ and $\widetilde{Q}^\infty = A^{\mathit{eff}} X$. By (21.15), for all $\varphi \in C_c(\Omega)$ with $\varphi \geq 0$,

$$\liminf_{m\to\infty} \int_\Omega F(P^m X, A^m P^m X)\,\varphi\,dx \geq \int_\Omega F(X, A^{\mathit{eff}} X)\,\varphi\,dx, \tag{21.18}$$

and since $F(P^m X, A^m P^m X) \leq g(A^m)$ by (21.16), one deduces that for all $X \in C^1\big(\Omega; \mathcal{L}(\mathbb{R}^N; \mathbb{R}^N)\big)$, and all $\varphi \in C_c(\Omega)$, $\varphi \geq 0$,

$$\liminf_{m\to\infty} \int_\Omega g(A^m)\,\varphi\,dx \geq \int_\Omega F(X, A^{\mathit{eff}} X)\,\varphi\,dx. \tag{21.19}$$

For $X \in L^\infty\big(\Omega; \mathcal{L}(\mathbb{R}^N; \mathbb{R}^N)\big)$, there exists a sequence $X^k \in C^1\big(\Omega; \mathcal{L}(\mathbb{R}^N; \mathbb{R}^N)\big)$, bounded in $L^\infty\big(\Omega; \mathcal{L}(\mathbb{R}^N; \mathbb{R}^N)\big)$ and converging a.e. to X; by the Lebesgue dominated convergence theorem $F(X^k, A^{\mathit{eff}} X^k)$ converges in $L^1(\Omega)$ strong to $F(X, A^{\mathit{eff}} X)$, so that (21.19) is true for all $X \in L^\infty\big(\Omega; \mathcal{L}(\mathbb{R}^N; \mathbb{R}^N)\big)$.

For $r < \infty$, define g_r on $\mathcal{L}(\mathbb{R}^N; \mathbb{R}^N)$ by

$$g_r(A) = \sup_{||P|| \leq r} F(P, A\,P) \text{ for all } A \in \mathcal{L}(\mathbb{R}^N; \mathbb{R}^N), \tag{21.20}$$

which is continuous, since F is uniformly continuous on bounded sets. For $\varepsilon > 0$, choose M_ε and X_ε to be step functions with values in $\mathcal{L}(\mathbb{R}^N; \mathbb{R}^N)$, such that

$$||M_\varepsilon - A^{\mathit{eff}}|| \leq \varepsilon \text{ a.e. in } \Omega, \tag{21.21}$$

$$||X_\varepsilon|| \leq r \text{ and } F(X_\varepsilon, M_\varepsilon X_\varepsilon) = g_r(M_\varepsilon) \text{ a.e. in } \Omega. \tag{21.22}$$

Since $g_r(M_\varepsilon)$ converges uniformly to $g_r(A^{\mathit{eff}})$ as ε tends to 0, one deduces from (21.19) used for X_ε that, for all $\varphi \in C_c(\Omega)$, $\varphi \geq 0$,

$$\liminf_{m\to\infty} \int_\Omega g(A^m)\,\varphi\,dx \geq \int_\Omega g_r(A^{\mathit{eff}})\,\varphi\,dx \text{ for all } r < +\infty. \tag{21.23}$$

Then, one deduces (21.17) by a theorem of B. LEVI,[6] since $g_r(A^{\mathit{eff}})$ increases and converges to $g(A^{\mathit{eff}})$ as r increases to $+\infty$. □

Of course, I had in mind to use our *compensated compactness* theorem 17.3 for F quadratic, since it characterizes the homogeneous quadratic part F_0 such that (21.14) implies (21.15): it is true if and only if

---

[6] Beppo LEVI, Italian-born mathematician, 1875–1961. He worked in Cagliari, Parma, and Bologna, Italy, and in Rosario, Argentina. The Instituto de Matematicas "Beppo Levi" of the National University of Rosario, Argentina, is named after him.

$$F_0(\xi \otimes \eta, Q) \geq 0 \text{ for all } \xi, \eta \in \mathbb{R}^N \text{ and all } Q \in \mathcal{L}(\mathbb{R}^N; \mathbb{R}^N) \text{ with } Q^T \xi = 0, \tag{21.24}$$

since the characteristic set Λ corresponds to $P\,e$ being parallel to ξ for all e, i.e., $P = \xi \otimes \eta$ for some η, and $Q\,e$ being orthogonal to ξ for all $e \in \mathbb{R}^N$.

Which functions F to choose for obtaining good bounds on A^{eff} was not so clear, and in June 1980, while I was visiting the Courant Institute in New York, NY, I chose to look for bounds in the case of mixing two isotropic materials of conductivity α, β, when A^{eff} is isotropic, and for that I decided to look at quadratic functions F which are invariant under a change of orthonormal basis, i.e., linear combinations of $Trace(P)$, $Trace(Q)$, $Trace(P^T P)$, $Trace(Q^T Q)$, and $Trace(Q^T P)$. By the *div–curl lemma*, for $i, j = 1, \ldots, N$,

$$F^{\pm}_{i,j}(P, Q) = \pm(Q^T P)_{i,j} = \pm \sum_k Q_{k,i}\,P_{k,j} \text{ satisfies (21.24)}, \tag{21.25}$$

so that

$$F^{\pm}(P, Q) = \pm Trace(Q^T P) \text{ satisfies (21.24)}, \tag{21.26}$$

and I observed that

$$\begin{aligned} &F_1(P) = Trace(P^T\,P) - Trace^2(P) \text{ satisfies (21.24)},\\ &F_2(Q) = (N-1)Trace(Q^T\,Q) - Trace^2(Q) \text{ satisfies (21.24)}, \end{aligned} \tag{21.27}$$

by Lemma 21.5, since one has $rank(P) \leq 1$ on Λ, and $rank(Q) \leq N - 1$.[7]

Lemma 21.5. *If* $M \in \mathcal{L}(\mathbb{R}^N; \mathbb{R}^N)$ *then*

$$rank(M)\,Trace(M^T\,M) - Trace^2(M) \geq 0. \tag{21.28}$$

Proof. If $rank(M) = k$, one chooses an orthonormal basis such that $R(M)$ (the range of M) is spanned by the first k vectors of the basis, and then $Trace(M) = \sum_i M_{i,i}$ and $Trace(M^T\,M) = \sum_{i,j} M_{i,j}^2 \geq \sum_i M_{ii}^2$, which by the Cauchy–Schwarz inequality is $\geq \frac{1}{k}(\sum_i M_{i,i})^2$. □

I tried general combinations of these particular functions, but the computations were too technical, and I selected two simple ones, corresponding to Lemma 21.6 and Lemma 21.7. In June 1980 I only computed $g(A)$ for $A = \lambda\,I$, but in the fall, François MURAT suggested to use the same functionals for an anisotropic A^{eff}, and we then did the computations shown below.

[7] For F_1, it is just that when $P = \xi \otimes \eta$, one has $F_1(P) = |\xi|^2|\eta|^2 - (\xi, \eta)^2 \geq 0$.

Lemma 21.6. *One chooses*

$$F_1(P,Q) = \alpha\left(Trace(P^T P) - Trace^2(P)\right) - Trace(Q^T P) + 2Trace(P). \tag{21.29}$$

For $A \in \mathcal{L}_{sym}(\mathbb{R}^N;\mathbb{R}^N)$ with $A \geq \alpha\, I$, $\lambda_j(A)$ denoting the eigenvalues of A,

$$g_1(A) = \frac{\tau}{1+\alpha\,\tau}, \text{ with } \tau = \sum_{j=1}^N \frac{1}{\lambda_j(A) - \alpha}. \tag{21.30}$$

Proof. Of course, if α is an eigenvalue of A then $\tau = \infty$ and $g_1(A) = \frac{1}{\alpha}$. On an orthonormal basis where A is diagonal, the form of $F_1(P, A\,P)$ is unchanged, and one must compute the supremum over all $P \in \mathcal{L}(\mathbb{R}^N;\mathbb{R}^N)$ of

$$\alpha \sum_{i,j=1}^N P_{i,j}^2 - \alpha\left(\sum_{i=1}^N P_{i,i}\right)^2 - \sum_{i,j=1}^N \lambda_i(A)\, P_{i,j}^2 + 2\sum_{i=1}^N P_{i,i}, \tag{21.31}$$

and for $i \neq j$ a good choice for $P_{i,j}$ is 0 (it does not really matter what $P_{i,j}$ is if $\lambda_i(A) = \alpha$), and one must then compute the supremum over all $P_{i,i}$ of

$$\sum_{i=1}^N (\alpha - \lambda_i(A))\, P_{i,i}^2 - \alpha\left(\sum_{i=1}^N P_{i,i}\right)^2 + 2\sum_{i=1}^N P_{ii}. \tag{21.32}$$

If $\sum_i P_{i,i} = t$ is fixed, then in the case where $\lambda_i(A) > \alpha$ for all i, maximizing $\sum_i (\alpha - \lambda_i(A))\, P_{i,i}^2$ is obtained by taking $P_{i,i} = \frac{C}{\lambda_i(A)-\alpha}$ for all i, for a Lagrange multiplier C, so that $t = C\,\tau$; one then finds t by maximizing $-C^2\,\tau - \alpha\, t^2 + 2t$, i.e., by maximizing $-\frac{t^2}{\tau} - \alpha\, t^2 + 2t$, which gives the value of t and the maximum equal to $\frac{\tau}{1+\alpha\,\tau}$. If $\lambda_i = \alpha$ for some i, then it is best to take $P_{i,i} = t$ and $P_{j,j} = 0$ for $j \neq i$, which gives the value of t and the maximum equal to $\frac{1}{\alpha}$. □

Lemma 21.7. *One chooses*

$$F_2(P,Q) = (N-1)Trace(Q^T Q) - Trace^2(Q) - \beta\,(N-1)Trace(Q^T P) + 2Trace(Q). \tag{21.33}$$

For $A \in \mathcal{L}_{sym}(\mathbb{R}^N;\mathbb{R}^N)$ with $A \leq \beta\, I$, $\lambda_j(A)$ denoting the eigenvalues of A,

$$g_2(A) = \frac{\sigma}{\sigma + N - 1}, \text{ with } \sigma = \sum_{j=1}^N \frac{\lambda_j(A)}{\beta - \lambda_j(A)}. \tag{21.34}$$

Proof. Of course, if β is an eigenvalue of A then $\sigma = \infty$ and $g_2(A) = 1$. On an orthonormal basis where A is diagonal, the form of $F_2(P, A\,P)$ is unchanged, and one must compute the supremum over all $P \in \mathcal{L}(\mathbb{R}^N;\mathbb{R}^N)$ of

$$(N-1)\sum_{i,j=1}^{N}\big(\lambda_i^2(A)-\beta\,\lambda_i(A)\big)\,P_{i,j}^2-\Big(\sum_{i=1}^{N}\lambda_i(A)\,P_{i,i}\Big)^2+2\sum_{i=1}^{N}\lambda_i(A)\,P_{i,i}, \tag{21.35}$$

and for $i\neq j$ a good choice for $P_{i,j}$ is 0 (it does not really matter what $P_{i,j}$ is if $\lambda_i(A)=\beta$), and one must then compute the supremum over all $P_{i,i}$ of

$$(N-1)\sum_{i=1}^{N}(\lambda_i(A)-\beta)\,\lambda_i(A)\,P_{i,i}^2-\Big(\sum_{i=1}^{N}\lambda_i(A)\,P_{i,i}\Big)^2+2\sum_{i=1}^{N}\lambda_i(A)\,P_{i,i}. \tag{21.36}$$

If $\sum_i\lambda_i(A)\,P_{i,i}=s$ is fixed, then in the case where $\lambda_i(A)<\beta$ for all i, maximizing $\sum_i(\lambda_i(A)-\beta)\lambda_i(A)\,P_{i,i}^2$ is obtained by taking $P_{i,i}=\frac{C}{\beta-\lambda_i(A)}$ for all i, for a Lagrange multiplier C, so that $s=C\,\sigma$; one then finds s by maximizing $-(N-1)C^2\,\sigma-s^2+2s$, i.e., by maximizing $-\frac{N-1}{\sigma}s^2-s^2+2s$, which gives the value of s and the maximum equal to $\frac{\sigma}{\sigma+N-1}$. If $\lambda_i(A)=\beta$ for some i, then it is best to take $P_{i,i}=\frac{s}{\lambda_i(A)}$ and $P_{j,j}=0$ for $j\neq i$, which gives the value of s and the maximum equal to 1. □

In June 1980, I first considered more general combinations

$$\begin{aligned}F_3(P,Q)&=-Trace(Q^TP)+a\,\big(Trace(P^TP)-Trace^2(P)\big)\\&+\;b\,\big((N-1)Trace(Q^TQ)-Trace^2(Q)\big)+2c\,Trace(P)+2d\,Trace(Q),\end{aligned} \tag{21.37}$$

with $a,b\geq 0$, and $g_3(\gamma\,I)$ requires the maximization in $P\in(\mathbb{R}^N;\mathbb{R}^N)$ of

$$(-\gamma+a+b\,(N-1)\,\gamma^2)Trace(P^TP)-(a+b\,\gamma^2)Trace^2(P)+2(c+\gamma\,d)Trace(P). \tag{21.38}$$

To have $g_3(\gamma\,I)<+\infty$, one needs $-\gamma+a+b(N-1)\gamma^2\leq 0$, and one then chooses all non diagonal coefficients of P equal to 0; for $Trace(P)$ given one wants to minimize $Trace(P^TP)$, so that one only considers $P=p\,I$, and one then wants to maximize $\big(-\gamma+a+b(N-1)\gamma^2-N(a+b\gamma^2)\big)p^2+2(c+\gamma\,d)p$, and one obtains

$$g_3(\gamma\,I)=\frac{(c+\gamma\,d)^2}{(N-1)a+\gamma+b\,\gamma^2}\ \text{ if } a,b\geq 0 \text{ and } -\gamma+a+b(N-1)\gamma^2\leq 0. \tag{21.39}$$

It was not easy to handle, so I chose the simplifications of either $b=d=0$, corresponding to Lemma 21.6, or $a=c=0$, corresponding to Lemma 21.7.

François MURAT and myself wanted to characterize A^{eff} for mixtures using proportions θ and $1-\theta$ of isotropic materials with tensors $\alpha\,I$ and $\beta\,I$, so in the fall of 1980 we found the necessary conditions of Lemma 21.8.

Lemma 21.8. *If a sequence of characteristic functions $\chi_n \rightharpoonup \theta$ in $L^\infty(\Omega)$ weak $\star$, with $A^n = \big(\alpha\,\chi_n + \beta\,(1-\chi_n)\big)\,I$ H-converges to A^{eff} in Ω, then*

$$\lambda_-(\theta) \le \lambda_j(A^{\mathit{eff}}) \le \lambda_+(\theta), j = 1,\ldots,N \text{ a.e. in } \Omega, \tag{21.40}$$

$$\sum_{j=1}^N \frac{1}{\lambda_j(A^{\mathit{eff}})-\alpha} \le \frac{(N-\theta)\alpha+\theta\,\beta}{(1-\theta)\alpha(\beta-\alpha)} = \frac{1}{\lambda_-(\theta)-\alpha} + \frac{N-1}{\lambda_+(\theta)-\alpha}, \tag{21.41}$$

$$\sum_{j=1}^N \frac{1}{\beta-\lambda_j(A^{\mathit{eff}})} \le \frac{(1-\theta)\alpha+(N+\theta-1)\beta}{\theta\,\beta(\beta-\alpha)} = \frac{1}{\beta-\lambda_-(\theta)} + \frac{N-1}{\beta-\lambda_+(\theta)}. \tag{21.42}$$

Proof. By Lemma 6.7, $A^{\mathit{eff}} \in \mathcal{B}(\theta)$, i.e., (21.40). Property (21.17) means

$$g(A^{\mathit{eff}}) \le \theta\,g(\alpha\,I) + (1-\theta)g(\beta\,I) \text{ a.e. in } \Omega, \tag{21.43}$$

whenever g is defined by (21.16) for a function F such that (21.14) implies (21.15). For g_1 given by (21.30) in Lemma 21.6, one has

$$g_1(\alpha\,I) = \frac{1}{\alpha}, g_1(\beta\,I) = \frac{N/(\beta-\alpha)}{1+\alpha\,N/(\beta-\alpha)} = \frac{N}{(N-1)\alpha+\beta}, \tag{21.44}$$

so that (21.43) means

$$\frac{\tau^{\mathit{eff}}}{1+\alpha\,\tau^{\mathit{eff}}} \le \frac{\theta}{\alpha} + \frac{(1-\theta)N}{(N-1)\alpha+\beta} = \frac{(N-\theta)\alpha+\theta\,\beta}{\alpha\big((N-1)\alpha+\beta\big)}, \tag{21.45}$$

which gives for τ^{eff} the bound (21.41); an explicit computation gives equality in the laminated case. For g_2 given by (21.34) in Lemma 21.7, one has

$$g_2(\alpha\,I) = \frac{N\,\alpha/(\beta-\alpha)}{N\,\alpha/(\beta-\alpha)+N-1} = \frac{N\,\alpha}{\alpha+(N-1)\beta}, g_2(\beta\,I) = 1, \tag{21.46}$$

so that (21.43) means

$$\frac{\sigma^{\mathit{eff}}}{\sigma^{\mathit{eff}}+N-1} \le \frac{\theta\,N\,\alpha}{\alpha+(N-1)\beta} + (1-\theta) = \frac{(\theta\,N+1-\theta)\alpha+(1-\theta)(N-1)\beta}{\alpha+(N-1)\beta}, \tag{21.47}$$

which gives for σ^{eff} the upper bound

$$\sigma^{\mathit{eff}} = \sum_{j=1}^N \frac{\lambda_j}{\beta-\lambda_j} \le \frac{(\theta\,N+1-\theta)\alpha+(1-\theta)(N-1)\beta}{\theta(\beta-\alpha)}, \tag{21.48}$$

and since $\sigma^{eff} = -N + \beta \sum_{j=1}^N \frac{1}{\beta-\lambda_j(A^{eff})}$ it gives the bound (21.42); an explicit computation gives equality in the laminated case. □

In June 1980, I showed my bounds for the case $A^{eff} = a_{eff} I$ to George PAPANICOLAOU, and he told me to compare them to bounds which I then heard about for the first time, those which Zvi HASHIN obtained with SHTRIKMAN almost 20 years before. I went to the library and I checked their article, and my bounds were the same as the Hashin–Shtrikman bounds, which I was probably the first to prove, since their "reasoning" had a step with no mathematical meaning, so that it could only be a conjecture. Their argument why the bounds must hold does not correspond to any physical principle that I know, so I cannot guess if it corresponds to a physical intuition that they had. When I introduced *H-measures* in the late 1980s, the situation became similar to that of Laurent SCHWARTZ explaining some curious computations of DIRAC, since the step which did not make any sense in their "proof" can be explained using arguments about *H-measures*.

Some people pretended to give a proof by considering a periodic situation, but the argument of Zvi HASHIN and SHTRIKMAN which is problematic is precisely that they do computations as in a periodic case, although they deal with a nonperiodic situation, since their understanding of material science was good enough to avoid a non-physical hypothesis of periodicity! It is precisely one of the properties of my *H-measures* to give a meaning to lots of computations made in continuum mechanics or physics in this way, i.e., to imitate a proof from a periodic situation in a situation without periodicity.

However, I had no difficulty adapting the argument of Zvi HASHIN and SHTRIKMAN showing that the bounds are attained, and actually one may consider that they proved that part, long before a general definition of homogenization was introduced, since they observed a special property, which is not satisfied in the general theory.

I shall describe their construction in Chap. 25, and I shall describe in Chap. 26 the generalization that I made with François MURAT, which I presented in June 1981 with Lemma 21.8 at a conference in New York, NY.

Finally, I want to mention a computation done with Gilles FRANCFORT and François MURAT in the mid 1980s [31],[8] giving a characterization of the effective properties of all mixtures of two anisotropic materials (i.e., with all possible orientations), only valid in the case $N = 2$, since we used Lemma 20.1; we did not impose the proportion θ of one of the two materials, i.e., we did not identify each of the sets $\mathcal{K}(\theta; M^1, M^2)$ for $0 < \theta < 1$, of all effective tensors A^{eff} obtained by mixing M^1 in proportion θ and M^2 in proportion

[8] I did not accept the proposition of Gilles FRANCFORT and François MURAT to be a coauthor of [30], since I felt that they did much of the computational work for obtaining the first version of the proofs, while my work was more about simplifying their proofs. For [31], I did provide an important idea, and Gilles FRANCFORT and François MURAT state in the acknowledgments of their article that the results were obtained in collaboration with me.

$1-\theta$, but only their union $\bigcup_{0\le\theta\le 1}\mathcal{K}(\theta;M^1,M^2)$. A characterization of all the sets which are stable by H-convergence followed. The constructions relied on simple laminations as in Chap. 12 (and not on the more general results of Chap. 27), and the necessary conditions used Lemma 20.1 and Lemma 6.7.[9]

[9] I have shown a similar idea in the proof of Lemma 20.8.

Chapter 22
Functions Attached to Geometries

During my talk at a conference in New York, NY, in June 1981, when I was presenting my result with François MURAT on the characterization of effective properties of mixtures of two isotropic conductors, proving and extending the Hashin–Shtrikman bounds, of which I presented the necessary part in Lemma 21.8, David BERGMAN asked me an interesting question, about the meaning of the sequence A^n that I used. I answered him as a joke, that it is the same thing as the thermodynamic limit that he used![1] Since I think that many do not perceive well what is homogenization and what is not, and it is important not to confuse his approach with homogenization, I shall be more precise.

David BERGMAN is a physicist, and what interested him was a number, for example the current going through two plates with a difference of potential of 1, the zone between the plates being filled with a mixture of two conductors, or the energy stored in the mixture. Although the domain filled with his mixture was fixed and bounded, he invoked a thermodynamic limit, supposed to be the limit of averages on arbitrarily large balls, so I found it a curious idea for a finite domain! However, I guessed that one could make a correct statement by considering a sequence of mixtures using shorter and shorter characteristic lengths, and I considered that his "argument using a thermodynamic limit" meant that he also considered a sequence of mixtures! Of course, what he was doing is not homogenization, since he did not speak of local properties, and of effective coefficients given by symmetric matrices!

David BERGMAN used an idea special to the case of mixing two materials, and Graeme MILTON had the same idea independently, which consists in using the same geometry with different isotropic conductivities, which may be given by complex numbers; they studied which functions of the ratio of the two conductivities one may obtain. I once heard someone attribute a similar

[1] Of course, using a sequence serves in learning about the topology of H-convergence, the natural one for comparing a fine mixture with very small pieces of different materials and a material with smoothly varying properties.

L. Tartar, *The General Theory of Homogenization*, Lecture Notes of the Unione Matematica Italiana 7, DOI 10.1007/978-3-642-05195-1_22,

idea to PRAGER.[2] In discussing with François MURAT after the conference, we easily applied this idea to homogenization, obtaining Lemma 22.1.

Lemma 22.1. *If a sequence χ_n of characteristic functions converges to θ in $L^\infty(\Omega)$ weak $\star$, there exists a subsequence indexed by m such that*

$$\begin{aligned}&A^m = \chi_m M^1 + (1-\chi_m)M^2 \textit{ H-converges to } A^{eff} = F(\cdot; M^1, M^2)\\ &\textit{for all matrices } M^1, M^2 \textit{ satisfying}\\ &(M^1\xi,\xi),(M^2\xi,\xi) \ge \alpha\,|\xi|^2 \textit{ for all } \xi\in\mathbb{R}^N, \textit{ for some } \alpha>0,\end{aligned}\tag{22.1}$$

and the Carathédory function F is analytic in M^1 and M^2, and satisfies

$$\begin{aligned}&F(\cdot; M, M) = M \textit{ for all } M,\\ &\nabla^1 F(\cdot; M,M).\widetilde{M} = \theta\,\widetilde{M} \textit{ for all } M \textit{ and all directions } \widetilde{M},\\ &\nabla^2 F(\cdot; M,M).\widetilde{M} = (1-\theta)\widetilde{M} \textit{ for all } M \textit{ and all directions } \widetilde{M},\end{aligned}\tag{22.2}$$

where ∇^j is the Fréchet derivative with respect to M^j, i.e., at order 1 and near the diagonal one has $F(\cdot; M^1, M^2)\approx \theta\,M^1 + (1-\theta)M^2$.

Proof. One uses a Cantor diagonal procedure for extracting a subsequence for M^1, M^2 in a countable dense subset of $\mathcal{L}_+(\mathbb{R}^N;\mathbb{R}^N)$,[3] and by Lemma 10.9 H-convergence holds on $\mathcal{L}_+(\mathbb{R}^N;\mathbb{R}^N)$; by Lemma 10.10, the H-limit inherits of regularity properties, in particular the analyticity in M^1 and M^2.

For $M^1 = M+\varepsilon\,\widetilde{M}$, $M^2 = M$, and $f\in H^{-1}(\Omega)$, one has

$$\begin{aligned}&-div\big(\big[\chi_m(M+\varepsilon\,\widetilde{M}) + (1-\chi_m)M)grad(u_m\big]\big) = f \text{ in } \Omega,\\ &grad(u_m) = grad(v) + \varepsilon\, grad(w_m) + o(\varepsilon) \in H^1_0(\Omega)\\ &-div\big(M\, grad(v)\big) = f \text{ in } \Omega, v\in H^1_0(\Omega)\\ &-div\big(M\, grad(w_m) + \chi_m\widetilde{M}\, grad(v)\big) = 0 \text{ in } \Omega, w_m\in H^1_0(\Omega),\end{aligned}\tag{22.3}$$

[2] Ennio DE GIORGI once said "Chi cerca trova, chi ricerca ritrova," a play on the words of the gospels, "Ask and it will be given to you; seek and you will find; knock and the door will be opened to you" (Matthew 7:7, Luke 11:9), which gave the French saying "Qui cherche trouve," or the Italian one, "Chi cerca trova." The play on the prefix "ri" in Italian, does not work as well in English, but Ennio DE GIORGI's remark means that searching leads to discovery and doing research leads to discovering again some results which are already "known." There is nothing wrong about having independently the same idea as someone else, but one should be aware of the possibility of plagiarism, or organized misattribution of ideas: I insist in describing the conditions which led me to the ideas that I had in order to help students and researchers understand about the creative process of the mathematical discovery, but also because my political opponents usually attribute my ideas to their friends, who cannot follow my example.

[3] $\mathcal{L}_+(\mathbb{R}^N;\mathbb{R}^N) = \{M\in\mathcal{L}(\mathbb{R}^N;\mathbb{R}^N) \mid (M\,\xi,\xi)>0 \text{ for all nonzero } \xi\in\mathbb{R}^N\}$.

so that,

$$\begin{aligned}&grad(u_\infty) = grad(v) + \varepsilon\, grad(w_\infty) + o(\varepsilon)\\&-div\big(M\, grad(w_\infty) + \theta\,\widetilde{M}\, grad(v)\big) = 0 \text{ in } \Omega\\&(\chi_m(M + \varepsilon\,\widetilde{M}) + (1-\chi_m)M)grad(u_m) \rightharpoonup A^{eff} grad(u_\infty)\\&\quad = \quad M\,\big(grad(v) + \varepsilon\, grad(w_\infty)\big) + \varepsilon\,\theta\,\widetilde{M}\, grad(v) + o(\varepsilon)\end{aligned} \tag{22.4}$$

and using $A^{eff} = M + \varepsilon\, B + o(\varepsilon)$, with $B = \nabla^1 F$, one deduces that

$$B\, grad(v) = \theta\,\widetilde{M}\, grad(v), \tag{22.5}$$

so that $B = \theta\,\widetilde{M}$ by varying f (or v), and similarly for $\nabla^2 F$, by exchanging the roles of θ and $1-\theta$. □

One uses G-convergence if M^1, M^2 are symmetric, and

$$F(\cdot;\alpha\, I, \beta\, I) = \beta\, G\Big(\cdot;\frac{\alpha}{\beta}\Big), \text{ with } G(\cdot;z) \text{ defined for } z \in \mathbb{C}\setminus(-\infty,0], \tag{22.6}$$

by using the Lax–Milgram lemma for the complex case.[4] Then, if

$$G(x;z) = g(x;z)I \text{ a.e. } x \in \Omega \text{ and all } z \in \mathbb{C}\setminus(-\infty,0], \tag{22.7}$$

which seems a very particular case, one may compare the properties of $g(\cdot;z)$ to those of the functions used by David BERGMAN and Graeme MILTON.

One needs an analogue of Lemma 22.1 for the extension to $z \in \mathbb{C}\setminus(-\infty,0]$ in the isotropic case, and it is Lemma 22.4, for which I use an idea of Eduardo ZARANTONELLO, concerning the convexity of the numerical range (the Hausdorff–Toeplitz theorem) of an operator in a complex Hilbert space.[5]

Definition 22.2. *If H is a complex Hilbert space and $M \in \mathcal{L}(H;H)$, then*

$$num(M) = \Big\{\frac{(M\,v,v)}{||v||^2} \in \mathbb{C} \mid v \neq 0, v \in H\Big\} \tag{22.8}$$

is called the numerical range *of M.*

---

[4] For a complex Hilbert space V, and a sesqui-linear form b on $V\times V$, one assumes that for some $\gamma > 0$ one has $|b(v,v)| \geq \gamma\,||v||^2$ for all $v \in V$, so that by Lemma 22.3 there exists a unimodular ζ_0 such that $\Re\big(\zeta_0 b(v,v)\big) \geq \gamma\,||v||^2$ for all $v \in V$. In the case where A^n only takes the values I and $z\,I$ it consists in separating strictly 1 and $z \in \mathbb{C}$ from 0 by a line, so the only z to avoid are the real ≤ 0.

[5] Otto TOEPLITZ, German-born mathematician, 1881–1940. He worked in Kiel, and in Bonn, Germany, and emigrated to Palestine in 1939, where he helped in the building up of Jerusalem University.

Lemma 22.3. *If H is a complex Hilbert space and $M \in \mathcal{L}(H;H)$, then the numerical range $num(M)$ is convex.*[6]

Proof. If $Z_1, Z_2 \in num(M)$ are distinct, there exist unit vectors $v_1, v_2 \in H$ with $Z_j = (M\,v_j, v_j)$, $j = 1,2$. For $\theta \in (0,1)$, one wants $z \in \mathbb{C}$ with

$$\big(M\,(v_1 + z\,v_2), (v_1 + z\,v_2)\big) - (\theta\,Z_1 + (1-\theta)Z_2)\,||v_1 + z\,v_2||^2 = 0, \quad (22.9)$$

and $Z_1 \neq Z_2$ implies $v_1 + z\,v_2 \neq 0$. For some $a, b \in \mathbb{C}$ (22.9) is

$$\theta\,(Z_2 - Z_1)\,|z|^2 - (1-\theta)\,(Z_2 - Z_1) = a\,z + b\,\overline{z}, \quad (22.10)$$

and after dividing by $Z_2 - Z_1$, the real and imaginary parts give

$$\begin{aligned} \theta\,(x^2 + y^2) - (1-\theta) &= C_1 x + C_2 y \\ 0 &= C_3 x + C_4 y, \end{aligned} \quad (22.11)$$

where $z = x + i\,y$. The first equation is a circle Γ with 0 in its interior, and the second is either a line going through 0, intersecting Γ at two distinct solutions, or the complex plane and every point in Γ is a solution. □

Lemma 22.4. *If χ_n is a sequence of characteristic functions, one can extract a subsequence such that $A^m = \chi_m M^1 + (1-\chi_m)M^2$ H-converges to $A^{eff} = F(\cdot; M^1, M^2)$ for all complex matrices M^1, M^2 satisfying*

$$0 \notin K(M^1, M^2) = conv\big(num(M^1) \cup num(M^2)\big), \quad (22.12)$$

where $conv(X)$ is the convex envelope of X. One has

$$num\big(F(x; M^1, M^2)\big) \subset \bigcup_{s\geq 1} s\,K(M^1, M^2) \text{ a.e. in } \Omega, \quad (22.13)$$

$$num\big(F(x; M^1, M^2)^{-1}\big) \subset \bigcup_{s\geq 1} s\,K\big((M^1)^{-1}, (M^2)^{-1}\big) \text{ a.e. in } \Omega. \quad (22.14)$$

Proof. Condition (22.12) serves for the existence of $\alpha > 0$ and ζ_0 unimodular such that $\Re\big(\zeta_0(M_j\eta, \eta)\big) \geq \alpha\,|\eta|^2$ for all $\eta \in \mathbb{C}^N$, and $j = 1,2$. For each pair (M^1, M^2) satisfying (22.12), one extracts a subsequence which H-converges; for a diagonal subsequence it is true for a countable dense set of such pairs, for example those having entries with real part and imaginary part $\in \mathbb{Q}$; by an argument of equicontinuity it is true for all matrices satisfying (22.12).

[6] If $\lambda \notin \overline{num(M)}$ the Lax–Milgram lemma in the complex case applies to $\lambda\,I - M$, so that $spectrum(M) \subset \overline{num(M)}$. In finite dimension with M normal (i.e., having an orthonormal basis of eigenvectors), $num(M) = conv\big(spectrum(M)\big)$.

To find where the effective matrix is, one needs to generalize the sets $\mathcal{M}(\alpha,\beta;\Omega)$ which I introduced with François MURAT for the real case. The analogue of an inequality $(A(x)\xi,\xi) \geq \alpha\,|\xi|^2$ for all $\xi \in \mathbb{R}^N$ in the real case is $\Re\big(\zeta_0(A(x)\eta,\eta)\big) \geq \alpha\,|\eta|^2$ for all $\eta \in \mathbb{C}^N$ for a ζ_0 unimodular; by homogeneity it is $\Re\big(\zeta_0(M_j e,e)\big) \geq \alpha$ for $|e|_{\mathbb{C}^N} = 1$ and $j = 1,2$, implying $\Re\big(\zeta_0(A^{eff}(x)e,e)\big) \geq \alpha$ for $|e|_{\mathbb{C}^N} = 1$ a.e. in Ω,[7] i.e., $(A^{\mathit{eff}}(x)e,e)$ takes its values in every half-space (of $\mathbb{C}$ considered as $\mathbb{R}^2$) containing $num(M^1)$ and $num(M^2)$ but not 0, i.e., which contains $K(M^1,M^2)$ but not 0, i.e., (22.13). Similarly, the analogue of an inequality $(A(x)\xi,\xi) \geq \frac{|A(x)\xi|^2}{\beta}$ for all $\xi \in \mathbb{R}^N$ in the real case is $\Re\big(\zeta_0(A(x)\eta,\eta)\big) \geq \frac{|A(x)\eta|^2}{\beta}$ for all $\eta \in \mathbb{C}^N$ for a ζ_0 unimodular, and becomes by homogeneity $\Re\big(\zeta_0(e,(M_j)^{-1}e)\big) \geq \frac{1}{\beta}$ for $|e|_{\mathbb{C}^N} = 1$ and $j = 1,2$, implying $\Re\big(\zeta_0(e,(A^{\mathit{eff}})^{-1}(x)e)\big) \geq \frac{1}{\beta}$ for all $\xi \in \mathbb{C}^N$, a.e. in Ω, and taking complex conjugates, one sees that $\big((A^{\mathit{eff}})^{-1}(x)e,e\big)$ takes its values in every half-space containing $num\big((M^1)^{-1}\big)$ and $num\big((M^2)^{-1}\big)$ but not 0, i.e., which contains $K\big((M^1)^{-1},(M^2)^{-1}\big)$ but not 0, i.e., (22.14). □

The proof of Lemma 22.4 gives the more general Lemma 22.5.

Lemma 22.5. *If $K_1, K_2 \subset \mathbb{C}$ are nonempty bounded closed convex subsets of $\mathbb{C}$, with $0 \notin K_1 \cup K_2$, and $A^n \in L^\infty\big(\Omega;\mathcal{L}(\mathbb{C}^N;\mathbb{C}^N)\big)$ satisfies*

$$num\big(A^n(x)\big) \subset K_1, num\big((A^n)^{-1}(x)\big) \subset K_2, \text{ a.e. } x \in \Omega, \text{ for all } n, \tag{22.15}$$

and A^n H-converges to A^{eff}, then

$$num\big(A^{\mathit{eff}}(x)\big) \subset K_1^*, num\big((A^{\mathit{eff}})^{-1}(x)\big) \subset K_2^* \text{ a.e. } x \in \Omega, \tag{22.16}$$

where K_j^* is the intersection of all closed half-spaces containing K_j but not 0.

Lemma 10.2 for $(A^n)^T$ gives the same property for $(A^n)^* = \overline{(A^n)^T}$ in the complex case; obviously, $\overline{A^n}$ H-converges to $\overline{A^{\mathit{eff}}}$, so $F(\cdot;M^1,M^2)$ satisfies

$$\begin{aligned} &F(x;(M^1)^T,(M^2)^T) = \big(F(x;M^1,M^2)\big)^T \\ &F(x;\overline{M^1},\overline{M^2}) = \overline{F(x;M^1,M^2)} \\ &\text{a.e. in } \Omega, \text{ and for all } M^1,M^2 \text{ satisfying (22.12).} \end{aligned} \tag{22.17}$$

The analogue of Lemma 10.7 for the pre-order $A \geq B$, meaning $\Re(A\,\eta,\eta) \geq \Re(B\,\eta,\eta)$ for all $\eta \in \mathbb{C}^N$, is Lemma 22.6.

---

[7] This uses the *div–curl lemma*, here for $E^m = grad(u_m), D^m = A^m grad(u_m) \in L^2(\Omega;\mathbb{C}^N)$ and (E^m,D^m) is their Hermitian scalar product in $\mathbb{C}^N$; one deduces the complex case from the real case by working on real and imaginary parts.

Lemma 22.6. *If $M^1 \geq H^1$, $M^2 \geq H^2$, and H^1, H^2 are Hermitian positive definite, then condition (22.12) is satisfied and*

$$F(x; M^1, M^2) \geq F(x; H^1, H^2) \text{ a.e. } \Omega. \tag{22.18}$$

Proof. One applies the analogue of Lemma 10.7 to $A^m = \chi_m M^1 + (1 - \chi_m) M^2$ and $B^m = \chi_m H^1 + (1 - \chi_m) H^2$, so that one has $A^m \geq B_m$ for all m and a.e. in Ω. The symmetry of B^m in the real case is replaced by the Hermitian symmetry and the positivity of B^m in the complex case, in order to apply the analogue of Lemma 10.6 proven by Ennio DE GIORGI and Sergio SPAGNOLO, i.e., if $grad(u_m) \rightharpoonup grad(u_\infty)$ in $L^2_{loc}(\Omega; \mathbb{C}^N)$ weak, then for all $\varphi \in C_c(\Omega)$, $\varphi \geq 0$, one has $\liminf_{n\to\infty} \big(\int_\Omega \varphi\, (B^m grad(u_m), grad(u_m))\, dx\big) \geq \int_\Omega \varphi\, (B^{eff} grad(u_\infty), grad(u_\infty))\, dx$. □

In their context, which is not homogenization, David BERGMAN and Graeme MILTON observed that their function g satisfies $g(1) = 1$ and $g'(1) = 1 - \theta$, and extends into a holomorphic function in the upper half-space $\Im z > 0$, where it satisfies $\Im\big(g(z)\big) > 0$.

In our homogenization context, the numerical range of $F(\cdot; I, z\, I)$ is constrained by Lemma 22.4;[8] if $F(\cdot; I, z\, I) = g(\cdot, z) I$ for $z \in C \setminus (-\infty, 0]$,[9] $g(\cdot, z)$ takes its values in the region limited by the segment with end-points 1 and z, and a piece of the circle going through 0, 1 and z (the arc between 1 and z that does not contain 0); the localization is then more precise, apart from the possibility (ruled out if $\theta < 1$) that $\Im(g)$ could vanish at some points.

Definition 22.7. *A* Pick function *g is a holomorphic function in the open upper half-space $\Im(z) > 0$ with $\Im\big(g(z)\big) \geq 0$.*[10] *A* Herglotz function *h is a holomorphic function in the open unit disc D with $\Re(h) \geq 0$.*[11]

If g is a Pick function and $\Im\big(g(z_0)\big) = 0$ for some z_0 with $\Im(z_0) > 0$, then g is constant, and if h is a Herglotz function and $\Re\big(h(z_0)\big) = 0$ for some z_0 with $|z_0| < 1$, then h is constant, by the maximum principle for holomorphic functions, applied to e^{-g} or to $e^{i\,h}$, which have modulus ≤ 1.

---

[8] If $A^n = c_n I$ with $\Im(c_n) \geq 0$ a.e. in Ω, $\Im\Big[\Big(A^m grad(u_m), grad(u_m)\Big)\Big] = \Im(c_m)\, |grad(u_m)|^2 \geq 0$ a.e. in Ω, so that $\Im\Big[\Big(A^{eff} grad(u_\infty), grad(u_\infty)\Big)\Big] \geq 0$ a.e. in Ω by the *div–curl lemma*, hence $num\Big(A^{eff}(x)\Big) \subset \{z \in \mathbb{C} \mid \Im(z) \geq 0\}$.

[9] For an $x \in \Omega$ such that $F(x; I, z\, I)$ is holomorphic in z in a connected open set $\Delta \subset \mathbb{C}$, if $z_\infty \in \Delta$ and $z_m \to z_\infty$ with $z_m \neq z_\infty$ and $F(x; I, z_m I) = \kappa_m I$ for all m, then $F(x; I, z_m I)_{i,j}$ must vanish in Δ if $i \neq j$, and $F(x; I, z_m I)_{i,i} - F(x; I, z_m I)_{j,j}$ must vanish in Δ for all i, j, so that $F(x; I, z\, I) = \kappa(z)\, I$ in Δ.

[10] These functions are also named after NEVANLINNA, and after STIELTJES.

[11] Gustav HERGLOTZ, Austrian-born mathematician, 1881–1953. He worked at Georg-August-Universität, Göttingen, Germany.

Lemma 22.8. *For a Herglotz function h, there exists a nonnegative Radon measure μ on $\mathbb{S}^1 = \partial D$ such that*[12]

$$h(z) = i\,\Im\big(h(0)\big) + \int_{\mathbb{S}^1} \frac{1+z\,e^{i\,\theta}}{1-z\,e^{i\,\theta}}\,d\mu(\theta) \text{ for } |z|<1. \tag{22.19}$$

Proof. From

$$h(z) = \sum_{n\geq 0} c_n z^n = P + i\,Q \text{ with } c_n = a_n + i\,b_n \text{ for } n \geq 0, \tag{22.20}$$

one deduces that

$$P(r\,e^{i\,\theta}) = \sum_{n\geq 0} a_n r^n \cos n\,\theta - \sum_{n\geq 1} b_n r^n \sin n\,\theta \text{ for } r<1, \tag{22.21}$$

and by results on Fourier series, for $r<1$,

$$\begin{aligned} a_0 &= \frac{1}{2\pi}\int_0^{2\pi} P(r\,e^{i\,\theta})\,d\theta, a_n r^n = \frac{1}{\pi}\int_0^{2\pi} P(r\,e^{i\,\theta})\cos n\,\theta\,d\theta \text{ for } n\geq 1\\ b_n r^n &= \frac{1}{\pi}\int_0^{2\pi} P(r\,e^{i\,\theta})\sin n\,\theta\,d\theta \text{ for } n\geq 1. \end{aligned} \tag{22.22}$$

Since $P\geq 0$, and $\int_0^{2\pi} P(r\,e^{i\,\theta})\,d\theta$ is independent of r, the restrictions of P on circles centred at 0 are bounded in L^1, and for a sequence r_m tending to 1 the restrictions converge in $\mathcal{M}(\mathbb{S}^1)$ weak $\star$ to $2\pi\,\mu\geq 0$, and (22.22) gives

$$c_0 = \langle \mu, 1\rangle + i\,\Im\big(h(0)\big), c_n = 2\big\langle \mu, e^{i\,n\,\theta}\big\rangle \text{ for } n\geq 1, \tag{22.23}$$

$$h(z) = i\,\Im\big(h(0)\big) + \Big\langle \mu, 1 + 2\sum_{n=1}^{\infty} e^{i\,n\,\theta} z^n\Big\rangle, \tag{22.24}$$

and $1 + 2\sum_{n=1}^{\infty} e^{i\,n\,\theta} z^n = -1 + \frac{2}{1-z\,e^{i\,\theta}} = \frac{1+z\,e^{i\,\theta}}{1-z\,e^{i\,\theta}}$. □

Lemma 22.9. *For a Pick function g, there exists $\gamma\geq 0$ and a nonnegative Radon measure $\nu\in\mathcal{M}_b(\mathbb{R})$ such that*

$$g(z) = \Re\big(g(i)\big) + \gamma\,z + \int_{\mathbb{R}} \frac{t\,z+1}{t-z}\,d\nu(t) \text{ for } \Im(z)>0. \tag{22.25}$$

---

[12] Traditional notation must always be interpreted in context: θ was a proportion before, but now it becomes an angle.

Proof. For $\Im(z) > 0$ one has $\xi = \frac{z-i}{z+i} \in D$, $z = \frac{i(1+\xi)}{1-\xi}$. One defines h on D by

$$h(\xi) = -i\, g\Big(\frac{i(1+\xi)}{1-\xi}\Big) \text{ for } |\xi| < 1, g(z) = i\, h\Big(\frac{z-i}{z+i}\Big) \text{ for } \Im(z) > 0. \quad (22.26)$$

One has $h(0) = -i\, g(i)$, so that $i\, \Im\big(h(0)\big) = -i\, \Re\big(g(i)\big)$. For μ defined by (22.19) for h, let $\gamma = \mu(\{1\}) \geq 0$, so that

$$h(\xi) = -i\, \Re\big(g(i)\big) + \gamma\, \frac{1+\xi}{1-\xi} + \int_{(0,2\pi)} \frac{1+\xi\, e^{i\,\theta}}{1-\xi\, e^{i\,\theta}}\, d\mu(\theta), \quad (22.27)$$

$$g(z) = \Re\big(g(i)\big) + \gamma\, z + i \int_{(0,2\pi)} \frac{z+i+(z-i)\, e^{i\,\theta}}{z+i-(z-i)\, e^{i\,\theta}}\, d\mu(\theta). \quad (22.28)$$

Using $1 + e^{i\,\theta} = 2\cos\frac{\theta}{2} e^{i\,\theta/2}$, $i(1 - e^{i\,\theta}) = 2\sin\frac{\theta}{2} e^{i\,\theta/2}$, one has

$$i\, \frac{z+i+(z-i)\, e^{i\,\theta}}{z+i-(z-i)\, e^{i\,\theta}} = \frac{z\, \cos(\theta/2) + \sin(\theta/2)}{\cos(\theta/2) - z\, \sin(\theta/2)}, \quad (22.29)$$

and one uses $t = \cot\frac{\theta}{2}$ and $d\nu(t) = d\mu(\theta)$. □

From the construction of μ in Lemma 22.8, one sees that if there is an open arc $\Gamma_0 \subset \mathbb{S}^1$ where h extends into a continuous function of real part 0, then $\mu = 0$ on Γ_0. Then, from the construction of ν in Lemma 22.9, one sees that if there is an open interval $J \subset \mathbb{R}$ where g extends into a continuous function taking real values, then $\nu = 0$ on J. For the functions $g(\cdot; z)$ arising in the particular case of Lemma 22.4 where $F(\cdot; I, z\, I) \equiv g(\cdot; z) I$, one uses $J = (0, \infty)$, and ν must have its support in $(-\infty, 0]$.

When ν has support in $(-\infty, 0]$, the condition $g(1) = 1$ gives

$$\Re\big(g(i)\big) = 1 - \gamma - \int_{(-\infty,0]} \frac{t+1}{t-1}\, d\nu(t), \quad (22.30)$$

so that, using $\frac{t\,z+1}{t-z} - \frac{t+1}{t-1} = \frac{(t^2+1)(z-1)}{(t-1)(t-z)}$,

$$g(z) = 1 + \gamma\,(z-1) + (z-1) \int_{(-\infty,0]} \frac{t^2+1}{(t-1)(t-z)}\, d\nu(t) \text{ for } \Im(z) > 0, \quad (22.31)$$

and the condition $g'(1) = 1 - \theta$ becomes

$$\gamma + \int_{(-\infty,0]} \frac{t^2+1}{(t-1)^2}\, d\nu(t) = 1 - \theta, \quad (22.32)$$

and since $\gamma \geq 0$, $\nu \geq 0$ and $\frac{1}{2} \leq \frac{t^2+1}{(t-1)^2} \leq 1$ for $t \in (-\infty, 0]$, it gives bounds for γ and ν.[13]

Since $\frac{t\,z+1}{t-z} = -t + \frac{1+t^2}{t-z}$, one deduces from (22.25) that if $\rho = (1+t^2)\,\nu \in \mathcal{M}_b(\mathbb{R})$, then $\delta = \Re\big(g(i)\big) - \int_{\mathbb{R}} t\,d\nu(t) \in \mathbb{R}$, and one has

$$g(z) = \gamma\,z + \delta + \int_{\mathbb{R}} \frac{d\rho(t)}{t-z} \text{ for } \Im(z) > 0, \gamma \geq 0, \delta \in \mathbb{R}, \rho \geq 0, \tag{22.33}$$

but not all Pick functions can be written in the form (22.33), even if $support(\nu) \subset (-\infty, 0]$, since it implies that at ∞ (in a direction in the upper half plane) one has $g(z) = \gamma\,z + \delta + O\big(\frac{1}{|z|}\big)$, which is not true for the Pick functions z^λ for $0 < \lambda < 1$.[14]

In the first article where he introduced his Pick function g in $\mathbb{C} \setminus (-\infty, 0]$, David BERGMAN (wrongly?) concluded that g is a rational fraction, i.e., that (22.33) holds with ν a finite combination of Dirac masses. In order to show that it is false, Graeme MILTON used an analogy with electrical circuits resembling Wheatstone bridges,[15,16] but I could not see how the argument of Graeme MILTON would be anything else than a conjecture that David BERGMAN made a mistake, since he used another type of limit, and inverting two limits is not always valid. It is often difficult for a mathematician to interpret some rules used by physicists: in 1981, when David BERGMAN asked me the question that I mentioned at the beginning of this chapter, I thought that he believed that his "definition" uses no sequence or limit!

In our homogenization framework of Lemma 22.4, one may have $g(x; z) = \sqrt{z}$ for x in a plane open set $\omega \subset \Omega \subset \mathbb{R}^2$ and $z \in \mathbb{C} \setminus (-\infty, 0]$, by using a checkerboard pattern in ω, since (20.4) holds in the complex case, as a consequence of Lemma 22.10, which extends Lemma 20.1.

Lemma 22.10. *If $N = 2$, and under the hypotheses of Lemma 22.5, one has*

$$R_{\pi/2}(A^n)^{-1}R_{\pi/2} = \frac{(A^n)^T}{\det(A^n)} \text{ H-converges to } R_{\pi/2}(A^{\mathit{eff}})^{-1}R_{\pi/2} = \frac{(A^{\mathit{eff}})^T}{\det(A^{\mathit{eff}})}. \tag{22.34}$$

Proof. Identical to that of Lemma 20.1, by considering $\widetilde{E}^n = R_{\pi/2}D^n$, and $\widetilde{D}^n = R_{\pi/2}E^n$, so that $(\widetilde{D}^n, \widetilde{E}^n) = \overline{(D^n, E^n)}$. □

---

[13] It is natural to compactify $\mathbb{R}$ as $\overline{\mathbb{R}} = \mathbb{R} \cup \{\infty\}$, and extend ν to $\overline{\mathbb{R}}$ into ν^* such that $\nu^*(\{\infty\}) = \gamma$, i.e., come back using the Radon measure μ on $\mathbb{S}^1$.

[14] Of course, for $z = r\,e^{i\,\theta}$ and $-\pi < \theta < \pi$, $z^\lambda = r^\lambda e^{i\,\lambda\,\theta}$.

[15] Sir Charles WHEATSTONE, English physicist, 1802–1875. He worked at King's College, London, England.

[16] As I was taught in my physics courses in 1963–1965, a Wheatstone bridge is used for making precise measurements of unknown resistances. However, the device was proposed 10 years before WHEATSTONE, by S.H. CHRISTIE.

Corollary 22.11. *If $N = 2$, and if (22.12) holds, one has*

$$\frac{F(\cdot; M^1, M^2)^T}{\det\big(F(\cdot; M^1, M^2)\big)} = F\Big(\cdot; \frac{(M^1)^T}{\det(M^1)}, \frac{(M^1)^T}{\det(M^1)}\Big), \tag{22.35}$$

and in particular if $F(\cdot; I, z\,I) = g(\cdot, z)\,I$ for $z \in \mathbb{C} \setminus (-\infty, 0]$, then

$$g(x; z) g\Big(x; \frac{1}{z}\Big) = 1, \text{ a.e. } x \in \Omega \subset \mathbb{R}^2, \text{ for all } z \in \mathbb{C} \setminus (-\infty, 0]. \tag{22.36}$$

In his framework without x variable,[17] Graeme MILTON showed that for $N = 2$ one can obtain all Pick functions g with ν a finite combination of Dirac masses in (22.33), so that g is a rational fraction, and satisfying (22.36), by using the two particular functions of the Hashin–Shtrikman coated spheres, and an argument of iteration valid for all $N \geq 1$, that from three realizable functions f, g, and h, one can create k, given by the formula

$$k(z) = f(z) h\Big(\frac{g(z)}{f(z)}\Big) \text{ for all } z \in C \setminus (-\infty, 0]. \tag{22.37}$$

This iteration procedure in our homogenization framework gives Lemma 22.12.

Lemma 22.12. *If one knows two functions F^1, F^2 independent of $x \in \Omega$, then for each $F(\cdot; M^1, M^2)$ one can construct $\widetilde{F}(\cdot; M^1, M^2)$, given by*

$$\begin{aligned}&\widetilde{F}(x; M^1, M^2) = F\big(x; F^1(M^1, M^2), F^2(M^1, M^2)\big), \text{ a.e. in } \Omega,\\ &\text{for all matrices } M^1, M^2 \text{ satisfying (22.12).}\end{aligned} \tag{22.38}$$

Proof. For $j = 1, 2,$[18] F^j is defined by a sequence of characteristic functions χ_n^j of measurable sets ω_n^j, converging in $L^\infty(\mathbb{R}^N)$ weak $\star$ to a constant θ^j. F is defined by a sequence of characteristic functions χ_m of sets ω_m, and

$$\begin{aligned}A^m &= \chi_m F^1(M^1, M^2) + (1 - \chi_m) F^2(M^1, M^2) \text{ H-converges to}\\ &\quad F\big(x; F^1(M^1, M^2), F^2(M^1, M^2)\big) \text{ in } \Omega,\end{aligned} \tag{22.39}$$

for all M^1, M^2 satisfying (22.12), since $F^1(M^1, M^2)$ and $F^2(M^1, M^2)$ satisfy (22.12), by (22.13) and (22.14). One then defines

$$\begin{aligned}A^{m,n} &= \chi_m\big(\chi_n^1 M^1 + (1 - \chi_n^1) M^2\big) + (1 - \chi_m)\big(\chi_n^2 M^1 + (1 - \chi_n^2) M^2\big)\\ &= \chi_{m,n} M^1 + (1 - \chi_{m,n}) M^2, \text{ with } \chi_{m,n} = \chi_m \chi_n^1 + (1 - \chi_m)\chi_n^2,\end{aligned} \tag{22.40}$$

[17] Since it is a global question of attaching a complex number to the mixture and its container, for precise boundary conditions, so that it is not homogenization!

[18] Using the local character of H-convergence, one may restrict F^j to a cube $Q^j \subset \Omega$, repeat the same pattern periodically, and obtain a similar function on $\mathbb{R}^N$.

and one observes that for all m

$$A^{m,n} \text{ H-converges to } A^m \text{ in } \Omega, \text{ as } n \to \infty, \tag{22.41}$$

for all M^1, M^2 satisfying (22.12), by application of the local character of H-convergence to ω_m and to $\Omega \setminus \omega_m$.[19] Then, one uses a metrizability argument for H-convergence, valid when one restricts oneself to sets of the type $\mathcal{M}(\alpha, \beta; \Omega)$,[20] and since one must choose subsequences independent of M^1 and M^2, one needs to pay a little attention, but as I wrote a detailed proof in [118], I just want to point out that the intuition is to fill ω_m with the first mixture giving F^1, and $\Omega \setminus \omega_m$ with the second mixture giving F^2. □

For $N = 1$, the only realizable functions have the form $\frac{1}{f(z)} = \theta + \frac{1-\theta}{z}$ for a value of $\theta \in [0,1]$, and using (22.37) does not enlarge the class.

A characterization of realizable functions f if $N \geq 3$ is not known. One does not know the function g for the "three-dimensional checkerboard," but in the early 1990s, Joe KELLER explained to me his argument showing that $g(z)$ behaves in $\sqrt{z}$ for z near 0, and it seems to apply to all $N \geq 3$.

Not much is known about realizable functions $F(\cdot; M^1, M^2)$.

David BERGMAN used an argument of regularity for deducing that g is regular near 0. In the periodic case, Stefano MORTOLA and Sergio STEFFÈ showed such a property for materials not necessarily isotropic: one may use one non elliptic material, under the condition that the lack of ellipticity is not too important, and this depends upon the regularity of the interface [67]; because of the example of the checkerboard, the regularity hypothesis is crucial.[21]

David BERGMAN also linked the behavior of g in 0 to a question of percolation, and this is not so clear. Percolation is a physical question related to mixing a conductor with a perfect insulator, which creates partial differential

---

[19] This uses the version of Sergio SPAGNOLO valid for measurable sets, but the same result can be obtained with the simpler version for open sets which I used with François MURAT. For $\varepsilon > 0$, there is a compact K_m and an open O_m with $K_m \subset \omega_m \subset O_m \subset \Omega$ and $measure(O_m \setminus K_m) < \varepsilon$; one then covers K_m by a finite number of open cubes included in O_m, and one replaces ω_m by a finite union of open cubes $\widetilde{\omega}_m$ so that $\chi_{\omega_m} - \chi_{\widetilde{\omega}_m}$ tends to 0 in $L^1(\Omega)$ strong, and thus in $L^p(\Omega)$ strong for all $p < \infty$ by the Hölder inequality; this does not change any of the H-limits considered, and since $\widetilde{\omega}_m$ is open and its boundary has measure 0, one then applies the local character of H-convergence to $\widetilde{\omega}_m$ and to $\Omega \setminus \overline{\widetilde{\omega}_m}$.

[20] I warned against using diagonal subsequences if one wants to let α tend to 0, since one no longer works on a bounded set, and the usual weak topologies and H-convergence are not metrizable. Some former mathematicians who decided not to prove what they say anymore did not listen to my advice, increasing the number of their articles with incomplete proofs, and maybe some false statements.

[21] One could use the technique that I developed in the fall of 1975 for homogenization in domains with holes, for avoiding the hypothesis of periodicity, but the smoothness of the interfaces would play a role, of course.

equations with Neumann condition on the boundary of the insulator, which one should not confuse with a game that physicists invented, which I call "percolation" in order to distinguish it from the physical problem.[22] My 1975 approach for proving a theorem of homogenization with Neumann conditions on the boundary of the holes, which is what one obtains if the holes are filled with a perfect insulator, described in Chap. 16, assumed a Lipschitz regularity of the interface between the insulator and the other conductors used,[23] but although this regularity hypothesis was weakened later by Grégoire ALLAIRE and François MURAT, this type of idea does not apply to the physical problem of percolation, which needs irregular interfaces.

The game of "percolation" consists in working on a discrete lattice for which one cuts bonds independently with a probability p, and one asks the probability with which the current will go through.

Experimentalists report that they observe a "critical proportion" θ_c of the insulator, such that for $\theta > \theta_c$ the current does not go through, and for $\theta < \theta_c$ the current goes through and one measures a current $i(\theta)$ which tends to 0 at θ_c, which they conjecture to be like $(\theta_c - \theta)^\gamma$ for a "universal exponent" γ. Of course, a procedure for creating mixtures was chosen, the creation of a mixture was repeated many times, and the measured values of the proportion of the insulator θ and the current i going through were made. Mathematicians who studied homogenization know that effective properties do not depend only on proportions used, and it only makes things worse that one material is an insulator, but some physicists are still stubborn enough to believe that there must be a formula, and they interpret the experimental measures by pretending that there is a curve $i(\theta)$ and that the dispersion of measured points is due to experimental errors! Of course, there is twisting of the scientific method here, but what interests me is rather to ask the question to explain how a function g, linked to a particular geometry with a particular proportion θ_0, which is $1 - g'(1)$, could tell one something on percolation, which consists experimentally of making measurements on a lot of different realizations with various proportions θ! George PAPANICOLAOU once told me that physicists do not like the Hashin–Shtrikman bounds since they do not show a percolation effect! It is a curious remark, since no one ever said than an experimental process could produce mixtures exhibiting the Hashin–Shtrikman bounds as effective conductivities! They are bounds, that one cannot trespass, even with unlikely mixtures of enormous fabrication cost. Physicists should abandon their naive idea that a procedure of fabrication

[22] "Percolation" is a physicists' problem, percolation is a physics problem.

[23] I used an extension $P_n u_n$ inside the insulator, and I wanted P_n bounded in $\mathcal{L}\Big(H^1(\Omega_n); H^1(\Omega)\Big)$ in order to have $P_m u_m \rightharpoonup u_*$ in $H^1(\Omega)$ weak (and in $L^2(\Omega)$ strong), so that $\widetilde{u_m} = \chi_{\Omega_m} P_m u_m \rightharpoonup \theta\, u_*$ in $L^2(\Omega)$ weak; then, if $\widetilde{u_n} \rightharpoonup u_\infty$ in $L^2(\Omega)$ weak, one has the surprise to have $\frac{u_\infty}{\theta} \in H^1(\Omega)$, although $\theta \in L^\infty(\Omega)$ is not necessarily regular (which the fanatics of periodicity have trouble imagining, since they believe that only mixtures at θ constant exist).

which contains non-clearly defined steps, like reducing in powder, shaking, compressing, will always give the same value for an effective coefficient; it is perfectly possible that the experimental span for a given fabrication procedure be very small compared to the ratio of the Hashin–Shtrikman bounds, but in order for mathematicians to study this question, one must tell them which fabrication procedure is chosen. Of course, the dogma that results of measurements are independent of the method used for mixing has no scientific interest!

I shall say more on this subject in Chap. 25, for some explicit constructions, and for obtaining bounds on effective coefficients, following ideas used by David BERGMAN and by Graeme MILTON, in particular his use of Padé approximants, and in Chap. 29 for obtaining information about second-order derivatives of F on the diagonal, using *H-measures* for "small-amplitude" homogenization questions.

Additional footnotes: CHRISTIE S. H. & J.,[24] NEVANLINNA,[25] STIELTJES.[26]

[24] Samuel Hunter CHRISTIE, English physicist, 1784–1865. He worked at the Royal Military Academy, Woolwich, England. The famous Christie's auction house in London, England, was founded in 1766 by his father, James CHRISTIE, 1730–1803.

[25] Rolf Herman NEVANLINNA, Finnish mathematician, 1895–1980. He worked in Helsinki, Finland.

[26] Thomas Jan STIELTJES, Dutch-born mathematician, 1856–1894. He worked in Leiden, The Netherlands, and in Toulouse, France.

Chapter 23
Memory Effects

In the early 1970s, Évariste SANCHEZ-PALENCIA studied the appearance of memory effects by homogenization (in the periodic case), explaining in this way the frequency dependence of the dielectric permittivity ε of a mixture as a consequence of the Ohm law,[1] and the visco-elastic behaviour of a solid containing inclusions of liquid.

I had not much intuition about continuum mechanics or physics at the time, and I did not study his articles, but I knew about the possibility, and I was not surprised when Jacques-Louis LIONS discussed a purely academic question leading to an effective equation using some kind of "pseudo-differential operator",[2] although it is just a fancy name for discussing memory effects (or relaxation effects), and the causality principle, expressing that only the present and the past are used in deducing the future. The basic idea is to get rid of time by using the Laplace transform, before studying an elliptic homogenization problem with the parameter p of the Laplace transform; it is because the coefficients of the effective equation are not polynomial in p that one mentions pseudo-differential operators!

In some way it is a surprising result that a weak limit of inverses of second-order elliptic operators is also the inverse of a second-order elliptic operator. The theorem of Sergio SPAGNOLO that starting from problems of the form $-div\big(a\,grad(u)\big) = f$ with a scalar a satisfying $\alpha \le a(x) \le \beta$ a.e. $x \in \Omega$, one needs to introduce the class of problems of the form $-div\big(A\,grad(u)\big) = f$ with a symmetric tensor A satisfying $\alpha\,I \le A(x) \le \beta\,I$ a.e. $x \in \Omega$, may be

[1] Around 1990, I asked Joe KELLER if he knew a similar reason for explaining the frequency dependence of the magnetic susceptibility μ of a mixture. He replied that magnetism is much more complex than electricity, and difficult to explain without something like quantum mechanics.

[2] A theory of pseudo-differential operators was first developed in the 1960s by Joseph KOHN and Louis NIRENBERG, but examples like the (M.) Riesz operators or the class of Calderón–Zygmund operators were used by specialists of singular integrals before, and a new idea was to develop a calculus on symbols, which are functions of x and ξ, although no effort was made concerning the regularity of coefficients, which were taken to be C^∞ functions in (x,ξ).

L. Tartar, *The General Theory of Homogenization*, Lecture Notes of the Unione Matematica Italiana 7, DOI 10.1007/978-3-642-05195-1_23,

found natural to a physicist using an interpretation of (stationary) diffusion of heat or electricity, by saying that the scalar case a corresponds to using locally isotropic materials, while the symmetric tensor case A corresponds to a locally anisotropic material, but although physicists used the notion of effective coefficients long before there was any mathematical definition of what they were, one must not confuse a mathematical proof that anisotropic materials can be constructed as limits of fine mixtures of isotropic materials (and that nothing more general can be obtained), with the fact that anisotropic materials like crystals were observed, prompting physicists to consider more general constitutive relations corresponding to anisotropic materials!

In a book that Garrett BIRKHOFF edited [8],[3] where one finds translations into English of a few important nineteenth century articles in analysis, he focused on a mistake of POINCARÉ, who wrote that the existence of a solution was not a problem since it was a physical problem, and he wondered how POINCARÉ could make such a mistake,[4] which he actually corrected in his next article, which included a proof of existence of that solution.

Physical intuition can be misleading, and physicists are not always aware of the defects of some of the equations that they use (like diffusion equations!),[5] so that mathematicians must be careful in analysing what they are told.

Although I find it questionable to mix homogenization and linearized elasticity, there is an interesting observation of Graeme MILTON concerning (linearized) elastic materials with negative Poisson ratio: pulling on a bar of such a material makes its cross-section expand. This may seem counter-intuitive, but Graeme MILTON first observed how to create such an effect by using almost rigid inclusions, and then he checked that such a material can be created by successive laminations, if one starts from two isotropic materials with properties sufficiently far apart. Nature does not seem to produce such materials, so that one sees the difference between observation and proofs.

In order to explain some observed natural processes, physicists invented rules which are quite different than the partial differential equations commonly used in continuum mechanics, and they proposed (or imposed!) silly

---

[3] Garrett BIRKHOFF, American mathematician, 1911–1996. He worked at Harvard University, Cambridge, MA.

[4] The only way to reject a mathematical model of "reality" is to prove that it has a property that contradicts what is observed, in which case it may only mean that reality is more complex than one thought, and the model could be good for a part of reality! Postulating a property is acceptable if one stresses which conjecture one uses, which must be checked later: I did not take the time to read POINCARÉ's articles, to ascertain the precise meaning of his words!

[5] What kind of deluded physicists would coin an expression like "anomalous diffusion" for expressing the fact that nature does not follow the equations which they like? In the fall of 1990, I went to a conference at IMA, Minneapolis, MN, and Dan JOSEPH offered to have a group visit his lab, where among other things he used extremely viscous fluids, and someone started a question "doesn't it contradict", but Dan JOSEPH did not let him finish his sentence and said that he did not care if it contradicted anything, because it was there!

games for the only reason that their output looked like something that one observed, i.e. they put in their hypotheses what they wanted in the conclusion, hardly an acceptable behaviour to someone who follows the scientific method!

Mathematicians should help in clarifying this unfortunate situation, but not by playing the games invented by physicists, which would be to confuse physicists' problems with physics problem, but in finding mathematical questions to study which agree with basic principles, and permit one to deduce a general class of equations to consider. For example, one may start with one of these partial differential equations commonly used in continuum mechanics, and while keeping in mind that it may already possess some unphysical property, one may look for the more general class of problems, which may be *beyond partial differential equations*, which follows in a mathematical way by homogenization. Of course, it is not an easy task, and one must find simpler training grounds, and not make the mistake of confusing the training ground with the real world.

From Évariste SANCHEZ-PALENCIA's explanations on visco-elastic effects, I understood that the movement of the solid puts the fluid in motion, and it is a way to store energy at a mesoscopic level, which may be given back later to the solid, but only in part since the equation for the fluid is dissipative.

In the spring of 1977, when I was spending some time at EPFL in Lausanne,[6] Switzerland, I was thinking that, because of weak convergence, the conservation of energy would have a part of the energy hidden at an intermediate level, which some Young measures may describe in a static way without the possibility to describe how it moves around, so that the first principle of thermodynamics was clear, but not the second principle. For describing the movement of energy hidden at a mesoscopic level, I thought of constructing a better mathematical tool, by splitting the Young measures in directions ξ, and in the summer of 1978 (while at ICM78 in Helsinki, Finland), I was trying functions of x, u_n and $\frac{grad(u_n)}{|grad(u_n)|}$ for that, without success, and it would take a few years before I looked in a quite different direction, for describing *H-measures*.

In the summer of 1979, I went to a conference mixing mathematicians and physicists in Cargèse, France, and having heard so curious things in the talks of "physicists", I looked at the first volume of FEYNMAN's course during the rest of the summer; I do not think that it had any effect on the ripening of this idea which I was looking for, and it probably was in the spring of 1980 that I finally understood something crucial, although obvious!

An experiment in spectroscopy consists in sending a wave in a gas, but the properties of the gas vary on small scales, because of things that one calls molecules, atoms, and electrons, so that there are resonance effects, and part of the energy of the wave is absorbed and given back later, i.e. the effective

[6] At the time, EPFL (École Polytechnique Fédérale de Lausanne) was still located in Lausanne, and it moved later a few kilometres away to Ecublens, Switzerland.

equation must be like an equation with added integral term used for memory effects: this is what the strange laws of absorption and emission invented by physicists must be about, to describe an effective equation with a nonlocal term! Again I found that this homogenization problem has no probabilities involved, and without probabilities one could at last start thinking about getting rid of the silly rules of quantum mechanics![7] My conjecture was that homogenization of hyperbolic equations or systems generate effective equations with nonlocal effects, but hyperbolic situations being notoriously more difficult than others, one probably needs to create better mathematical tools for settling this conjecture. Actually, the problem of homogenization should be studied for a semi-linear hyperbolic system.[8]

My analysis in 1980 was not as precise, and I thought that a good problem would be to identify the class of effective equations for

$$\sum_{i=1}^{N} a_i^n \frac{\partial u_n}{\partial x_i} + b_n u_n = f, \tag{23.1}$$

using a reasonable assumption for $div(a^n)$ so that (23.1) can be written in conservative form. This was too ambitious, since one would almost understand what turbulence is on the way.[9] I started with the simpler equation

$$\begin{aligned} &\tfrac{\partial u_n(x,t)}{\partial t} + a_n(x) u_n(x,t) = f(x,t), \text{ a.e. in } \Omega \times (0,\infty), \\ &u_n(x,0) = v(x), \text{ a.e. in } \Omega, \end{aligned} \tag{23.2}$$

assuming that $\alpha \le a_n \le \beta$ a.e. $x \in \Omega \subset \mathbb{R}^N$, and one may assume $\alpha > 0$,[10] that a_n corresponds to a Young measure $\nu_x, x \in \Omega$, and that v and f are bounded. I conjectured an effective equation of the form

$$\begin{aligned} &\tfrac{\partial u_\infty(\cdot,t)}{\partial t} + a_\infty u_\infty(\cdot,t) = f(\cdot,t) + \textstyle\int_0^t K(\cdot,t-s) u_\infty(\cdot,s)\,ds, \\ &u_\infty(\cdot,0) = v, \end{aligned} \tag{23.3}$$

[7] The rules of quantum mechanics were invented for giving lists of numbers, but these numbers do not exist: with improved measuring devices, physicists observed a density of absorption of light at all wavelengths with some sharp peaks, which do not look like Gaussians but like Lorentzians, $\frac{1}{1+x^2}$ rescaled, as for the real part of a meromorphic function with single poles very near the real axis!

[8] One of the dogmas in quantum mechanics is that the equations are linear, as if one wanted to hide from some of the students that their teacher might do research in quantum field theory, the nonlinear aspect of quantum mechanics!

[9] A purely Lagrangian point of view seems useless (like that of geometers who rewrite the Euler equation using affine connections), since the class of first-order hyperbolic equations is not stable by homogenization, a question forced upon us by the fluctuations in the velocity field a^n; a purely Eulerian point of view might not be much better either, but all this should become clearer when the effective equation will be understood.

[10] One writes $u_n = e^{\lambda t} u_n^*$, $f = e^{\lambda t} f^*$, and a_n is replaced by $\lambda + a_n$.

where $a_n \rightharpoonup a_\infty$ in $L^\infty(\Omega)$ weak $\star$. I conjectured $K \geq 0$, a *sufficient* condition for solutions of (23.3) to satisfy $u_\infty \geq 0$ whenever $v \geq 0$ and $f \geq 0$. As Évariste SANCHEZ-PALENCIA and Jacques-Louis LIONS did, I used the Laplace transform, defined if $support(h) \subset [0,\infty)$ by

$$\mathcal{L}h(p) = \int_0^\infty h(t)e^{-pt}\,dt, \text{ for } \Re p > \gamma \text{ for some } \gamma \in \mathbb{R}, \tag{23.4}$$

if h does not grow too much at infinity; assuming that $\mathcal{L}K$ exists, I identified it. Of course, x is a parameter but no differential structure of Ω is used (to use Young measures, one only needs a Radon measure without atoms).

Lemma 23.1. *Weak limits of solutions u_n of (23.2) satisfy (23.3) (for every f, v) if and only if the Laplace transform of K is given by*

$$\mathcal{L}K(x,p) = p + \int_{[\alpha,\beta]} a\,d\nu_x(a) - \Big(\int_{[\alpha,\beta]} \frac{d\nu_x(a)}{p+a}\Big)^{-1} \textit{ for } \Re p > -\alpha, \textit{ a.e. } x \in \Omega. \tag{23.5}$$

Proof. Taking the Laplace transform of (23.2) and (23.3), one has

$$\begin{aligned}&\big(p + a_n(x)\big)\mathcal{L}u_n(x,p) = \mathcal{L}f(x,p) + v(x)\\ &\big(p + a_\infty(x) - \mathcal{L}K(x,p)\big)\mathcal{L}u_\infty(x,p) = \mathcal{L}f(x,p) + v(x),\end{aligned} \tag{23.6}$$

so that, by varying $\mathcal{L}f + v$, one must have

$$\frac{1}{p + a_\infty - \mathcal{L}K(\cdot,p)} = \text{weak limit of } \frac{1}{p+a_n} = \int_{[\alpha,\beta]} \frac{d\nu.(a)}{p+a}, \tag{23.7}$$

and the right side is well defined in $\Re p > -\alpha$ (and is holomorphic with positive real part), and this is the same as (23.5). □

The kernel K then depends only upon the Young measure ν_x, but one must first show that the right side of (23.5) is indeed the Laplace transform of a nice function K. Since I was conjecturing that one has $K \geq 0$, I tried to apply the Bernstein theorem,[11] characterizing the Laplace transform g of a nonnegative measure on $[0,\infty)$, which is that g is a nonnegative C^∞ function defined on $(0,\infty)$ satisfying $(-1)^m \frac{d^m g}{dp^m} \geq 0$ on $(0,\infty)$ and all $m \geq 1$.[12]

---

[11] Sergei Natanovich BERNSTEIN, Russian mathematician, 1880–1968. He worked in Kharkov, in Leningrad, and in Moscow, Russia.

[12] I wonder why Laurent SCHWARTZ, who taught the Bochner theorem for the Fourier transform, did not mention the Bernstein theorem for the Laplace transform. I noticed it from the talk of a physicist, Daniel BESSIS, at a conference in Imbours, France, in June 1977, but I actually heard about it before, in a discussion with John NOHEL in the spring of 1971.

Lemma 23.2. *One defines h by*

$$\begin{aligned}&h(x,t) = \int_{[\alpha,\beta]} a\, e^{-t\,a}\, d\nu_x(a) \text{ for } t > 0, 0 \text{ for } t < 0, \text{ a.e. } x \in \Omega,\\ &\psi = \mathcal{L}h, \text{ so that } \psi(x,p) = \int_{[\alpha,\beta]} \tfrac{a}{p+a}\, d\nu_x(a) \text{ for } \Re(p) \geq 0, \text{ a.e. } x \in \Omega,\end{aligned} \tag{23.8}$$

and, using $\star$ for convolution in t,[13] *one defines H by*

$$\begin{aligned}&H = h + (h\star h) + (h\star h\star h) + \ldots \geq 0, H(x,0+) = a_\infty(x) \text{ a.e. } x \in \Omega,\\ &\mathcal{L}H = \tfrac{\psi}{1-\psi} \text{ in } \Re(p) > \beta - \alpha, \text{ a.e. in } \Omega,\end{aligned} \tag{23.9}$$

and one defines K by

$$K = -\frac{\partial H}{\partial t} \text{ for } t > 0, 0 \text{ for } t < 0 \tag{23.10}$$

and then (23.5) holds for $\Re(p) > \beta - \alpha$.

Proof. For $t \in [0,\infty)$, one has

$$\begin{aligned}&0 \leq h(x,t) \leq a_\infty(x)\, e^{-\alpha\, t},\\ &0 \leq h\star h \leq a_\infty^2 t\, e^{-\alpha\, t},\\ &0 \leq h\star h\star h \leq a_\infty^3 \tfrac{t^2}{2!}\, e^{-\alpha\, t},\\ &\ldots\\ &0 \leq H(x,t) \leq a_\infty(x)\, e^{(a_\infty(x)-\alpha)t},\end{aligned} \tag{23.11}$$

so that the Laplace transform of H is well defined if $\Re(p + \alpha - a_\infty) > 0$, a consequence of $\Re(p) > \beta - \alpha$. The series defining H converges uniformly on any interval $[0,T]$, but not in $L^1(\mathbb{R})$, since H is not integrable.[14] A similar result holds for the series of derivatives in the sense of distributions, because

$$D_t h = a_\infty \delta_0 - k, \text{ with } k(\cdot,t) = \int_{[\alpha,\beta]} a^2 e^{-t\,a}\, d\nu_x(a) \text{ for } t > 0, 0 \text{ for } t < 0, \tag{23.12}$$

and one deduces that

$$D_t H = a_\infty \delta_0 - k + a_\infty H - k\star H, \tag{23.13}$$

[13] I first based my proof on the remark that if φ satisfies the hypothesis of the Bernstein theorem, then $\frac{1}{1-\varphi}$ also does if $\varphi(0) < 1$. I then asked Yves MEYER if this was classical, and he gave me a simple proof by convolution, and with his idea one can then avoid using the Bernstein theorem.

[14] Since $1 = \int_{\mathbb{R}} h\, dt = \int_{\mathbb{R}} h\star h\, dt = \ldots$.

and similar estimates hold for the term of the series giving $k \star H$. Then for $\Re(p) > \beta - \alpha$, one does an integration by parts

$$\begin{array}{c}\mathcal{L}K(\cdot,p) = \int_0^\infty K(\cdot,t)e^{-pt}\,dt = -\int_0^\infty \frac{\partial H}{\partial t}\,e^{-pt}\,dt = H(\cdot,0+)\\ -p\int_0^\infty H\,e^{-pt}\,dt = a_\infty - p\,\mathcal{L}H(\cdot,p) = a_\infty - \frac{p\,\psi}{1-\psi},\end{array} \tag{23.14}$$

and since

$$\psi = 1 - p\,M_1, \text{ with } M_1 = \int_{[\alpha,\beta]} \frac{d\nu_x(a)}{p+a}, -\frac{p\,\psi}{1-\psi} = p - \frac{1}{M_1}, \tag{23.15}$$

one deduces that (23.14)–(23.15) gives (23.5). □

To have $K = 0$, one needs $M_1 = \frac{1}{p+a_\infty}$; since $\frac{1}{p+a}$ is strictly convex in a, it only happens if ν_x is a Dirac mass at a_∞, i.e. if a_n converges in $L^1_{loc}(\Omega)$ strong, which implies $L^q_{loc}(\Omega)$ strong for all $q < \infty$ since a_n is bounded in $L^\infty(\Omega)$. If then a_n converges weakly but not strongly, (23.2) gives an example where a limit of semi-groups is not a semi-group, and in 1980 I did not know of such a possibility.

Since one can write an explicit formula for u_n, and deduce what u_∞ is, it is important to observe that one should not play the game of finding an equation that a given function satisfies without understanding about a natural class of equations. Here, one deals with *linear* equations *invariant by translation in t*, and Laurent SCHWARTZ showed that if they map $C_c^\infty(\mathbb{R})$ continuously into the space of distributions $\mathcal{D}'(\mathbb{R})$ they are convolution operators: one then looks for an effective equation in the class of convolution operators (in t), but because of the *causality principle*, that only the present and the past should be used for predicting the future, i.e. one then looks for the convolution with a distribution with support in $[0,\infty)$![15] The nonscientific attitude of putting in one's hypotheses what one wants in the conclusion has unfortunately become quite common, and mathematicians should warn other scientists not to play whatever game they imagine, even if it gives a result looking like something which is observed; instead, they should think about ways nature could implement their game, and for imagining how "particles" behave, I find it instructive to learn what nature does for animal behaviour.[16]

---

[15] The Laplace transform is natural since it transforms convolution into multiplication, but not all distributions with support in $[0,\infty)$ have a Laplace transform.

[16] I am thinking about what LORENZ and TINBERGEN discovered in studying an instinctive reaction of some geese: if they see an egg outside their nest, they go fetch it and roll it back into the nest. A goose does not know how many eggs it is sitting upon, and it cannot stop fetching eggs even when the nest is full; it even fetches other objects, if they are smooth and not brightly coloured! If nature uses geese with almost no idea about what an egg is, why would it use "intelligent particles" which understand when to switch rules in a game depending upon the situation, or "angels playing dice" for telling "particles" what to do? Actually, there are no "particles"!

The kernel H is C^∞ and corresponds to an integrated form of (23.3),

$$u_\infty + H \star u_\infty = F, \text{ with } F(\cdot,t) = v + \int_0^t f(\cdot,s)\,ds. \tag{23.16}$$

All the preceding results can be obtained without Laplace transform, using convolution and (23.2). From the representation formula for the solutions

$$u_n(\cdot,t) = E_n(\cdot,t)v + \int_0^t E_n(\cdot,t-s)f(\cdot,s)\,ds, \text{ with } E_n(\cdot,t) = e^{-t\,a_n}, \tag{23.17}$$

one obtains

$$\begin{aligned} u_\infty(\cdot,t) &= E_\infty(\cdot,t)v + \textstyle\int_0^t E_\infty(\cdot,t-s)f(\cdot,s)\,ds, \\ E_\infty(\cdot,t) &= \textstyle\int_{[\alpha,\beta]} e^{-t\,a}\,d\nu.(a), \end{aligned} \tag{23.18}$$

and (23.3) is true for every bounded data v, f, if and only if K satisfies

$$\int_0^t K(\cdot,t-s)E_\infty(\cdot,s)\,ds = \frac{\partial E_\infty}{\partial t}(\cdot,t) + a_\infty E_\infty(\cdot,t) \text{ a.e. for } t \geq 0, \tag{23.19}$$

corresponding to the previous equation $M_1 \mathcal{L} K = a_\infty M_1 - \psi$, since $M_1 = \mathcal{L} E_\infty$. One finds that $K = 0$ only happens if $E_\infty(\cdot,t) = e^{-t\,a_\infty}$, which, by the strict convexity of exponentials, means that $a_n \to a_\infty$ in $L^1_{loc}(\mathbb{R})$ strong (and therefore in $L^q_{loc}(\mathbb{R})$ strong for every $q \in [1,\infty)$). Since $E_\infty(\cdot,0) = 1$ and $\frac{\partial E_\infty}{\partial t}(\cdot,0) = -a_\infty$, the right side of (23.19) tends to 0 as t tends to 0,[17] and (23.19) is then equivalent to the equation obtained by derivation in t, i.e.

$$K + K \star \frac{\partial E_\infty}{\partial t} = \frac{\partial^2 E_\infty}{\partial t^2} + a_\infty \frac{\partial E_\infty}{\partial t}, \tag{23.20}$$

where all functions are extended by 0 for $t < 0$; (23.20) has a unique solution which is 0 for $t < 0$, given by the formula

$$\begin{aligned} K &= \left(\delta_0 - \tfrac{\partial E_\infty}{\partial t} + \tfrac{\partial E_\infty}{\partial t} \star \tfrac{\partial E_\infty}{\partial t} - \ldots\right) \star \left(\tfrac{\partial^2 E_\infty}{\partial t^2} + a_\infty \tfrac{\partial E_\infty}{\partial t}\right) \\ &= (\delta_0 + h + (h \star h) + \ldots) \star \left(-\tfrac{\partial h}{\partial t} - a_\infty h\right), \end{aligned} \tag{23.21}$$

as $h = -\frac{\partial E_\infty}{\partial t}$, and this is in essence the content of (23.9) and (23.10).[18]

---

[17] In footnote 10, I used $f = f^* e^{\lambda\,t}$ for changing a_n into $a_n + \lambda$, in order to assume $\alpha > 0$, and the new f may grow exponentially, but it is not a restriction to assume that f has compact support, since the solution on $[0,T]$ only needs f on $(0,T)$.

[18] Although the convolution equation $K - (K \star h) = k$ satisfies the maximum principle, i.e. $k \geq 0$ implies $K \geq 0$, here $k = \int_{[\alpha,\beta]} a(a - a_\infty)e^{-t\,a}\,d\nu(a)$ satisfies $k(0) > 0$ if ν is not a Dirac mass, but k may change sign.

Then, I found a different way to analyse (23.5), by using Pick functions. Actually, I did not know at the time that these functions which David BERGMAN and Graeme MILTON used were called Pick functions, and after explaining my new idea to my student Luísa MASCARENHAS, I wanted to give her a precise mathematical reference, by asking someone, since I never read much.[19] Due to my firm opposition to a method of inventing results of votes sent to the minister in charge of the universities, I had no good relations with most of my colleagues in Orsay, France, and I thought of asking Ciprian FOIAS about that,[20] because I sat in a course that he gave during the year when we were colleagues, and he mentioned interpolation results of PICK and NEVANLINNA, which I thought related.[21]

Lemma 23.3. *There is a nonnegative Radon measure ρ_x with support in $[-\beta,-\alpha]$, weakly $\star$ measurable in $x \in \Omega$, such that*

$$\left(\int_{[\alpha,\beta]} \frac{d\nu_x(a)}{p+a}\right)^{-1} = p + a_\infty(x) + \int_{[-\beta,-\alpha]} \frac{d\rho_x(\lambda)}{\lambda - p} \text{ for } \Re p > -\alpha, \text{ a.e. } x \in \Omega, \tag{23.22}$$

so that

$$\begin{aligned} \mathcal{L}K(x,p) &= \int_{[-\beta,-\alpha]} \tfrac{d\rho_x(\lambda)}{p-\lambda} \text{ for } \Re p > -\alpha, \text{ a.e. } x \in \Omega \\ K(x,t) &= \int_{[-\beta,-\alpha]} e^{\lambda t}\, d\rho_x(\lambda) \text{ for } t > 0, \text{ a.e. } x \in \Omega. \end{aligned} \tag{23.23}$$

Proof. The function M_1 given by (23.15) is holomorphic for $p \notin -I_x \subset [-\beta,-\alpha]$, where $I_x = [a_-(x), a_+(x)]$ is the smallest closed interval containing the support of ν_x, i.e. $conv\big(support(\nu_x)\big)$, and $\Im M_1 < 0$ when $\Im p > 0$, so that $\Phi = \frac{1}{M_1}$ is a Pick function. Since $M_1 > 0$ on $(-a_-(x), +\infty)$ and $M_1 < 0$ on $\big(-\infty, -a_+(x)\big)$, one deduces that Φ is real on $\mathbb{R} \setminus -I_x$, so that (22.33) holds with $support(\rho_x) \subset -I_x$ a.e. $x \in \Omega$. The values of $\gamma \geq 0$ and δ real are deduced from the Taylor expansion of Φ near infinity,[22] and since

[19] I do not recommend acting as I did, but I was a student in Paris in the late 1960s, and I acquired a reasonable mathematical culture by just thinking about the various mathematical results which I heard in seminars, often given by well-known mathematicians. Also, I had difficulties writing and reading at the time.

[20] Ciprian Ilie FOIAS, Romanian-born mathematician. He worked in Bucharest, Romania, at University of Indiana, Bloomington, IN, and at Texas A&M University, College Station, TX. He was my colleague in 1978–1979, when he visited Université de Paris Sud, Orsay, France.

[21] I do not remember when I asked him, but I expected Ciprian FOIAS to give me an early reference, and he gave a relatively recent one [46]. In the mid 1990s, my late colleague Victor MIZEL mentioned a book for another reason [24], and it has a section on Pick functions; I understood how the hypothesis is used by looking at a proof, but I follow a slightly different approach in my proof of Lemma 22.8.

[22] Brook TAYLOR, English mathematician, 1685–1731. He worked in London, England.

$$\begin{aligned}&\tfrac{1}{p+a}=\tfrac{1}{p}\big(1-\tfrac{a}{p}+\tfrac{a^2}{p^2}+O\big(\tfrac{1}{p^3}\big)\big)\\&M_1(x,p)=\tfrac{1}{p}\big(1-\tfrac{a_\infty(x)}{p}+\tfrac{1}{p^2}\textstyle\int_{[\alpha,\beta]}a^2d\nu_x(a)+O\big(\tfrac{1}{p^3}\big)\big)\\&\varPhi(x,p)=p+a_\infty(x)+\tfrac{1}{p}\big(a_\infty^2(x)-\textstyle\int_{[\alpha,\beta]}a^2d\nu_x(a)\big)+O\big(\tfrac{1}{p^2}\big),\end{aligned}\tag{23.24}$$

one deduces that $\gamma=1$, $\delta=a_\infty(x)$, proving (23.22), but also

$$\int_{[-\beta,-\alpha]}d\rho_x=\Big(\int_{[\alpha,\beta]}a^2d\nu_x(a)\Big)-a_\infty^2(x).\tag{23.25}$$

One deduces the value of $\mathcal{L}K(x,p)$ in (23.23), and then the value of K. The weak $\star$ measurability for ρ_x means that for all continuous functions φ,[23] the quantity $\langle\rho_x,\varphi\rangle$, which is written as $\int\varphi(a)\,d\rho_x(a)$ for simplification, is measurable in x. Since the supports of all ρ_x are included in $[-\beta,-\alpha]$, it is equivalent to prove this when φ is a polynomial, by the Weierstrass theorem of approximation;[24] by using the Taylor expansion of $\varPhi$ at a higher order than for (23.24)–(23.25), one then finds that the moment of order m of ρ_x is a polynomial in the moments of order $\leq m+1$ of ν_x, and the moments of ν_x are measurable in x (as the $L^\infty(\Omega)$ weak $\star$ limits of powers of a_n). □

My second method not only shows that $K\geq 0$, but (23.23) also implies that K is the Laplace transform of a nonnegative Radon measure.

If ν_x is a combination of r distinct Dirac masses at points $n_1(x)<\ldots<n_r(x)$ (with positive weights of sum 1), then $\varPhi$ is a rational fraction and ρ_x is a combination of $r-1$ distinct Dirac masses at points $m_1(x)<\ldots<m_{r-1}(x)$ with $n_j(x)<m_j(x)<n_{j+1}(x)$ for $j=1,\ldots,r-1$, which are the roots of a polynomial. Giving ρ_x and trying to find ν_x leads to Lemma 23.4.

Lemma 23.4. *Given $\alpha,\beta,a_\infty(x)$ (and accepting all compatible ν_x), then ρ_x may be any nonnegative Radon measure with support in $[-\beta,-\alpha]$ satisfying*

$$\alpha-\int_{[-\beta,-\alpha]}\frac{d\rho_x(\lambda)}{\lambda+\alpha}\leq a_\infty(x)\leq\beta-\int_{[-\beta,-\alpha]}\frac{d\rho_x(\lambda)}{\lambda+\beta}.\tag{23.26}$$

Given α,β (and accepting all compatible $a_\infty(x),\nu_x$), ρ_x may be any nonnegative Radon measure with support in $[-\beta,-\alpha]$ satisfying

$$-\int_{[-\beta,-\alpha]}\frac{d\rho_x(\lambda)}{(\lambda+\alpha)(\lambda+\beta)}\leq 1.\tag{23.27}$$

---

[23] In principle, considering Radon measures in $\mathcal{M}(\mathbb{R})$, one should take $\varphi\in C_c(\mathbb{R})$, but the support of all the ρ_x being in a fixed interval $[-\beta,-\alpha]$, one may take $\varphi\in C(\mathbb{R})$, since only the restriction to $[-\beta,-\alpha]$ counts.

[24] Karl Theodor Wilhelm WEIERSTRASS, German mathematician, 1815–1897. He worked in Berlin, Germany.

Proof. Of course, the statements must be understood as valid a.e. $x \in \Omega$. If ρ_x is a nonnegative Radon measure with support in $[-\beta,-\alpha]$, one wants to find a nonnegative Radon measure ν_x with support in $[-\beta,-\alpha]$ with integral 1 (i.e. a probability measure) and centre of mass $a_\infty(x)$ such that (23.22) holds. However, one should just check that (23.22) holds for a nonnegative ν_x with support in $[-\beta,-\alpha]$, since the Taylor expansion at infinity then implies $\int_{[-\beta,-\alpha]} d\nu_x(a) = 1$ and $\int_{[-\beta,-\alpha]} a\, d\nu_x(a) = a_\infty(x)$. One defines Ψ by

$$\Psi(x,p) = p + a_\infty(x) + \int_{[-\beta,-\alpha]} \frac{d\rho_x(\lambda)}{\lambda - p} \text{ in } \Omega \times (\mathbb{C} \setminus [-\beta,-\alpha]), \tag{23.28}$$

so that Ψ is a Pick function, and since

$$\frac{\partial \Psi(x,p)}{\partial p} = 1 + \int_{[-\beta,-\alpha]} \frac{d\rho_x(\lambda)}{(\lambda - p)^2} > 0 \text{ for } p \in \mathbb{R} \setminus [-\beta,-\alpha], \tag{23.29}$$

one has $\Psi(x,p) \neq 0$ for $p \in \mathbb{R} \setminus [-\beta,-\alpha]$ if and only if $\Psi(x,-\beta-0) \leq 0$ and $\Psi(x,-\alpha+0) \geq 0$,[25] i.e. (23.26); in this case, $-\frac{1}{\Psi}$ is a Pick function holomorphic outside $[-\beta,-\alpha]$ and $O(\frac{1}{p})$ at infinity, so that (22.33) holds with $\gamma = \delta = 0$,

$$-\frac{1}{\Psi} = \int_{[-\beta,-\alpha]} \frac{d\pi_x(\lambda)}{\lambda - p} \text{ for } p \in \mathbb{R} \setminus [-\beta,-\alpha], \tag{23.30}$$

for a nonnegative measure π_x, and comparing to (23.22), one must have

$$\int_{[\alpha,\beta]} \frac{d\nu_x(a)}{p + a} = -\int_{[-\beta,-\alpha]} \frac{d\pi_x(\lambda)}{\lambda - p} \text{ for } p \in \mathbb{R} \setminus [-\beta,-\alpha], \tag{23.31}$$

i.e. ν_x is obtained from π_x by changing p into $-p$. Then, (23.27) follows from (23.26) by writing that one can find $a_\infty(x)$ so that (23.26) holds, i.e. the left side of (23.26) should be less or equal than the right side of (23.26). $\square$

The model (23.2) could be given an interpretation of a mixture of radioactive materials with different rates of decay, if $f = 0$, but with the curious property that one starts without any oscillations in u_n; the lesson is that, unlike for the elliptic case (like the stationary diffusion equation), the effective equation for the hyperbolic case is not in general a partial differential equation, since it may contain nonlocal terms; in my example, one finds a term involving the past, but other terms will be found in Chap. 24.

This suggests that there is something wrong about the rules of thermodynamics concerning the evolution of the internal energy, since the energy hidden at a mesoscopic level is made of different modes which behave in

---

[25] $f(x-0)$ and $f(x+0)$ are the limits of f at x, from the left and from the right.

various ways, and their sum may not in general satisfy a partial differential equation.

It is a puzzling fact that numerical simulations for Hamiltonian systems with a large number of degrees of freedom show irreversibility, forward or backward in time, but it is a mistake to advocate the Boltzmann equation in this situation, since the Boltzmann equation is obtained by postulating a kind of irreversibility by introducing probabilities, which always destroy physical reality.[26] I am not sure who pointed out this problem first, but I read about it in a book by Clifford TRUESDELL and Robert MUNCASTER [120]: starting from a reversible Hamiltonian system of forces acting at a distance,[27] it is impossible to deduce the Boltzmann equation, since an observer looking at time flowing backward would also deduce the Boltzmann equation, which is not reversible because of the Boltzmann H-theorem.[28]

My observation is that if an effective equation has a memory term using the past, it means that the forward observer adds a term $\int_{-\infty}^{t}$, but the backward observer adds a term $\int_{t}^{+\infty}$, so that it could be possible that they capture the same solutions. In other words, changing t into $-t$ is not an obvious matter for equations which contain integral terms![29]

Equation (23.2) generates a group of operators, but it is not a good model for describing a physical phenomenon, since changing t into $-t$ creates a problem with growth conditions;[30] however, one can still learn from it.

---

[26] The reason is that the rules imposed come from the short list of processes which probabilists are used to, and none of these seems adapted to the partial differential equations of continuum mechanics or physics. Understanding waves at the surface of the sea is certainly quite challenging if the wind is strong enough to make waves break, but it is hopeless to describe this in a classical way, even by using equations from kinetic theory, if one does not know of all the defects of the Boltzmann equation for example.

[27] Already not physical by POINCARÉ's principle of relativity.

[28] It is that $\int_{\mathbb{R}^3\times\mathbb{R}^3} f \log f \,dx\,dv$ is nonincreasing in time, and it is constant only for a finite list of solutions (Gaussians in v solving a free transport equation).

[29] I think that FEYNMAN used the future as well as the past for equations like the wave equation. The rules of his game were probably different than mine, but the physical questions behind might be similar.

[30] One must avoid a time of "creation" like $t = 0$ before which everything is 0, and it would be better if one could look instead at solutions defined for all $t \in \mathbb{R}$, so that "forward" observers (like us) could interpret the situation by saying that the creation took place at $-\infty$, while "backward" observers could interpret it by saying that the creation took place at the other end of $\mathbb{R}$, which is $+\infty$ for us, but the forward observer would select the solutions with $e^{\alpha\, t}||u(\cdot,t)||_{L^2(\Omega)} \to 0$ as $t \to -\infty$, and the backward observer the solutions with $e^{\beta\, t}||u(\cdot,t)||_{L^2(\Omega)} \to 0$ as $t \to +\infty$, and these two solutions are usually different. The same discrepancy would occur for the Laplace transform, the forward observer using it for $\Re(p) > -\alpha$, and the backward observer for $\Re p < -\beta$, and one sees the advantage of my second method using Pick functions.

In the case where ν_x is a combination of $m+1$ Dirac masses, so that ρ_x is a combination of m Dirac masses, one can easily restore the semi-group property by introducing an adequate system. One assumes that

$$\nu_x = \sum_{i=0}^{m+1} \theta_i(x)\delta_{\gamma_i(x)},$$
$$\rho_x = \sum_{i=0}^{m} w_i(x)\delta_{-c_i(x)}, \tag{23.32}$$
$$\text{so that } K(x,t) = \sum_{i=0}^{m} w_i(x)e^{-c_i(x)\,t} \text{ for } t > 0. \tag{23.33}$$

One introduces the auxiliary functions Z_i, $i = 1, \ldots, m$,[31] defined in Ω by

$$Z_i(x,t) = \int_0^t e^{-c_i(x)(t-s)} u_\infty(x,s)\,ds, \tag{23.34}$$

and using (23.33) the effective (23.3) is equivalent to the system

$$\begin{aligned} &\tfrac{\partial u_\infty}{\partial t} + a_\infty u_\infty - \textstyle\sum_{i=1}^m w_i Z_i = f, u_\infty(\cdot,0) = v,\\ &\tfrac{\partial Z_i}{\partial t} + c_i Z_i - u_\infty = 0, Z_i(\cdot,0) = 0, i = 1, \ldots, m. \end{aligned} \tag{23.35}$$

One could eliminate the $Z_i, i = 1, \ldots, m$, by applying the operator $\prod_{i=1}^m \big(\frac{\partial}{\partial t} + c_i\big)$ to the first equation in (23.35), and obtain an equation of order $m+1$ for u_∞, with Cauchy data at $t = 0$, but it would not be useful for the general case where ν_x is non-atomic; however, it helps understanding that if one observes u_∞ and its time derivatives up to order m at time $t_0 > 0$, then one can deduce what u_∞ is for $t > t_0$,[32] and the "irreversibility" of (23.3) in that case is not too different from the "fake irreversibility" that one would deduce for the usual wave equation $u_{tt} - c^2\Delta u = 0$, if one was giving the value of u at time 0, without giving the value of u_t at time 0.

Since $w_i \geq 0, i = 1, \ldots, m$, (23.35) satisfies the maximum principle, i.e. if $f \geq 0, v \geq 0$, then for $t \geq 0$ one has $u_\infty \geq 0$ and $Z_i \geq 0$, $i = 1, \ldots, m$. One could invent a probabilistic game generating (23.35) for expectations of various densities of particles,[33] but it is worth emphasizing once more that (23.35)

---

[31] I follow here an observation of Youcef AMIRAT, Kamel HAMDACHE and Hamid ZIANI for another problem, which I describe in Chap. 24.

[32] It is not realistic: for non-atomic ν_x one needs to assume that u_∞ is C^∞ in t, limiting f to be C^∞ in t. It is better to use the system (23.35), or to keep the integral term in (23.3) and observe that one needs the past information on u_∞.

[33] Defining u_∞ as the density of particles of type 0, Z_i as the density of particles of type i, $i = 1, \ldots, m$: particles of type 0 disappear at rate a_∞ and create spontaneously

is obtained in a purely deterministic fashion, and that the sequences and the weak $\star$ convergences serve in the mathematical purpose of identifying what kind of distance to use in asserting that a real solution presenting short-scale variations (in x) can be considered near an effective solution presenting no such variations; identifying what effective equation (or systems) the limiting solution must satisfy is then of great importance.

If one lets M be the matrix appearing in (23.35), and one looks for its eigenvalues, $M\,\xi = \lambda\,\xi$, with $\xi \neq 0$ a vector with $m+1$ components denoted $\xi_0, \ldots \xi_m$, one finds that λ is not one of the c_i, that $\xi_j = \frac{\xi_0}{c_j - \lambda}$ for $j = 1, \ldots, m$, and that $a_\infty - \lambda - \sum_{j=1}^m \frac{w_j}{c_j - \lambda} = 0$. Using (23.32) and (23.22) with $p = -\lambda$ shows that the right side of (23.22) is 0, so p is a pole of M_1, i.e. $\lambda = \gamma_j$ for some j, so that $\lambda \geq 0$ if $\alpha \geq 0$. If $c_i > 0$, $i = 1, \ldots, m$, $M\,\xi = \eta$ gives $\xi_j = \frac{\xi_0 + \eta_j}{c_j}$ for $j = 1, \ldots, m$, so that $\big(a_\infty - \sum_{j=1}^m \frac{w_j}{c_j}\big)\xi_0 = \eta_0 + \sum_{j=1}^m \frac{w_j \eta_j}{c_j}$, and using (23.22) with $p = 0$, $\xi_0 = \big(\sum_{j=1}^m \frac{\theta_j}{\gamma_j}\big)\big(\eta_0 + \sum_{j=1}^m \frac{w_j \eta_j}{c_j}\big)$ and M^{-1} has nonnegative entries.

For a general nonnegative ρ_x, (23.34)–(23.35) becomes (23.36)–(23.37):

$$F(x,c,t) = \int_0^t e^{-c(t-s)} u_\infty(x,s)\, ds \text{ on } \Omega \times [\alpha,\beta] \times (0,\infty), \qquad (23.36)$$

$$\begin{aligned} &\tfrac{\partial u_\infty}{\partial t} + a_\infty u_\infty - \textstyle\int_{[\alpha,\beta]} F(\cdot,c,t)\, d\rho_x(-c) = f, u_\infty(\cdot,0) = v,\\ &\tfrac{\partial F}{\partial t} + c\,F - u_\infty = 0, F(\cdot,c,0) = 0, c \in [\alpha,\beta], \end{aligned} \qquad (23.37)$$

and (23.29) resembles a transport equation used in kinetic theory: one should then wonder if "particles" used by physicists are just mathematical objects for describing in usual terms effective equations containing nonlocal terms! Additional footnotes: Albert of Prussia,[34] Daniel BESSIS,[35] BOCHNER,[36]

---

one particle of type i for each i at rate w_i, and particles of type i disappear at rate c_i and create spontaneously one particle of type 0 at rate 1.

[34] Albert (Albrecht), 1st duke of Prussia, 1490–1568. He was the 37th grand master of the Teutonic knights (1510–1525). In 1544, he founded a university in Königsberg, (in Prussia, then Germany, now Kaliningrad, Russia), which was named after him until its destruction in 1945, Albertus Universität.

[35] Daniel BESSIS, French-born physicist, born in 1933. He worked at CEA (Commissariat à l'Énergie Atomique), Saclay, France, at Clark-Atlanta University, Atlanta, GA, and at Texas Southern University, Houston, TX.

[36] Salomon BOCHNER, Polish-born mathematician, 1899–1982. He worked in München (Munich), Germany, and at Princeton University, Princeton, NJ.

BRANDEIS,[37] CLARK D.W.,[38] Joseph KOHN,[39] LORENZ,[40] TINBERGEN,[41] VON FRISCH.[42]

---

[37] Louis Dembitz BRANDEIS, American lawyer, 1856–1939. He was Justice of the United States Supreme Court, 1916–1939. Brandeis University, Waltham, MA, is named after him, as well as the Louis D. Brandeis School of Law of University of Louisville, Louisville, KY.

[38] Davis Wasgatt CLARK, American clergyman, 1812–1871. He was elected bishop (Methodist Episcopal Church) in 1864. Clark College (1869), which became Clark University (1877), and Clark Atlanta University (1988), is named after him.

[39] Joseph John KOHN, Czech-born mathematician, born in 1932. He worked at Brandeis University, Waltham, MA, and at Princeton University, Princeton, NJ.

[40] Konrad LORENZ, Austrian ethologist, 1903–1989. He received the Nobel Prize in Physiology or Medicine in 1973, jointly with Karl VON FRISCH and Nikolaas TINBERGEN, for their discoveries concerning organization and elicitation of individual and social behaviour patterns. He worked in Vienna, Austria, at the Albertus University in Königsberg (then in Germany, now Kaliningrad, Russia), in Buldern, Germany, and in Seewiesen, Austria. The Konrad Lorentz Institute for evolution and cognition research, Altenberg, Austria, is named after him.

[41] Nikolaas TINBERGEN, Dutch-born ethologist, 1907–1988. He received the Nobel Prize in Physiology or Medicine in 1973, jointly with Karl VON FRISCH and Konrad LORENZ, for their discoveries concerning organization and elicitation of individual and social behaviour patterns. He worked in Leiden, The Netherlands, and Oxford, England.

[42] Karl Ritter VON FRISCH, Austrian-born ethologist, 1886–1982. He received the Nobel Prize in Physiology or Medicine in 1973, jointly with Konrad LORENZ and Nikolaas TINBERGEN, for their discoveries concerning organization and elicitation of individual and social behaviour patterns. He worked in Rostock, in Breslau, and in München (Munich), Germany.

Chapter 24
Other Nonlocal Effects

The analysis of (23.2) relied on linearity and translation invariance in t, and there are natural questions to ask, one about linear equations with coefficients depending upon t, so that translation invariance in t is lost,

$$\frac{\partial u_n}{\partial t} + a_n(x,t)\,u_n = f \text{ in } \Omega \times (0,\infty), u_n(\cdot,0) = v \text{ in } \Omega, \tag{24.1}$$

another one about nonlinear equations which are translation invariant in t,[1]

$$\frac{\partial u_n}{\partial t} + a_n(x)\,u_n^2 = f \text{ in } \Omega \times (0,\infty), u_n(\cdot,0) = v \text{ in } \Omega. \tag{24.2}$$

I thought that studying time-dependent problems was a first step towards the nonlinear problems, and I asked my student Luísa MASCARENHAS to investigate (24.1); I conjectured an effective equation of the form

$$\frac{\partial u_\infty}{\partial t} + a_\infty(\cdot,t)\,u_\infty = f + \int_0^t K(\cdot,t,s)\,u_\infty(\cdot,s)\,ds \text{ in } \Omega \times (0,\infty), u_\infty(\cdot,0) = v, \tag{24.3}$$

and she used a time discretization [60], in order to use the linear case (which at the time was obtained by the convolution method (23.8)–(23.10)), but this approach did not help much for the nonlinear case. Even for $f = 0$, the solution u_n of (24.2) and its weak $\star$ limit u_∞ can be given explicitly by

$$\begin{aligned} u_n(x,t) &= \tfrac{v(x)}{1+t\,v(x)a_n(x)},\\ u_\infty(x,t) &= \int_{[\alpha,\beta]} \tfrac{v(x)\,d\nu_x(a)}{1+t\,v(x)\,a}, \end{aligned} \tag{24.4}$$

[1] The third one is to use nonlinear equations which are not translation invariant in t, but one should wait for a good answer to the second question before that.

L. Tartar, *The General Theory of Homogenization*, Lecture Notes of the Unione Matematica Italiana 7, DOI 10.1007/978-3-642-05195-1_24,

if $\alpha \geq 0$ and $v \geq 0$ in order to avoid blow-up, but it is not clear what a reasonable class of equations one should consider for this problem.[2]

It was at the end of the 1980s, while preparing an article for Bernard COLEMAN's 60th birthday [107],[3] that I found a way to handle both the time-dependent case and the nonlinear case, at least in principle, since I did not push the computations very far in the nonlinear case. My "new" idea was to use an old method, a perturbation expansion, and the form of the computations helped me understand why FEYNMAN probably introduced diagrams.[4]

I assumed that a_n is bounded, uniformly equicontinuous in t,[5]

$$\begin{aligned}&\alpha \leq a_n(x,t) \leq \beta \text{ a.e. in } \Omega \times (0,\infty),\\&|a_n(x,t) - a_n(x,s)| \leq \varepsilon(|t-s|) \text{ a.e. } (x,t,s) \in \Omega \times (0,\infty) \times (0,\infty),\end{aligned} \tag{24.5}$$

with $\varepsilon(\sigma) \to 0$ as $\sigma \to 0$; I defined b_n by

$$b_n(x,t) = a_n(x,t) - a_\infty(x,t) \text{ in } \Omega \times (0,\infty) \tag{24.6}$$

in order to consider an equation with a parameter γ:

$$\frac{\partial U_n(\cdot;\gamma)}{\partial t} + (a_\infty + \gamma\, b_n)U_n(\cdot;\gamma) = f \text{ in } \Omega \times (0,\infty), U_n(\cdot,0;\gamma) = v \text{ in } \Omega, \tag{24.7}$$

so that (24.1) corresponds to $\gamma = 1$, and I based my analysis on analyticity properties in γ. Using the equicontinuity property (24.5), I extracted a subsequence such that for each $k \geq 1$ and for every $s_1, \ldots, s_k \in (0,\infty)$

$$b_n(\cdot,s_1)\cdots b_n(\cdot,s_k) \rightharpoonup M_k(\cdot,s_1,\ldots,s_k) \text{ in } L^\infty(\Omega) \text{ weak } \star. \tag{24.8}$$

---

[2] Besides convolutions, the transformations $v \mapsto F(v)$ for F real continuous on $\mathbb{R}$ commute with translation in t, but these generate too large a class.

[3] Bernard COLEMAN proposed an intuitive explanation of the need for a memory effect: taking weak limits is like averaging, here the results of independent equations, and remembering all the averages from 0 to t gives some information on the (ill-posed) inverse problem of identifying the oscillating data which created the averages, and one then expects to be able to deduce the future averages.

[4] I once asked James GLIMM if there was a good reference for learning about diagrams, and his answer was negative, and he said that it is a state of mind! For a mathematician, it means that physicists are used to playing games without learning their rules, so that it would be a good service to the scientific community to clarify all that, but the first step is to distinguish between physicists' problems and physics problems. What are the physics problems behind all that? I guess that they are the formulation of effective equations showing nonlocal effects!

[5] Luísa MASCARENHAS assumed the a_n uniformly Lipschitz continuous in t.

Some integrals of M_k over particular domains appeared in the computations, and I defined new functions N_k and P_k on $\Omega\times(0,\infty)\times(0,\infty)$ by $N_1 = P_1 = 0$, $P_2 = M_2$, and for $k \geq 2$, $x \in \Omega$ and $0 \leq s \leq t$,

$$\begin{aligned} &N_k(x,s,t) = \textstyle\int_{\Delta(s,t)} M_k(x,s,s_2,\ldots,s_k)\,ds_2\ldots ds_k,\\ &P_{k+1}(x,s,t) = \textstyle\int_{\Delta(s,t)} M_{k+1}(x,s,s_2,\ldots,s_k,t)\,ds_2\ldots ds_k,\\ &\Delta(s,t) = \{s_2,\ldots,s_k \mid 0 \leq s \leq s_2 \leq \ldots \leq s_k \leq t\}, \end{aligned} \tag{24.9}$$

and I obtained Lemma 24.1:

Lemma 24.1. *The kernel K in (24.3) is obtained with $\gamma = 1$ in the following power series expansion with infinite radius of convergence*

$$\begin{aligned} &K(x,s,t;\gamma) = \textstyle\sum_{k=2}^\infty \gamma^k K_k(x,s,t),\\ &K_2(x,s,t) = M_2(x,s,t)\,exp\big(-\textstyle\int_s^t a_\infty(x,\sigma)\,d\sigma\big),\\ &for\ k \geq 3, K_k(x,s,t) = (-1)^k P_k(x,s,t)\,exp\big(-\textstyle\int_s^t a_\infty(x,\sigma)\,d\sigma\big)\\ &\quad -\textstyle\sum_{j=2}^{k-1}\int_s^t K_j(x,z,t)N_{k-j}(x,s,z)\,exp\big(-\int_s^z a_\infty(x,\tau)d\tau\big)\,dz. \end{aligned} \tag{24.10}$$

Proof. U_n admits the expansion

$$U_n(x,t;\gamma) = \sum_{k=0}^\infty \gamma^k U_{n,k}(x,t) \text{ in } \Omega\times(0,\infty), \tag{24.11}$$

in which $U_{n,0} = U_*$ solves (24.12) and $U_{n,k}$ is defined by induction

$$\frac{\partial U_*}{\partial t} + a_\infty U_* = f \text{ in } \Omega\times(0,\infty), U_*(\cdot,0) = v \text{ in } \Omega, \tag{24.12}$$

$$\frac{\partial U_{n,k}}{\partial t} + a_\infty U_{n,k} + b_n U_{n,k-1} = 0, U_{n,k}(\cdot,0) = 0, \text{ for } k \geq 1. \tag{24.13}$$

Assuming $\alpha > 0$ (which is not a restriction), $U_{n,k}$ is bounded for each k, and one extracts a subsequence with $U_{n,k} \rightharpoonup U_{\infty,k}$ in $L^\infty\big(\Omega\times(0,\infty)\big)$ weak $\star$ for all k. One has $U_{\infty,0} = U_*$ and $U_{\infty,1} = 0$, since $b_n \rightharpoonup 0$ in $L^\infty\big(\Omega\times(0,\infty)\big)$ weak $\star$, so that the first correction to (24.12) is in γ^2. From (24.13) one has

$$\text{for } k \geq 1, U_{n,k}(x,t) = -\int_0^t b_n(x,s)U_{n,k-1}(x,s)\,e^{-\int_s^t a_\infty(x,\sigma)\,d\sigma}\,ds, \tag{24.14}$$

$$U_{n,k}(x,t) = (-1)^k \int_\Delta b_n(x,s_1)\cdots b_n(x,s_k)U_*(x,s_1)\,e^{-\int_{s_1}^t a_\infty(x,\sigma)\,d\sigma}\,ds_1\ldots ds_k, \tag{24.15}$$

for $k \geq 1$, with Δ defined by $0 \leq s_1 \leq \ldots \leq s_k \leq t$. Using (24.8), one can let n tend to ∞ in (24.15) and for $k \geq 1$ one obtains

$$U_{\infty,k}(x,t) = (-1)^k \int_{\Delta} M_k(x, s_1, \ldots, s_k)\, U_*(x, s_1) e^{-\int_{s_1}^t a_\infty(x,\sigma)\, d\sigma}\, ds_1 \ldots ds_k. \tag{24.16}$$

Similarly, the weak $\star$ limit $W_{\infty,k}$ of $b_n U_{n,k}$ is given for $k \geq 0$ by

$$W_{\infty,k}(x,t) = (-1)^k \int_{\Delta} M_{k+1}(x, s_1, \ldots, s_k, t)\, U_*(x, s_1)\, e^{-\int_{s_1}^t a_\infty(x,\sigma)\, d\sigma}\, ds_1 \ldots ds_k. \tag{24.17}$$

Using notation (24.9), (24.16) for $k \geq 1$ and (24.17) for $k \geq 0$ become

$$U_{\infty,k}(x,t) = (-1)^k \int_0^t N_k(x,s,t)\, U_*(x,s)\, e^{-\int_s^t a_\infty(x,\sigma)\, d\sigma}\, ds, \tag{24.18}$$

$$W_{\infty,k}(x,t) = (-1)^k \int_0^t P_{k+1}(x,s,t)\, U_*(x,s)\, e^{-\int_s^t a_\infty(x,\sigma)\, d\sigma}\, ds, \tag{24.19}$$

For $k \geq 1$, the limit of (24.13) then gives

$$\frac{\partial U_{\infty,k}}{\partial t} + a_\infty U_{\infty,k} + W_{\infty,k-1} = 0 \text{ in } \Omega \times (0,\infty), U_{\infty,k}(\cdot,0) = 0 \text{ in } \Omega, \tag{24.20}$$

so that by using (24.12) one deduces that

$$\frac{\partial(\sum_{k=0}^\infty \gamma^k U_{\infty,k})}{\partial t} + a_\infty \Big(\sum_{k=0}^\infty \gamma^k U_{\infty,k}\Big) + \sum_{k=1}^\infty \gamma^{k+1} W_{\infty,k} = f \text{ in } \Omega \times (0,\infty), \tag{24.21}$$

and one must then show that

$$\sum_{k=1}^\infty \gamma^{k+1} W_{\infty,k}(x,t) = -\int_0^t K(x,s,t;\gamma) \Big(\sum_{k=0}^\infty \gamma^k U_{\infty,k}(x,s)\Big)\, ds \tag{24.22}$$

with K being analytic in γ and having the expansion (24.10). The identification of the first terms in the expansion gives

$$K(x,s,t;\gamma) = \gamma^2 M_2(x,s,t)\, e^{-\int_s^t a_\infty(x,\sigma)\, d\sigma} + O(\gamma^3) \tag{24.23}$$

giving K_2 as in (24.10). For the following terms, one must check that

$$\begin{aligned}(-1)^{k-1} \textstyle\int_0^t P_k(x,s,t)\, U_*(x,s) e^{-\int_s^t a_\infty(x,\sigma)\, d\sigma}\, ds = -\int_0^t K_k(x,t,s)\, U_*(x,s)\, ds \\ -\textstyle\sum_{j=2}^{k-1} \int_0^t K_j(x,s,t)\Big((-1)^{k-j} \int_0^s N_{k-j}(x,\sigma,s)\, U_*(x,\sigma)\, e^{-\int_\sigma^t a_\infty(x,\tau)\, d\tau}\, d\sigma\Big)\, ds,\end{aligned} \tag{24.24}$$

and this holds for every choice of smooth functions U_* if K_k satisfies the condition in (24.10) for $k \geq 3$. For taking $\gamma = 1$, it remains to check the radius of convergence of the power series.

From (24.5) and (24.8) one deduces that M_k satisfy the bounds

$$|M_k(x, s_1, \ldots, s_k)| \leq (\beta - \alpha)^k \text{ for } k \geq 2, \tag{24.25}$$

and from notation (24.9) one then deduces the bounds

$$|N_k(x, s, t)| \leq (\beta - \alpha)^k \frac{(t-s)^{k-1}}{(k-1)!} \text{ for } k \geq 2, \tag{24.26}$$

$$|P_k(x, s, t)| \leq (\beta - \alpha)^k \frac{(t-s)^{k-2}}{(k-2)!} \text{ for } k \geq 2. \tag{24.27}$$

Then (24.10) implies for K_2 the bound

$$|K_2(x, s, t)| \leq (\beta - \alpha)^2 e^{-\alpha(t-s)} \tag{24.28}$$

and when $k > 2$ one looks for a bound of K_k of the form

$$|K_k(x, s, t)| \leq C_k (\beta - \alpha)^k \frac{(t-s)^{k-2}}{(k-2)!} e^{-\alpha(t-s)}. \tag{24.29}$$

Putting this bound into (24.10) implies the inequalities

$$C_k \leq 1 + \sum_{j=2}^{k-1} C_j \text{ for } k > 2, \tag{24.30}$$

$$C_2 = 1 \text{ implies } C_k \leq 2^{k-2} \text{ for } k \geq 2, \tag{24.31}$$

giving the power series (24.10) an infinite radius of convergence. □

When a_n depends upon t, (24.10) does not imply that M_2 is nonnegative, and so the kernel K may change sign.[6] I am wondering if in such a situation, physicists would invent an explanation involving "anti-particles".

When a_n is independent of t, the power series algorithm for computing K differs from the one using the convolution equation, or the one using Pick functions; it might be more efficient from a practical point of view, but it does not seem to imply easily the theoretical characterization obtained before.

Formula (24.10) for the first term of the expansion of K reminds one of small-amplitude homogenization, which I describe with *H-measures* in Chap. 29.

[6] K being nonnegative is only *sufficient* for the maximum principle to hold.

I also applied my method using Pick functions in a computation done with François MURAT, on a degenerate elliptic problem.[7] The question was asked if our general theory of homogenization extends to some degenerate elliptic problems,[8] since one cannot carry out one step in our proof, relying on the compact embedding of $H^1(\Omega)$ into $L^2_{loc}(\Omega)$; we then looked for a case where the computations could be made explicit. For $\Omega = \mathbb{R} \times \omega$, one considers

$$-a_n(y)\frac{\partial^2 u_n}{\partial x^2} + b_n(y)u_n = f(\cdot, y) \text{ in } \Omega, u_n(\cdot, y) \in H^1(\mathbb{R}) \text{ a.e. } y \in \omega, \quad (24.32)$$

where the coefficients a_n, b_n satisfy

$$\begin{aligned} &0 < \alpha \le a_n \le \beta \text{ a.e. in } \omega, \\ &0 < \alpha' \le \frac{b_n}{a_n} \le \beta' \text{ a.e. in } \omega, \end{aligned} \quad (24.33)$$

so that for $f \in L^2(\Omega)$ there is a unique solution u_n of (24.32), and the sequence u_n stays bounded in $L^2\big(\omega; H^1(\mathbb{R})\big)$. One assumes that (a_n, b_n) corresponds to a Young measure ν_y, although the analysis will show that the complete knowledge of ν_y is not necessary for identifying the effective equation satisfied by u_∞: one only needs to apply ν_y to functions of the form $\frac{1}{a}\varphi\big(\frac{b}{a}\big)$, or equivalently one uses the list $A_-, B_1, B_2, \ldots$, defined by

$$\begin{aligned} &\tfrac{1}{a_n} \rightharpoonup \tfrac{1}{A_-} = \textstyle\int \frac{1}{a}\, d\nu_y(a,b) \text{ in } L^\infty(\omega) \text{ weak } \star, \\ &\tfrac{b_n^m}{a_n^{m+1}} \rightharpoonup \tfrac{B_m}{A_-^{m+1}} = \textstyle\int \frac{b^m}{a^{m+1}}\, d\nu_y(a,b) \text{ in } L^\infty(\omega) \text{ weak } \star, \text{ for } m \ge 1. \end{aligned} \quad (24.34)$$

Lemma 24.2. *$u_n \rightharpoonup u_\infty$ in $L^2(\Omega)$ weak, where u_∞ satisfies*

$$\begin{aligned} &-A_-(y)\tfrac{\partial^2 u_\infty}{\partial x^2} + B_1(y)u_\infty - \textstyle\int_{\mathbb{R}} H(\cdot - x', y)u_\infty(x', y)\, dx' = f(\cdot, y) \text{ in } \Omega, \\ &u_\infty(\cdot, y) \in H^1(\mathbb{R}) \text{ a.e. } y \in \omega, \end{aligned} \quad (24.35)$$

and H can be determined from $A_-, B_1, \ldots, B_m, \ldots$ by

$$\psi(y, z) = \int \frac{1}{z\,a + b}\, d\nu_y(a,b) \text{ for } z \in \mathbb{C} \setminus [-\beta', -\alpha'], \quad (24.36)$$

$$\frac{1}{\psi(\cdot, z)} = \gamma\, z + \delta + \int_{[-\beta', -\alpha']} \frac{d\mu.(\lambda)}{\lambda - z}, \text{ for } z \in \mathbb{C} \setminus [-\beta', -\alpha'], \quad (24.37)$$

---

[7] A scalar first-order partial differential equation is a case of elliptic degeneracy, but the example that I shall study now is partially elliptic.

[8] Jean-Claude NÉDÉLEC asked this question.

$$H(x,y) = \int_{[-\beta',-\alpha']} \frac{1}{2\sqrt{|\lambda|}}\, e^{-\sqrt{|\lambda|}\,|x|}\, d\mu_y(\lambda). \tag{24.38}$$

Proof. One defines the partial Fourier transform $\mathcal{F}_x$, with respect to x,[9] by

$$\mathcal{F}_x g(\xi,y) = \int_{\mathbb{R}} g(x,y)\, e^{-2i\,\pi\,x\,\xi}\, dx, \tag{24.39}$$

and one applies it to (24.32), which gives

$$\mathcal{F}u_n(\xi,y) = \frac{\mathcal{F}f(\xi,y)}{4\pi^2\xi^2 a_n(y) + b_n(y)} \text{ in } \Omega, \tag{24.40}$$

$$\mathcal{F}u_\infty(\xi,y) = \mathcal{F}f(\xi,y) \int \frac{1}{4\pi^2\xi^2 a + b}\, d\nu_y(a,b) \text{ in } \Omega. \tag{24.41}$$

By (24.34), ψ is holomorphic for $z \in \mathbb{C}\setminus[-\beta',-\alpha']$, and $\frac{1}{\psi}$ is a Pick function, giving (22.37) with $\gamma \geq 0$, $\delta \in \mathbb{R}$, and μ_y a nonnegative Radon measure with support in $[-\beta',-\alpha']$, weakly $\star$ measurable in y. Using (24.34), one has

$$\psi(y,z) = \frac{1}{A_-(y)z} - \frac{B_1(y)}{A_-^2(y)z^2} + \frac{B_2(y)}{A_-^3(y)z^3} - \ldots \text{ for } |z| > \beta', \tag{24.42}$$

and a Taylor expansion of ψ at ∞ gives $\gamma = A_-(y)$ and $\delta = B_1(y)$ in ω. Using (24.41), (24.36) and (24.37) with $z = 4\pi^2\xi^2$, one obtains

$$\Big(4\pi^2\xi^2 A_-(y) + B_1(y) + \int_{[-\beta',-\alpha']} \frac{d\mu_y(\lambda)}{\lambda - 4\pi^2\xi^2}\Big)\mathcal{F}u_\infty(\xi,y) = \mathcal{F}f(\xi,y) \text{ in } \Omega, \tag{24.43}$$

so that u_∞ satisfies (24.35) for every $f \in L^2(\Omega)$ if and only if H satisfies

$$\mathcal{F}H(\xi,y) = \int_{[-\beta',-\alpha']} \frac{d\mu_y(\lambda)}{4\pi^2\xi^2 - \lambda} \text{ in } \Omega, \tag{24.44}$$

which, after inverting the Fourier transform, gives (24.38). □

One has $H \geq 0$, a sufficient condition for the maximum principle to hold for (24.35). Moreover, H decays exponentially, since it is a linear combination with nonnegative weights of the elementary solutions of the operators $-\frac{d^2}{dx^2}+c$ for values of c in the interval where $\frac{b_n}{a_n}$ takes its values, and applying the same

[9] The variable y belongs to ω, which may only be endowed with a measure without atoms for describing Young measures.

observation of Youcef AMIRAT, Kamel HAMDACHE, and Hamid ZIANI, which I used for (23.36)–(23.37), one may rewrite the effective equation as

$$F(x,y;c) = \int_{\mathbb{R}} \frac{1}{2\sqrt{c}}\, e^{-\sqrt{c}\,(x-x')}\, u_\infty(x',y)\, dx' \text{ on } \mathbb{R}\times\omega, \text{ for } c>0, \quad (24.45)$$

$$\begin{aligned} &-A_-(y)\tfrac{\partial^2 u_\infty}{\partial x^2} + B_1(y)u_\infty = f(\cdot,y) + \textstyle\int_{[-\beta',-\alpha']} F(x,y;|\lambda|)\, d\mu_y(\lambda)\\ &-\tfrac{\partial^2 F(\cdot,y;|\lambda|)}{\partial x^2} + |\lambda|\, F(\cdot,y;|\lambda|) = u_\infty. \end{aligned} \quad (24.46)$$

Around 1990, I asked my PhD student Nenad ANTONIĆ to generalize the result of Lemma 24.2 to the case where there are terms of order 1 in the equation,[10] which he did; I also asked him to investigate the effect of adding boundary conditions if one works in a bounded domain, and he obtained results in a special case. I think that it is important to understand this question, both from a mathematical point of view, since the use of pseudo-differential operators is not so simple when one does not work on the whole space, and from a physical point of view, because I think that it is related to the question of size effects which physicists sometimes mention.

It was because I could not handle the general first-order scalar hyperbolic equation (23.1) that I started with (23.2), which has a constant characteristic velocity,[11] but Youcef AMIRAT, Kamel HAMDACHE, and Hamid ZIANI later studied a more interesting example [1, 2], where they applied the Laplace transform and the partial Fourier transform,[12]

$$\begin{aligned} &\tfrac{\partial u_n(x,y,t)}{\partial t} + a_n(y)\tfrac{\partial u_n(x,y,t)}{\partial x} = 0 \text{ in } \mathbb{R}\times\omega\times(0,\infty),\\ &u_n\big|_{t=0} = v \in L^\infty(\mathbb{R}\times\omega), v \text{ having compact support in } x, \end{aligned} \quad (24.47)$$

where $a_n \rightharpoonup a_\infty$ in $L^\infty(\omega)$ weak $\star$, and

$$\alpha \le a_n(y) \le \beta \text{ a.e. in } \omega, \text{ and defines a Young measure } \nu_y, y\in\omega. \quad (24.48)$$

Lemma 24.3. *The solutions of (24.47) satisfy $u_n \rightharpoonup u_\infty$ in $L^\infty\big(\mathbb{R}\times\omega\times(0,\infty)\big)$ weak $\star$, and u_∞ solves the effective equation*

$$\begin{aligned} &\tfrac{\partial u_\infty(x,y,t)}{\partial t} + a_\infty(y)\tfrac{\partial u_\infty(x,y,t)}{\partial x} - \textstyle\int_0^t\int_{[-\beta,-\alpha]} \tfrac{\partial^2 u_\infty(x+\lambda(t-s),y,s)}{\partial x^2}\, d\mu_y(\lambda)\, ds = 0\\ &\text{in } \mathbb{R}\times\omega\times(0,\infty), u_\infty(x,y,0) = v(x,y) \text{ in } \mathbb{R}\times\omega. \end{aligned} \quad (24.49)$$

---

[10] Nenad ANTONIĆ, Croatian mathematician. He works in Zagreb, Croatia. He was my PhD student (1992) at CMU (*Carnegie Mellon University*), Pittsburgh, PA.

[11] By moving at this velocity, I got rid of derivatives in x, and I studied the effect of a non-homogeneous absorption, since my plan was to explain the effects observed in spectroscopy as the appearance of nonlocal effects by homogenization.

[12] I describe the case with right-hand side $f = 0$ as a simplification, and where $\mathcal{F}_x v$ appears one should think that it means $\mathcal{F}_x\mathcal{L}f + \mathcal{F}_x v$ in the general case.

where the nonnegative Radon measures μ_y, $y \in \omega$, with support in $[-\beta,-\alpha]$, are defined by

$$\left(\int_{[\alpha,\beta]} \frac{d\nu_y(\lambda)}{z+\lambda}\right)^{-1} = z + a_\infty(y) + \int_{[-\beta,-\alpha]} \frac{d\mu_y(\lambda)}{\lambda - z} \text{ on } \omega \times (\mathbb{C} \setminus [-\beta,-\alpha]). \tag{24.50}$$

Proof. One applies the Laplace transform $\mathcal{L}$ in t, and the partial Fourier transform $\mathcal{F}_x$ in x, and one obtains

$$\mathcal{L}\mathcal{F}u_n(\xi,y,p) = \frac{\mathcal{F}v(\xi,y)}{p + 2i\,\pi\,\xi\,a_n(y)}, \tag{24.51}$$

which is valid for $\Re p > 0$, and letting $n \to \infty$ gives

$$\mathcal{L}\mathcal{F}u_\infty(\xi,y,p) = \int_{[\alpha,\beta]} \frac{\mathcal{F}v(\xi,y)}{p + 2i\,\pi\,\xi\,\lambda}\,d\nu_y(\lambda) = \frac{\mathcal{F}v(\xi,y)}{2i\,\pi\,\xi}\Psi\Big(y, \frac{p}{2i\,\pi\,\xi}\Big), \tag{24.52}$$

$$\Psi(y,z) = \int_{[\alpha,\beta]} \frac{d\nu_y(\lambda)}{z+\lambda} \text{ on} \omega \times (\mathbb{C} \setminus [-\beta,-\alpha]). \tag{24.53}$$

$\frac{1}{\Psi}$ is a Pick function, and (24.50) follows as usual, from (22.33) and a Taylor expansion at ∞. Then, for $\frac{p}{2i\,\pi\xi} \in \mathbb{C} \setminus [-\beta,-\alpha]$, using (24.50) gives

$$\begin{aligned} \frac{2i\,\pi\xi}{\Psi\big(y, \frac{p}{2i\,\pi\xi}\big)} &= 2i\,\pi\xi\big(\tfrac{p}{2i\,\pi\xi} + a_\infty(y) + \textstyle\int_{[-\beta,-\alpha]} \frac{d\mu_y(\lambda)}{\lambda - \frac{p}{2i\,\pi\xi}}\big) \\ &= p + 2i\,\pi\xi\,a_\infty(y) - 4\pi^2\xi^2 \textstyle\int_{[-\beta,-\alpha]} \frac{d\mu_y(\lambda)}{2i\,\pi\xi\lambda - p}. \end{aligned} \tag{24.54}$$

Using (24.54) in (24.52), which are valid for $\Re p > 0$, one inverts the Laplace transform, and one obtains a delay equation for $\mathcal{F}u_\infty$

$$\begin{aligned} &\tfrac{\partial \mathcal{F}u_\infty(\xi,y,t)}{\partial t} + 2i\,\pi\xi\,a_\infty(y)\mathcal{F}u_\infty(\xi,y,t) + \textstyle\int_0^t K(\xi,y,t-s)\mathcal{F}u_\infty(\xi,y,s)\,ds = 0 \\ &\text{in } \mathbb{R} \times \omega \times (0,\infty), \mathcal{F}u_\infty(\xi,y,0) = \mathcal{F}v(\xi,y) \text{ in } \mathbb{R} \times \omega, \end{aligned} \tag{24.55}$$

where

$$K(\xi,y,t) = 4\pi^2\xi^2 \int_{[-\beta,-\alpha]} e^{2i\,\pi\xi\lambda\,t}\,d\mu_y(\lambda) \text{ in } \mathbb{R} \times \omega \times (0,\infty); \tag{24.56}$$

then, one inverts the partial Fourier transform and one obtains (24.49). □

The preceding proof works for a right side in (24.47) satisfying $f \in L^\infty\big(\mathbb{R} \times \omega \times (0,\infty)\big)$. Due to the finite propagation speed property of (24.47), it is not a restriction to assume that v or f has compact support in x, but it helps avoid some technical difficulties with the Fourier transform.

Of course, (24.49) inherits the finite propagation speed property of (24.47), and Youcef AMIRAT, Kamel HAMDACHE, and Hamid ZIANI actually studied directly the finite propagation speed property for (24.49) for a nonnegative measure μ_y not necessarily introduced as a consequence of (24.47).

Since so many mistakenly believe that diffusion terms automatically appear in any problem for describing effects occurring at a smaller scale, it is worth emphasizing that (24.49) is not a diffusion equation.[13] I am even tempted to suggest that viscosity effects in real fluids should not be described by the usual term that NAVIER derived before STOKES, but by terms similar in nature to those occurring in (24.49);[14] not knowing at the moment about a good way to describe the nonlocal effects appearing by homogenization of the general equation (23.1),[15] I cannot yet suggest a replacement.

Youcef AMIRAT, Kamel HAMDACHE, and Hamid ZIANI also had the original idea, which I already used twice, of transforming (24.49) into a system looking like an equation from kinetic theory,

$$F(x,y,t;\lambda) = \int_0^t \frac{\partial u_\infty(x+\lambda(t-s),y,s)}{\partial x}\,ds, \tag{24.57}$$

$$\begin{aligned}
&\tfrac{\partial u_\infty(x,y,t)}{\partial t} + a_\infty(y)\tfrac{\partial u_\infty(x,y,t)}{\partial x} - \textstyle\int_{[-\beta,-\alpha]} \tfrac{\partial F(x,y,t;\lambda)}{\partial x}\,d\mu_y(\lambda)\,ds = 0,\\
&\tfrac{\partial F(x,y,t;\lambda)}{\partial t} + \lambda\tfrac{\partial F(x,y,t;\lambda)}{\partial x} = \tfrac{\partial u_\infty}{\partial x},\\
&\quad\text{in } \mathbb{R}\times\omega\times(0,\infty), u_\infty(x,y,0) = v(x,y), F(x,y,0;\lambda) = 0 \text{ in } \mathbb{R}\times\omega.
\end{aligned} \tag{24.58}$$

Youcef AMIRAT, Kamel HAMDACHE, and Hamid ZIANI studied (24.47) as a model for fluid flow in a porous medium, and I once mentioned their result to James GLIMM,[16] who said that their result is false, and that they should obtain the Burgers equation![17] I wondered by what kind of mistake

[13] Diffusion occurs from unphysical games of instantaneous jumps in position, or if one lets a characteristic speed (like the speed of light c) tend to ∞.

[14] Dan JOSEPH studied elastic effects in fluids, experimentally and theoretically. In a seminar that he gave at CMU (*Carnegie Mellon University*), Bill PRITCHARD showed a simple experiment, which he attributed to G.I. TAYLOR, where a drop of ink in a highly viscous fluid is spread by turning a crank, and then by turning the crank backward all the ink comes back to its original position, so that it seems that reversibility occurs.

[15] The scalar equation (23.1) serves as a first step for a similar question about the Euler equation, and a different approach to turbulence, where one avoids any probabilistic jargon and one concentrates on understanding properties of fluids.

[16] James G. GLIMM, American mathematician, born in 1934. He worked at MIT (Massachusetts Institute of Technology), Cambridge, MA, at NYU (New York University), New York, NY, and at SUNY (State University of New York) at Stony Brook, NY.

[17] I also mentioned to James GLIMM my conjecture about the "noise" in $1/f$, inverse of frequency, that it must correspond to the creation of triangular-shaped solutions by the Burgers equation, natural for a small nonlinear effect in conductivity.

he could imagine that starting from a linear equation the limit would be a nonlinear equation, but I then remembered that James GLIMM started as a physicist, and I interpreted what he said to mean that they worked on an incorrect model for flow in a porous medium. However, when I mentioned to him the question of anisotropic diffusion, I was surprised that he had no intuition about that question, quite natural when one knows about an instability called fingering, which he expects to occur! On another occasion, at a conference in October 1990 at MSRI in Berkeley, CA, James GLIMM said that he appreciated a talk since it used something that he likes (the game of statistical mechanics), but my feeling was that the speaker, Alexandre CHORIN,[18] said clearly that the classical ideas of statistical mechanics do not work for fluids!

One should not underestimate the pleasure felt when hearing something that one knows, and it is probably the main reason why people go to synagogues, churches, mosques, or temples of any kind, but usually only the ones corresponding to the faith of their parents, without realizing that the good part of a religion is almost the same whatever the religion is, and it is the bad part, political or nationalistic, which varies. I then find it strange that scientists would make the same mistakes that religious fanatics have done over and over, and be unable to analyse for what reason they believe something.

For studying the effective equation corresponding to a nonlinear equation

$$\frac{\partial u_n}{\partial t} + a_n u_n^2 = f \text{ in } \Omega \times (0,\infty), u_n(\cdot,0) = v \text{ in } \Omega, \tag{24.59}$$

I first chose a_n depending only upon x, in order to keep the invariance by translation in t. However, since all the nonlinear transformations $v \mapsto F(v)$ for F real continuous on $\mathbb{R}$ commute with translation in t, but do not commute with convolutions, the group generated by all these mappings is quite large. Even if one restricts attention to non-anticipative mappings, since $u_\infty(\cdot,t)$ only depends upon v and $f(\cdot,s)$ for $0 \leq s \leq t$, there is no clear simplification in the case $a_n(x)$ for defining the effective equation for (24.59), so I consider the general case $a_n(x,t)$.

In order to avoid questions of blow-up of u_n, which would force one to work on a finite interval $(0,T)$ adapted to the data, I chose

$$\alpha \geq 0, 0 \leq v \leq M_0 \text{ in } \Omega, 0 \leq f \leq F_0 \text{ a.e. in } \Omega \times (0,\infty), \tag{24.60}$$

so that the solution

$$\begin{aligned} &u_n = \Phi_n(v;f) \text{ satisfies } 0 \leq u_n \leq \max\Big\{M_0, \tfrac{\sqrt{F_0}}{\sqrt{\alpha}}\Big\} \text{ if } \alpha > 0,\\ &\quad 0 \leq u_n \leq M_0 + t\,F_0, \text{ if } \alpha = 0. \end{aligned} \tag{24.61}$$

[18] Alexandre J. CHORIN, Polish-born mathematician, born in 1938. He works at UCB (University California in Berkeley), Berkeley, CA.

I assumed that

$$\begin{aligned} &a_n \rightharpoonup a_\infty \text{ in } L^\infty\big(\Omega \times (0,\infty)\big) \text{ weak } \star,\\ &a_n \text{ defines a Young measure } \nu_{x,t}, \text{ a.e. } (x,t) \in \Omega \times (0,\infty). \end{aligned} \tag{24.62}$$

For each v, f one can extract a subsequence such that u_n converges weakly $\star$ to a function u_∞, but in order to extract a subsequence such that $\Phi_n(v; f)$ converges weakly $\star$ to $u_\infty = \Phi_\infty(v; f)$ for every v, f satisfying (24.60), one wants to show that Φ_n is uniformly continuous and that a countable dense set of v, f can be found. Since L^∞ is not separable, one uses the (strong) L^1 topology for v and f, adding the constraint $0 \le t \le T < \infty$; for all T, the restriction of Φ_n to the set of v, f satisfying (24.60) is Lipschitz continuous with values in $L^\infty\big(\Omega \times (0,T)\big)$, the Lipschitz constant depending only on $M_0, F_0, \alpha, \beta, T$. Φ_∞ is non-anticipative, since the value of each $u_n(\cdot,t)$ and thus the value of $u_\infty(\cdot,t)$ depends only upon the values of $f(\cdot,s)$ for $s \in (0,t)$.

Among the various ideas that I described for linear equations, I could only see how to use the method of power series for (24.59), using notation (24.6) and considering the following nonlinear analogue of (24.7)

$$\frac{\partial U_n(\cdot;\gamma)}{\partial t} + (a_\infty + \gamma\, b_n)U_n^2(\cdot;\gamma) = f \text{ in } \Omega \times (0,\infty), U_n(\cdot,0;\gamma) = v \text{ in } \Omega, \tag{24.63}$$

and looking for an expansion

$$U_n(\cdot;\gamma) = U_* + \sum_{k=1}^\infty \gamma^k U_{n,k} \text{ in } \Omega \times (0,\infty), \tag{24.64}$$

where U_* is independent of n and the two following terms satisfy

$$\frac{\partial U_*}{\partial t} + a_\infty U_*^2 = f \text{ in } \Omega \times (0,\infty), U_*(\cdot,0) = v \text{ in } \Omega. \tag{24.65}$$

$$\begin{aligned} &\tfrac{\partial U_{n,1}}{\partial t} + 2a_\infty U_* U_{n,1} + b_n U_*^2 = 0 \text{ in } \Omega \times (0,\infty),\\ &\tfrac{\partial U_{n,2}}{\partial t} + 2a_\infty U_* U_{n,2} + a_\infty U_{n,1}^2 + 2b_n U_* U_{n,1} = 0 \text{ in } \Omega \times (0,\infty),\\ &U_{n,1}(\cdot,0) = U_{n,2}(\cdot,0) = 0 \text{ in } \Omega, \end{aligned} \tag{24.66}$$

and more generally, by induction for $k \ge 1$,

$$\begin{aligned} &\tfrac{\partial U_{n,k}}{\partial t} + a_\infty V_{n,k} + b_n V_{n,k-1} = 0 \text{ in } \Omega \times (0,\infty),\\ &U_{n,k}(\cdot,0) = 0 \text{ in } \Omega, \end{aligned} \tag{24.67}$$

$$\begin{aligned} &V_{n,0} = U_*^2, V_{n,1} = 2U_* U_{n,1}\\ &V_{n,k} = 2U_* U_{n,k} + \textstyle\sum_{j=1}^{k-1} U_{n,j} U_{n,k-j} \text{ for } k \ge 2. \end{aligned} \tag{24.68}$$

Since b_n converges weakly $\star$ to 0, one deduces that $U_{n,1}$ converges weakly $\star$ to 0, by (24.66), and $V_{n,1}$ converges weakly $\star$ to 0, by (24.68). For computing other weak $\star$ limits, one uses more precise expressions: from (24.66), one has[19]

$$\begin{aligned} U_{n,1}(x,t) &= -\int_0^t R(x;s,t)\,b_n(x,s)\,U_*^2(x,s)\,ds \text{ in } \Omega\times(0,\infty)\\ &= -\int_0^t S(x;s,t)\,b_n(x,s)\,ds \end{aligned} \tag{24.69}$$

$$R(x;s,t) = e^{-\int_s^t 2a_\infty(x,\sigma)\,U_*(x,\sigma)\,d\sigma} \text{ in } \Omega\times(0,\infty)\times(0,\infty), \tag{24.70}$$

$$S(x;s,t) = R(x;s,t)\,U_*^2(x,s) \text{ in } \Omega\times(0,\infty). \tag{24.71}$$

Using (24.69) and the functions $M_k, k = 2,\ldots,$ defined as in (24.8), one can compute the weak $\star$ limit $U_{\infty,k}$ of $U_{n,k}$, the weak $\star$ limit $V_{\infty,k}$ of $V_{n,k}$, and the weak $\star$ limit $W_{\infty,k}$ of $b_n V_{n,k}$ for all k, related by

$$\begin{aligned} &\tfrac{\partial U_{\infty,k}}{\partial t} + a_\infty V_{\infty,k} + W_{\infty,k-1} = 0 \text{ in } \Omega\times(0,\infty),\\ &U_{\infty,k}(\cdot,0) = 0 \text{ in } \Omega, \end{aligned} \tag{24.72}$$

but one stumbles on a problem of *bookkeeping* of the computations to do. For example, dropping the variable x, one deduces easily from (24.69) that

$$\begin{aligned} &W_{\infty,1}(t) = -2U_*(t)\int_0^t S(s,t)\,M_2(s,t)\,ds,\\ &V_{\infty,2}(t) = 2U_*(t)U_{\infty,2}(t) + \int_0^t\int_0^t S(s_1,t)\,S(s_2,t)\,M_2(s_1,s_2)\,ds_1\,ds_2, \end{aligned} \tag{24.73}$$

and one can derive an explicit formula for $U_{\infty,2}$ using (24.72), but in order to go further one still needs the precise expression of $U_{n,2}$ obtained from (24.66)

$$U_{n,2}(t) = -\int_0^t R(s,t)\big(a_\infty(s)U_{n,1}^2(s) + 2b_n(s)U_*(s)U_{n,1}(s)\big)\,ds, \tag{24.74}$$

and one deduces that

$$\begin{aligned} U_{\infty,2}(t) = &-\int_{0\le s_1,s_2\le s\le t} a_\infty(s)R(s,t)S(s_1,s)S(s_2,s)M_2(s_1,s_2)\,ds_1\,ds_2\,ds\\ &-2\int_{0\le s_1\le s\le t} R(s,t)S(s_1,s)M_2(s_1,s)U_*(s)\,ds_1\,ds, \end{aligned} \tag{24.75}$$

and one may create a notation for simplifying some expressions,

$$\begin{aligned} &\int_{0\le s_1,s_2\le s\le t} h\,ds_1\,ds_2\,ds \text{ is written } \int_{D(s_1,s_2;s;t)} h\\ &\int_{0\le s_1\le s\le t} h\,ds_1\,ds \text{ is written } \int_{D(s_1;s;t)} h, \end{aligned} \tag{24.76}$$

---

[19] The function R of (24.70) appears because if $\frac{dw}{dt} + 2a_\infty(x,t)\,U_*(x,t)w = g(t)$ and $w(0) = w_0$, then $w(t) = R(x;0,t)w_0 + \int_0^t R(x;s,t)g(s)\,ds$.

and then one has

$$
\begin{aligned}
W_{\infty,2}(t) &= -2U_*(t) \int_{D(s_1,s_2;s;t)} a_\infty(s)R(s,t)S(s_1,s)S(s_2,s)M_3(s_1,s_2,t) \\
&\quad -4U_*(t) \int_{D(s_1;s;t} R(s,t)S(s_1,s)M_3(s_1,s,t)U_*(s) \\
&\quad + \int_{D(s_1,s_2;t)} S(s_1,t)S(s_2,t)M_3(s_1,s_2,t).
\end{aligned}
\tag{24.77}
$$

Without pushing further these computations, one sees easily that $U_{\infty,k}$ and $V_{\infty,k}$ are expressed in terms of multiple integrals involving M_k, and $W_{\infty,k}$ is expressed in terms of multiple integrals involving M_{k+1}, with an argument equal to t, but it would be a little futile to continue these computations without a good bookkeeping method for keeping track of all the terms that will appear; this is probably a part of what FEYNMAN did with his diagrams.

It is often not the case that the power series has a radius of convergence > 1, because if $a_\infty + \gamma\, b_n$ changes signs one may have a blow-up of the solutions in finite time. One must then identify an adapted summation procedure, and many physicists seem to like Padé approximants for this step.

One can use the preceding computations for writing a nonlinear delay equation satisfied at order 2 in γ for a truncated expansion

$$
\widetilde{U}(x,t) = U_*(x,t) + \gamma^2 U_{\infty,2}(x,t) \tag{24.78}
$$

then $\widetilde{U}$ satisfies an equation

$$
\begin{aligned}
&\tfrac{\partial \widetilde{U}}{\partial t} + a_\infty \widetilde{U}^2 + \gamma^2 a_\infty \int_0^t \int_0^t M^*(\cdot,s_1,s_2)R^*(\cdot,s_1,t)R^*(\cdot,s_2,t)\,ds_1\,ds_2 \\
&\quad -\gamma^2 \int_0^t 2M^*(\cdot,s,t)R^*(\cdot,s,t)\,ds = f + O(\gamma^3) \text{ in } \Omega \times (0,\infty) \\
&\widetilde{U}(\cdot,0) = v \text{ in } \Omega,
\end{aligned}
\tag{24.79}
$$

where M^* and R^* are functions of $\widetilde{U}$ given in $\Omega \times (0,\infty) \times (0,\infty)$ by

$$
\begin{aligned}
M^*(x,s,t) &= M_2(x,s,t)\widetilde{U}(x,s)\widetilde{U}(x,t), \\
R^*(x,s,t) &= e^{-2\int_s^t a_\infty(x,\sigma)\widetilde{U}(x,\sigma)\,d\sigma}\, \widetilde{U}(x,s).
\end{aligned}
\tag{24.80}
$$

I think that the study of homogenization of hyperbolic equations with general oscillating coefficients is the key to understanding many of the strange rules invented by physicists, and talking about spontaneous creation of "particles" and their interaction could be no more than a language for describing the nonlinear nonlocal effects created by homogenization.

Additional footnotes: Jean-Claude NÉDÉLEC,[20] Bill PRITCHARD,[21] TAYLOR G.I.[22]

[20] Jean-Claude NÉDÉLEC, French mathematician, born in 1943. He worked at Université de Rennes 1, Rennes, and at École Polytechnique, Palaiseau, France.

[21] William Gardiner PRITCHARD, Australian-born mathematician, 1942–1994. He worked at PennState (The Pennsylvania State University), State College, PA, where the William G. Pritchard Fluid Mechanics Laboratory is named after him.

[22] Sir Geoffrey Ingram TAYLOR, English mathematician, 1886–1975. He worked in Cambridge, England.

Chapter 25
The Hashin–Shtrikman Construction

I described in Chap. 21 my method for obtaining bounds on effective coefficients, which I introduced in the fall of 1977, a natural improvement of the initial method which François MURAT and myself devised in the early 1970s.

Using it, I proved bounds for effective isotropic mixtures of two isotropic conductors, while I was visiting NYU in June 1980, and George PAPANICOLAOU pointed out to me the Hashin–Shtrikman "bounds": there was a gap in the "proof" of Zvi HASHIN and SHTRIKMAN in [38],[1] so that I was the first to prove that they are valid bounds, but their proof that there exist mixtures exhibiting these extreme conductivities made sense to me. Their construction with a coated spheres geometry did not need a theory of homogenization, like the G-convergence developed by Sergio SPAGNOLO 15 years after their article, since they observed the special property of Lemma 25.2.

Definition 25.1. *For $\omega \subset \mathbb{R}^N$ bounded open with Lipschitz boundary, $A \in \mathcal{M}(\alpha, \beta; \omega)$ is said to be* equivalent *to $M \in \mathcal{L}_+(\mathbb{R}^N; \mathbb{R}^N)$ if, after extending A by $A(x) = M$ for $x \in \mathbb{R}^N \setminus \omega$, for all $\lambda \in \mathbb{R}^N$ there exists $w_\lambda \in H^1_{loc}(\mathbb{R}^N)$ with*

$$- div\big(A\, grad(w_\lambda)\big) = 0 \text{ in } \mathbb{R}^N, w_\lambda(x) = (\lambda, x) \text{ in } \mathbb{R}^N \setminus \omega. \tag{25.1}$$

Equivalently,[2]

$$\begin{aligned} &-div\big(A\, grad(w_\lambda)\big) = 0 \text{ in } \omega,\\ &w_\lambda - (\lambda, \cdot) \in H^1_0(\omega), \big(A\, grad(w_\lambda), \nu\big) = (M\,\lambda, \nu) \text{ on } \partial\omega. \end{aligned} \tag{25.2}$$

[1] I filled this gap by developing the theory of *H-measures* in the late 1980s. My proof of 1980 followed a different approach, and most of the proofs which others proposed afterward were either incomplete or not as general.

[2] I should use the normal trace on $H(div; \omega)$ as defined by Jacques-Louis LIONS, and instead of an integral on $\partial\omega$ the duality product between $H^{1/2}(\partial\omega)$ and its dual $H^{-1/2}(\partial\omega)$ should appear.

L. Tartar, *The General Theory of Homogenization*, Lecture Notes of the Unione Matematica Italiana 7, DOI 10.1007/978-3-642-05195-1_25,

Lemma 25.2. *If ω is a sphere of centre C_ω with*

$$A(x) = \begin{cases} \alpha_1 I \text{ if } |x - C_\omega| < r_1 \\ \alpha_2 I \text{ if } r_1 < |x - C_\omega| < r_2 \end{cases}, \tag{25.3}$$

then A is equivalent to γI with

$$\frac{\gamma - \alpha_2}{\gamma + (N-1)\alpha_2} = \theta \frac{\alpha_1 - \alpha_2}{\alpha_1 + (N-1)\alpha_2}, \text{ where } \theta = \frac{r_1^N}{r_2^N}. \tag{25.4}$$

Proof. Taking C_ω as origin, $r = |x|$, and adding $w_\lambda(0) = 0$, one notices that one must have $w_\lambda = (\lambda, x) f(r)$, with f of the form $b + \frac{c}{r^N}$ in $r < r_1$, as well as in $r_1 < r < r_2$, so that

$$f(r) = \begin{cases} b_1 \text{ if } r < r_1 \\ b_2 + \frac{c_2}{r^N} \text{ if } r_1 < r < r_2 \\ 1 \text{ if } r_2 < r \end{cases} \tag{25.5}$$

and at r_1 and r_2 one has continuity of w_λ and of $a(r)\big(grad(w_\lambda), \frac{x}{r}\big)$, i.e. continuity of f and, using $' = \frac{d}{dr}$, continuity of $a(r)\big(f(r) + r\,f'(r)\big)$, which gives

$$\begin{aligned} &b_1 = b_2 + \tfrac{c_2}{r_1^N}, \text{ and } \alpha_1 b_1 = \alpha_2\big(b_2 + \tfrac{(1-N)c_2}{r_1^N}\big), \\ &b_2 + \tfrac{c_2}{r_2^N} = 1, \text{ and } \alpha_2\big(b_2 + \tfrac{(1-N)c_2}{r_2^N}\big) = \gamma. \end{aligned} \tag{25.6}$$

Eliminating b_1 gives

$$(\alpha_1 - \alpha_2) b_2 + (\alpha_1 + (N-1)\alpha_2) \frac{c_2}{r_1^N} = 0, \tag{25.7}$$

and expressing $\gamma - \alpha_2$ and $\gamma + (N-1)\alpha_2$ gives

$$\gamma - \alpha_2 = -N\,\alpha_2 \frac{c_2}{r_2^N}, \text{ and } \gamma + (N-1)\alpha_2 = N\,\alpha_2 b_2, \tag{25.8}$$

and comparing $\frac{\gamma - \alpha_2}{\gamma + (N-1)\alpha_2}$ from (25.8) to $\frac{\alpha_1 - \alpha_2}{\alpha_1 + (N-1)\alpha_2}$ from (25.7) makes the ratio $\theta = \frac{r_1^N}{r_2^N}$ appear, giving (25.4). □

Lemma 25.3. *If $A \in \mathcal{M}(\alpha, \beta; \omega)$ is equivalent to $M \in \mathcal{L}_+(\mathbb{R}^N; \mathbb{R}^N)$, then there exists a sequence $A^n \in \mathcal{M}(\alpha, \beta; \Omega)$ which H-converges to M, and uses the same proportions as A in ω.*

Proof. One uses a sequence of Vitali coverings of Ω by reduced copies of ω,[3]

$$meas\big(\Omega \setminus \cup_{k\in K}(\varepsilon_{k,n}\omega + y^{k,n})\big) = 0, \text{ with } \eta_n = \sup_{k\in K} \varepsilon_{k,n} \to 0, \tag{25.9}$$

for a finite or countable K; then, a.e. $x \in \Omega$, one defines

$$A^n(x) = A\Big(\frac{x - y^{k,n}}{\varepsilon_{k,n}}\Big) \text{ in } \varepsilon_{k,n}\omega + y^{k,n}, \tag{25.10}$$

which makes sense since for each n the sets $\varepsilon_{k,n}\omega + y^{k,n}$, $k \in K$ are disjoint. For $\lambda \in \mathbb{R}^N$, and w_λ as in (25.1), one defines $u_n \in H^1(\Omega)$ by

$$u_n(x) = \varepsilon_{k,n} w_\lambda\Big(\frac{x - y^{k,n}}{\varepsilon_{k,n}}\Big) + (\lambda, y^{k,n}) \text{ in } \varepsilon_{k,n}\omega + y^{k,n}, k \in K, \tag{25.11}$$

and one wants to show that

$$grad(u_n) \rightharpoonup \lambda, A^n grad(u_n) \rightharpoonup M\,\lambda \text{ in } L^2(\Omega; \mathbb{R}^N) \text{ weak}, \tag{25.12}$$

and the first part of (25.12) follows from

$$\begin{aligned} &\textstyle\int_\Omega |grad(u_n)|^2\,dx = \sum_{k\in K} \int_{\varepsilon_{k,n}\omega+y^{k,n}} \big|grad(w_\lambda)\big(\frac{x-y^{k,n}}{\varepsilon_{k,n}}\big)\big|^2\,dx\\ &\quad= \big(\textstyle\int_\omega |grad(w_\lambda)|^2\,dx\big) \sum_{k\in K} \varepsilon_{k,n}^N, \text{ and } \sum_{k\in K} \varepsilon_{k,n}^N = \frac{meas(\Omega)}{meas(\omega)}, \end{aligned} \tag{25.13}$$

$$\begin{aligned} &\textstyle\int_\Omega |u_n - (\lambda,\cdot)|^2\,dx = \sum_{k\in K} \varepsilon_{k,n}^N \int_{\varepsilon_{k,n}\omega+y^{k,n}} \Big|w_\lambda\big(\frac{x-y^{k,n}}{\varepsilon_{k,n}}\big) - \big(\lambda, \frac{x-y^{k,n}}{\varepsilon_{k,n}}\big)\Big|^2\,dx\\ &\quad= \Big(\textstyle\int_\omega |w_\lambda - (\lambda,\cdot)|^2\,dx\Big) \sum_{k\in K} \varepsilon_{k,n}^{N+2}, \text{ and } \sum_{k\in K} \varepsilon_{k,n}^{N+2} \le \eta_n^2 \frac{meas(\Omega)}{meas(\omega)}. \end{aligned} \tag{25.14}$$

In proving the second part of (25.12), one notices that (25.2) implies[4]

$$\begin{aligned} &\textstyle\int_\omega \big(A\,grad(w_\lambda), grad(v)\big)\,dx = \int_{\partial\omega} (A\,grad(w_\lambda), \nu)\,v\,dH^{N-1}\\ &\quad= \textstyle\int_{\partial\omega} (M\,\lambda, \nu)\,v\,dH^{N-1} = \int_\omega \big(M\,\lambda, grad(v)\big)\,dx \text{ for } v \in H^1(\omega), \end{aligned} \tag{25.15}$$

and taking $grad(v)$ constant, one deduces that

$$\int_\omega A\,grad(w_\lambda)\,dx = \int_\omega M\,\lambda\,dx, \tag{25.16}$$

and for ψ Lipschitz on Ω one deduces from rescaling that

---

[3] Giuseppe VITALI, Italian mathematician, 1875–1932. He worked in Modena, in Padova (Padua), and in Bologna, Italy. The department of pure and applied mathematics of Università degli Studi di Modena e Reggio Emilia is named after him.

[4] Like for (25.2), I use integrals of the normal trace on $\partial\omega$ instead of the duality product between $H^{1/2}(\partial\omega)$ and its dual $H^{-1/2}(\partial\omega)$.

$$\Big|\int_\Omega (A^n grad(u_n)-M\,\lambda)\,\psi\,dx\Big| \leq C\,\eta_n||A^n grad(u_n)-M\,\lambda||_{L^1(\Omega;\mathbb{R}^N)}, \quad (25.17)$$

with $2C > diam(\omega)\,||\psi||_{Lip(\Omega)}$, implying $A^n grad(u_n) \rightharpoonup M\,\lambda$ in $L^2(\Omega;\mathbb{R}^N)$ weak, since $A^n grad(u_n)$ is bounded in $L^2(\Omega;\mathbb{R}^N)$. □

Corollary 25.4. *The bounds of Lemma 21.8 are optimal, attained by the Hashin–Shtrikman coated spheres geometry, the lower bound with the best conductor as core, the upper bound with the worst conductor as core.*

Proof. The lower bound of (21.41) corresponds to $\frac{N}{\gamma_- - \alpha} = \frac{(N-\theta)\alpha+\theta\,\beta}{(1-\theta)\alpha(\beta-\alpha)}$, so that

$$\begin{aligned}\gamma_- - \alpha &= N\,\alpha\,\tfrac{(1-\theta)\,(\beta-\alpha)}{(N-\theta)\,\alpha+\theta\,\beta},\\ \gamma_- + (N-1)\,\alpha &= N\,\alpha\,\tfrac{\beta+(N-1)\,\alpha}{(N-\theta)\,\alpha+\theta\,\beta},\end{aligned} \quad (25.18)$$

$$\frac{\gamma_- - \alpha}{\gamma_- + (N-1)\,\alpha} = (1-\theta)\,\frac{\beta-\alpha}{\beta+(N-1)\,\alpha}, \quad (25.19)$$

which is the analogue of (25.4) for a core $\alpha_1 = \beta$ coated by $\alpha_2 = \alpha$, the proportion of the core being $1-\theta$. The upper bound of (21.42) corresponds to $\frac{N}{\beta-\gamma_+} = \frac{(1-\theta)\alpha+(N+\theta-1)\beta}{\theta\,\beta(\beta-\alpha)}$, so that

$$\begin{aligned}\gamma_+ - \beta &= N\,\beta\,\tfrac{\theta\,(\alpha-\beta)}{(1-\theta)\alpha+(N+\theta-1)\beta},\\ \gamma_+ + (N-1)\,\beta &= N\,\beta\,\tfrac{\alpha+(N-1)\,\beta}{(1-\theta)\alpha+(N+\theta-1)\beta},\end{aligned} \quad (25.20)$$

$$\frac{\gamma_+ - \beta}{\gamma_+ + (N-1)\,\beta} = \theta\,\frac{\alpha-\beta}{\alpha+(N-1)\,\beta}, \quad (25.21)$$

which is the analogue of (25.4) for a core $\alpha_1 = \alpha$ coated by $\alpha_2 = \beta$, the proportion of the core being θ. □

Formulas like (25.4) come with various combinations of names in the literature. MOSSOTTI apparently wrote about it in 1850 for dielectric materials,[5] LORENZ,[6] LORENTZ, and CLAUSIUS apparently wrote about it for questions of refraction of light,[7] independently, in 1869, 1870 and 1879, although the formula is given the names Clausius–Mossotti or Lorenz–Lorentz,[8] but

[5] Ottaviano Fabrizio MOSSOTTI, Italian mathematician, 1791–1863. He worked in Milano, Italy, in Buenos Aires, Argentina, in Corfu, Greece, and in Pisa, Italy.

[6] Ludvig Valentin LORENZ, Danish physicist, 1829–1891. He worked in Copenhagen, Denmark.

[7] Rudolf Julius Emmanuel CLAUSIUS, German physicist, 1822–1888. He worked in Berlin, Germany, at ETH (Eidgenössische Technische Hochschule), Zürich, Switzerland, in Würzburg and in Bonn, Germany.

[8] L.V. LORENZ also introduced a gauge in electromagnetism, often wrongly attributed to H.A. LORENTZ!

MAXWELL apparently found it also, for questions of conductivity, although the relation is usually called Maxwell–Garnett,[9] either because one thought that GARNETT had something to do with it,[10] or because one erroneously attributes it to his son,[11] not so much because he wrote some articles in optics (transparent films), but because his father named him James Clerk Maxwell in memory of his great teacher.[12]

Corollary 25.5. *If a binary mixture has* $F(\cdot; I, z\,I) = g(\cdot, z)\,I$ *for* z *near 1,*

$$g'(\cdot,1) = 1-\theta \text{ with } 0<\theta<1 \text{ implies } g''(\cdot,1) = -\frac{2\theta\,(1-\theta)}{N}. \tag{25.22}$$

Proof. If $\alpha = 1$ and $\beta = 1+\varepsilon$ with $\varepsilon > 0$, then $g(1+\varepsilon) \in [\gamma_-, \gamma_+]$, with γ_- given by (25.18)–(25.19) and γ_+ given by (25.20)–(25.21), i.e.

$$\gamma_- = 1 + \frac{N\,(1-\theta)\,\varepsilon}{N+\theta\,\varepsilon} = 1 + (1-\theta)\,\varepsilon\left(1 - \frac{\theta\,\varepsilon}{N} + \ldots\right), \tag{25.23}$$

$$\gamma_+ = 1+\varepsilon - \frac{N\,(1+\varepsilon)\,\theta\,\varepsilon}{N + (N+\theta-1)\varepsilon} = 1+\varepsilon-\theta\,\varepsilon\left(1 - \frac{(N+\theta-1)\,\varepsilon}{N} + \ldots\right). \tag{25.24}$$

Since γ_- and γ_+ coincide at order ε^2, $g(1+\varepsilon) = 1 + (1-\theta)\,\varepsilon - \frac{\theta\,(1-\theta)\,\varepsilon^2}{N} + \ldots$. ∎

One should not confuse Corollary 25.5 with an argument of David BERGMAN, that $g''(1) = -\frac{2\theta\,(1-\theta)}{N}$ for microstructures with cubic symmetry. Using *H-measures* (for the sequence χ_n), I shall show in Chap. 29 the Taylor expansion at order 2 of $F(\cdot; M^1, M^2)$ on the diagonal; without any symmetry assumption, it gives $g''(\cdot,1) = -\frac{2\theta\,(1-\theta)}{N}$, using the hypothesis that $F(\cdot; I, z\,I) = g(\cdot, z)\,I$ for z real near 1 (and by analyticity in $\mathbb{C}\setminus(-\infty,0]$), of course. In my proof of Corollary 25.5, I use my proof that the Hashin–Shtrikman bounds hold.

From his hypothesis of cubic symmetry David BERGMAN derived the Hashin–Shtrikman bounds on the positive real axis (in his framework which is not homogenization), and Graeme MILTON analysed the case $z \in \mathbb{C}\setminus(-\infty,0]$; their proofs use results like Lemma 25.6, but by different methods.

Lemma 25.6. *If a Pick function in* $\mathbb{C}\setminus(-\infty,0]$*, real on* $(0,\infty)$*, satisfies* $g(1) = 1$, $g'(1) = 1-\theta$, *and* $g''(1) = -\frac{2\theta\,(1-\theta)}{d}$ *with* $d > \theta$*, then using the*

---

[9] William GARNETT, English physicist, 1850–1932. He worked in Nottingham, England.

[10] William GARNETT started his career as MAXWELL's demonstrator at the Cavendish Laboratory in Cambridge, England.

[11] James Clerk Maxwell GARNETT (son of William GARNETT), English educationalist and peace activist, 1880–1958. He was the secretary of the League of Nations Union in England at one time.

[12] So that some people thought that his last name was MAXWELL-GARNETT!

representation formula (22.31), the extreme values of $g(z)$ are obtained for a measure ν having at most one Dirac mass in $(-\infty,0)$, and either a Dirac mass at 0 and $\gamma=0$, or no Dirac mass at 0 and $\gamma\geq 0$.

Proof. Using Lemma 22.9 and $g(1)=1$ led to (22.31) for $z\in\mathbb{C}\setminus(-\infty,0]$, so that for $\gamma\geq 0$ and a nonnegative Radon measure ν,

$$\begin{aligned} g(z) &= 1+\gamma\,(z-1)+\int_{(-\infty,0]}\frac{(t^2+1)(z-1)}{(1-t)(z-t)}\,d\nu(t),\\ g'(z) &= \gamma+\int_{(-\infty,0]}\frac{t^2+1}{(z-t)^2}\,d\nu(t),\\ g''(z) &= -2\int_{(-\infty,0]}\frac{t^2+1}{(z-t)^3}\,d\nu(t), \end{aligned}\tag{25.25}$$

and the constraints on $g'(1)$ and $g''(1)$ become

$$\begin{aligned} \gamma+\int_{(-\infty,0]}\frac{t^2+1}{(1-t)^2}\,d\nu(t) &= 1-\theta,\\ \int_{(-\infty,0]}\frac{t^2+1}{(1-t)^3}\,d\nu(t) &= \frac{\theta\,(1-\theta)}{d}. \end{aligned}\tag{25.26}$$

On $\overline{\mathbb{R}}$, the Aleksandrov compactification of $\mathbb{R}$,[13] one defines $\widetilde{\nu}=\nu+\gamma\,\delta_\infty\in\mathcal{M}(\overline{\mathbb{R}})$, and the first constraint in (25.26) defines a bounded weak $\star$ closed convex set of $\mathcal{M}(\overline{\mathbb{R}})$,[14] which is weak $\star$ compact by the Banach–Alaoglu theorem,[15] and (25.26) defines a weak $\star$ compact convex set Γ of $\mathcal{M}(\overline{\mathbb{R}})$. Since the mapping G defined by $G(\widetilde{\nu})=g(z)$ is ($\mathbb{R}$-) linear continuous from $\mathcal{M}(\overline{\mathbb{R}})$ with values in $\mathbb{C}$ (considered as $\mathbb{R}^2$), the range $G(\Gamma)$ is a compact convex K of $\mathbb{C}$. For $\zeta\in K$, $\Gamma_\zeta=G^{-1}(\{\zeta\})=\{\widetilde{\nu}\mid g(z)=\zeta\}$ is a nonempty weak $\star$ compact convex subset of Γ, so that it has an extreme point by the Krein–Milman theorem.[16,17] Then, one uses an argument of Zvi ARTSTEIN [3]: since Γ_ζ is defined by four ($\mathbb{R}$-) linear constraints (three if z is real), an extreme point of Γ_ζ is on a face of the convex cone P of all nonnegative Radon measures in $\mathcal{M}(\overline{\mathbb{R}})$, of dimension at most four, so that it is an atomic measure

---

[13] Pavel Sergeevich ALEKSANDROV, Russian mathematician, 1896–1982. He worked in Smolensk, and in Moscow, Russia.

[14] In (25.26), $\widetilde{\nu}$ is applied to functions in $C(\overline{\mathbb{R}})$, since $\frac{t^2+1}{(1-t)^2}\to 1$ and $\frac{t^2+1}{(1-t)^3}\to 0$ as $t\to\infty$, and the coefficients of γ are 1 and 0.

[15] Leonidas ALAOGLU, Canadian-born mathematician, 1914–1981. He worked at Pennsylvania State College (to become in 1953 The Pennsylvania State University, known as Penn State), State College, PA, at Harvard University, Cambridge, MA, and at Purdue University, West Lafayette, IN, before working for the United States Air Force, and the Lockheed Corporation.

[16] Mark Grigorievich KREIN, Ukrainian mathematician, 1907–1989. He received the Wolf Prize in 1982, for his fundamental contributions to functional analysis and its applications, jointly with Hassler WHITNEY. He worked in Moscow, in Kazan, Russia, and in Odessa and Kiev, Ukraine.

[17] David Pinhusovich MILMAN, Ukrainian-born mathematician, 1912–1982. He worked in Tel Aviv, Israel.

with at most four Dirac masses.[18] For reducing the number of Dirac masses to use, my idea consists in moving the positions of the Dirac masses, and to select measures which give a point on the boundary of K,[19] and the implicit function theorem tells one which measures give a point in the interior of K.

For example, if $z = x + i\,y$ with $y \neq 0$, and $\widetilde{\nu}$ contains two Dirac masses at $t_1, t_2 \in (-\infty, 0)$ (with coefficients $c_1, c_2 > 0$), then $g(z)$ is not on the boundary of K; for showing this, let f_1, f_2, f_3, f_4 be defined by

$$\begin{aligned} f_1(t) &= \tfrac{t^2+1}{(1-t)}\,\Re\big(\tfrac{1}{z-t}\big), f_2(t) = \tfrac{t^2+1}{(1-t)}\,\Im\big(\tfrac{1}{z-t}\big),\\ f_3(t) &= \tfrac{t^2+1}{(1-t)^2}, f_4(t) = \tfrac{t^2+1}{(1-t)^3}, \end{aligned} \tag{25.27}$$

for $t \in (-\infty, 0]$, or $t = \infty$, with $f_1(\infty) = f_3(\infty) = 1$, $f_2(\infty) = f_4(\infty) = 0$. With $'$ denoting the derivative in t, the following matrix

$$\begin{pmatrix} f_1(t_1) & f_1'(t_1) & f_1(t_2) & f_1'(t_2)\\ f_2(t_1) & f_2'(t_1) & f_2(t_2) & f_2'(t_2)\\ f_3(t_1) & f_3'(t_1) & f_3(t_2) & f_3'(t_2)\\ f_4(t_1) & f_4'(t_1) & f_4(t_2) & f_4'(t_2) \end{pmatrix} \tag{25.28}$$

appears if one varies c_1, t_1, c_2, t_2; if it is invertible, the implicit function theorem applies, and one finds a curve in the (c_1, t_1, c_2, t_2) space with $\langle \widetilde{\nu}, f_3\rangle$ and $\langle \widetilde{\nu}, f_4\rangle$ constant, and $\langle \widetilde{\nu}, f_1\rangle$ and $\langle \widetilde{\nu}, f_2\rangle$ vary in an arbitrary direction, so that (25.26) is true, and by (25.25) $g(z)$ varies in an arbitrary direction, i.e. $g(z)$ belongs to the interior of Γ. If $g(z) \in \partial\Gamma$, the matrix in (25.28) is not invertible, so that there exist $\lambda_1, \lambda_2, \lambda_3, \lambda_4$ not all 0, such that

$$h = \lambda_1 f_1 + \lambda_2 f_2 + \lambda_3 f_3 + \lambda_4 f_4 \text{ satisfies } h(t_1) = h'(t_1) = h(t_2) = h'(t_2) = 0, \tag{25.29}$$

and one wants to show that h cannot have two double zeros. The condition does not change if one multiplies h by a C^1 function which does not vanish at t_1 and t_2, and using the factor $\frac{t^2+1}{(1-t)^3[(x-t)^2+y^2]}$, one has a combination of $(x-t)\,(1-t)^2$, $y\,(1-t)^2$, $(1-t)\,\big((x-t)^2+y^2\big)$, and $(x-t)^2+y^2$, which are polynomials of degree ≤ 3, and the combination must be identically 0; one then checks easily that the four polynomials are linearly independent. The same method shows that one cannot have three Dirac masses at ∞,

[18] If $\widetilde{\nu}$ has a diffuse part it is on a face of P of infinite dimension. If $\widetilde{\nu}$ is a combination of Dirac masses at k distinct points, it is on a face of P of dimension $\geq k$ (by changing the coefficients of the Dirac masses).

[19] I developed it in 1990 for a question which I then studied with Gilles FRANCFORT and François MURAT [32], to identify the convex set (in $\mathbb{R}^{15}$) of moments of order 4 of nonnegative measures on $\mathbb{S}^2$, and to reduce the number of Dirac masses from 15 (as given by the argument of Zvi ARTSTEIN) to 5 for measures giving moments which are boundary points (and the number is reduced to 6 in general).

$t_1 \in (-\infty, 0)$ and 0: the combination h must have a simple zero at ∞ and 0 (as one cannot change these positions), and a double zero at t_1; the condition at ∞ implies that one must have a combination of $f_1 - f_3$, f_2, and f_4, and using the same factor, one has a combination of $y\,(1-t)^2$, $(x-t)^2+y^2$, and $(x-t)\,(1-t)^2 - (1-t)\,\big((x-t)^2+y^2\big) = (1-t)\big((x-t)\,(1-x)-y^2\big)$, which are polynomials of degree ≤ 2, and the combination must be identically 0; one then checks easily that the three polynomials are linearly independent.

The boundary of K then corresponds to $\widetilde{\nu}$ having only two Dirac masses, one of them at 0 or ∞; each case corresponds to three parameters and the two constraints (25.26) leave one parameter for describing a curve. One intersection of the two curves corresponds to a single Dirac mass, at $t_1 = 1 - \frac{d}{\theta}$ (and one must have $d \geq \theta$, of course),[20] since one must have

$$c_1 \frac{t_1^2+1}{(1-t_1)^2} = 1-\theta, c_1 \frac{t_1^2+1}{(1-t_1)^3} = \frac{\theta\,(1-\theta)}{d}, \tag{25.30}$$

and the other intersection of the two curves corresponds to a Dirac mass at 0 and a Dirac mass at ∞, with weights γ and c_1 satisfying

$$\gamma + c_1 = 1-\theta, c_1 = \frac{\theta\,(1-\theta)}{d}. \tag{25.31}$$

In the case where $z \in \mathbb{R}$, one only deals with f_1, f_3, and f_4, and the matrix which appears is that of (25.26) with the second row removed; it should not have rank 3, so that one must rule out a pattern of zeros for a nonzero combination $h = \lambda_1 f_1 + \lambda_3 f_3 + \lambda_4 f_4$. Putting in the factor $\frac{t^2+1}{(1-t)^3(x-t)}$, one has a combination of $(1-t)^2$, $(1-t)(x-t)$, and $(x-t)$, which cannot have a double zero in two distinct negative values t_1, t_2, or a double zero at $t_1 < 0$ and a single zero at 0, unless the combination is identically 0, and if $x \neq 1$ the three polynomials are linearly independent. A single zero at ∞ implies a combination of $(1-t)^2 - (1-t)(x-t) = (1-t)(1-x)$, and $(x-t)$, which cannot have a double zero at $t_1 < 0$, so that the combination giving the extreme values are either a single Dirac mass at $t_1 = 1 - \frac{d}{\theta}$ with (25.30), or a Dirac mass at 0 and at ∞ with (25.31). □

In the real case with $x \neq 1$, the extreme value corresponding to $\widetilde{\nu} = c_1 \delta_{t_1}$ and $t_1 = 1 - \frac{d}{\theta}$, gives $g(z) = 1 + c_1 \frac{1+t_1^2}{1-t_1}\frac{z-1}{z-t_1}$, and by (25.30) $c_1 \frac{1+t_1^2}{1-t_1} = (1-\theta)(1-t_1) = \frac{d\,(1-\theta)}{\theta}$, and one deduces that

$$g(z) = 1 + \frac{d\,(1-\theta)\,(z-1)}{\theta\,z + d - \theta}, \tag{25.32}$$

and this corresponds to

---

[20] In practice, d is a positive integer, but Lemma 25.6 is stated in general, so that if $d < \theta$, no function g exists, while if $d = \theta$ only one function g exists.

$$\frac{g(z)-1}{g(z)+d-1} = (1-\theta)\,\frac{z-1}{z+d-1}, \tag{25.33}$$

which is one of the Hashin–Shtrikman bounds in "dimension" d. However, the other extreme value is not the other Hashin–Shtrikman bound! This lack of symmetry is related to the fact that not all Pick functions considered in Lemma 25.6 can be associated to a geometry. By exchanging the two materials, $z\,g\big(\frac{1}{z}\big)$ should also be a Pick function, so that if $d = N$ (25.33) gives a lower bound if $z > 1$ and an upper bound if $0 < z < 1$, and the similar bounds for the function $z\,g\big(\frac{1}{z}\big)$ give the missing bounds for g.

The case $\nu = c_1\delta_0$ gives $g(z) = 1+\gamma\,(z-1)+c_1\frac{z-1}{z}$, and $z\,g\big(\frac{1}{z}\big) = z+\gamma\,(1-z)+c_1 z\,(1-z)$, which is not a Pick function if $c_1 > 0$, or if $c_1 = 0$ and $\gamma > 1$. Lemma 25.7 gives a more general condition.

Lemma 25.7. *If $g(z)$ is a Pick function, with $g(1) = 1$ and formula (25.25) defining γ and ν, and if $z\,g\big(\frac{1}{z}\big)$ is also a Pick function, then*

$$\nu(\{0\}) = 0, \ \ \textit{and}\ 0 \le \int_{(-\infty,0]} \frac{t^2+1}{-t\,(1-t)}\,d\nu(t) \le 1-\gamma. \tag{25.34}$$

Proof. Formula (25.25) shows that

$$\frac{g(z)-1}{z-1} - \gamma = G(z) = \int_{(-\infty,0]} \frac{t^2+1}{(1-t)(z-t)}\,d\nu(t), \tag{25.35}$$

and when one restricts G to $\mathbb{R}_+$, G is non-increasing and one has

$$\begin{aligned} &\lim_{x\to+\infty} G(x) = 0,\\ &\lim_{x\to 0} x\,G(x) = \nu(\{0\}), \end{aligned} \tag{25.36}$$

since for $x \ge 1$ one has $\frac{t^2+1}{(1-t)(x-t)} \le \frac{t^2+1}{(1-t)^2} \in L^1(\nu)$ and $\frac{t^2+1}{(1-t)(x-t)} \to 0$ for all $t \le 0$ as $x \to \infty$, so that the Lebesgue dominated convergence theorem gives the first part of (25.36); for $0 < x \le 1$ one has $x\,\frac{t^2+1}{(1-t)(x-t)} \le \frac{t^2+1}{(1-t)^2} \in L^1(\nu)$ and $x\,\frac{t^2+1}{(1-t)(x-t)} \to 0$ for all $t < 0$ as $x \to 0$, and is equal to 1 for $t = 0$, so that the Lebesgue dominated convergence theorem gives the second part of (25.36). If $g(z)$ and $z\,g\big(\frac{1}{z}\big)$ are Pick functions, one has

$$\begin{aligned} &g(z) = 1+\gamma\,(z-1)+\textstyle\int_{(-\infty,0]} \frac{(t^2+1)(z-1)}{(1-t)(z-t)}\,d\nu(t),\\ &z\,g\big(\tfrac{1}{z}\big) = z+\gamma\,(1-z)+z\textstyle\int_{(-\infty,0]} \frac{(t^2+1)(1-z)}{(1-t)(1-t\,z)}\,d\nu(t),\\ &\tfrac{z\,g(1/z)-1}{z-1} = K(z) = 1-\gamma-z\textstyle\int_{(-\infty,0]} \frac{t^2+1}{(1-t)(1-t\,z)}\,d\nu(t). \end{aligned} \tag{25.37}$$

Since $K(x)$ is non-increasing for $x > 0$ and $\lim_{x\to+\infty} K \ge 0$, one has

$$x\int_{(-\infty,0]}\frac{t^2+1}{(1-t)(1-t\,x)}\,d\nu(t)\leq 1-\gamma \text{ for } x>0, \tag{25.38}$$

and since $0\leq x\,\frac{t^2+1}{(1-t)(1-t\,x)}$ increases from 0 to $\frac{t^2+1}{-t\,(1-t)}$ for $t\leq 0$ as x increases from 0 to $+\infty$, one applies the Fatou theorem,[21] which implies

$$0\leq\int_{(-\infty,0]}\frac{t^2+1}{-t\,(1-t)}\,d\nu(t)\leq 1-\gamma,\ \text{ so that } \nu(\{0\})=0, \tag{25.39}$$

and $\lim_{x\to 0}K(x)=0$, so that $\lim_{x\to 0}x\,K(x)=0$. □

At a conference in New York, NY, in June 1981, I described the optimal bounds for anisotropic effective mixtures of two isotropic conductors obtained with François MURAT, and I shall describe in Chap. 26 our construction using confocal ellipsoids. At the end of my talk, I conjectured that in mixing three or more isotropic materials the optimal bounds were probably similar, with a construction using spheres in the effective isotropic case, with materials of increasing conductivity from inside out or from outside in, depending upon which bound one wants, but Graeme MILTON, still a graduate student at the time, said that even for three materials it is not always so; when I visited him at NYU in the early 1990s, he gave me a physical intuition for that: if the worst conductor is almost an insulator, it is natural to put it in the core, and since the lines of currents avoid the core, there is a high density of current just outside the core, and it is there that one needs the best conductor.

In order to understand this question in a quantitative way, one studies a more general radial symmetric situation, described in Lemma 25.8.

Lemma 25.8. *If $A(x)=a(r)\,I$, the restriction of A to the ball of radius r is equivalent to $a_{eff}(r)\,I$ (according to Definition 25.1), and $a_{eff}(r)$ satisfies*

$$r\,a'_{eff}(r)+\frac{[a_{eff}(r)-a(r)]\,[a_{eff}(r)+(N-1)\,a(r)]}{a(r)}=0. \tag{25.40}$$

Proof. One looks at solutions u of

$$-\,div\big(a(r)\,grad(u)\big)=0, \tag{25.41}$$

of the form $u=x_j f(r)$,[22]

[21] Pierre Joseph Louis FATOU, French mathematician, 1878–1929. He worked at the Observatory in Paris, France.

[22] For the case of confocal ellipsoids, described in Chap. 26, one looks at solutions $u=x_j f_j(r)$. There are also solutions depending only upon r, which appear in calculations of (electrostatic) capacity, for example.

$$\begin{aligned}
&E = grad\big(x_j f(r)\big) = x_j f'(r)\,\tfrac{x}{r} + f(r)\,e_j,\\
&D_k = a(r)\,E_k = x_j x_k \tfrac{a(r)\,f'(r)}{r} + a(r)\,f(r)\,\delta_{j,k}, k = 1,\ldots,N,\\
&\big(D,\tfrac{x}{r}\big) = x_j a(r)\big(f'(r) + \tfrac{f(r)}{r}\big),\\
&div(D) = (N+1)x_j \tfrac{a(r)\,f'(r)}{r} + x_j r\,\big(\tfrac{a(r)\,f'(r)}{r}\big)' + x_j \tfrac{[a(r)\,f(r)]'}{r},
\end{aligned} \tag{25.42}$$

if f and a are smooth. If a has discontinuities, then u and $\big(D,\frac{x}{r}\big)$ are continuous, i.e. $f(r)$ and $a(r)\big(f'(r)+\frac{f(r)}{r}\big)$ are continuous, and only the derivatives of these quantities should appear, so that (25.41) must be written as

$$\Big[a(r)\Big(f'(r) + \frac{f(r)}{r}\Big)\Big]' + \frac{N\,a(r)\,f'(r)}{r} + \frac{a(r)\,f(r)}{r^2} = 0. \tag{25.43}$$

If $f(r) = \kappa$ and $a(r) = \gamma$ for $r > r_0$, then $f(r) \to \kappa$ as r increases to r_0 and $a(r)\big(f'(r) + \frac{f(r)}{r}\big)$ tends to $\frac{\gamma\,\kappa}{r_0}$, so that a_{eff} is characterized by

$$a_{\mathit{eff}}(r) = \frac{a(r)[r\,f'(r) + f(r)]}{f(r)}, \tag{25.44}$$

and using $a(r)\big(f'(r) + \frac{f(r)}{r}\big) = a_{\mathit{eff}}(r)\,\frac{f(r)}{r}$ in (25.43) gives (25.40). □

(25.40) is a Riccati equation,[23] and for a general (scalar) Riccati equation

$$\frac{dv}{dx} = m(x)\,v^2 + n(x)\,v + p(x), \tag{25.45}$$

if one knows one solution v_1 of (25.45), then

$$w = \frac{1}{v - v_1} \text{ satisfies } \frac{dw}{dx} = -(2m\,v_1 + n)\,w - m; \tag{25.46}$$

if one knows two distinct solutions v_1, v_2 of (25.45), then

$$z = \frac{v - v_2}{v - v_1} \text{ satisfies } \frac{dz}{dx} = m\,(v_2 - v_1)\,z; \tag{25.47}$$

---

[23] Jacopo Francesco RICCATI, Italian mathematician, 1676–1754. The equation named after him, which he wrote about in 1724, was studied before him by Jacob BERNOULLI.

if one knows three distinct solutions v_1, v_2, v_3 of (25.45), then

$$\text{the cross ratio } (v, v_1; v_2, v_3) = \frac{v - v_2}{v - v_3}\,\frac{v_1 - v_3}{v - v_2} \text{ is constant.} \tag{25.48}$$

In the case where $a(r)$ is constant in an open interval, then (25.40) has two constant solutions, $a_{\mathit{eff}} = a$ and $a_{\mathit{eff}} = -(N-1)a$ (which is not a physical solution) and the quantity appearing in (25.47) is then $\frac{a_{\mathit{eff}} - a}{a_{\mathit{eff}} + (N-1)a}$, and one can deduce (25.4) by integrating (25.47), since $m = \frac{-1}{r\,a}$ and $v_1 - v_2 = N\,a$.

If one uses (25.40) on $(0, R)$, with $0 < \alpha \le a(r) \le \beta < +\infty$ on $(0, R)$, then there is no need for an initial condition at $r = 0$, i.e. there is only one solution a_{eff} satisfying $0 < \alpha \le a_{\mathit{eff}}(r) \le \beta < +\infty$ on $(0, R)$,[24] but if one uses a core $a = \alpha$ for $r < r_0$, one takes the initial condition $a_{\mathit{eff}}(r_0) = \alpha$ for (25.40).

Lemma 25.9. *If $N \ge 2$ and $0 < \alpha < \beta < \gamma$, the isotropic mixture with highest effective coefficient for proportions $\theta_\alpha, \theta_\beta, \theta_\gamma > 0$ is* never *given by the coated spheres construction with α inside, β in the middle, and γ outside.*

Proof. One assumes that the configuration with $a^*(r) = \alpha$ for $0 < r < r_0$, $a^*(r) = \beta$ for $r_0 < r < r_1$, and $a^*(r) = \gamma$ for $r_1 < r < r_2$, is better than all the radial configurations with $a_n = \alpha$ on $(0, r_0)$, $a_n = \beta\,\chi_n + \gamma\,(1 - \chi_n)$ on (r_0, r_2), and $\int_{r_0}^{r_2} \chi_n(r)\,r^{N-1}\,dr = \frac{r_1^N - r_0^N}{N}$. One uses (25.40) for a_n, and if $\chi_n \rightharpoonup \theta$ in L^∞ weak $\star$, with $\int_{r_0}^{r_2} \theta\,r^{N-1}\,dr = \frac{r_1^N - r_0^N}{N}$, one has

$$r\,a'_{\mathit{eff}} + a^2_{\mathit{eff}}\left(\frac{\theta}{\beta} + \frac{1-\theta}{\gamma}\right) + (N-2)\,a_{\mathit{eff}} - (N-1)\big(\theta\,\beta + (1-\theta)\,\gamma\big) = 0. \tag{25.49}$$

A necessary condition of optimality is that the derivative δa_{eff} in every admissible direction $\delta\theta$ satisfies $\delta a_{\mathit{eff}}(r_2) \le 0$, and one has

$$\delta\theta \le 0 \text{ in } (r_0, r_1), \delta\theta \ge 0 \text{ in } (r_1, r_2), \int_{r_0}^{r_2} \delta\theta\,r^{N-1}\,dr = 0$$

$$r\,\delta a'_{\mathit{eff}} + \delta a_{\mathit{eff}}\,\varphi^* + \delta\theta\,(\gamma - \beta)\,\psi^* = 0, \text{ and } \delta a_{\mathit{eff}}(r_0) = 0, \tag{25.50}$$

$$\varphi^* = \frac{2a^*_{\mathit{eff}}}{\beta} + N - 2 \text{ on } (r_0, r_1), \varphi^* = \frac{2a^*_{\mathit{eff}}}{\gamma} + N - 2 \text{ on } (r_1, r_2),$$

$$\psi^* = \frac{a^*_{\mathit{eff}}}{\beta\,\gamma} + N - 1, \tag{25.51}$$

---

[24] Replacing a by α for $r < \varepsilon$ gives a smaller a^-_{eff} on (ε, R), and replacing a by β for $r < \varepsilon$ gives a larger a^+_{eff} on (ε, R); $b = a^+_{\mathit{eff}} - a^-_{\mathit{eff}}$ satisfies $r\,b' + \psi\,b = 0$ and $\psi = \frac{a^+_{\mathit{eff}} + a^-_{\mathit{eff}}}{a} + (N-2)$ is bounded below (if $N \ge 2$), and b tends to 0 when ε tends to 0. If $N = 1$, $\frac{1}{a_{\mathit{eff}}}$ satisfies a linear equation, and the result is easily seen.

$$e^{\int_{r_0}^{r_2} \frac{\varphi^*(r)}{r}\,dr}\,\delta a_{\mathit{eff}}(r_2) = -(\gamma-\beta)\int_{r_0}^{r_2} \frac{\psi^*(r)}{r}\, e^{\int_{r_0}^{r} \frac{\varphi^*(s)}{s}\,ds}\,\delta\theta\, dr \le 0, \quad (25.52)$$

and using a Lagrange multiplier, it is necessary that

$$H = \frac{\psi^*(r)}{r^N}\, e^{\int_{r_0}^{r} \frac{\varphi^*(s)}{s}\,ds} \text{ satisfies } H(r) \le H(r_1) \text{ on } (r_0, r_1). \quad (25.53)$$

It is false, since H is not constant on (r_0, r_1) and $H(r_1) \le H(r_0)$: on (r_0, r_1),

$$\frac{a^*_{\mathit{eff}}(r)-\beta}{a^*_{\mathit{eff}}(r)+(N-1)\,\beta} = -X, X = \frac{\kappa\, r_0^N}{r^N}, \kappa = \frac{\beta-\alpha}{\alpha+(N-1)\,\beta}, \quad (25.54)$$

$$a^*_{\mathit{eff}}(r) = \frac{\beta\,(1-(N-1)\,X)}{1+X}, \varphi^* = \frac{N\,(1-X)}{1+X}, X \le \kappa < \frac{1}{N-1}, \quad (25.55)$$

$$\int_{r_0}^{r} \frac{\varphi^*(s)}{s}\,ds = -\int_{\kappa}^{X} \frac{1-Y}{1+Y}\frac{dY}{Y} = \log\frac{\kappa\,(1+X)^2}{(1+\kappa)^2 X}, \quad (25.56)$$

$$(1+\kappa)^2 r_0^N H = (1+X)^2\psi^* = \frac{\beta\,(1-(N-1)\,X)^2}{\gamma} + (N-1)\,(1+X)^2, \quad (25.57)$$

and, using $X \le \kappa$, it is $\le \frac{\beta\,(1-(N-1)\,\kappa)^2}{\gamma} + (N-1)\,(1+\kappa)^2$ if

$$\frac{2\beta\,(N-1)}{\gamma}\,((N-1)\,\kappa + (N-1)\,X - 2) + 2(N-1)\,(\kappa + X + 2) \ge 0, \quad (25.58)$$

which is true if $\beta < \gamma$. □

The computations (25.54)–(25.58) actually show that H is decreasing on (r_0, r_1) and suggest keeping the computations in differential form, and using them for characterizing the optimal radial solution.

Lemma 25.10. *If $N \ge 2$, and $0 < \alpha < \beta < \gamma$, and $\theta_\alpha, \theta_\beta, \theta_\gamma > 0$, the isotropic mixture with radial geometry of the highest effective coefficient using all the material α as core is* always *obtained by coating it with γ, until one reaches an effective conductivity β or one uses all the γ; one then uses all the material of conductivity β, and finally what eventually remains of the material of conductivity γ.*

Proof. There is an optimal θ^* satisfying the constraint $\int_{r_0}^{r_2} \theta^* r^{N-1}\,dr = \frac{r_1^N - r_0^N}{N}$ and such that the corresponding solution a^*_{eff} of (25.49) with $a^*_{\mathit{eff}}(r_0) = \alpha$ achieves the maximum of $a_{\mathit{eff}}(r_2)$. The derivative δa_{eff} in every admissible direction $\delta\theta$ satisfies $\delta a_{\mathit{eff}}(r_2) \le 0$, and one has

$$r\,\delta a'_{\mathit{eff}} + \varphi^*\,\delta a_{\mathit{eff}} + (\gamma-\beta)\,\psi^*\,\delta\theta = 0, \text{ and } \delta a_{\mathit{eff}}(r_0) = 0, \quad (25.59)$$

with $\varphi^* = 2a^*_{\mathit{eff}}\Big(\frac{\theta^*}{\beta} + \frac{1-\theta^*}{\gamma}\Big) + N - 2$, and $\psi^* = \frac{(a^*_{\mathit{eff}})^2}{\beta\,\gamma} + N - 1$, (25.60) and the necessary condition of optimality of θ^* and a^*_{eff} is (25.52), i.e.

$$\frac{e^{\int_{r_0}^{r_2}\frac{\varphi^*(r)}{r}\,dr}}{\gamma-\beta}\,\delta a_{eff}(r_2) = -\int_{r_0}^{r_2} r^{N-1} H\,\delta\theta\,dr \le 0, \tag{25.61}$$

but more precise information than (25.53) for $H = \frac{\psi^*(r)}{r^N}\,e^{\int_{r_0}^{r}\frac{\varphi^*(s)}{s}\,ds}$ is

$$\frac{r\,H'}{H} = \frac{-2(N-1)(a^*_{eff}-\beta)\,(a^*_{eff}-\gamma)}{\beta\,\gamma\,\psi^*} \text{ on } (r_0,r_1). \tag{25.62}$$

Indeed,

$$\begin{aligned}
&\beta\,\gamma\,\psi^*\,\tfrac{r\,H'}{H} = \beta\,\gamma\,r\,(\psi^*)' + \beta\,\gamma\,\psi^*\,(\varphi^*-N)\\
&= 2a^*_{eff}r\,(a^*_{eff})' + 2\beta\,\gamma\,\psi^*\big[a^*_{eff}\big(\tfrac{\theta^*}{\beta}+\tfrac{1-\theta^*}{\gamma}\big)-1\big]\\
&= -2a^*_{eff}\big[(a^*_{eff})^2\big(\tfrac{\theta^*}{\beta}+\tfrac{1-\theta^*}{\gamma}\big)+(N-2)\,a^*_{eff}-(N-1)\big(\theta^*\beta+(1-\theta^*)\,\gamma\big)\big]\\
&\quad +2\big((a^*_{eff})^2+(N-1)\,\beta\,\gamma\big)\big[a^*_{eff}\big(\tfrac{\theta^*}{\beta}+\tfrac{1-\theta^*}{\gamma}\big)-1\big]\\
&= -2(N-1)\,(a^*_{eff})^2+2(N-1)\,a^*_{eff}(\beta+\gamma)-2(N-1)\,\beta\,\gamma.
\end{aligned} \tag{25.63}$$

Notice that $\theta^*,\varphi^*\in L^\infty(r_0,r_2)$, but a^*_{eff}, ψ^*, and H are Lipschitz continuous. Since $\int_{r_0}^{r_2} r^{N-1}H\,\delta\theta\,dr \ge 0$ for all admissible $\delta\theta$ satisfying $\int_{r_0}^{r_2} r^{N-1}\delta\theta\,dr = 0$, one deduces that there exists a Lagrange multiplier $\lambda\in\mathbb{R}$ such that

$$H\ge\lambda \text{ where } \theta^*=0, H=\lambda \text{ where } 0<\theta^*<1, H\le\lambda \text{ where } \theta^*=1. \tag{25.64}$$

An optimal solution a^*_{eff} cannot use $0<\theta^*<1$ on a set of positive measure, since it implies H constant so that $H'=0$ a.e., i.e. a^*_{eff} must be β, but since a mixture has effective conductivity $>\beta$, a^*_{eff} cannot stay equal to β. If $a^*_{eff}(r_2)\le\beta$, then $H'\le 0$, and one uses γ as long as $H\ge\lambda$ and β afterwards. If $a^*_{eff}>\beta$, then H decreases, then stays constant with a^*_{eff} necessarily equal to β and then increases; the value λ must be that for which $a^*_{eff}=\beta$, since a^*_{eff} is continuous. □

I believe that Graeme MILTON constructed these solutions long ago, but I only did the computations of the last two lemmas when writing this book.

Additional footnotes: BERNOULLI Ja.,[25] LOCKHEED A.H. & M.,[26] WHITNEY.[27]

[25] Jacob BERNOULLI, Swiss mathematician, 1654–1705. He worked in Basel, Switzerland.

[26] Allan Haines LOCKHEED (LOUGHEAD), American businessman, 1889–1969. With his brother, Malcolm LOCKHEED (LOUGHEAD), American businessman, 1887– 1958, he formed the Alco Hydro-Aeroplane Company, that became Lockheed Corporation.

[27] Hassler WHITNEY, American mathematician, 1907–1989. He received the Wolf Prize in 1982, for his fundamental work in algebraic topology, differential geometry and differential topology, jointly with Mark Grigorievich KREIN. He worked at Harvard University, Cambridge, MA, and at IAS (Institute for Advanced Study), Princeton, NJ.

Chapter 26
Confocal Ellipsoids and Spheres

In the fall of 1980, I showed to François MURAT my optimal bounds for isotropic effective mixtures of two isotropic conductors, obtained during the summer, and he proposed to follow the same strategy for anisotropic effective mixtures, i.e., use my method from the fall of 1977, shown in Lemma 21.4, with the same functionals. I doubted myself that the necessary bounds would be optimal, thinking that my choice was adapted to the isotropic case, but we performed the computations, shown in Lemma 21.6 and 21.7.

There was a natural construction to try for showing that our bounds were optimal, which was to replace the family of concentric spheres used by Zvi HASHIN and SHTRIKMAN by a family of confocal ellipsoids,

$$\sum_{j=1}^{N} \frac{x_j^2}{c_j^2 + t} = 0, \text{ for } t > 0 \text{ if } \min\{c_1, \ldots, c_N\} = 0. \tag{26.1}$$

I mentioned our plan to Edward FRAENKEL, who visited Paris in the fall of 1980, and he gave me some advice about computing with ellipsoids in $\mathbb{R}^3$,[1] writing for me a one-page review on ellipsoidal coordinates, which I misplaced soon after, so that I did not use it when I resumed my computations with François MURAT. Another reason why we tried another approach was that we wanted to work in $\mathbb{R}^N$; we started by considering a general family of hypersurfaces, and in order to go through a quite technical computation, we made a simplifying assumption, and that gave us the case of confocal ellipsoids, using different formulas than the ones that I saw before.

Lemma 26.1. *For $m_1, \ldots, m_N \in \mathbb{R}$, and $\rho + m_j > 0$ for $j = 1, \ldots, N$, the family of confocal ellipsoids S_ρ of equation*

$$\sum_{j=1}^{N} \frac{x_j^2}{\rho + m_j} = 1, \tag{26.2}$$

[1] Graeme MILTON told me later that he also made computations with ellipsoids.

L. Tartar, *The General Theory of Homogenization*, Lecture Notes of the Unione Matematica Italiana 7, DOI 10.1007/978-3-642-05195-1_26,

defines implicitly a real function ρ*, outside a possibly degenerate ellipsoid in a subspace of dimension* $< N$*, which satisfies*

$$\begin{aligned} &\frac{\partial \rho}{\partial x_j} = \frac{1}{\sigma}\frac{2x_j}{\rho+m_j}, j = 1,\dots,N, \textit{ with } \sigma = \textstyle\sum_{k=1}^{N} \frac{x_k^2}{(\rho+m_k)^2}, \\ &|grad(\rho)|^2 = \tfrac{4}{\sigma}, \end{aligned} \tag{26.3}$$

$$\Delta\,\rho = \frac{1}{\sigma}\sum_{k=1}^{N}\frac{2}{\rho+m_k}. \tag{26.4}$$

Proof. One defines the singular set Σ by

$$x_j = 0 \text{ for } j \in J = \{j \mid m_j = \min_k m_k\}, \sum_{i \notin J} \frac{x_i^2}{m_i - \min_k m_k} \le 1. \tag{26.5}$$

For $x \notin \Sigma$ and $\rho + \min_k m_k > 0$, the left side of (26.2) is a C^∞ function, whose partial derivative in ρ is different from 0, and the implicit function theorem applies: for $x \notin \Sigma$ there is only one ellipsoid S_ρ containing x, and ρ is a C^∞ function near x. Defining σ in (26.3), and deriving (26.2) in x_j gives

$$-\,\sigma\frac{\partial \rho}{\partial x_j} + \frac{2x_j}{\rho+m_j} = 0, j = 1,\dots,N, \tag{26.6}$$

from which one deduces the rest of (26.3). Then, for $j,k = 1,\dots,N$, one has

$$\frac{\partial \sigma}{\partial x_j} = -2\tau\frac{\partial \rho}{\partial x_j} + \frac{2x_j}{(\rho+m_j)^2}, \text{ with } \tau = \sum_{k=1}^{N}\frac{x_k^2}{(\rho+m_k)^3}, \tag{26.7}$$

$$\frac{\partial^2 \rho}{\partial x_j \partial x_k} = \frac{2\delta_{j,k}}{\sigma\,(\rho+m_j)} - \frac{4x_j x_k}{\sigma^2(\rho+m_j)(\rho+m_k)}\Big(\frac{1}{\rho+m_j} + \frac{1}{\rho+m_k} - \frac{2\tau}{\sigma}\Big), \tag{26.8}$$

and taking $k = j$ and summing in j gives (26.4). □

Lemma 26.2. *If* $0 < \alpha \le a(\rho) \le \beta < +\infty$, $\rho^* + \min_k m_k > 0$, *the equation*

$$-\,div\big(a(\rho)\,grad(u)\big) = 0 \textit{ in } E_{\rho^*} = \Big\{x \mid \sum_{j=1}^{N}\frac{x_j^2}{\rho^*+m_j} \le 1\Big\}, \tag{26.9}$$

has particular solutions of the following form, where $' = \frac{d}{d\rho}$,

$$\begin{aligned} &u = f_0(\rho) \textit{ with } f_0 \textit{ and } a\,f_0' \textit{ continuous, and} \\ &(a\,f_0')' + \frac{a\,f_0'}{2}\sum_{k=1}^{N}\frac{1}{\rho+m_k} = 0, \end{aligned} \tag{26.10}$$

$$u = x_j f_j(\rho) \text{ with } f_j \text{ and } a\, f_j' + \frac{a\, f_j}{2(\rho+m_j)} \text{ continuous, and}$$

$$\Big(a\, f_j' + \frac{a\, f_j}{2(\rho+m_j)}\Big)' + \frac{a\, f_j'}{2}\Big(\frac{1}{\rho+m_j} + \sum_{k=1}^N \frac{1}{\rho+m_k}\Big) + \frac{a\, f_j}{2(\rho+m_j)^2} = 0. \quad (26.11)$$

Proof. If $div(D) = 0$, then (D, ν) is continuous on S_ρ, but ν being parallel to $grad(\rho)$, $\big(D, grad(\rho)\big)$ is continuous. If $u = f_0(\rho)$, then $D = a(\rho)\, f_0'(\rho)\, grad(\rho)$, and $\frac{\sigma}{4}\,\big(D, grad(\rho)\big) = a\, f_0'$ is continuous, and $div(D) = 0$ gives

$$\big(a(\rho)\, f_0'(\rho)\big)'|grad(\rho)|^2 + a(\rho)\, f_0'(\rho)\, \Delta\,\rho = 0, \quad (26.12)$$

which implies (26.10) by using (26.3) and (26.4). If $u = x_j f_j(\rho)$, then

$$D = x_j a\, f_j' grad(\rho) + a\, f_j e_j, \text{ and } \frac{\sigma}{4}\,\big(D, grad(\rho)\big) = x_j a\, f_j' + x_j \frac{a\, f_j}{2(\rho+m_j)}, \quad (26.13)$$

so that $a\, f_j' + \frac{a\, f_j}{2(\rho+m_j)}$ is continuous, and $div(D) = 0$ gives

$$x_j (a\, f_j')'|grad(\rho)|^2 + x_j a\, f_j' \Delta\,\rho + a\, f_j' \frac{\partial \rho}{\partial x_j} + (a\, f_j)' \frac{\partial \rho}{\partial x_j} = 0, \quad (26.14)$$

and putting in the factor $\frac{4 x_j}{\sigma}$ gives

$$(a\, f_j')' + \frac{a\, f_j'}{2}\Big(\sum_{k=1}^N \frac{1}{\rho+m_k}\Big) + \frac{a\, f_j'}{2(\rho+m_j)} + \frac{(a\, f_j)'}{2(\rho+m_j)} = 0, \quad (26.15)$$

and combining the first and last terms for making the derivative of $a\, f_j' + \frac{a\, f_j}{2(\rho+m_j)}$ appear, gives (26.11). □

The existence of solutions $u = f_0(\rho)$ relies on (26.12) and uses $\Delta\,\rho = g(\rho)\,|grad(\rho)|^2$ (and $grad(\rho) \neq 0$), while the existence of solutions $u = x_j f_j(\rho)$ relies on (26.14) and also uses $\frac{\partial \rho}{\partial x_j} = x_j g_j(\rho)\,|grad(\rho)|^2$.

Lemma 26.3. *If $A(x) = a(\rho)\, I$, the restriction of A to E_{ρ^*} is equivalent to $A^{\mathit{eff}}(\rho^*)$, which is diagonal, and for $j = 1, \ldots, N$, $A^{\mathit{eff}}_{j,j}(\rho)$ satisfies*

$$(A^{\mathit{eff}}_{j,j})' + \frac{(A^{\mathit{eff}}_{j,j} - a)^2}{2a\,(\rho+m_j)} + \frac{A^{\mathit{eff}}_{j,j} - a}{2}\Big(\sum_{k=1}^N \frac{1}{\rho+m_k}\Big) = 0. \quad (26.16)$$

Proof. One uses the solutions $x_j f_j(\rho)$ computed at (26.11), and one has

$$a\, f_j' + \frac{a\, f_j}{2(\rho+m_j)} = \frac{A^{\mathit{eff}}_{j,j}\, f_j}{2(\rho+m_j)}, \quad (26.17)$$

so that (26.11) becomes 510.5

$$\Big(\frac{A^{\text{eff}}_{j,j}\,f_j}{2(\rho+m_j)}\Big)'+\frac{(A^{\text{eff}}_{j,j}-a)\,f_j}{4(\rho+m_j)}\Big(\frac{1}{\rho+m_j}+\sum_{k=1}^{N}\frac{1}{\rho+m_k}\Big)+\frac{a\,f_j}{2(\rho+m_j)^2}=0, \tag{26.18}$$

$$\Big(\frac{A^{\text{eff}}_{j,j}\,f_j}{2(\rho+m_j)}\Big)'=\frac{(A^{\text{eff}}_{j,j})'\,f_j}{2(\rho+m_j)}-\frac{A^{\text{eff}}_{j,j}\,f_j}{2(\rho+m_j)^2}+\frac{A^{\text{eff}}_{j,j}}{2(\rho+m_j)}\,\frac{(A^{\text{eff}}_{j,j}-a)f_j}{2a\,(\rho+m_j)}, \tag{26.19}$$

so that, putting in the factor $\frac{f_j}{2(\rho+m_j)}$ one has

$$(A^{\text{eff}}_{j,j})'-\frac{A^{\text{eff}}_{j,j}-a}{2(\rho+m_j)}+\frac{A^{\text{eff}}_{j,j}\,(A^{\text{eff}}_{j,j}-a)}{2a\,(\rho+m_j)}+\frac{A^{\text{eff}}_{j,j}-a}{2}\Big(\sum_{k=1}^{N}\frac{1}{\rho+m_k}\Big)=0, \tag{26.20}$$

which gives (26.16). □

In all m_i are equal, the ellipsoids are concentric spheres, but $\rho+m_1>0$ corresponds to r^2; also, $'=\frac{d}{dr}$ in (25.40), but $'=\frac{d}{d\rho}$ in (26.16); multiplying (26.16) by $2(\rho+m_j)$ which is $2r^2$, the first term is $2r^2\frac{dA^{\text{eff}}_{j,j}}{dr^2}$, i.e., $r\,\frac{dA^{\text{eff}}_{j,j}}{dr}$.

Corollary 26.4. *If one defines $V(\rho)$ by*

$$V(\rho)=\prod_k\sqrt{\rho+m_k}, \tag{26.21}$$

if $a(\rho)=\alpha$ for $\rho<\rho_1$ and $a(\rho)=\beta$ for $\rho_1<\rho<\rho_2$,

$$\frac{1}{\beta-A^{\text{eff}}_{j,j}(\rho_2)}=\frac{V(\rho_2)}{(\beta-\alpha)\,V(\rho_1)}-\frac{V(\rho_2)}{2\beta}\int_{\rho_1}^{\rho_2}\frac{d\rho}{(\rho+m_j)V(\rho)}, \tag{26.22}$$

$$\sum_j\frac{1}{\beta-A^{\text{eff}}_{j,j}}=\frac{(1-\theta)\alpha+(N+\theta-1)\beta}{\theta\,\beta(\beta-\alpha)}. \tag{26.23}$$

Proof. One has $A^{\text{eff}}_{j,j}(\rho_1)=\alpha$, and (26.16) for $a=\beta$ for $\rho>\rho_1$ gives

$$\Big(\frac{1}{\beta-A^{\text{eff}}_{j,j}}\Big)'=\frac{1}{2(\beta-A^{\text{eff}}_{j,j})}\Big(\sum_k\frac{1}{\rho+m_k}\Big)-\frac{1}{2\beta\,(\rho+m_j)}, \tag{26.24}$$

and since the solution of the homogeneous equation is proportional to $V(\rho)$,

$$\frac{1}{\beta-A^{\text{eff}}_{j,j}}=C\,V(\rho),\ \text{and}\ C'=\frac{1}{2\beta\,(\rho+m_j)\,V(\rho)}, \tag{26.25}$$

which gives (26.22) by using $C(\rho_1)=\frac{1}{V(\rho_1)}$. For obtaining (26.23), one notices that $\sum_j\frac{1}{2\beta\,(\rho+m_j)\,V(\rho)}=\frac{V'(\rho)}{\beta\,V^2(\rho)}$, whose integral is $\frac{1}{\beta\,V(\rho_1)}-\frac{1}{\beta\,V(\rho_2)}$. □

The volume of E_ρ is $\omega_N V(\rho)$, ω_N being the volume of the unit ball in $\mathbb{R}^N$, so that $\frac{V(\rho_1)}{V(\rho_2)} = \theta$, and (26.23) corresponds to the equality in (21.42). The coated confocal ellipsoids construction then gives effective materials satisfying the bounds that I obtained with François MURAT, and the other bound is obtained by exchanging the roles of α and β and replacing θ by $1-\theta$.

I refer to [101] for checking that all tensors with (21.40)–(21.42) are effective tensors of a mixture using proportions θ of α and $1-\theta$ of β.

Although I thought that my choice of functionals of June 1980 for my method from the fall of 1977 (Lemma 21.4) was adapted to isotropic effective materials, François MURAT was right to propose to check what they imply for anisotropic effective materials (Lemma 21.6 and 21.7). We then found a natural way to make explicit constructions with confocal ellipsoids (Lemma 26.1 and 26.2), different from what Edward FRAENKEL showed to me, and it was nice that for mixtures of two isotropic materials the two approaches fit well together and Corollary 26.4 showed that we characterized which effective tensors can be obtained with given proportions.

One should remember that research is about discovery, but that well-organized development helps research; it is useful to put old and new results in order, and to simplify proofs, so that one understands better what was done and where to go. Explicit constructions of solutions is an eighteenth century point of view in ordinary differential equations, before the development in the beginning of the nineteenth century of a general method of existence by CAUCHY, and it is a nineteenth century point of view in partial differential equations, before the development of the general methods of existence of functional analysis in the beginning of the twentieth century. I showed the way in developing new mathematical tools for twentieth century continuum mechanics and physics, and if I use old methods of looking for explicit solutions, it is not for a reason of sabotage like advocating fake mechanics or physics which is done by important groups, but because it helps checking if the new tools that I developed are good enough.

One of my goals was to create mathematical tools for describing effective properties of mixtures, and their evolution; in the late 1980s, when I introduced *H-measures*, which I describe in Chap. 28, I was only making one step in the right direction, and for the moment no one sees how to go further, but by describing all the aspects of what I did and by choosing to describe other pieces of mathematics which I expect to be relevant for my quest, I hope to help a few researchers acquire a broad knowledge. I mentioned before the motto of Hugo of Saint Victor "Learn everything, and you will see afterward that nothing is useless," with the advice of taking the time to understand fully what one learned, but few received a good general training like I did, in algebra, analysis, and geometry, as well as in classical and continuum mechanics, and various aspects of physics, and so it is my duty to show the way.

In this book I discuss homogenization as an important subject in continuum mechanics and physics, and it is important to realize that the many aspects of the problems require using various mathematical tools. However, one should be aware that it is not so easy to learn about some domains, because of a few kinds of behavior. There are domains where many acquired a racist tendency to despise others, in part since they feel superior because what others did and not because of their own accomplishments or their deep understanding, in part since their refusing to answer questions is a way to avoid one discovering that they are mediocre mathematicians, in part for political reasons to create havoc in the scientific arena, by introducing class struggle (choosing to pretend to be upper class, of course!). One should remember that it is not the field which decides if one is a good mathematician or not, but the importance of the new ideas that one introduced, and the scientific importance of the problems that one considered, and one should not let mediocre mathematicians brainwash students for obviously political reasons. It is then not so easy for those trained in analysis to learn what they may need in algebra or in geometry, and it would be the same for topology, which I cannot yet see as being useful.[2] Another obstacle to learning is that powerful political groups organize fashions by having their adepts follow them and advocate that others follow the same trend,[3] and since the organizers have not a good brain for mathematical ideas, they resurrect old ideas to which they give new names, chaos for the work of POINCARÉ, fractals for the work of FATOU and of JULIA,[4] fractional derivatives for some classical work that Laurent SCHWARTZ taught in the 1960s and which some pretended to be new in the 1980s, wavelets from the work of HAAR and of GABOR,[5,6] a subject

[2] General topology is used as a part of functional analysis, but is algebraic topology of any use? POINCARÉ started such questions of topology, motivated by classical mechanics, but they play no role in continuum mechanics, although some want to see topology around "invariants," forgetting that conserved quantities may hide at a mesoscopic level, and it is so crucial a fact that it is nonsense to only work before the time when it happens. Physicists are prone to pseudo-logic, playing any game if it looks like something observed, but instead of playing their games with strings, they should recall POINCARÉ's principle of relativity: nature uses no games with instantaneous forces at distance, and everything results from semi-linear hyperbolic systems with only the speed of light c as characteristic speed!

[3] In my student days, someone told me in this way to read a book on foundations of mechanics, and I read it, but although it was a good book for learning questions on manifolds, it contained no mechanics at all! A friend told me about René THOM's ideas in a less directive way, and I wondered if these ideas were useful in biology, of which I learned too little; however, it looked too much like considering a world described by ordinary differential equations, quite a naive approach.

[4] Gaston Maurice JULIA, French mathematician, 1893–1978. He worked at École Polytechnique, Paris, France.

[5] Alfréd HAAR, Hungarian mathematician, 1885–1933. He worked at Georg-August-Universität, Göttingen, Germany, in Kolozsvár (then in Hungary, now Cluj-Napoca, Romania), in Budapest and in Szeged, Hungary.

[6] Dénes (Dennis) GÁBOR, Hungarian-born physicist, 1900–1979. He received the Nobel Prize in Physics in 1971 for his invention and development of the holographic method. He worked at British Thomson–Houston in Reading, and at Imperial College, London, England.

which disappointed a few when it started becoming useful,[7] mass transport for the work of MONGE and of KANTOROVICH,[8,9] correctly describing the motivation of MONGE but not that of KANTOROVICH.[10]

More generally, all the fashions that I witnessed pushed the researchers away from understanding about sciences other than mathematics, and engineering. For what concerns homogenization, besides hiding the names of the pioneers, Sergio SPAGNOLO in Italy, François MURAT and myself in France, they either discussed periodic structures without any real engineering applications, forgetting also to mention the pioneers like Évariste SANCHEZ-PALENCIA in France or Ivo BABUŠKA in United States, and even imposed a periodic analysis on general sequences when it could be of no use; they also ignored the defects of the second principle which result from the simple observation that the effective properties cannot be deduced from proportions alone if $N \geq 2$, and some even used the term Young measures to say something more silly (pretending that Young measures characterize microstructures), and many advocated fake mechanics principles like minimization of potential energy, showing a complete disdain for elementary classical mechanics, or just the first principle.

The confocal ellipsoids form a unified family of geometries, which contains the concentric spheres, by taking all the m_i equal, and which contains the laminations, by fixing one m_i and letting the other m_j tend to $+\infty$, and it also contains a cylindrical geometry with base consisting of confocal ellipsoids, by keeping a few of the m_j fixed and letting the others tend to $+\infty$. However, we must go further for what concerns explicit formulas. In the proof of Lemma 25.9, we saw the importance of using laminates for necessary conditions: they have $N-1$ equal eigenvalues along the tangent hyperplane and a different one along the radial direction, but the last eigenvalue is not arbitrary, and we must

---

[7] However, many who play with these special bases of Hilbert spaces were not trained well enough to avoid some theoretical computations by using the Lions–Peetre theory of interpolation spaces, which I described in the second part of [117].

[8] Gaspard MONGE, French mathematician, 1746–1818. He was made count by Napoléon I in 1808. He worked in Mézières, and in Paris, France.

[9] Leonid Vitalyevich KANTOROVICH, Russian mathematician, 1912–1986. He received the Sveriges Riksbank Prize in Economic Sciences in Memory of Alfred Nobel in 1975, jointly with Tjalling C. KOOPMANS, for their contributions to the theory of optimum allocation of resources. He worked in Leningrad, at the Siberian branch of the Russian Academy of Sciences, Novosibirsk, and in Moscow, Russia.

[10] MONGE considered the cost of building a road, by transporting soil from one place to another, while KANTOROVICH described allocating and transferring resources from one sector of the economy to another, and for that he invented linear programming, and the duality method.

check a more general formula, from a construction of SCHULGASSER,[11,12] which Lemma 26.5 presents; Lemma 26.6 gives the extension to confocal ellipsoids, which I checked with Gilles FRANCFORT in the fall of 1994, while he worked for 1 year at CMU (*Carnegie Mellon University*), Pittsburgh, PA.

Lemma 26.5. *If $A(x)\,x = \lambda_{rad}(r)\,x$ and $A(x)\,v = \lambda_{tan}(r)\,v$ for all $v\perp x$, the restriction of A to the ball of radius r is equivalent to $a_{\mathit{eff}}(r)\,I$, and one has*

$$r\,a'_{\mathit{eff}} + \frac{a^2_{\mathit{eff}}}{\lambda_{rad}} + (N-2)\,a_{\mathit{eff}} - (N-1)\,\lambda_{tan} = 0. \tag{26.26}$$

Proof. One looks for solutions $u = x_j f(r)$ of $div\big(A\,grad(u)\big) = 0$, so that

$$\begin{aligned}
&grad(u) = x_j f'(r)\,\tfrac{x}{r} + f(r)\,e_j,\\
&D = A\,grad(u) = \lambda_{rad} x_j f'\,\tfrac{x}{r} + f\,\big(\lambda_{tan} e_j + (\lambda_{rad}-\lambda_{tan})\,\tfrac{x_j}{r}\,\tfrac{x}{r}\big),\\
&(D,x) = x_j \lambda_{rad}(r\,f' + f),
\end{aligned} \tag{26.27}$$

and both f and $\lambda_{rad}(r\,f' + f)$ must be continuous in r. The coefficient of λ_{rad} in D is $x_j f' \frac{x}{r} + \frac{x_j}{r}\,f\,\frac{x}{r}$, so that the term containing λ_{rad} in $div(D)$ is $x_j(\lambda_{rad} f')' + \lambda_{rad} f' \frac{N\,x_j}{r} + (\lambda_{rad} f)' \frac{x_j}{r} + \lambda_{rad} f\,\frac{(N-1)\,x_j}{r^2}$; the coefficient of λ_{tan} in D is $f\,e_j - x_j f\,\frac{x}{r^2}$, so that the term containing λ_{tan} in $div(D)$ is $-(N-1)\lambda_{tan} f\,\frac{x_j}{r^2}$; adding, putting in the factor $\frac{x_j}{r}$, one obtains

$$\big(\lambda_{rad}(r\,f' + f)\big)' + (N-1)\lambda_{rad} f' + (N-1) f\,\frac{\lambda_{rad} - \lambda_{tan}}{r} = 0. \tag{26.28}$$

Then one uses

$$\lambda_{rad}(r)\,\big(r\,f'(r) + f(r)\big) = a_{\mathit{eff}}(r)\,f(r), \tag{26.29}$$

so that $\lambda_{rad} f' = \frac{(a_{\mathit{eff}} - \lambda_{rad})\,f}{r}$, and

$$\big(\lambda_{rad}(r\,f' + f)\big)' = (a_{\mathit{eff}} f)' = a'_{\mathit{eff}} f + a_{\mathit{eff}}\,\frac{(a_{\mathit{eff}} - \lambda_{rad})\,f}{r\,\lambda_{rad}}, \tag{26.30}$$

and putting in the factor $\frac{f}{r}$ gives (26.26). □

Lemma 26.6. *If $A(x)\,grad(\rho) = \lambda_n(\rho)\,grad(\rho)$ and $A(x)v = \lambda_{tan}(\rho)\,v$ for all $v\perp grad(\rho)$, the restriction of A to the E_{ρ^*} is equivalent to $A^{\mathit{eff}}(\rho^*)$, which is diagonal, and for $j = 1,\ldots,N$, $A^{\mathit{eff}}_{j,j}(\rho)$ satisfies*

---

[11] Kalman M. SCHULGASSER, American–born physicist, born in 1938. He worked at Ben-Gurion University of the Negev, Beer-Sheva, Israel.

[12] For creating an isotropic conductor out of a sole anisotropic conductor, with conductivities $(\alpha_1,\alpha_2,\alpha_3)$, he first used lamination in the plane for obtaining $(\sqrt{\alpha_1\,\alpha_2},\sqrt{\alpha_1\,\alpha_2},\alpha_3)$ and then he applied his construction.

$$\begin{array}{ll}(A^{\textit{eff}}_{j,j})' + \frac{(A^{\textit{eff}}_{j,j})^2}{\lambda_n}\,\frac{1}{2(\rho+m_j)} + A^{\textit{eff}}_{j,j}\left(-\frac{1}{\rho+m_j} + \sum_{k=1}^N \frac{1}{2(\rho+m_k)}\right) \\ +\quad \lambda_{tan}\left(\frac{1}{2(\rho+m_j)} - \sum_{k=1}^N \frac{1}{2(\rho+m_k)}\right) = 0.\end{array} \tag{26.31}$$

Proof. One looks for solutions $u = x_j f_j(\rho)$ of $div\big(A\, grad(u)\big) = 0$, so that

$$\begin{array}{l} grad(u) = x_j f_j'(\rho)\, grad(\rho) + f_j(\rho)\, e_j, \\ D = A\, grad(u) = x_j f_j' \lambda_n grad(\rho) + f_j\left(\lambda_{tan} e_j + (\lambda_n - \lambda_{tan})\,\frac{\partial \rho}{\partial x_j}\,\frac{grad(\rho)}{|grad(\rho)|^2}\right), \\ \frac{\sigma}{4}\,\big(D, grad(\rho)\big) = x_j \lambda_n\left(f_j' + \frac{f_j}{2(\rho+m_j)}\right). \end{array} \tag{26.32}$$

Both f_j and $\lambda_n\left(f_j' + \frac{f_j}{2(\rho+m_j)}\right)$ are continuous in ρ, and as $\frac{\frac{\partial \rho}{\partial x_j}}{|grad(\rho)|^2} = \frac{x_j}{2(\rho+m_j)}$,

$$D = x_j \lambda_n f_j' grad(\rho) + \lambda_{tan} f_j e_j + x_j(\lambda_n - \lambda_{tan})\,\frac{f_j}{2(\rho+m_j)}\, grad(\rho). \tag{26.33}$$

In $div(D)$, the coefficient of $\Delta\,\rho$ is $x_j \lambda_n f_j' + x_j(\lambda_n - \lambda_{tan})\,\frac{f_j}{2(\rho+m_j)}$, the coefficient of $|grad(\rho)|^2$ is $x_j(\lambda_n f_j')' + x_j\left((\lambda_n - \lambda_{tan})\,\frac{f_j}{2(\rho+m_j)}\right)'$, and the coefficient of $\frac{\partial \rho}{\partial x_j}$ is $\lambda_n f_j' + (\lambda_{tan} f_j)' + (\lambda_n - \lambda_{tan})\,\frac{f_j}{2(\rho+m_j)}$. Adding and using $\frac{4 x_j}{\sigma}$ as a factor,

$$\begin{array}{ll}(\lambda_n f_j')' + \left(\frac{(\lambda_n - \lambda_{tan})\, f_j}{2(\rho+m_j)}\right)' + \frac{\lambda_n f_j' + (\lambda_{tan} f_j)'}{2(\rho+m_j)} + \frac{(\lambda_n - \lambda_{tan})\, f_j}{4(\rho+m_j)^2} \\ +\quad \left(\lambda_n f_j' + \frac{(\lambda_n - \lambda_{tan})\, f_j}{2(\rho+m_j)}\right)\left(\sum_k \frac{1}{2(\rho+m_k)}\right) = 0,\end{array} \tag{26.34}$$

and using

$$\lambda_n\left(f_j' + \frac{f_j}{2(\rho+m_j)}\right) = \frac{A^{\textit{eff}}_{j,j}\, f_j}{2(\rho+m_j)}, \tag{26.35}$$

$$\begin{array}{ll}\left(\frac{A^{\textit{eff}}_{j,j}\, f_j}{2(\rho+m_j)}\right)' + \frac{A^{\textit{eff}}_{j,j}\, f_j}{4(\rho+m_j)^2} + \frac{A^{\textit{eff}}_{j,j}\, f_j}{2(\rho+m_j)}\left(\sum_k \frac{1}{2(\rho+m_k)}\right) \\ +\quad \frac{\lambda_{tan}\, f_j}{4(\rho+m_j)^2} - \frac{\lambda_{tan}\, f_j}{2(\rho+m_j)}\left(\sum_k \frac{1}{2(\rho+m_k)}\right) = 0,\end{array} \tag{26.36}$$

$$\left(\frac{A^{\textit{eff}}_{j,j}\, f_j}{2(\rho+m_j)}\right)' = \frac{(A^{\textit{eff}}_{j,j})' f_j}{2(\rho+m_j)} - \frac{A^{\textit{eff}}_{j,j}\, f_j}{2(\rho+m_j)^2} + \frac{A^{\textit{eff}}_{j,j}}{2(\rho+m_j)}\,\frac{(A^{\textit{eff}}_{j,j} - \lambda_n) f_j}{2(\rho+m_j)\,\lambda_n}, \tag{26.37}$$

so that putting $\frac{f_j}{2(\rho+m_j)}$ as a factor gives (26.31). □

Corollary 26.7. *If (26.26) is valid for fixed $\lambda_{rad}, \lambda_{tan}$, in $0 < r_1 < r < r_2$,*

$$\begin{array}{l} z = \frac{a_{\textit{eff}} - v_+}{a_{\textit{eff}} - v_-} \textit{ satisfies } r\, z' + z\,\sqrt{(N-2)^2 \lambda_{rad}^2 + 4(N-1)\,\lambda_{rad}\lambda_{tan}} = 0, \\ v_\pm = \frac{-(N-2)\,\lambda_{rad} \pm \sqrt{(N-2)^2 \lambda_{rad}^2 + 4(N-1)\,\lambda_{rad}\lambda_{tan}}}{2}, \end{array} \tag{26.38}$$

and if r_1 tends to 0, then $a_{\textit{eff}} = v_+$.

Proof. One applies (25.47) for a general Riccati equation (25.45), with two explicit solutions $v_\pm$, with $v_- < 0 < v_+$, and the variable is actually $\log r$. □

Corollary 26.8. *If (26.31) is valid for fixed λ_n, λ_{tan}, in $0 < \rho_1 < \rho < +\infty$,*

$$\lim_{\rho\to+\infty} A^{eff}_{j,j}(\rho) = \frac{-(N-2)\,\lambda_n + \sqrt{(N-2)^2\lambda_n^2 + 4(N-1)\,\lambda_n\lambda_{tan}}}{2}. \quad (26.39)$$

Proof. Defining $v_\pm$ as in (26.38) but with λ_n replacing λ_{rad}, then

$$z = \frac{A^{eff}_{j,j} - v_+}{A^{eff}_{j,j} - v_-} \text{ satisfies } \rho\, z' + \kappa\, z = O\Big(\frac{1}{\rho}\Big), \quad (26.40)$$

with $\kappa = \sqrt{(N-2)^2\lambda_n^2 + 4(N-1)\,\lambda_n\lambda_{tan}} > 0$; as ρ tends to $+\infty$, (26.40) implies that z tends to 0 as $O\big(\frac{1}{\rho}\big)$. □

If one considers (26.26) with $\int_{r_0}^{r_2} \theta\, r^{N-1} dr = \frac{r_1^N - r_0^N}{N}$ and

$$\begin{aligned} &\lambda_-(\theta) = \big(\tfrac{\theta}{\beta} + \tfrac{1-\theta}{\gamma}\big)^{-1} \le \lambda_{rad}, \lambda_{tan} \le \lambda_+(\theta) = \theta\,\beta + (1-\theta)\,\gamma,\\ &\tfrac{1}{\lambda_{rad}-\beta} + \tfrac{N-1}{\lambda_{tan}-\beta} \le \tfrac{1}{\lambda_-(\theta)-\beta} + \tfrac{N-1}{\lambda_+(\theta)-\beta},\\ &\tfrac{1}{\gamma-\lambda_{rad}} + \tfrac{N-1}{\gamma-\lambda_{tan}} \le \tfrac{1}{\gamma-\lambda_-(\theta)} + \tfrac{N-1}{\gamma-\lambda_+(\theta)}, \end{aligned} \quad (26.41)$$

can the Schulgasser geometry of Lemma 26.5 improve Lemma 25.10?

Lemma 26.9. *The maximum effective coefficient $a_{eff}(r_2)$ for $(\theta, \lambda_{rad}, \lambda_{tan})$ satisfying (26.41) and $\int_{r_0}^{r_2} \theta\, r^{N-1} dr = \frac{r_1^N - r_0^N}{N}$ is that of Lemma 25.10.*

Proof. One compares a general $(\theta, \lambda_{rad}, \lambda_{tan})$ to a candidate for optimality indexed by $*$, by considering $0 \le \eta \le 1$ on (r_0, r_2) and

$$r\, a'_{eff} + a^2_{eff}\Big(\frac{1-\eta}{\lambda^*_{rad}} + \frac{\eta}{\lambda_{rad}}\Big) + (N-2)\, a_{eff} - (N-1)\,\big((1-\eta)\lambda^*_{tan} + \eta\,\lambda_{tan}\big) = 0, \quad (26.42)$$

with $a_{eff}(r_0) = \alpha$ and η constrained by $\int_{r_0}^{r_2} \big((1-\eta)\,\theta^* + \eta\,\theta\big)\, r^{N-1} dr = \frac{r_1^N - r_0^N}{N}$.[13] That $a_{eff}(r_2)$ is minimum at $\eta = 0$ gives $\delta a_{eff}(r_2) \le 0$, where

$$\begin{aligned} &r\,\delta a'_{eff} + \varphi^*\,\delta a_{eff} + \widetilde{\psi}\,\delta\eta = 0, \delta a_{eff}(r_0) = 0,\\ &\widetilde{\psi} = (a^*_{eff})^2\big(\tfrac{1}{\lambda_{rad}} - \tfrac{1}{\lambda^*_{rad}}\big) - (N-1)\,(\lambda_{tan} - \lambda^*_{tan}), \end{aligned} \quad (26.43)$$

so that with the constraint $\int_{r_0}^{r_2} (\theta - \theta^*)\, r^{N-1}\delta\eta\, dr = 0$, and $\delta\eta \ge 0$,

[13] The characterization (26.41) implies that $\Big(\frac{1}{\lambda_{rad}}, \lambda_{tan}, \theta\Big)$ belongs to a convex.

$$\int_{r_0}^{r_2} e^{\int_{r_0}^{r} \frac{\varphi^*(s)}{s}\,ds} \frac{\widetilde{\psi}}{r}\,\delta\eta\,dr \geq 0, \tag{26.44}$$

and using a Lagrange multiplier $\kappa \in \mathbb{R}$, depending upon $(\theta, \lambda_{rad}, \lambda_{tan})$, one has

$$e^{\int_{r_0}^{r} \frac{\varphi^*(s)}{s}\,ds} \frac{\widetilde{\psi}}{r} + \kappa\,(\theta - \theta^*)\,r^{N-1} \geq 0 \text{ a.e. } r \in (r_0, r_2). \tag{26.45}$$

For $\theta = \theta^*$, $\widetilde{\psi} \geq 0$ a.e. $r \in (r_0, r_2)$, i.e.,[14]

$$\frac{(a^*_{\mathit{eff}})^2}{\lambda_{rad}} - (N-1)\,\lambda_{tan} \geq \frac{(a^*_{\mathit{eff}})^2}{\lambda^*_{rad}} - (N-1)\,\lambda^*_{tan} \text{ a.e. } r \in (r_0, r_2), \tag{26.46}$$

so that one maximizes $a_{\mathit{eff}}(r_2)$ by maximizing a'_{eff} a.e. $r \in (r_0, r_2)$.

The necessary condition (26.46) is satisfied for the optimal radial construction of Lemma 25.10,[15] but one must find all solutions for which it holds, by looking for what $a^*_{\mathit{eff}} \in (\alpha, \gamma)$, and $\theta \in (0,1)$, the minimum of $\frac{(a^*_{\mathit{eff}})^2}{\lambda_{rad}} - (N-1)\,\lambda_{tan}$ is attained for $\lambda_{rad} \in \big(\lambda_-(\theta), \lambda_+(\theta)\big)$. Since it only happens for $a^*_{\mathit{eff}} > \beta$, one cannot improve Lemma 25.10 by using the Schulgasser geometry,[16] but it is worth sketching a way to do the computations.

For $\lambda_-(\theta) < \lambda_{rad} < \lambda_+(\theta)$, the maximum value of λ_{tan} is given by

$$\frac{1}{\gamma - \lambda_{rad}} + \frac{N-1}{\gamma - \lambda_{tan}} = Z(\theta) = \frac{1}{\gamma - \lambda_-(\theta)} + \frac{N-1}{\gamma - \lambda_+(\theta)}, \tag{26.47}$$

so that the variations $\delta\lambda_{rad}$ and $\delta\lambda_{tan}$ satisfy

$$\frac{\delta\lambda_{rad}}{(\gamma - \lambda_{rad})^2} + \frac{(N-1)\delta\lambda_{tan}}{(\gamma - \lambda_{tan})^2} = 0, \tag{26.48}$$

while at the minimum of $\frac{(a^*_{\mathit{eff}})^2}{\lambda_{rad}} - (N-1)\,\lambda_{tan}$, one has

$$\frac{(a^*_{\mathit{eff}})^2\delta\lambda_{rad}}{(\lambda_{rad})^2} + (N-1)\delta\lambda_{tan} = 0. \tag{26.49}$$

The minimum is attained between $\lambda_{rad} = 0$ and $\lambda_{rad} = \gamma - \frac{1}{Z(\theta)}$, where the function tends to $+\infty$, and (26.48) and (26.49) give

---

[14] One uses a countable dense set of $(\theta^*, \lambda_{rad}, \lambda_{tan})$ satisfying (26.41).

[15] It has either $\theta^* = 0$, and $\lambda_{rad} = \lambda_{tan} = \gamma$, or $\theta^* = 1$, and $\lambda_{rad} = \lambda_{tan} = \beta$.

[16] Once a ball of radius r_* is equivalent to $\beta\,I$, one may replace it by the material β, and one is then led to create a geometry with the highest effective coefficient using only β and γ, one answer being a Hashin–Shtrikman coated sphere.

$$\frac{1}{\gamma-\lambda_{tan}} = \frac{\lambda_{rad}}{a^*_{\mathit{eff}}(\gamma-\lambda_{rad})}, Z(\theta) = \frac{1}{\gamma-\lambda_{rad}}\Big(1+\frac{(N-1)\,\lambda_{rad}}{a^*_{\mathit{eff}}}\Big), \tag{26.50}$$

and as $t \mapsto \frac{a^*_{\mathit{eff}}+(N-1)\,t}{\gamma-t}$ is increasing, it happens in $\big(\lambda_-(\theta),\lambda_+(\theta)\big)$ if

$$\begin{aligned}&\tfrac{1}{\gamma-\lambda_-(\theta)}+\tfrac{(N-1)\,\lambda_-(\theta)}{a^*_{\mathit{eff}}[\gamma-\lambda_-(\theta)]} < Z(\theta) < \tfrac{1}{\gamma-\lambda_+(\theta)}+\tfrac{(N-1)\,\lambda_+(\theta)}{a^*_{\mathit{eff}}[\gamma-\lambda_+(\theta)]},\\ &\tfrac{\lambda_-(\theta)}{a^*_{\mathit{eff}}[\gamma-\lambda_-(\theta)]} < \tfrac{1}{\gamma-\lambda_+(\theta)} \text{ and } \tfrac{1}{\gamma-\lambda_-(\theta)}+\tfrac{N-2}{\gamma-\lambda_+(\theta)} < \tfrac{(N-1)\,\lambda_+(\theta)}{a^*_{\mathit{eff}}[\gamma-\lambda_+(\theta)]},\end{aligned} \tag{26.51}$$

so that $a^*_{\mathit{eff}} > \frac{\lambda_-(\theta)\,[\gamma-\lambda_+(\theta)]}{\gamma-\lambda_-(\theta)} = \beta$, and $a^*_{\mathit{eff}} < \frac{(N-1)\,\lambda_+(\theta)}{N-2+\theta+(1-\theta)\,\frac{\beta}{\gamma}} < \gamma$. □

After the intuitive explanation of Graeme MILTON, I started a computation with François MURAT for the case $N=2$, using symmetric tensors whose eigenvectors make an angle with the radial direction, depending only upon r. It was a way to embed the radial laminates and the Schulgasser geometry into a larger family of geometries, and my hope was to find a way for the current to turn efficiently around a non-conducting core.

The idea was to look for solutions $u(x_1,x_2) = x_1 f(r) + x_2 g(r)$, to derive a differential system for the pair (f,g), and to use a complex notation for $f + i\,g$, expecting to find a complex Riccati equation for the effective equivalent conductivity; we stopped short of that, but with my students Sergio GUTIÉRREZ and Gregor WEISKE we carried the computations further in the fall of 1994.

In the Euclidean plane outside the origin, one uses as basis vectors $e_r = \frac{x}{r}$ and $e_\theta = R_{\pi/2}e_r$. Using polar coordinates and variational formulations,

$$\mathit{grad}(u) = \frac{\partial u}{\partial r}\,e_r + \frac{1}{r}\,\frac{\partial u}{\partial\theta}\,e_\theta, \tag{26.52}$$

$$\begin{aligned}&-\mathit{div}\big(A(x)\,\mathit{grad}(u)\big) = f \text{ in } r_1<r<r_2 \text{ means}\\ &\int_{r_1<r<r_2}\big(A\,\mathit{grad}(u),\mathit{grad}(v)\big)\,r\,dr\,d\theta = \int_{r_1<r<r_2} f\,v\,r\,dr\,d\theta\\ &\qquad\text{for all } v \text{ with compact support in } r_1<r<r_2.\end{aligned} \tag{26.53}$$

$$\begin{aligned}&\int_{r_1<r<r_2}\Big(\big(A\,\mathit{grad}(u),e_r\big)\,\frac{\partial v}{\partial r} + \big(A\,\mathit{grad}(u),e_\theta\big)\,\frac{1}{r}\,\frac{\partial v}{\partial\theta}\Big)\,r\,dr\,d\theta\\ &\qquad = \int_{r_1<r<r_2} f\,v\,r\,dr\,d\theta,\end{aligned} \tag{26.54}$$

$$-\frac{\partial\big(r\,(A\,\mathit{grad}(u),e_r)\big)}{\partial r} - \frac{\partial(A\,\mathit{grad}(u),e_\theta)}{\partial\theta} = r\,f \text{ in } r_1<r<r_2. \tag{26.55}$$

If A is discontinuous on a circle $r = \mathit{constant}$, both u and $(A\,\mathit{grad}(u),e_r)$ are continuous there. We considered tensors A depending only upon r when expressed in the basis (e_r,e_θ), i.e.,

$$A(x) = \begin{pmatrix} a_{11}(r) & a_{12}(r) \\ a_{21}(r) & a_{22}(r) \end{pmatrix} \text{ in the basis } (e_r, e_\theta), \tag{26.56}$$

$$\begin{aligned} A\,grad(u) &= \tfrac{\partial u}{\partial r}\,(a_{1,1}e_r + a_{2,1}e_\theta) + \tfrac{1}{r}\,\tfrac{\partial u}{\partial \theta}\,(a_{1,2}e_r + a_{2,2}e_\theta), \\ (A\,grad(u), e_r) &= a_{1,1}\,\tfrac{\partial u}{\partial r} + \tfrac{a_{1,2}}{r}\,\tfrac{\partial u}{\partial \theta}, \\ (A\,grad(u), e_\theta) &= a_{2,1}\,\tfrac{\partial u}{\partial r} + \tfrac{a_{2,2}}{r}\,\tfrac{\partial u}{\partial \theta}. \end{aligned} \tag{26.57}$$

We considered complex solutions u of the special form

$$u(r,\theta) = g_m(r)\,e^{i\,m\,\theta} \text{ in } r_1 < r < r_2, \tag{26.58}$$

whose real and imaginary parts give the solutions for the boundary conditions $u = \cos(m\,\theta)$ or $u = \sin(m\,\theta)$,[17] and for $f(x) = F(r)\,e^{i\,m\,\theta}$ the equation is

$$-\,(r\,a_{1,1}g'_m + i\,m\,a_{1,2}g_m)' - i\,m\left(a_{2,1}g'_m + i\,m\,a_{2,2}\,\frac{g_m}{r}\right) = r\,F, \tag{26.59}$$

where $'$ denotes $\frac{\partial}{\partial r}$, recalling that $r\,a_{1,1}g'_m + i\,m\,a_{1,2}g_m$ is continuous. If $F = 0$, one uses the unknown $A_{m,\mathit{eff}}$ defined by

$$A_{m,\mathit{eff}} = \frac{r\,a_{1,1}g'_m + i\,m\,a_{1,2}g_m}{g_m}, \text{ or } g'_m = \frac{A_{m,\mathit{eff}} - i\,m\,a_{1,2}}{r\,a_{1,1}}\,g_m, \tag{26.60}$$

and the equation in $r_1 < r < r_2$ becomes

$$\begin{aligned} 0 &= A'_{m,\mathit{eff}}\,g_m + A_{m,\mathit{eff}}\,g'_m + i\,m\,\big(a_{2,1}g'_m + i\,m\,\frac{a_{2,2}}{r}\,g_m\big) \\ &= A'_{m,\mathit{eff}}\,g_m + (A_{m,\mathit{eff}} + i\,m\,a_{2,1})\,g'_m - m^2\,\frac{a_{2,2}}{r}\,g_m, \end{aligned} \tag{26.61}$$

$$A'_{m,\mathit{eff}} + \frac{(A_{m,\mathit{eff}} + i\,m\,a_{2,1})\,(A_{m,\mathit{eff}} - i\,m\,a_{1,2})}{r\,a_{1,1}} - m^2\,\frac{a_{2,2}}{r} = 0. \tag{26.62}$$

Lemma 26.10. *Assuming that (26.56) holds with $a_{12} = a_{21}$ and $\alpha_0 I \le A(x) \le \beta_0 I$ in $r_1 < r < r_2$, with $0 < \alpha_0 \le \beta_0 < +\infty$, the restriction of A to the disc of radius r is equivalent to $a_{\mathit{eff}}(r)\,I$ with*

$$r\,a'_{\mathit{eff}} + \frac{a^2_{\mathit{eff}} - det(A)}{a_{1,1}} = 0 \text{ in } r_1 < r < r_2. \tag{26.63}$$

Proof. Due to the symmetry assumption, (26.61) becomes

$$A'_{m,\mathit{eff}} + \frac{A^2_{m,\mathit{eff}} - m^2 det(A)}{r\,a_{1,1}} = 0 \text{ in } r_1 < r < r_2, \tag{26.64}$$

[17] They are the traces of homogeneous harmonic polynomials of degree m.

and (26.63) is the case $m = \pm 1$, and as the same coefficient appears for data in cos θ or in sin θ, the effective tensor is isotropic. □

Turning the eigenvectors changes $a_{1,1}$ and in order to have $a_{eff}(r_2)$ maximum it seems natural to maximize a'_{eff} and in the case were $det(A) > a_{eff}^2$ one wants to minimize $a_{1,1}$, i.e., use a radial laminate.

Equation (26.64) for $A_{m,eff}$ can be solved for $r_1 = 0$ without a value for $A_{m,eff}(0)$, either assuming the uniform ellipticity of $A(r)$, or simply

$$\alpha^2 \leq det\big(A(r)\big) \leq \beta^2 \text{ in } 0 < r < r_2 \text{ with } 0 < \alpha \leq \beta < \infty, \tag{26.65}$$

since

$$|m|\,\alpha \leq A_{m,eff}(r_1) \leq |m|\,\beta \text{ implies } |m|\,\alpha \leq A_{m,eff}(r) \leq |m|\,\beta \text{ in } r_1 < r < r_2, \tag{26.66}$$

and once this uniform bound is obtained, the influence of the condition at r_1 becomes negligible when $r_1 \to 0$ as soon as

$$\int_0^{r_2} \frac{dr}{r\,a_{1,1}} = +\infty, \tag{26.67}$$

which tells one what degenerescence is allowed under (26.65).

Only the case $m = 1$ is of direct interest for computing the effective conductivity, but knowing $A_{m,eff}$ for $m \in \mathbb{Z}$ serves in computing the solution of a general Dirichlet problem. In order to compute the solution corresponding to $u(r_2, \theta) = e^{i\,m\,\theta}$ one solves (26.60) with the condition $g_m(r_2) = 1$. In order to compute the solution corresponding to a general function $u(r_2, \theta) = h(\theta)$ one decomposes h in Fourier series and one sums the corresponding solutions; for the convergence of the series, the general variational theory says that under the uniform ellipticity assumption the series converges in $H^1(r < r_2)$ if $h \in H^{1/2}(r = r_2)$, i.e.,

$$h(\theta) = \sum_{m \in \mathbb{Z}} h_m e^{i\,m\,\theta} \text{ with } \sum_{m \in \mathbb{Z}} |m|\,|h_m|^2 < \infty. \tag{26.68}$$

There are explicit solutions in the special case where $det(A)$ is a positive constant in $r_1 < r < r_2$, by using the new unknown $B_m(r)$ defined by

$$B_m = \frac{A_{m,eff} - m\sqrt{det(A)}}{A_{m,eff} + m\sqrt{det(A)}}, \tag{26.69}$$

so the equation becomes

$$
\begin{aligned}
B'_m &= \frac{2m\,\sqrt{det(A)}}{\big(A_{m,\mathit{eff}}+m\,\sqrt{det(A)}\big)^2}\,A'_{m,\mathit{eff}}\\
&= \frac{-2m\,\sqrt{det(A)}}{\big(A_{m,\mathit{eff}}+m\,\sqrt{det(A)}\big)^2}\,\frac{A^2_{m,\mathit{eff}}-m^2 det(A)}{r\,a_{1,1}} = -\frac{2m\,\sqrt{det(A)}}{r\,a_{1,1}}B_m,
\end{aligned}
\tag{26.70}
$$

giving, in that special case where $det(A)$ is a positive constant,

$$
\frac{A_{m,\mathit{eff}}(r_2) - m\,\sqrt{det(A)}}{A_{m,\mathit{eff}}(r_2) + m\,\sqrt{det(A)}} = \frac{A_{m,\mathit{eff}}(r_1) - m\,\sqrt{det(A)}}{A_{m,\mathit{eff}}(r_1) + m\,\sqrt{det(A)}}\,e^{-2m\,\sqrt{det(A)}\,\int_{r_1}^{r_2}\frac{dr}{r\,a_{11}}}.
\tag{26.71}
$$

Finally, I want to show a computation done with Gilles FRANCFORT during the year 1994–1995, which he spent at CMU (*Carnegie Mellon University*).[18] We considered a family of hypersurfaces indexed by $\rho\in(\rho_-,\rho_+)\subset\mathbb{R}$, $\Phi(x_1,\ldots,x_N)=\rho$, assuming that $grad(\rho)$ does not vanish, so that the normal $n=\frac{grad(\rho)}{|grad(\rho)|}$ is well defined. For $\rho_0\in(\rho_-,\rho_+)$, one fills the region $\rho\in(\rho_-,\rho_0)$ with an isotropic conductor of conductivity $a(\rho)$, and the region $\rho\in(\rho_0,\rho_+)$ with a possibly anisotropic conductor of conductivity $A^{\mathit{eff}}(\rho_0)$, and writing

$$
A(\rho)=\begin{cases} a(\rho)\,I \text{ for } \rho\in(\rho_-,\rho_0)\\ A^{\mathit{eff}}(\rho_0) \text{ for } \rho\in(\rho_0,\rho_+)\end{cases}
\tag{26.72}
$$

we wondered which functions Φ have the property that for any smooth positive function a there exists $A^{\mathit{eff}}(\rho_0)$ (symmetric positive definite) and N independent solutions of

$$
\begin{aligned}
&div\big(A(\rho)\,grad(u_j)\big)=0\\
&u_j(x)=x_j f_j^{\rho_0}(\rho),\ \text{with } f_j^{\rho_0}=1 \text{ for } \rho\in(\rho_0,\rho_+),
\end{aligned}
\tag{26.73}
$$

but we also added the condition that $e_1,\ldots,e_N$ is an orthonormal basis of eigenvectors of $A^{\mathit{eff}}(\rho_0)$,

$$
A^{\mathit{eff}}(\rho_0)\,e_j=\lambda_j^{\mathit{eff}}(\rho_0)e_j, j=1,\ldots,N,
\tag{26.74}
$$

and since we then let ρ_0 vary, a more general condition remains to be checked.[19] Forgetting the superscript ρ_0 for $f_j, j=1,\ldots,N$, one has

$$
\begin{aligned}
&grad(u_j)=f_j(\rho)\,e_j+x_j f'_j(\rho)\,grad(\rho)\\
&div\big(a(\rho)f_j(\rho)\,e_j+x_j a(\rho)f'_j(\rho)\,grad(\rho)\big)=0 \text{ for } \rho\in(\rho_0,\rho_+)\\
&x_j[(a\,f'_j)'\,|grad(\rho)|^2+a\,f'_j\Delta\,\rho]+[(a\,f_j)'+a\,f'_j]\,\frac{\partial\rho}{\partial x_j}=0
\end{aligned}
\tag{26.75}
$$

[18] I forgot the detail of these computations, but Gilles FRANCFORT sent me some handwritten notes (in French), dated January 1995.

[19] Like for the computations shown in (26.52)–(26.71), one could consider N independent solutions of the form $u_j(x)=\sum_{k=1}^N x_k f_{j,k}(\rho)$, and accept that the basis of eigenvectors of $A^{\mathit{eff}}(\rho)$ varies with ρ.

where $' = \frac{d}{d\rho}$, and for the interface condition at $\rho = \rho_0$, one has

$$\begin{aligned}&(a(\rho)grad(u_j), n) = (A^{\mathit{eff}}(\rho_0)e_j, n) = \lambda_j^{\mathit{eff}}(e_j, n) \text{ at } \rho = \rho_0\\&a(\rho_0)\Big[f_j(\rho_0)\,\tfrac{\partial\rho}{\partial x_j} + x_j f_j'(\rho_0)\,|grad(\rho)|^2\Big] = \lambda_j^{\mathit{eff}}\,\tfrac{\partial\rho}{\partial x_j},\end{aligned} \tag{26.76}$$

so that

$$\begin{aligned}&\tfrac{\partial\rho}{\partial x_j} = x_j|grad(\rho)|^2F_j(\rho), j = 1,\ldots,N,\\&F_j(\rho_0) = \tfrac{a(\rho_0)f_j'(\rho_0)}{\lambda_j^{\mathit{eff}} - a(\rho_0)f_j(\rho_0)},\end{aligned} \tag{26.77}$$

and since $|grad(\rho)|$ is assumed finite, F_j must be finite, but it could be indeterminate if f_j' vanishes with $\lambda_j^{\mathit{eff}} - a\,f_j$, and in that case one uses (26.75), which gives $F_j = -\frac{a\,f_j''}{a'f_j}$, so that by choosing a strictly monotone one obtains (26.77) for some functions $F_1,\ldots,F_N$, whose precise form is not needed (and it will be determined). One deduces from (26.77) that $\frac{\partial\rho}{\partial x_j} = x_j G\,F_j$ with

$$G = |grad(\rho)|^2 = \frac{1}{\sum_{k=1}^N x_k^2F_k^2}, \tag{26.78}$$

and it gives

$$\begin{aligned}&\tfrac{\partial^2\rho}{\partial x_i\partial x_j} = \delta_{i,j}G\,F_j + x_jG\,F_j'x_iG\,F_i\\&\quad -x_jF_jG^2\big(2x_iF_i^2 + \big(\textstyle\sum_{k=1}^N 2x_k^2F_kF_k'\big)x_iG\,F_i\big),\end{aligned} \tag{26.79}$$

which must be symmetric in i and j; as $\delta_{i,j}G\,F_j = \delta_{i,j}G\,F_i$, one deduces that

$$\begin{aligned}&F_j'F_i - 2F_jF_i^2 = F_i'F_j - 2F_iF_j^2 \text{ for } i \neq j,\\&\tfrac{F_j'}{F_j} + 2F_j = \tfrac{F_i'}{F_i} + 2F_i = H\\&\text{independent of } i,j = 1,\ldots,N.\end{aligned} \tag{26.80}$$

One deduces that $\varphi_i = \frac{1}{F_i}$ satisfies

$$\varphi_i' + H\,\varphi_i = 2, i = 1,\ldots,N, \tag{26.81}$$

so that if $K' = H$, and $L' = e^K$, one multiplies by e^K and one obtains

$$\begin{aligned}&\varphi_i = c_ie^{-K} + 2L\,e^{-K}, i = 1,\ldots,N,\\&F_i = \tfrac{e^K}{c_i+2L},\\&\tfrac{1}{G} = e^{2K}\big(\textstyle\sum_{k=1}^N \frac{x_k^2}{(c_k+2L)^2}\big).\end{aligned} \tag{26.82}$$

Finally

$$\begin{aligned}\frac{\partial}{\partial x_j}\left(\sum_{k=1}^{N}\frac{x_k^2}{c_k+2L}\right) &= \frac{2x_j}{c_j+2L}-\sum_{k=1}^{N}\frac{2x_k^2}{(c_k+2L)^2}L'x_jG\,F_j,\\ &= \frac{2x_j}{c_j+2L}-2\frac{e^{-2K}}{G}\,e^K x_jG\,\frac{e^K}{c_j+L},\\ &= 0 \text{ for } j=1,\ldots,N,\end{aligned} \tag{26.83}$$

showing that

$$\sum_{k=1}^{N}\frac{x_k^2}{c_k+2L} = constant, \tag{26.84}$$

and the family of hypersurfaces are actually confocal ellipsoids (with a different parametrization).
Additional footnotes: BEN-GURION,[20] HOUSTON,[21] KOOPMANS,[22] René THOM,[23] THOMSON E.[24]

---

[20] David BEN-GURION, Polish-born Israeli statesman, 1886–1973. He was the first Prime Minister of Israel in 1948. Ben-Gurion University of the Negev, and Ben-Gurion International Airport, Lod (near Tel-Aviv), Israel, are named after him.

[21] Edwin James HOUSTON, American engineer, 1847–1914. With E. THOMSON, he founded the Thomson–Houston Electric Company in 1879.

[22] Tjalling Charles KOOPMANS, Dutch-born economist, 1910–1985 He received the Sveriges Riksbank Prize in Economic Sciences in Memory of Alfred Nobel in 1975, jointly with Leonid Vitalyevich KANTOROVICH, for their contributions to the theory of optimum allocation of resources. He worked at Yale University, New Haven, CT.

[23] René Frédéric THOM, French mathematician, 1923–2002. He received the Fields Medal in 1958 for his work in topology. He worked in Grenoble, in Strasbourg, and at IHES (Institut des Hautes Études Scientifique) at Bures-sur-Yvette, France.

[24] Elihu THOMSON, English-born engineer, 1853–1937. With E. J. HOUSTON, he founded the Thomson–Houston Electric Company in 1879.

Chapter 27
Laminations Again, and Again

In the spring of 1982, I gave an introductory course to homogenization for researchers at École Polytechnique, Palaiseau, France; students from École Polytechnique could follow it as an optional course, and I needed to give an assignment to two students, Philippe BRAIDY and Didier POUILLOUX.[1,2] I asked them to make a numerical study of the set attainable by successive laminations for comparing it with the optimal set characterized by François MURAT and myself. I thought the set obtained by repeated laminations to be different, i.e., strictly smaller, and our construction with confocal ellipsoids unavoidable,[3] so that I was surprised that they reported that the two sets looked alike; a few days after, they provided a proof that the two sets are equal.

Lemma 27.1. *If one laminates A in proportion θ and $\beta\,I$ in proportion $1-\theta$ in direction e_j and A is diagonal with eigenvalues $a_1,\ldots,a_N$, the result B is diagonal with eigenvalues $b_1,\ldots,b_N$, and*

$$\begin{aligned} &b_i = \theta\,a_i + (1-\theta)\,\beta \text{ for } i \neq j, \text{ i.e., } \tfrac{1}{\beta-b_i} = \tfrac{1}{\theta}\,\tfrac{1}{\beta-a_i},\\ &b_j = \left(\tfrac{\theta}{a_j} + \tfrac{1-\theta}{\beta}\right)^{-1}, \text{ i.e., } \tfrac{1}{\beta-b_j} = \tfrac{1}{\theta}\,\tfrac{1}{\beta-a_j} - \tfrac{1-\theta}{\theta\,\beta}\\ &\textstyle\sum_k \tfrac{1}{\beta-b_k} - \tfrac{1}{\beta} = \tfrac{1}{\theta}\left(\sum_k \tfrac{1}{\beta-a_k} - \tfrac{1}{\beta}\right). \end{aligned} \tag{27.1}$$

Proof. Of course, it means that for a sequence of characteristic functions χ_n depending only upon x_j and such that $\chi_n \rightharpoonup \theta$ in L^∞ weak $\star$, $\chi_n A + (1-\chi_n)\,\beta\,I$ H-converges to B. The formulas for $b_1,\ldots,b_N$ are those of Lemma 5.2, and one just checks what $\frac{1}{\beta-b_k}$ is. □

[1] Philippe BRAIDY, French engineer, born in 1960.

[2] Didier POUILLOUX, French engineer.

[3] It was silly of me to think that, since there is a high degree of nonuniqueness in the construction, which uses an arbitrary Vitali covering.

L. Tartar, *The General Theory of Homogenization*, Lecture Notes of the Unione Matematica Italiana 7, DOI 10.1007/978-3-642-05195-1_27,

Corollary 27.2. *If one starts with αI and one laminates successively with βI in proportion $1-\theta_j$ in direction e_j, for $j=1,\ldots,N$, and $\theta=\theta_1\cdots\theta_N$, the result M is diagonal with eigenvalues $\lambda_1,\ldots,\lambda_N$ with*

$$\begin{aligned}&\lambda_-(\theta)\le\lambda_j\le\lambda_+(\theta), j=1,\ldots,N,\\ &\textstyle\sum_k\frac{1}{\beta-\lambda_k}=\frac{(N-1+\theta)\,\beta+(1-\theta)\,\alpha}{\theta\,\beta\,(\beta-\alpha)}.\end{aligned}\tag{27.2}$$

Conversely, if (27.2) holds, there is a unique choice of $\theta_1,\ldots,\theta_N\in(0,1)$.

Proof. Notice that the second line of (27.2) corresponds to the equality in (21.42). By repeated applications of Lemma 27.1, one has

$$\begin{aligned}\tfrac{1}{\beta-\lambda_k}&=\tfrac{1}{\theta\,(\beta-\alpha)}-\tfrac{1-\theta_k}{\beta\,\theta_k\cdots\theta_N}, k=1,\ldots,N\\ \textstyle\sum_k\frac{1}{\beta-\lambda_k}&=\tfrac{1}{\beta}+\tfrac{1}{\theta}\left(\tfrac{N}{\beta-\alpha}-\tfrac{1}{\beta}\right),\end{aligned}\tag{27.3}$$

and the second line of (27.3) gives the second line of (27.2). Furthermore

$$d_k=\frac{1-\theta_k}{\beta\,\theta_k\cdots\theta_N}\ge 0, k=1,\ldots,N,\text{ and }\sum_k d_k=\frac{1-\theta}{\theta\,\beta},\tag{27.4}$$

the value of the sum following from (27.3). $\frac{1}{\beta-\lambda_k}\le\frac{1}{\theta\,(\beta-\alpha)}$ means $\lambda_k\le\lambda_+(\theta)$, and $\frac{1}{\beta-\lambda_k}\ge\frac{1}{\theta\,(\beta-\alpha)}-\frac{1-\theta}{\theta\,\beta}$ means $\lambda_k\ge\lambda_-(\theta)$. Conversely, given $d_1,\ldots,d_N\ge 0$ with sum $\frac{1-\theta}{\theta\,\beta}$, one defines $\theta_1,\ldots,\theta_k$ by

$$\theta\,\beta\,(d_1+\cdots+d_k)=1-\theta_1\cdots\theta_k, k=1,\cdots,N,\tag{27.5}$$

and this defines $\theta_1,\ldots,\theta_N\in(0,1)$, with product θ. □

The proof of Philippe BRAIDY and Didier POUILLOUX is elementary, and could have been found by Antonio MARINO and Sergio SPAGNOLO or by François MURAT and myself when we first computed with laminations in the early 1970s. However, the fact that the mapping $\lambda\mapsto\frac{1}{\beta-\lambda}$ is useful to consider comes from the form of the optimality condition, which we only found in 1980. One may ponder if I could have found this proof in June 1980 if I was asked to show that my bounds are attained, since I only knew the lamination method for constructing new materials at the time. This shows that discovery is a very sensitive issue, and that being told a line of proof may become a handicap.

My advisor told me to read articles by only looking at statements of theorems and trying to supply my own proofs, and there is an interesting anecdote in that respect about Alberto CALDERÓN, and Antoni ZYGMUND.[4]

---

[4] The MacTutor History of Mathematics archive mentions that Antoni ZYGMUND gave a talk in 1948 at the University of Buenos Aires, Argentina, and asked a question

I noticed that for the case of an isotropic result, i.e., when one considers the (upper) Hashin–Shtrikman bound,[5] the repeated lamination construction corresponds to a function $g(z)$ in $\mathbb{C}\setminus(-\infty,0]$, identical to that of the Hashin–Shtrikman coated spheres construction. This comes from the fact that all the d_k are equal in the case of an isotropic material, so that the common value is $\frac{1-\theta}{\theta\,\beta\,N}$, which is independent of α, so that the θ_k do not depend upon α, which can then be taken in $\mathbb{C}\setminus(-\infty,0]$.

I shall show in Chap. 33 the functions $F(\cdot,M^1,M^2)$ for more general repeated lamination constructions, and the result in this case is not independent of which orthonormal basis is used; similar functions for the Hashin–Shtrikman coated sphere constructions (or the coated ellipsoids) are not known, and they may actually depend upon which Vitali covering is used.

During the spring of 1983, while I visited MSRI in Berkeley, CA, I decided to compute the formula for laminating general materials in arbitrary directions, having in mind to reiterate the procedure.[6] I wanted to use tensors A and B in proportions θ and $1-\theta$ and laminate perpendicularly to e, and the formula of Lemma 5.2 for the effective tensor C corresponds to

$$\begin{aligned}
\frac{1}{(C\,e,e)} &= \frac{\theta}{(A\,e,e)} + \frac{1-\theta}{(B\,e,e)},\\
\frac{(C\,f,e)}{(C\,e,e)} &= \frac{\theta\,(A\,f,e)}{(A\,e,e)} + \frac{(1-\theta)\,(B\,f,e)}{(B\,e,e)} \text{ for all } f\perp e,\\
\frac{(C\,e,f)}{(C\,e,e)} &= \frac{\theta\,(A\,e,f)}{(A\,e,e)} + \frac{(1-\theta)\,(B\,e,f)}{(B\,e,e)} \text{ for all } f\perp e,\\
(C\,f,g) - \frac{(C\,f,e)\,(C\,e,g)}{(C\,e,e)} &= \theta\left((A\,f,g) - \frac{(A\,f,e)\,(A\,e,g)}{(A\,e,e)}\right)\\
&+ (1-\theta)\left((B\,f,g) - \frac{(B\,f,e)\,(B\,e,g)}{(B\,e,e)}\right) \text{ for all } f\perp e, g\perp e,
\end{aligned} \tag{27.6}$$

and I looked for an intrinsic formulation. I observed that for θ small it implies $C = B + \theta\,F(A,B,e) + o(\theta)$, which suggests to writing a differential equation $M' = F(A,M,e)$ with $M(0) = B$, and hopefully integrating it explicitly. In other words, for e fixed, increasing the proportion of A from 0 to 1 creates

which puzzled Alberto CALDERÓN, who told the speaker that his own book *Trigonometric Series* contained the answer, but Antoni ZYGMUND disagreed that he had a proof of that in his book. After discussion, it appeared that Alberto CALDERÓN only read the statement of a theorem, and supplied his own proof, obviously more general than that in the book since it also answered the question that Antoni ZYGMUND just asked; without knowing the proof in the book, he (wrongly) assumed that it was the same as his.

[5] Of course, one may exchange the roles of α and β and consider the lower Hashin–Shtrikman bound.

[6] The cases used by Philippe BRAIDY and Didier POUILLOUX are the simplest: commuting symmetric positive definite tensors, with the direction of lamination being a common eigenvector.

a curve going from B to A in the space of tensors, and I first computed this curve as the solution of a differential equation, easy to write down. This gives[7]

$$\begin{aligned}
(C\,e,e) &= (B\,e,e)+\theta\left((B\,e,e)-\tfrac{(B\,e,e)^2}{(A\,e,e)}\right)+o(\theta),\\
(C\,f,e) &= (B\,f,e)+\theta\left(\tfrac{(A\,f,e)\,(B\,e,e)}{(A\,e,e)}-\tfrac{(B\,f,e)\,(B\,e,e)}{(A\,e,e)}\right)+o(\theta),\\
(C\,e,g) &= (B\,e,g)+\theta\left(\tfrac{(A\,e,g)\,(B\,e,e)}{(A\,e,e)}-\tfrac{(B\,e,g)\,(B\,e,e)}{(A\,e,e)}\right)+o(\theta),\\
(C\,f,g) &= (B\,f,g)+\theta\left((A\,f,g)-(B\,f,g)\right)\\
&\quad-\theta\,\tfrac{[(B\,f,e)-(A\,f,e)]\,[(B\,e,g)-(A\,e,g)]}{(A\,e,e)}+o(\theta),
\end{aligned} \tag{27.7}$$

and the form of $(C\,f,g)$ suggests the formula

$$C = B+\theta\Big(A-B-(B-A)\frac{e\otimes e}{(A\,e,e)}(B-A)\Big)+o(\theta), \tag{27.8}$$

which one checks to be compatible with the rest of (27.7). For e and A given, (27.8) corresponds to the differential equation

$$M' = A-M-(M-A)\frac{e\otimes e}{(A\,e,e)}(M-A), \tag{27.9}$$

and if $M-A$ is invertible,[8] (27.9) can be written as

$$\big((M-A)^{-1}\big)' = -(M-A)^{-1}M'(M-A)^{-1} = (M-A)^{-1}+\frac{e\otimes e}{(A\,e,e)}, \tag{27.10}$$

which is a linear equation in $(M-A)^{-1}$. Using τ as variable, and assuming that $\tau=0$ corresponds to B, the solution is

$$(M-A)^{-1} = -\frac{e\otimes e}{(A\,e,e)}+e^\tau\Big((B-A)^{-1}+\frac{e\otimes e}{(A\,e,e)}\Big), \tag{27.11}$$

and if M corresponds to using proportion $\eta(\tau)$ of A and $1-\eta(\tau)$ of B, then for θ small $\eta(\tau+\theta)=\theta+(1-\theta)\,\eta(\tau)+o(\theta)$ gives $\eta'=1-\eta$ and therefore $\eta=1-e^{-\tau}$ or equivalently $e^\tau=\frac{1}{1-\eta}$ for proportion η of A, giving

$$(M-A)^{-1} = \frac{(B-A)^{-1}}{1-\eta}+\frac{\eta}{1-\eta}\,\frac{e\otimes e}{(A\,e,e)} \text{ for proportion } \eta \text{ of } A. \tag{27.12}$$

---

[7] I use the Euclidean structure of $\mathbb{R}^N$, but it can be avoided by denoting $\mathcal{E}$ the ambient vector space, taking e as an element of the dual $\mathcal{E}'$, where $grad(u)$ lives, and considering the tensors A, B, as elements of $\mathcal{L}(\mathcal{E}',\mathcal{E})$, which is also the case of $e\otimes e$, which appears in some formulas.

[8] If $(B-A)\,z=0$ for a nonzero vector z, (27.9) implies $(M-A)\,z=0$, and one must reinterpret the equations involving $(M-A)^{-1}$.

Of course, exchanging the role of A and B and changing η into $1-\eta$,

$$(M-B)^{-1} = \frac{(A-B)^{-1}}{\eta} + \frac{1-\eta}{\eta}\,\frac{e\otimes e}{(B\,e,e)} \text{ for proportion } \eta \text{ of } A. \quad (27.13)$$

With (27.12), I easily reiterated the lamination process with various directions of lamination, with each lamination using the material with tensor A,[9] and it gave Lemma 27.3, generalizing the formula obtained by Philippe BRAIDY and Didier POUILLOUX in the special case where A and B have a common basis of eigenvectors and each e is one of these common eigenvectors.

Lemma 27.3. *For $\eta \in (0,1)$, let $\xi_1,\ldots,\xi_p > 0$ with $\sum_j \xi_j = 1-\eta$, let $e_1,\ldots,e_p \in \mathbb{R}^N \setminus \{0\}$, then using materials with tensors A and B in proportions η and $1-\eta$, one can construct by repeated lamination the material with tensor M such that*

$$(M-B)^{-1} = \frac{(A-B)^{-1}}{\eta} + \frac{1}{\eta}\Big(\sum_{j=1}^{p} \xi_j\,\frac{e_j\otimes e_j}{(B\,e_j,e_j)}\Big). \quad (27.14)$$

Proof. Of course, one assumes $B-A$ invertible, since the formula must be reinterpreted if $B-A$ is not invertible. One starts from $M_0 = A$ and by induction one constructs M_j by laminating M_{j-1} and B in proportions η_j and $1-\eta_j$, with lamination orthogonal to e_j. Formula (27.13) gives

$$(M_j-B)^{-1} = \frac{(M_{j-1}-B)^{-1}}{\eta_j} + \frac{1-\eta_j}{\eta_j}\,\frac{e_j\otimes e_j}{(B\,e_j,e_j)} \text{ for } j=1,\ldots,p, \quad (27.15)$$

which is adapted to reiteration and provides (27.14) with

$$\begin{aligned} &\eta = \eta_1\cdots\eta_p \\ &\xi_1 = 1-\eta_1, \xi_j = \eta_1\cdots\eta_{j-1}(1-\eta_j) \text{ for } j=1,\ldots,p, \end{aligned} \quad (27.16)$$

which gives $\xi_1+\ldots+\xi_j = 1-\eta_1\cdots\eta_j$ for $j=1,\ldots,p$, and this defines in a unique way η_j for $j=1,\ldots,p$. ◻

The preceding computations do not require A or B to be symmetric. The characterization of the sum $\sum_j \xi_j \frac{e_j\otimes e_j}{(B\,e_j,e_j)}$ for all $\xi_j>0$ with sum $1-\eta$ and all nonzero vectors e_j only depends upon the symmetric part of B (and η).

[9] With (27.13), each lamination must use the material with tensor B.

Lemma 27.4. *If B is symmetric positive definite then for $\xi_1, \ldots, \xi_p > 0$ and nonzero vectors $e_1, \ldots, e_p$, one has*

$$\begin{aligned}\textstyle\sum_{j=1}^p \xi_j \frac{e_j \otimes e_j}{(B\,e_j, e_j)} = B^{-1/2}\,K\,B^{-1/2}, \text{ with}\\ K \geq 0 \text{ symmetric, and } trace(K) = \textstyle\sum_{j=1}^p \xi_j,\end{aligned} \tag{27.17}$$

and conversely, any such K can be obtained in this way.

Proof. Putting $e_j = B^{-1/2} f_j$ for $j = 1, \ldots, p$, one has $K = \sum_j \xi_j \frac{f_j \otimes f_j}{|f_j|^2}$, and each $\frac{f_j \otimes f_j}{|f_j|^2}$ is a nonnegative symmetric tensor with trace 1, and (27.17) follows. Conversely, if K is a symmetric nonnegative tensor with trace equal to S, then there is an orthonormal basis of eigenvectors $f_1, \ldots, f_N$, with $K\,f_j = \kappa_j\,f_j$ and $\kappa_j \geq 0$ for $j = 1, \ldots, N$, and $\sum_j \kappa_j = S$, so that $K = \sum_j \kappa_j\,f_j \otimes f_j$. □

Using Lemma 27.3 and Lemma 27.4, with $A = \alpha\,I$ and $B = \beta\,I$, one can construct materials with a symmetric tensor M with eigenvalues $\lambda_1, \ldots, \lambda_N$, and (27.14) with Lemma 27.4 mean

$$\begin{aligned}\tfrac{1}{\lambda_j - \beta} \geq \tfrac{1}{\eta(\alpha-\beta)} \text{ for } j = 1, \ldots, N\\ \textstyle\sum_{j=1}^N \frac{1}{\lambda_j - \beta} = \frac{N}{\eta\,(\alpha-\beta)} + \frac{1-\eta}{\eta\,\beta},\end{aligned} \tag{27.18}$$

i.e., $\lambda_j \leq \lambda_+(\eta)$ for $j = 1, \ldots, N$, and the second part of (27.18) is the same as the second part of (27.2), which implies $\lambda_j \geq \lambda_-(\eta)$ for $j = 1, \ldots, N$. Of course, by exchanging the roles of A and B one can obtain another part of the boundary of possible effective tensors.

Formula (27.14) corresponds to a special case of

$$(M - B)^{-1} = \frac{(A - B)^{-1}}{\eta} + \frac{1 - \eta}{\eta} \int_{\mathbb{S}^{N-1}} \frac{e \otimes e}{(B\,e, e)}\,d\nu(e), \tag{27.19}$$

for a probability measure ν on the unit sphere $\mathbb{S}^{N-1}$, obtained as a limit of (27.14) for the atomic measures $\sum_{j=1}^p \frac{\xi_j}{1-\eta}\,\delta_{e_j}$. Such integrals over $\mathbb{S}^{N-1}$ appear naturally in the theory of *H-measures*, and I describe them in Chap. 33.

I noticed afterward that it is related to using relaxation for the differential equation (27.9), considering the choice of $e \in \mathbb{S}^{N-1}$ as a control, but then one may also consider A as part of the control, so that if one considers a set $\mathcal{A}$ of possible A, one can use a probability measure on $\mathbb{S}^{N-1} \times \mathcal{A}$ and consider

$$M' = \int_{\mathbb{S}^{N-1}\times\mathcal{A}} \Big(A - M - (M - A)\frac{e \otimes e}{(A\,e, e)}(M - A)\Big)\,d\mu(e, A), \tag{27.20}$$

for a probability measure μ on $\mathbb{S}^{N-1} \times \mathcal{A}$; one can let μ vary with time τ, keeping track of how much of each $A \in \mathcal{A}$ one uses. I hoped that this trick would give more characterizations of effective coefficients, and I described this method at a meeting at IMA, Minneapolis, MN, in the spring of 1985.[10]

After arriving at Lemma 27.1 by differential equations, I proved similar results directly (still in the spring of 1983, while at MSRI in Berkeley, CA).

Lemma 27.5. *Laminating orthogonally to e materials with tensors A and B in proportions η and $1-\eta$ gives an effective tensor C given by*

$$C = \eta\,A + (1-\eta)\,B - \eta\,(1-\eta)(B-A)\,\frac{e\otimes e}{(1-\eta)\,(A\,e,e) + \eta\,(B\,e,e)}\,(B-A). \tag{27.21}$$

Proof. One constructs a sequence of characteristic functions χ_n depending on (x,e), with $\chi_n \rightharpoonup \eta$ in $L^\infty(\mathbb{R})$ weak $\star$, and $A^n = \chi_n\,A + (1-\chi_n)\,B$. For $E^\infty \in \mathbb{R}^N$, one constructs $E^n = grad(u_n)$, depending on (x,e), with $E^n \rightharpoonup E^\infty$ in $L^2_{loc}(\mathbb{R}^N;\mathbb{R}^N)$ weak and $div\big(A^n\,grad(u_n)\big) = 0$, and one computes the limit in $L^2_{loc}(\mathbb{R}^N;\mathbb{R}^N)$ weak of $D^n = A^n\,grad(u_n)$, which is $D^\infty = C\,E^\infty$, with C given by (27.21). For doing that, one looks for $E_A, E_B \in \mathbb{R}^N$ with

$$\begin{aligned} &E^n = \chi_n\,E_A + (1-\chi_n)\,E_B \text{ and } \eta\,E_A + (1-\eta)\,E_B = E^\infty,\\ &D^n = \chi_n\,A\,E_A + (1-\chi_n)\,B\,E_B, \end{aligned} \tag{27.22}$$

and the constraints $curl(E^n) = div(D^n) = 0$ become

$$E_B - E_A = c\,e \text{ and } (B\,E_B - A\,E_A, e) = 0, \tag{27.23}$$

and then one should have

$$\eta\,A\,E_A + (1-\eta)\,B\,E_B = C\,E^\infty. \tag{27.24}$$

One then chooses

$$E_A = E^\infty + c_A\,e;\; E_B = E^\infty + c_B\,e;\; \eta\,c_A + (1-\eta)\,c_B = 0, \tag{27.25}$$

and (27.23) requires that

$$\big((B-A)E^\infty, e\big) + c_B(B\,e,e) - c_A(A\,e,e) = 0, \tag{27.26}$$

and (27.25) and (27.26) give

$$\begin{aligned} &\big((1-\eta)(A\,e,e) + \eta\,(B\,e,e)\big)c_A = (1-\eta)\,\big((B-A)\,E^\infty, e\big)\\ &\big((1-\eta)(A\,e,e) + \eta\,(B\,e,e)\big)c_B = -\eta\,\big((B-A)\,E^\infty, e\big), \end{aligned} \tag{27.27}$$

[10] I only mentioned it in writing for a meeting at LANL (Los Alamos National Laboratory), Los Alamos, NM, in January 1987; it seems that what I wrote became a part of an internal report, whose reference I could not obtain.

so that (27.24) becomes

$$
\begin{aligned}
C\,E^\infty &= (\eta\,A + (1-\eta)\,B)\,E^\infty \\
&+ \tfrac{((B-A)\,E^\infty, e)}{(1-\eta)\,(A\,e,e)+\eta\,(B\,e,e)}\,(\eta\,(1-\eta)\,A\,e - \eta\,(1-\eta)\,B\,e),
\end{aligned}
\tag{27.28}
$$

and since (27.28) is true for every $E^\infty \in R^N$, one deduces (27.21). □

A result of linear algebra then proves (27.12) and (27.13) from (27.21).

Lemma 27.6. *If $M \in \mathcal{L}(\mathcal{E};\mathcal{F})$ is invertible, if $a \in \mathcal{F}, b \in \mathcal{E}'$ and $(M^{-1}a, b) \neq -1$, then $M + a \otimes b$ is invertible, with*

$$
(M + a \otimes b)^{-1} = M^{-1} - \frac{M^{-1}(a \otimes b)\,M^{-1}}{1 + (M^{-1}a, b)}.
\tag{27.29}
$$

Proof. One wants to solve $(M + a \otimes b)\,x = y$, i.e., $M\,x + a\,(b,x) = y$, so that $x = M^{-1}\,y - t\,M^{-1}\,a$ with $t = (b,x)$, but one then needs to have $t = (b, M^{-1}\,y) - t\,(M^{-1}\,a, b)$, which is possible since $(M^{-1}\,a, b) \neq -1$, and gives $x = M^{-1}\,y - M^{-1}\,a\,\frac{(b, M^{-1}\,y)}{1+(M^{-1}\,a,b)}$, and since y is arbitrary it gives (27.29). □

In the late 1980s, I characterized with François MURAT the possible *H-measures* for a weakly $\star$ converging sequence of characteristic functions, and we used the more general lamination formula of Lemma 27.7, and the formula for small-amplitude homogenization that I describe in Chap. 29. I cited our result in [105], and then I mentioned more general relations between *Young measures* and *H-measures* that we obtained afterward, for a conference in the fall of 1991 in Ferrara, Italy, for the 600th anniversary of the University of Ferrara [109], and for a conference in Udine, Italy, in the summer of 1994 [110]. The details of our construction of admissible pairs of a Young measure and an H-measure associated with a sequence was never published, and I shall sketch a little more about this question in Chap. 33, but the starting point is the analogue (27.30) of (27.21), when one laminates r different materials.

Lemma 27.7. *Laminating orthogonally to e materials with tensors $M^1, \ldots, M^r$, in proportions $\eta_1, \ldots, \eta_r$, gives an effective tensor M^{eff} given by*

$$
\begin{aligned}
M^{eff} &= \textstyle\sum_{i=1}^r \eta_i\,M^i - \sum_{1\leq i<j\leq r} \eta_i\eta_j(M^i - M^j)R^{i,j}(M^i - M^j) \\
R^{i,j} &= \tfrac{1}{(M^i\,e,e)}\,\tfrac{e\otimes e}{H}\,\tfrac{1}{(M^j\,e,e)} \textit{ for } i,j = 1,\ldots,r \\
H &= \textstyle\sum_{k=1}^r \frac{\eta_k}{(M^k\,e,e)}.
\end{aligned}
\tag{27.30}
$$

Proof. Like for the proof of Lemma 27.5, one uses

$$
E^i = E^\infty + c_i\,e \text{ in the material } i, \text{ for } i = 1,\ldots,r, \text{ and } \sum_{i=1}^r \eta_i\,c_i = 0,
\tag{27.31}
$$

and one must have $(M^i E^i, e) = (M^j E^j, e)$ if there is an interface between material i and material j, so that there exists a constant C with

$$(M^i E^i, e) = C \text{ for } i = 1, \ldots, r. \tag{27.32}$$

With the definition (27.31) of $E^i, i = 1, \ldots, r$, (27.32) implies

$$c_i = \frac{C - (M^i E^\infty, e)}{(M^i e, e)} \text{ for } i = 1, \ldots, r, \tag{27.33}$$

and the condition $\sum_i \eta_i c_i = 0$ gives

$$H\, C = \sum_{i=1}^{r} \eta_i \, \frac{(M^i E^\infty, e)}{(M^i e, e)}, \tag{27.34}$$

with H given in (27.30). Using (27.32) one obtains

$$\begin{aligned} H\,(M^i e, e)\, c_i &= \Big(\textstyle\sum_{j=1}^r \eta_j \, \frac{(M^j E^\infty, e)}{(M^j e, e)}\Big) - \frac{(M^i E^\infty, e)}{(M^i e, e)} \\ &= \textstyle\sum_{j=1}^r \eta_j \, \frac{\big((M^j - M^i)\, E^\infty, e\big)}{(M^j e, e)} \text{ for } i = 1, \ldots, r. \end{aligned} \tag{27.35}$$

This gives

$$\begin{aligned} M^{eff}\, E^\infty &= \textstyle\sum_{i=1}^r \eta_i \, M^i E^i = \big(\sum_{i=1}^r \eta_i \, M^i\big)\, E^\infty \\ &+ \textstyle\frac{1}{H} \sum_{i=1}^r \frac{\eta_i}{(M^i e, e)} \Big(\sum_{j=1}^r \eta_j \, \frac{((M^j - M^i)\, E^\infty, e)}{(M^j e, e)}\Big)\, M^i e \\ &= \textstyle\big(\sum_{i=1}^r \eta_i \, M^i\big)\, E^\infty - \frac{1}{2H} \sum_{i,j=1}^r \eta_i \eta_j \, \frac{\big((M^i - M^j) E^\infty, e\big)}{(M^i e, e)\, (M^j e, e)} \, (M^i - M^j)\, e \\ &= \textstyle\big(\sum_{i=1}^r \eta_i \, M^i\big)\, E^\infty - \frac{1}{H} \sum_{i<j} \eta_i \eta_j \, (M^i - M^j) \, \frac{e \otimes e}{(M^i e, e)\, (M^j e, e)} \, (M^i - M^j)\, E^\infty, \end{aligned} \tag{27.36}$$

proving (27.30). □

Chapter 28
Wave Front Sets, H-Measures

In the summer of 1972, I listened to a conference on partial differential equations in Jerusalem, Israel. It was the first time that I heard Lars HÖRMANDER talk,[1] and his work was related to lacunas,[2] which is about identifying the exact support of the elementary solution E of an hyperbolic equation having support in $t \geq 0$; he introduced a new notion, the wave front set of a distribution $T \in \mathcal{D}'(\Omega)$,[3] denoted $WF(T)$ and also called the essential singular support of T, for which he proved propagation results, which enabled him to identify $WF(E)$ (but not the support of E).[4]

Laurent SCHWARTZ defined the singular support of T as the complement of the largest open set $\omega \subset \Omega \subset \mathbb{R}^N$ such that the restriction of T to ω belongs to $C^\infty(\omega)$, and the existence of ω follows from the existence of C^∞ partitions of unity, but Lars HÖRMANDER introduced a more precise notion of microlocal regularity, which is not seen in Ω but in $\Omega \times (\mathbb{R}^N \setminus \{0\})$: T is microlocally regular at (x_0, ξ_0) if there exists $\varphi \in C_c^\infty(\Omega)$ with $\varphi(x_0) \neq 0$ such that $\mathcal{F}(\varphi T)$ decays rapidly in a conic neighborhood of ξ_0.[5] Using C^∞ partitions of unity on $\mathbb{S}^{N-1}$, one easily proves that if T is microlocally regular at (x_0, ξ) for all directions ξ, then T is C^∞ near x_0, so that the projection in Ω of $WF(T)$ is the singular support of T as defined by Laurent SCHWARTZ.

---

[1] I first heard about Lars HÖRMANDER's work in lectures that Salah BAOUENDI gave when I still was a student at École Polytechnique in Paris, France, in 1967, and a few years after I heard about his work on a class of hypo-elliptic operators at the Lions–Schwartz seminar at IHP (Institut Henri Poincaré), in Paris.

[2] I first heard about lacunas in the late 1960s, in a talk by Lars GÅRDING at the Lions–Schwartz seminar at IHP (Institut Henri Poincaré) in Paris, France.

[3] There seems to be something resembling a wave front set in the theory of hyperfunctions of SATO.

[4] Pippo (Giuseppe) GEYMONAT told me that the problem of lacunas was solved later by Michael ATIYAH, Raoul BOTT, and Lars GÅRDING.

[5] More precisely, for some $\psi \in C^\infty(\mathbb{S}^{N-1})$ with $\psi\big(\frac{\xi_0}{|\xi_0|}\big) \neq 0$, and $\chi \in C_c^\infty(\mathbb{R}^N)$ equal to 1 near 0, one has $(1-\chi)\,\psi\big(\frac{\xi}{|\xi|}\big)\,\mathcal{F}(\varphi T) \in \mathcal{S}(\mathbb{R}^N)$.

L. Tartar, *The General Theory of Homogenization*, Lecture Notes of the Unione Matematica Italiana 7, DOI 10.1007/978-3-642-05195-1_28,

By adapting the stationary-phase principle, Lars HÖRMANDER proved a localization result, that if $S \in \mathcal{D}'(\Omega)$ satisfies

$$\begin{aligned}&\textstyle\sum_{j=1}^N b_j \frac{\partial S}{\partial x_j} + c\,S = f \text{ in } \Omega,\\ &b_j (j = 1, \ldots, N), c, f \in C^\infty(\Omega),\end{aligned} \tag{28.1}$$

then

$$\begin{aligned}&WF(S) \subset \{(x, \xi) \in \Omega \times (\mathbb{R}^N \setminus \{0\}) \mid P(x, \xi) = 0\},\\ &P(x, \xi) = \textstyle\sum_{j=1}^N b_j(x)\xi_j.\end{aligned} \tag{28.2}$$

By developing a theory of Fourier integral operators, an extension of pseudo-differential operators,[6] Lars HÖRMANDER then proved results of propagation of microlocal regularity:[7] assuming that the coefficients b_j are real, if S is microlocally regular at (x_0, ξ_0), then it is microlocally regular along the whole bicharacteristic ray going through (x_0, ξ_0), defined by

$$\begin{aligned}&\tfrac{dx_j}{d\tau} = b_j\big(x(\tau)\big) = \tfrac{\partial P}{\partial \xi_j}, j = 1, \ldots, N,\\ &\tfrac{d\xi_j}{d\tau} = -\tfrac{\partial P}{\partial x_j}, j = 1, \ldots, N,\\ &(x, \xi) \mid_{\tau=0} = (x_0, \xi_0).\end{aligned} \tag{28.3}$$

With the mathematical tools that he developed, Lars HÖRMANDER showed that the propagation of microlocal regularity holds for a scalar wave equation,[8]

$$\varrho\,\frac{\partial^2 T}{\partial t^2} - \sum_{i,j=1}^N \frac{\partial}{\partial x_i}\Big(a_{i,j}\,\frac{\partial T}{\partial x_j}\Big) = f \text{ in } \mathbb{R}^N \times \mathbb{R}, \tag{28.4}$$

[6] Joseph KOHN and Louis NIRENBERG introduced pseudo-differential operators for questions about elliptic equations, but the mapping which to the initial data for the wave equation gives its solution at time t is not a pseudo-differential operator, and Lars HÖRMANDER developed the theory of Fourier integral operators in order to handle such operators.

[7] I observed an intention to mislead by some followers of Lars HÖRMANDER, who not only wrongly use the term propagation of singularities to refer to what is propagation of microlocal regularity, but also pretend that it is related to a result of HADAMARD about discontinuities in the gradient of solutions of a scalar wave equation: HADAMARD talked of something which he measured along a bicharacteristic ray (a jump in gradient), while followers of Lars HÖRMANDER are afraid to work on a wave front set, which is a no man's land for them, and they cannot measure anything there! Misleading students and researchers is the worst possible sin for a teacher, and besides teaching what is a singularity and what is not, one should mention that singularities are not of much use for physical problems anyway!

[8] Lars HÖRMANDER never seemed interested in developing mathematical tools for continuum mechanics or physics, as he did not work with systems! Did he learn that real light is polarized, so that it is not about the scalar wave equation but about the Maxwell–Heaviside system? Did he learn that what is important about a ray of light is that it transports energy and momentum?

$$WF(T) \subset \{(x,\xi) \in \mathbb{R}^{N+1} \times (\mathbb{R}^{N+1} \setminus \{0\}) \mid Q(x,\xi) = 0\},$$
$$Q(x,\xi) = \varrho(x)\xi_0^2 - \sum_{i,j=1}^N a_{i,j}(x)\xi_i\xi_j, \tag{28.5}$$

the microlocal regularity propagating along the bicharacteristic rays of Q

$$\frac{dx_j}{d\tau} = \frac{\partial Q}{\partial \xi_j}, j = 0,\ldots,N,$$
$$\frac{d\xi_j}{d\tau} = -\frac{\partial Q}{\partial x_j}, j = 0,\ldots,N, \tag{28.6}$$

where besides assuming that the coefficients ϱ and $a_{i,j}, i,j = 1,\ldots,N$, and the data f are C^∞ in $\mathbb{R}^N \times \mathbb{R}$, and defining $t = x_0$, one also assumes that the coefficients are real and independent of t, that $a_{i,j} = a_{j,i},\ i,j = 1,\ldots,N$, that $\varrho > 0$, and that $\sum_{i,j} a_{i,j}\lambda_i\lambda_j > 0$ for all $\lambda \in \mathbb{R}^N \setminus \{0\}$.

Without reading much about the work of Lars HÖRMANDER, I thought that it was not of much use for my purpose, but I first rejected it because of its description in terms of propagation of singularities, as I knew the concept of singularity to be almost useless for understanding continuum mechanics or physics,[9] and it was only after I introduced *H-measures* that Mike CRANDALL pointed out to me that it is microlocal regularity which is transported along bicharacteristic rays in Lars HÖRMANDER's work.[10]

It would be better if those who received a little talent for science, like Lars HÖRMANDER seems to have, would not bury it in the ground through fear, and use it for the benefit of the scientific community: was it so difficult for him to point out a simple example of pseudo-logic, that although bicharacteristic rays appear in the formal theory of geometrical optics, his own theory of propagation of microlocal regularity has not much to do with it?

In the late 1970s, I looked for a mathematical object more general than Young measures, and I thought of splitting Young measures in directions ξ in order to have a variable for the direction of propagation.[11] Although

---

[9] I immediately had a negative feeling when the fashion concerning solitons started, because its organizers also advocated eighteenth century mechanics instead of explaining what is nineteenth century mechanics and suggesting to work on twentieth century mechanics. It was only much later that I understood what is wrong with quantum mechanics, so that advocating solitons for the hope of understanding about elementary particles is necessarily doomed, as there are no particles! However, there could be interesting mathematical questions to study which mix partial differential equations and algebra or geometry, but there is no reason to lie about the motivation for working on such questions.

[10] Michael Grain CRANDALL, American mathematician, born in 1940. He worked at Stanford University, Stanford, CA, at UCLA (University of California at Los Angeles), Los Angeles, CA, at UW (University of Wisconsin), Madison, WI, and at UCSB (University of California at Santa Barbara), Santa Barbara, CA.

[11] In the summer of 1978, I investigated functionals $\int_\Omega F\big(x, v_n, grad(v_n)\big)\,dx$ with $F(x,v,q)$ homogeneous of degree 0 in q, but I wanted a tool for describing the evolution of mixtures, and regularization produced different limits, and I thought it pointless to use only mixtures with an interface of finite perimeter.

I advocated the study of propagation of oscillations and not of singularities, hearing my talk in the fall of 1980 at the Goulaouic–Meyer–Schwartz seminar [99] was not the best way for understanding the differences between my programme and that of Lars HÖRMANDER, and as he was visiting École Polytechnique in Palaiseau, France, at the time, he heard me describe my results concerning the oscillations for the Carleman model,[12] but in this case the Young measures are sufficient for explaining what happens.

When I wrote about the past and future of *compensated compactness* [100] in the summer of 1982, I could not see how to reformulate quantum mechanics and statistical mechanics inside my programme. In the fall of 1982, after my lost fight against inventing results of votes in the "academic" world, exhausted by the racist behavior of those who insisted that I should not have the right to vote,[13] I took leave from my university, thanks to the help of Robert DAUTRAY, who offered me a position at CEA; I felt the possibility that I would not come back to mathematics if there were no sanctions against those whom I opposed, showing that they had complete control of the French "academic" system, and I wrote down a few ideas to try if I remained in mathematics, and I gave a copy to François MURAT.

My idea was to go further than the functions of geometries that I described in Chap. 22, and to attach to a sequence U^n many sequences $A^n = \Phi(U^n)$ of coefficients of elliptic equations, not restricted to $div\big(A^n grad(u_n)\big) = f$,[14] and their H-limits, and that the joint knowledge of all the corresponding H-limits could be handled by purely algebraic methods; I guessed that it could be related to something that I had quite vaguely heard about, sheaf theory.[15]

I wondered about a possible unified character of mathematics.[16]

---

[12] Tage Gillis Torsten CARLEMAN, Swedish mathematician, 1892–1949. He worked in Lund and in Stockholm, Sweden.

[13] Did those who invented results of votes that were sent to the minister "in charge" of the French universities also invent results of other experiments (in mathematics, physics, chemistry, and biology), so that their "results" would fit with the obsolete laws which they wanted to keep?

[14] Unphysical questions like linearized elasticity would then be useful, and many others not related to continuum mechanics and physics, for describing more precisely something about the geometry of the pieces used in the mixtures considered.

[15] It was only in the fall of 1984 that Jean LERAY gave me some detail about his answer to my letter, addressed in the spring of 1984 to a few professors at Collège de France, in Paris. He told me about switching to do research in topology when he was a prisoner of war, about developing the basis of sheaf theory, about his election to a chair at Collège de France in 1948 against a prominent member of Bourbaki whose disgraceful attitude during the war worked against him, and about another member of Bourbaki openly plagiarizing his work on sheaf theory. In the fall of 1982, I was not aware that sheaf theory developed out of the work of Jean LERAY, and I only used guesses about ideas of Alexandre GROTHENDIECK, some of which heard in a talk by Jean-Pierre SERRE at the Bourbaki seminar, when he described the proof of the Weil conjectures by Pierre DELIGNE, who followed ideas of Alexandre GROTHENDIECK.

[16] For my thesis defence, in April 1971, Jean-Pierre SERRE gave me a second subject, so that I read about modular functions and described error estimates for the problem

The new tool of *H-measures*, which I first described in the beginning of 1987, corresponds to using only the Taylor expansion of Φ at order 2, and I shall describe more about this aspect in Chap. 29.

Was it really a new mathematical tool? It was also introduced, shortly after but independently, by Patrick GÉRARD,[17] for reasons related to kinetic theory (which I described in [119]), but I only heard about his work after giving a seminar talk on my work on *H-measures* in January 1989 at Collège de France in Paris. In November 1989, I gave a seminar talk at UCLA, and Gregory ESKIN mentioned that he saw measures in x and ξ used before,[18] by A.I. SHNIRELMAN,[19] who was followed by Y. COLIN DE VERDIÈRE.[20] In January 1990, I gave another seminar talk at Collège de France, where I mentioned my idea for using one characteristic length, and again Patrick GÉRARD shortly after sent me his work on the subject, using what he called semi-classical measures, and I shall describe in Chap. 32 the similarities and the differences with my idea, but one of his examples was precisely the question studied by A.I. SHNIRELMAN and Y. COLIN DE VERDIÈRE which Gregory ESKIN mentioned, so that it was about semi-classical measures and not about *H-measures.*[21]

The only earlier idea that I see as announcing *H-measures* is my own Theorem 17.3 of *compensated compactness*, because one may interpret it as saying that if $U^n \rightharpoonup 0$ in $L^2_{loc}(\Omega; \mathbb{R}^p)$ weak and $U^n \otimes U^n \rightharpoonup \nu$ in $\mathcal{M}(\Omega; \mathbb{R}^p \otimes \mathbb{R}^p)$ weak $\star$, then if $\nu \in L^1_{loc}(\Omega; \mathbb{R}^p \otimes \mathbb{R}^p)$ one deduces that $\nu(x)$ belongs to the convex hull of $\Lambda \otimes \Lambda$ a.e. in Ω,[22] so that it is a convex combination indexed by ξ of elements of the form $\lambda \otimes \lambda$ with $\lambda \in \Lambda_\xi$; an H-measure is precisely a

---

of the circle (counting integer solutions of $x^2 + y^2 \leq n$), but I felt frustrated that the analytical methods could not grasp the special algebraic nature of the problem. From the form of the Ramanujan conjecture, proven by Pierre DELIGNE, I thought many years after that there must be a similarity with the compensation effects that I studied [115], and it gave me the idea, which I only mentioned to Joel ROBBIN in the early 1990s, that the Riemann conjecture is about a compensation effect!

[17] I do not like the name that he chose, microlocal defect measure, because it has the defect of the approach of Lars HÖRMANDER, which induces people to confuse propagation of light with questions of microlocal regularity.

[18] Gregory I. ESKIN, Russian-born mathematician. He works at UCLA (University of California Los Angeles), Los Angeles, CA.

[19] Alexander I. SHNIRELMAN, Russian-born mathematician. He worked in Tel Aviv, Israel, in Hull, England, and in Montréal, Québec.

[20] Yves COLIN DE VERDIÈRE, French mathematician. He works at Université de Grenoble I (Joseph Fourier), Saint-Martin-d'Hères, France.

[21] I do not think that A.I. SHNIRELMAN and Y. COLIN DE VERDIÈRE made any attempt at finding a general framework, like that of semi-classical measures of Patrick GÉRARD.

[22] In the general case, one uses the Radon–Nikodym theorem.

nonnegative measure which gives this convex combination, but this approach may not show clearly the importance of using a calculus with symbols.[23]

For a scalar sequence u_n converging weakly to 0 in $L^2_{loc}(\Omega)$, and for $\varphi \in C_c(\Omega)$, $\varphi\, u_n \rightharpoonup 0$ in $L^2(\mathbb{R}^N)$ weak (by using the value 0 outside Ω), so that $\mathcal{F}(\varphi\, u_n) \rightharpoonup 0$ in $L^2(\mathbb{R}^N)$ weak, but $\mathcal{F}(\varphi\, u_n)$ staying bounded in $C_0(\mathbb{R}^N)$ and converging pointwise to 0, the Lebesgue dominated convergence theorem implies that $\mathcal{F}(\varphi\, u_n) \to 0$ in $L^2_{loc}(\mathbb{R}^N)$ strong. If $\varphi\, u_n$ does not converge strongly to 0 in $L^2(\mathbb{R}^N)$, it implies that the information on $|\mathcal{F}(\varphi\, u_n)|^2$ moves toward ∞, and I decided to see how much goes away in each direction by considering

$$\lim_{n\to\infty} \int_{\mathbb{R}^N} |\mathcal{F}(\varphi\, u_n)|^2 \psi\Big(\frac{\xi}{|\xi|}\Big)\, d\xi, \tag{28.7}$$

for $\psi \in C(\mathbb{S}^{N-1})$.[24] One may have to extract a subsequence, but as $|\mathcal{F}(\varphi\, u_n)|^2$ is bounded in $L^1(\mathbb{R}^N)$ and $C(\mathbb{S}^{N-1})$ is separable, a Cantor diagonal process gives a subsequence for which the limit exists for all $\psi \in C(\mathbb{S}^{N-1})$: this consists in projecting $|\mathcal{F}(\varphi\, u_n)|^2$ on the unit sphere $\mathbb{S}^{N-1}$, giving a bounded sequence in $L^1(\mathbb{S}^{N-1})$ by the Fubini theorem,[25] and extracting a subsequence which converges in $\mathcal{M}(\mathbb{S}^{N-1})$ weak $\star$, to a limit which depends upon φ, so that

$$\lim_{m\to\infty} \int_{\mathbb{R}^N} |\mathcal{F}(\varphi\, u_m)|^2 \psi\Big(\frac{\xi}{|\xi|}\Big)\, d\xi = \langle \mu_\varphi, \psi\rangle \text{ for all } \psi \in C(\mathbb{S}^{N-1}). \tag{28.8}$$

My intuition suggested that there exists a nonnegative $\mu \in \mathcal{M}(\Omega \times \mathbb{S}^{N-1})$, which I called the *H-measure associated to the subsequence*,[26] such that

$$\langle \mu_\varphi, \psi\rangle = \langle \mu, |\varphi|^2 \otimes \psi\rangle \text{ for all } \varphi \in C_c(\Omega), \psi \in C(\mathbb{S}^{N-1}), \tag{28.9}$$

but my intuition came from a situation with u_n bounded in $L^\infty(\Omega)$!

---

[23] It is understandable that engineers or physicists, who use formal manipulations without bothering if it makes any sense, may think that they already knew about *H-measures*. Why would mathematicians not realize that I gave definitions of mathematical objects which were apparently new (since no one pointed out that someone gave a precise definition earlier), and that I proved theorems about the way to use them which corresponds to interesting questions in continuum mechanics or physics? I think that I gave the first correct description of what geometrical optics is about, because I proved that *for all* sequences of solutions of the wave equation converging weakly to 0, the energy propagates along light rays in the limit (if the coefficients are C^2), and not only that *there exists* a sequence that one constructs for which it is true, only outside caustics! Are mathematicians so ill-trained nowadays, that they may confuse the quantifiers $\forall$ and $\exists$?

[24] I thought of using $\psi \in L^\infty(\mathbb{S}^{N-1})$, but $L^\infty(\mathbb{S}^{N-1})$ is not separable.

[25] Guido FUBINI, Italian-born mathematician, 1879–1943. He worked in Catania, in Genova (Genoa), in Torino (Turin), Italy, and in New York, NY.

[26] I was working on what I called small-amplitude homogenization questions, which I describe in Chap. 29, so that the prefix H reminds one of homogenization. I think that it is a reasonable name, but *H-measures* are only a piece of the mathematical apparatus needed for explaining continuum mechanics and physics.

Because coefficients in homogenization are usually bounded, no concentration effects occur, and although my Theorem 17.3 of *compensated compactness* applies to concentration effects, I neglected to study them directly in questions of continuum mechanics or physics before introducing *H-measures*. However, even with just an intuition of what an H-measure would be, I felt that the way to study concentration effects was not adapted to continuum mechanics or physics: for example, if $u_n \rightharpoonup 0$ in $L^2(\Omega)$ weak and $|u_n|^2 \rightharpoonup \nu$ in $\mathcal{M}(\Omega)$ weak $\star$, writing $\nu = f\,dx + \nu_0$ with $f \in L^1(\Omega)$ and ν_0 singular with respect to the Lebesgue measure looked of little interest to me, and maybe it corresponds to being stuck with ideas like Young measures;[27] I felt that concentration effects should be studied with microlocal tools, for understanding how they move, and this is what my theory of *H-measures* does, but it does not seem able to study propagation effects in semi-linear partial differential equations.[28]

Pierre-Louis LIONS also criticized the fact that my theory of *H-measures* uses L^2 bounds, and that there is no L^p theory, but I let the problems in continuum mechanics and physics lead me to what is needed.[29] Anyway, I am not sure if L^p spaces are natural, and I discussed in [117] a larger class of spaces, the Lorentz spaces,[30] which are studied as interpolation spaces with the theory developed by Jacques-Louis LIONS and Jaak PEETRE (to whom one owes the important simplifications which made the theory more easily applicable),[31] because I noticed situations in partial differential equations from continuum mechanics and physics which seem to require new spaces; it is not clear if these spaces will be interpolation spaces, but it might be useful to know how one created many spaces before. At a basic level, the equations of physics must be hyperbolic, because of the principle of relativity of POINCARÉ, and L^p spaces for $p \neq 2$ are not adapted to such equations, but conservation of energy involves quadratic quantities, so that using *H-measures* is not a bad idea, until one finds how to handle semi-linear hyperbolic systems!

---

[27] I heard Erik BALDER mention the possibility of defining Young measures with values in various compactifications.

[28] It is not the correct theory for quasi-linear equations, but such equations used in continuum mechanics have some defects, which are not corrected by postulating inadequate equations in kinetic theory, and I discussed these questions in [119].

[29] I was careful not to confuse reality with physicists' problems, and I knew enough defects of most theories used by physicists before introducing *H-measures*, and if some defects are corrected by the introduction of my ideas, some others remain, so that better mathematical tools must be developed.

[30] George Gunther LORENTZ, Russian-born mathematician, 1910–2006. He worked in Toronto, Ontario, at Wayne State University, Detroit, MI, in Syracuse, NY, and University of Texas, Austin, TX.

[31] Jacques-Louis LIONS was influenced by the work of Nachman ARONSZAJN and Emilio GAGLIARDO, and Jaak PEETRE probably by the work of M. RIESZ, who was the pioneer for questions of interpolation.

I found it natural to use $\mathbb{S}^{N-1}$, but what is really used is the quotient space of $\mathbb{R}^N \setminus \{0\}$ when one identifies x and $s\,x$ for $s > 0$,[32] and the unit sphere is a convenient way to choose one point in each equivalent class. I only understood this point after Patrick GÉRARD introduced his semi-classical measures, but not immediately, because at the beginning I saw his approach as very different from mine.[33] It was in explaining why the work of A.I. SHNIRELMAN and of Y. COLIN DE VERDIÈRE mentioned by Gregory ESKIN is not about using an H-measure but using a semi-classical measure that I understood that I used a quotient space, while they used the sphere inside $\mathbb{R}^N$! If one looks at the eigenvalues of $-div\big(A\,grad(u_n)\big) = \lambda_n u_n$, for a symmetric and smooth A, and one uses the characteristic length $\varepsilon_n = \lambda_n^{-1/2}$, their (semi-classical) measures live on $(A(x)\,\xi,\xi) = 1$ as a consequence of the *localization principle*: it was their use of the Laplacian which made the sphere appear!

I only understood how to prove (28.9) after considering the case of vector-valued sequences $U^n \rightharpoonup 0$, because I thought of localizing components U_j^n and U_k^n with two different test functions $\varphi_1, \varphi_2 \in C_c(\Omega)$: it is natural for a quadratic form to introduce the associated bilinear form, and for a Hermitian form one introduces the associated sesqui-linear form, and here it means to extract a subsequence such that, for all $\varphi_1, \varphi_2 \in C_c(\Omega)$ and all $\psi \in C(\mathbb{S}^{N-1})$

$$\lim_{m\to\infty} \int_{\mathbb{R}^N} \mathcal{F}(\varphi_1 u_m)\overline{\mathcal{F}(\varphi_2 u_m)}\psi\Big(\frac{\xi}{|\xi|}\Big)\,d\xi = L(\varphi_1,\varphi_2,\psi) \text{ exists,} \tag{28.10}$$

and then my conjecture was

$$L(\varphi_1,\varphi_2,\psi) = \langle \mu, \varphi_1\overline{\varphi_2} \otimes \psi\rangle. \tag{28.11}$$

It is not obvious that $L(\varphi_1,\varphi_2,\psi)$ only depends upon $\varphi_1\overline{\varphi_2}$, because a quantity

$$\int_{\Omega\times\Omega\times\mathbb{S}^{N-1}} \varphi_1(x)\,\overline{\varphi_2(y)}\,\psi(\xi)\,d\mu(x,y,\xi) \text{ for } \mu \in \mathcal{M}(\Omega\times\Omega\times\mathbb{S}^{N-1}) \tag{28.12}$$

satisfies the same bounds deduced from the definition (28.10), that for every compact $K \subset \Omega$, one has $|L(\varphi_1,\varphi_2,\psi)| \le C_K||\varphi_1||\,||\varphi_2||\,||\psi||$, where the

[32] I think that it is a mistake from a practical point of view to identify ξ and $-\xi$ and use the projective space $\mathbb{P}^{N-1}$, although for real sequences an H-measure charges in the same way ξ and $-\xi$; it is not reasonable to use only real sequences, because complex-valued sequences are created by the pseudo-differential operators (without traditional smoothness hypotheses) which one uses.

[33] I use the unit sphere $\mathbb{S}^N \subset \mathbb{R}^{N+1}$ while the space $\mathbb{R}^N$ that Patrick GÉRARD uses is in some way the tangent plane to $\mathbb{S}^N$ at e_{N+1}, and the information that he loses at ∞ is found in my approach on the equator of $\mathbb{S}^N$, but there are other differences in his approach.

norms are sup norms, and C_K is an upper bound for $\int_K |u_m|^2\,dx$. I then rewrote (28.10) using a class of "pseudo-differential operators."[34]

Definition 28.1. *For $\Omega \subset \mathbb{R}^N$ and $b \in L^\infty(\Omega)$ one defines M_b by*

$$\begin{aligned} &\text{for } v \in L^2(\Omega), M_b v = b\,v \in L^2(\Omega)\\ &||M_b||_{\mathcal{L}(L^2(\Omega);L^2(\Omega))} = ||b||_{L^\infty(\Omega)}, \end{aligned} \tag{28.13}$$

and for $a \in L^\infty(\mathbb{R}^N)$, one defines P_a by

$$\begin{aligned} &\text{for } w \in L^2(\mathbb{R}^N), \mathcal{F}(P_a w) = a\,\mathcal{F}w \in L^2(\mathbb{R}^N), \ \text{i.e., } P_a = \mathcal{F}^{-1}M_a\mathcal{F}\\ &||P_a||_{\mathcal{L}(L^2(\mathbb{R}^N);L^2(\mathbb{R}^N))} = ||a||_{L^\infty(\mathbb{R}^N)}. \end{aligned} \tag{28.14}$$

Using the same notation ψ for the function extended to $\mathbb{R}^N \setminus \{0\}$ as a homogeneous function of order 0, the left side of (28.10) can be rewritten

$$\int_{\mathbb{R}^N} \mathcal{F}(\varphi_1 u_m)\,\overline{\mathcal{F}(\varphi_2 u_m)}\,\psi\Big(\frac{\xi}{|\xi|}\Big)\,d\xi = \int_{\mathbb{R}^N} \mathcal{F}\,P_\psi M_{\varphi_1} u_m\,\overline{\mathcal{F}\,M_{\varphi_2} u_m}\,d\xi, \tag{28.15}$$

which, using the Plancherel theorem, gives

$$\begin{aligned} &\textstyle\int_{\mathbb{R}^N} \mathcal{F}(\varphi_1 u_m)\,\overline{\mathcal{F}(\varphi_2 u_m)}\,\psi\Big(\frac{\xi}{|\xi|}\Big)\,d\xi = \int_{\mathbb{R}^N} P_\psi M_{\varphi_1} u_m\,\overline{M_{\varphi_2} u_m}\,dx\\ &= \textstyle\int_{\mathbb{R}^N} (M_{\overline{\varphi_2}} P_\psi M_{\varphi_1} u_m)\,\overline{u_m}\,dx, \end{aligned} \tag{28.16}$$

and I observed that if $M_{\overline{\varphi_2}}P_\psi$ could be replaced by $P_\psi M_{\overline{\varphi_2}}$, then $M_{\overline{\varphi_2}}M_{\varphi_1} = M_{\overline{\varphi_2}\varphi_1}$ would appear, and the limit would only depend upon $\overline{\varphi_2}\varphi_1$;[35] I then wanted the commutator of $M_{\overline{\varphi_2}}$ and P_ψ to be a compact operator from $L^2(\mathbb{R}^N)$ into itself, for transforming the weakly converging sequence $M_{\varphi_1}u_m$ into a strongly converging sequence (to 0). I proved a first commutation lemma, with the hypothesis $\psi \in C(\mathbb{S}^{N-1})$ extended to be homogeneous of degree 0, but a few years later I found that my proof applied to a more general setting: I gave in [105] a formula using *H-measures* which expressed M^{eff} from Lemma 19.1, but in the evolution case [103] something different is needed, which I investigated with two postdoctoral students at CMU

---

[34] I use quotes for pointing out that I do not impose the regularity hypotheses of the classical theory. Interfaces between different materials arise for partial differential equations of continuum mechanics and physics, requiring discontinuous symbols. However, some theorems impose regularity for the coefficients, which one should check carefully; for example, I use C^1 or C^2 regularity for the transport of energy for the wave equation, so that refraction effects are not described correctly yet.

[35] It would not be the same to replace $P_\psi M_{\varphi_1}$ by $M_{\varphi_1}P_\psi$, because P_ψ does not act on u_m if $\Omega \neq \mathbb{R}^N$.

(*Carnegie Mellon University*), Konstantina TRIVISA and Chun LIU,[36] and a different scaling and quotient space appeared, where one identifies (τ, ξ) with $(s^2\tau, s\,\xi)$ for $s > 0$.

Lemma 28.2. *If $b \in C_0(\mathbb{R}^N)$, and if $a \in L^\infty(\mathbb{R}^N)$ satisfies*

$$\begin{aligned}&\text{for every } \rho > 0, \varepsilon > 0, \text{ there exists } \kappa \text{ such that}\\ &|\xi^1|, |\xi^2| \geq \kappa \text{ and } |\xi^1 - \xi^2| \leq \rho \text{ imply } |a(\xi^1) - a(\xi^2)| \leq \varepsilon,\end{aligned} \tag{28.17}$$

then $[M_b, P_a] = M_b P_a - P_a M_b$ is compact from $L^2(\mathbb{R}^N)$ into itself.

Proof: Because $||[M_b, P_a]||_{\mathcal{L}(L^2(\mathbb{R}^N); L^2(\mathbb{R}^N))} \leq 2||a||_{L^\infty(\mathbb{R}^N)}||b||_{L^\infty(\mathbb{R}^N)}$, one approaches b uniformly by a sequence $b_n \in S(\mathbb{R}^N)$ with $\mathcal{F}b_n$ having compact support, inside $|\xi| \leq \rho_n$; if one shows that each $C_n = [M_{b_n}, P_a]$ is compact, then $[M_b, P_a]$ is compact, as a uniform limit of compact operators. For constructing b_n, one chooses $f_n \in S(R^N)$ converging to b uniformly, and $g_{n,m} \in C_c^\infty(\mathbb{R}^N)$ converging to $\mathcal{F}f_n$ in $L^1(R^N)$, so that $\mathcal{F}^{-1}g_{n,m}$ approaches f_n uniformly; a diagonal subsequence of $\mathcal{F}^{-1}g_{n,m}$ serves as the sequence b_n. For $v \in L^2(\mathbb{R}^N)$,

$$\begin{aligned}\mathcal{F}(C_n v)(\xi) &= \int_{\mathbb{R}^N} \mathcal{F}b_n(\xi - \eta) a(\eta)\, \mathcal{F}v(\eta)\, d\eta - a(\xi) \int_{\mathbb{R}^N} \mathcal{F}b_n(\xi - \eta)\mathcal{F}\, v(\eta)\, d\eta\\ &= \int_{\mathbb{R}^N} \big(a(\eta) - a(\xi)\big)\, \mathcal{F}b_n(\xi - \eta)\, \mathcal{F}v(\eta)\, d\eta,\end{aligned} \tag{28.18}$$

so that $\mathcal{F}(C_n v)$ is obtained from $\mathcal{F}v$ by an operator with kernel $K_n(\xi, \eta) = \big(a(\eta) - a(\xi)\big)\, \mathcal{F}b_n(\xi - \eta)$, which only involves $|\xi - \eta| \leq \rho_n$. For $\varepsilon > 0$, one applies (28.17) to find κ associated to ρ_n and ε, and one cuts the kernel in two pieces, one for which $|\xi|, |\eta| \geq \kappa$, which gives a kernel bounded by $\varepsilon\, |\mathcal{F}b_n(\xi - \eta)|$, corresponding to an operator of norm $\leq \varepsilon\, ||\mathcal{F}b_n||_{L^1(\mathbb{R}^N)}$, and one with either $|\xi|$ or $|\eta| < \kappa$, and $|\xi - \eta| \leq \rho_n$, which corresponds to a bounded kernel with compact support, certainly in $L^2(\mathbb{R}^N \times \mathbb{R}^N)$, giving a Hilbert–Schmidt operator;[37] the operator C_n being a uniform limit of Hilbert–Schmidt operators, which are compact, is then compact. □

Corollary 28.3. *If $\psi \in C(\mathbb{S}^{N-1})$ and $a(\xi) = \psi(\frac{\xi}{|\xi|})$, then (28.17) holds.*

Proof: There exists $\delta > 0$ such that $|\psi(\eta^1) - \psi(\eta^2)| \leq \varepsilon$ for $\eta^1, \eta^2 \in \mathbb{S}^{N-1}$ with $|\eta^1 - \eta^2| \leq \delta$. One looks for κ such that $|\xi^1|, |\xi^2| \geq \kappa$ and $|\xi^1 - \xi^2| \leq \rho$

---

36 Konstantina TRIVISA and Chun LIU did not seem so interested in continuing this work, and later my former PhD student Nenad ANTONIĆ started investigating this parabolic variant of *H-measures*, with his own student Martin LAZAR.

37 Erhard SCHMIDT, German mathematician, 1876–1959. He worked in Bonn, Germany, in Zürich, Switzerland, in Erlangen, Germany, in Breslau (then in Germany, now Wrocław, Poland), and in Berlin, Germany.

imply $\left|\frac{\xi^1}{|\xi^1|}-\frac{\xi^2}{|\xi^2|}\right|\leq\delta$. One has $\left|\frac{\xi^1}{|\xi^1|}-\frac{\xi^2}{|\xi^2|}\right|\leq\left|\frac{\xi^1}{|\xi^1|}-\frac{\xi^2}{|\xi^1|}\right|+\left|\frac{\xi^2}{|\xi^1|}-\frac{\xi^2}{|\xi^2|}\right|$; the first term is $\leq\frac{\rho}{|\xi^1|}\leq\frac{\rho}{\kappa}$; the second term is the absolute value of $1-\frac{|\xi^2|}{|\xi^1|}$, which because $|\xi^1|-\rho\leq|\xi^2|\leq|\xi^1|+\rho$ is also $\leq\frac{\rho}{|\xi^1|}\leq\frac{\rho}{\kappa}$, and one takes $\kappa\geq\frac{2\rho}{\delta}$. □

Corollary 28.4. *If $a\in C\big((\mathbb{R}^N\times\mathbb{R})\setminus\{0\}\big)$ satisfies*

$$a(s\,\xi,s^2\tau)=a(\xi,\tau)\textit{ for all }s>0\textit{ and }|\xi|+|\tau|\neq 0, \tag{28.19}$$

then (28.17) holds, with N replaced by $N+1$.

Proof: One defines

$$\Sigma=\{(\xi,\tau)\mid |\xi|^4+\tau^2=1\}, \Phi(\xi,\tau)=(|\xi|^4+\tau^2)^{1/4}\text{ if }|\xi|+|\tau|\neq 0, \tag{28.20}$$

and for $|\xi|+|\tau|\neq 0$ one has

$$a(\xi,\tau)=a\Big(\frac{\xi}{\Phi(\xi,\tau)},\frac{\tau}{\Phi^2(\xi,\tau)}\Big),\text{ and }\eta=\Big(\frac{\xi}{\Phi(\xi,\tau)},\frac{\tau}{\Phi^2(\xi,\tau)}\Big)\in\Sigma. \tag{28.21}$$

There exists $\delta>0$ such that $|a(\eta^1)-a(\eta^2)|\leq\varepsilon$ whenever $\eta^1,\eta^2\in\Sigma$ with $|\eta^1-\eta^2|\leq\delta$. For $|\xi^1-\xi^2|^2+|\tau^1-\tau^2|^2\leq\rho^2$ one wants to take $\Phi(\xi^j,\tau^j)\geq K$ for $j=1,2$, with K large enough, and deduce $|\eta^1-\eta^2|\leq\delta$. Using $|\xi^j|\leq\Phi(\xi^j,\tau^j)$ and $|\tau^j|\leq\Phi^2(\xi^j,\tau^j)$ for $j=1,2$, one has

$$\begin{aligned}&\left|\tfrac{\xi^1}{\Phi(\xi^1,\tau^1)}-\tfrac{\xi^2}{\Phi(\xi^2,\tau^2)}\right|\leq\tfrac{|\xi^1-\xi^2|}{\Phi(\xi^1,\tau^1)}+\left|\tfrac{\xi^2}{\Phi(\xi^1,\tau^1)}-\tfrac{\xi^2}{\Phi(\xi^2,\tau^2)}\right|\\&\leq\tfrac{\rho}{K}+\left|1-\tfrac{\Phi(\xi^2,\tau^2)}{\Phi(\xi^1,\tau^1)}\right|,\end{aligned} \tag{28.22}$$

$$\begin{aligned}&\left|\tfrac{\tau^1}{\Phi^2(\xi^1,\tau^1)}-\tfrac{\tau^2}{\Phi^2(\xi^2,\tau^2)}\right|\leq\tfrac{|\tau^1-\tau^2|}{\Phi^2(\xi^1,\tau^1)}+\left|\tfrac{\tau^2}{\Phi^2(\xi^1,\tau^1)}-\tfrac{\tau^2}{\Phi^2(\xi^2,\tau^2)}\right|\\&\leq\tfrac{\rho}{K^2}+\left|1-\tfrac{\Phi^2(\xi^2,\tau^2)}{\Phi^2(\xi^1,\tau^1)}\right|,\end{aligned} \tag{28.23}$$

and because

$$|grad(\Phi)|=\frac{\sqrt{4|\xi|^6+|\tau|^2}}{2\Phi^3}\leq\frac{\sqrt{4\Phi^2+1}}{2\Phi}\leq 2\text{ if }\Phi\geq\frac{1}{\sqrt{12}}, \tag{28.24}$$

one deduces that if the segment joining (ξ^1,τ^1) to (ξ^2,τ^2) has $\Phi\geq\frac{1}{\sqrt{12}}$, which is true if K is large enough compared to ρ, one has

$$\Phi(\xi^1,\tau^1)-2\rho\leq\Phi(\xi^2,\tau^2)\leq\Phi(\xi^1,\tau^1)+2\rho,\text{ i.e., }1-\frac{2\rho}{K}\leq\frac{\Phi(\xi^2,\tau^2)}{\Phi(\xi^1,\tau^1)}\leq 1+\frac{2\rho}{K}, \tag{28.25}$$

and the desired inequalities hold for K large enough. □

Corollary 28.3 serves in defining the H-measure of a subsequence, and in proving what I called the *localization principle*, from which one deduces the *compensated compactness* theory. Often, the regularity hypothesis of a in ξ is not important but the regularity hypothesis of b in x is important, because a is a test function and b is a coefficient of a partial differential equation that one studies, and one should pay attention to the hypotheses of regularity needed in obtaining a particular result. My counter-example of Lemma 7.3 for the *div–curl lemma* shows that the commutator $[M_b, P_a]$ cannot be compact for all a smooth if b is the characteristic function of a nonempty smooth set. After writing my article [105] on *H-measures*, I became aware of an improvement on the regularity of b, which can be taken to be in $L^\infty(\mathbb{R}^N) \cap VMO(\mathbb{R}^N)$,[38] thanks to a result which Raphaël COIFMAN proved with ROCHBERG and WEISS [19],[39,40] that for $b \in BMO(\mathbb{R}^N)$ the commutators $[M_b, R_j]$ map $L^p(\mathbb{R}^N)$ into itself for $1 < p < \infty$, with a norm depending only on the $BMO(\mathbb{R}^N)$ semi-norm of b;[41] $b \in VMO(\mathbb{R}^N)$ means that it can be approached in the $BMO(\mathbb{R}^N)$ semi-norm by $b_k \in C_c^\infty(\mathbb{R}^N)$, and if $b \in L^\infty(\mathbb{R}^N) \cap VMO(\mathbb{R}^N)$ one may truncate these b_k so that they stay bounded in $L^\infty(\mathbb{R}^N)$ and the truncated functions may only be Lipschitz continuous with compact support, but they belong to $C_0(\mathbb{R}^N)$, and all $[M_{b_k}, R_j]$ are compact from $L^2(\mathbb{R}^N)$ into itself by Lemma 28.2 and Corollary 28.3, so that the uniform limit $[M_b, R_j]$ is then also compact; then one notices that $[M_b, P_a R_j] = [M_b, P_a]\, R_j + P_a[M_b, R_j]$, so that by induction one deduces that $[M_b, P_a]$ is compact if a is a polynomial in $\frac{\xi}{|\xi|}$ (as R_j corresponds to $a = \frac{i\,\xi_j}{|\xi|}$), and by the Weierstrass approximation theorem one deduces that $[M_b, P_a]$ is compact for all $a \in C(\mathbb{S}^{N-1})$. Not being knowledgeable enough, I sent a message a few years ago to ROCHBERG, asking him a few questions, and he answered that UCHIYAMA proved in [121] that $b \in VMO(\mathbb{R}^N)$ is necessary for the commutators to be compact;[42] he also said that for extending results to

[38] I knew it in November 1989, when I wrote a text for a seminar talk at CMU (*Carnegie Mellon University*), Pittsburgh, PA, where I mentioned various properties of the Hardy space $\mathcal{H}^1(\mathbb{R}^N)$, and the spaces $BMO(\mathbb{R}^N)$ and $VMO(\mathbb{R}^N)$ (BMO = bounded mean oscillation, VMO = vanishing mean oscillation), in particular that $[M_b, R_j]$ is compact if $b \in L^\infty(\mathbb{R}^N) \cap VMO(\mathbb{R}^N)$, and R_j is a (M.) Riesz operator. In July 1993, I mentioned that one can define *H-measures* with test functions in $L^\infty(\mathbb{R}^N) \cap VMO(\mathbb{R}^N)$ in my CBMS–NSF lectures in Santa Cruz, CA. In a discussion with my former PhD student Sergio GUTIÉRREZ, we deduced that the Gårding inequality holds with coefficients in $L^\infty(\Omega) \cap VMO(\Omega)$.

[39] Richard Howard ROCHBERG, American mathematician, born in 1943. He works at Washington University, St Louis, MO.

[40] Guido Leopold WEISS, Italian-born mathematician, born in 1928. He worked at DePaul University, and he works now at Washington University, St Louis, MO.

[41] They also prove the converse, that if the commutators map $L^p(\mathbb{R}^N)$ into itself, then b must belong to $BMO(\mathbb{R}^N)$.

[42] Akihito UCHIYAMA, Japanese mathematician, 1948–1997. He worked at The University of Chicago, Chicago, IL, and at Tohoku University, Sendai, Japan.

nonisotropic situations like that of Corollary 28.4, there is a theory of BMO adapted to other geometries, whose best introduction is an article by Raphaël COIFMAN and WEISS [20], but I never checked these references. Despite the possibility of using $b \in L^\infty(\Omega) \cap VMO(\Omega)$, I shall only describe my theory of *H-measures* using functions b belonging to $C_c(\Omega)$ (or to $C_0(\mathbb{R}^N)$).

Theorem 28.5. *If $u_n \rightharpoonup 0$ in $L^2_{loc}(\Omega)$ weak, there exists a subsequence u_m and $\mu \in \mathcal{M}(\Omega \times \mathbb{S}^{N-1})$, with $\mu \geq 0$, called the H-measure associated to the subsequence, such that for all $\varphi_1, \varphi_2 \in C_c(\Omega)$ and all $\psi \in C(\mathbb{S}^{N-1})$ one has*

$$\int_{\mathbb{R}^N} \mathcal{F}(\varphi_1 u_m)\,\overline{\mathcal{F}(\varphi_2 u_m)}\,\psi\Big(\frac{\xi}{|\xi|}\Big)\,d\xi \to \langle \mu, \varphi_1 \overline{\varphi_2} \otimes \psi\rangle. \tag{28.26}$$

Proof: Having extracted a subsequence such that (28.10) holds, one has

$$M_{\overline{\varphi_2}} P_\psi M_{\varphi_1} u_m - P_\psi M_{\overline{\varphi_2}\,\varphi_1} u_m \to 0 \text{ in } L^2(\mathbb{R}^N) \text{ strong}, \tag{28.27}$$

by Lemma 28.2 and Corollary 28.3, so that

$$\int_{\mathbb{R}^N} (M_{\overline{\varphi_2}} P_\psi M_{\varphi_1} u_m)\,\overline{u_m}\,dx - \int_{\mathbb{R}^N} (P_\psi M_{\overline{\varphi_2}\,\varphi_1} u_m)\,\overline{u_m}\,dx \to 0, \tag{28.28}$$

and $L(\varphi_1, \varphi_2, \psi)$ only depends upon the product $\overline{\varphi_2}\,\varphi_1$. One deduces that $L(\varphi_1, \varphi_2, \psi)$ is independent of φ_2 once it is equal to 1 on the support of φ_1, and this value $B(\varphi_1, \psi)$ is bilinear in φ_1 and ψ, and continuous in the sense that for every compact $K \subset \Omega$ there exists a constant $C(K)$, which one may take to be the upper bound for $\int_K |u_m|^2\,dx$,[43] such that

$$\begin{aligned} &|B(\varphi, \psi)| \leq C(K)\,||\varphi||_{L^\infty(K)}\,||\psi||_{C(\mathbb{S}^{N-1})} \text{ for all } \psi \in C(\mathbb{S}^{N-1})\\ &\quad \text{and all } \varphi \in C_K(\Omega) = \{\varphi \in C_c(\Omega) \mid \mathit{support}(\varphi) \subset K\}. \end{aligned} \tag{28.29}$$

It means that $B(\varphi, \psi)$ defines a linear continuous mapping from $C_c(\Omega)$ into $\mathcal{M}(\mathbb{S}^{N-1})$, which by the kernel theorem of Laurent SCHWARTZ is given by $\langle T, \varphi \otimes \psi\rangle$ for a distribution $T \in \mathcal{D}'(\Omega \times \mathbb{S}^{N-1})$; then, because $B(\varphi, \psi) \geq 0$ whenever $\varphi \geq 0$ and $\psi \geq 0$, T is a nonnegative distribution, thus a nonnegative Radon measure $\mu \in \mathcal{M}(\Omega \times \mathbb{S}^{N-1})$ by another theorem of Laurent SCHWARTZ, more elementary. □

In [105], I wrote a simple proof of what I needed, based on properties of Hilbert–Schmidt operators, whose kernels belong to L^2.[44] I do not remember why I did not mention that in my student days my advisor, Jacques-Louis

[43] So that one may actually take $C(K) = \limsup_{m\to\infty} \int_K |u_m|^2\,dx$.

[44] I first used a regularization, which gave a Hilbert–Schmidt operator, with a kernel in L^2, and I checked that there was a uniform bound for the L^1 norm of the kernel.

LIONS, told me that he wrote with Lars GÅRDING a simple proof of the kernel theorem, which I then read; maybe, I did not know how to find the reference of their article at that time, which is easy now with MathSciNet, the AMS online database for *Math Reviews*, as they only wrote one joint article [33], but the referee mentions that the exposition of the last section (the only one that I read, on the kernel theorem of Laurent SCHWARTZ) is based on a method of Leon EHRENPREIS [25]![45]

Corollary 28.6. *If $U^n \rightharpoonup 0$ in $L^2_{loc}(\Omega;\mathbb{C}^p)$ weak, there exists a subsequence U^m and a nonnegative $p\times p$ Hermitian symmetric $\mu\in\mathcal{M}(\Omega\times\mathbb{S}^{N-1};\mathbb{C}^{p\times p})$, called the H-measure associated to the subsequence, such that for all $j,k\in\{1,\ldots,p\}$, all $\varphi_1,\varphi_2\in C_c(\Omega)$ and all $\psi\in C(\mathbb{S}^{N-1})$ one has*

$$\int_{\mathbb{R}^N}\mathcal{F}(\varphi_1 U_j^m)\,\overline{\mathcal{F}(\varphi_2 U_k^m)}\,\psi\Big(\frac{\xi}{|\xi|}\Big)\,d\xi\to\langle\mu^{j,k},\varphi_1\overline{\varphi_2}\otimes\psi\rangle. \tag{28.30}$$

Proof: By the same arguments as for Theorem 28.5 repeated for all j,k, one may extract a subsequence and the limit is $\langle T^{j,k},\varphi_1\overline{\varphi_2}\otimes\psi\rangle$ for distributions $T^{j,k}\in\mathcal{D}'(\Omega\times\mathbb{S}^{N-1})$, $j,k=1,\ldots,p$; obviously, one has $T^{k,j}=\overline{T^{j,k}}$ for $j,k=1,\ldots,p$. Then, the H-measure of the sequence U_j^m is $T^{j,j}\in\mathcal{M}(\Omega\times\mathbb{S}^{N-1})$, the H-measure of the sequence $U_j^m+U_k^m$ is $T^{j,j}+T^{k,k}+2\Re T^{j,k}$, and the H-measure of the sequence $U_j^m+i\,U_k^m$ is $T^{j,j}+T^{k,k}+2i\,\Im T^{j,k}$, showing that $T^{j,k}\in\mathcal{M}(\Omega\times\mathbb{S}^{N-1})$. Then, for $w_1,\ldots,w_p\in C(\Omega)$ the H-measure of the sequence $\sum_j w_jU_j^m$ is $\sum_{j,k}\mu^{j,k}w_j\overline{w_k}\ge 0$ (in $\mathcal{M}(\Omega\times\mathbb{S}^{N-1})$).[46] □

I then proved the *localization principle*, Theorem 28.7, and a few consequences, among them an improved version of *compensated compactness*, Corollary 28.11. My reason for choosing the name was that if $u_n\rightharpoonup u_\infty$ in $L^2_{loc}(\Omega)$ weak, and satisfies

$$\sum_{j=1}^N b_j\frac{\partial u_n}{\partial x_j}=f_n \text{ stays in a compact of } H^{-1}_{loc}(\Omega) \text{ strong,} \tag{28.31}$$

for $b_1,\ldots,b_N$ of class C^1, and u_n-u_∞ defines an H-measure μ, then

$$\begin{gathered}P\,\mu=0, \text{ with } P(x,\xi)=\textstyle\sum_{j=1}^N b_j(x)\,\xi_j,\\ \text{i.e., } support(\mu)\subset \text{ zero set of } P,\end{gathered} \tag{28.32}$$

[45] Leon EHRENPREIS, American mathematician, and orthodox rabbi, born in 1930. He worked at NYU (New York University), New York, NY, and he works now at Temple University, Philadelphia, PA.

[46] Using a partition of unity, one can deduce that $\sum_{j,k}\mu^{j,k}\Phi_j\overline{\Phi_k}\ge 0$ for all $\Phi_1,\ldots,\Phi_p\in C(\Omega\times\mathbb{S}^{N-1})$, but it is useful to recall that by the Radon–Nikodym theorem, if $\nu=\sum_j\mu^{j,j}$ one has $\mu^{j,k}=M^{j,k}\nu$ with $M^{j,k}$ being ν-measurable, and $M(x,\xi)$ is Hermitian ≥ 0 (and $trace(M)=1$) for ν a.e. $(x,\xi)\in\Omega\times\mathbb{S}^{N-1}$.

which is Corollary 28.8. A consequence is that one has $\mu = 0$ if the zero set of P is empty,[47] which implies strong convergence in $L^2_{loc}(\Omega)$. In the vector case, the algebraic equations obtained by applying the localization principle do not always imply constraints for the support of μ.

Theorem 28.7. *Let $U^n \rightharpoonup U^\infty$ in $L^2_{loc}(\Omega; \mathbb{C}^p)$ weak and satisfy*

$$\sum_{j=1}^N \sum_{k=1}^p \frac{\partial(A_{j,k} U_k^n)}{\partial x_j} \text{ belongs to a compact of } H^{-1}_{loc}(\Omega) \text{ strong,} \tag{28.33}$$

with $A_{j,k} \in C(\Omega)$ for $j = 1, \ldots, N$, and $k = 1, \ldots, p$. Then, if $U^n - U^\infty$ defines an H-measure μ, one has

$$\sum_{j=1}^N \sum_{k=1}^p \xi_j A_{j,k} \mu^{k,\ell} = 0 \text{ in } \Omega \times \mathbb{S}^{N-1}, \text{ for } \ell = 1, \ldots, p. \tag{28.34}$$

Proof: For $\phi \in C^1_c(\Omega)$, $V^n = \varphi U^n - \varphi U^\infty \rightharpoonup 0$ in $L^2(\mathbb{R}^N; \mathbb{C}^p)$ weak, satisfies

$$\sum_{j=1}^N \sum_{k=1}^p \frac{\partial(A_{j,k} V_k^n)}{\partial x_j} \to 0 \text{ in } H^{-1}(\mathbb{R}^N) \text{ strong,} \tag{28.35}$$

and defines the H-measure $|\varphi|^2 \mu$. Using $\frac{\partial}{\partial x_j} = (-\Delta)^{1/2} R_j$ (where R_j is the (M.) Riesz operator), (28.35) is equivalent to

$$\sum_{j=1}^N \sum_{k=1}^p R_j A_{j,k} V_k^n \to 0 \text{ in } L^2(\mathbb{R}^N) \text{ strong,} \tag{28.36}$$

because (28.35) has the form $\sum_{j=1}^N \frac{\partial f_{j,n}}{\partial x_j} \to 0$ in $H^{-1}(\mathbb{R}^N)$ strong, or $\sum_{j=1}^N \frac{\xi_j \mathcal{F} f_{j,n}}{1+|\xi|} \to 0$ in $L^2(\mathbb{R}^N)$ strong, and for $|\xi| \geq \eta > 0$, it implies the strong convergence of $\sum_{j=1}^N \frac{\xi_j \mathcal{F} f_{j,n}}{|\xi|}$, but for $|\xi| \leq \eta$ one uses the fact that $\mathcal{F} f_{j,n}$ is bounded in L^∞, as $f_{j,n}$ converges weakly to 0 in $L^2(\mathbb{R}^N)$ and keeps its support in a fixed compact set of $\mathbb{R}^N$, and letting η tend to 0 one deduces that $\sum_{j=1}^N \frac{\xi_j \mathcal{F} f_{j,n}}{|\xi|} \to 0$ in $L^2(\mathbb{R}^N)$ strong; then, one uses $R_j = P_{a_j}$ with $a_j = \frac{i\,\xi_j}{|\xi|}$. By Corollary 28.6

$$\langle \xi_j A_{j,k} |\varphi|^2 \mu^{k,\ell}, \varphi_1 \overline{\varphi_2} \otimes \psi \rangle = \lim_m \int_{\mathbb{R}^N} (P_\psi M_{\varphi_1} R_j M_{A_{j,k}} V_k^n) \overline{M_{\varphi_2} V_\ell^n}\, dx, \tag{28.37}$$

[47] If the functions b_j take complex values, $j = 1, \ldots, N$, it may happen that the zero set of P is empty without all the b_j being 0.

so that summing in j and k makes $\sum_{j,k} R_j A_{j,k} V_k^n$ appear, and the sum of the limits is 0, i.e.,

$$\sum_{j=1}^{N}\sum_{k=1}^{p} \langle \xi_j A_{j,k} |\varphi|^2 \mu^{k,\ell}, \varphi_1 \overline{\varphi_2} \otimes \psi \rangle = 0 \tag{28.38}$$

and varying φ, φ_1, φ_2, and ψ gives (28.34). □

Corollary 28.8. *If $u_n \rightharpoonup u_\infty$ in $L^2_{loc}(\Omega)$ weak, and satisfies*

$$\sum_{j=1}^{N} \frac{\partial(b_j u_n)}{\partial x_j} = f_n \textit{ stays in a compact of } H^{-1}_{loc}(\Omega) \textit{ strong,} \tag{28.39}$$

with $b_1, \ldots, b_N \in C(\Omega)$, and if $u_n - u_\infty$ defines an H-measure μ, then

$$P\,\mu = 0, \textit{ with } P(x,\xi) = \sum_{j=1}^{N} b_j(x)\,\xi_j. \tag{28.40}$$

Proof: It is an obvious application of Theorem 28.7, and μ^{11} is denoted μ. □

Corollary 28.9. *If $w_n \rightharpoonup w_\infty$ in $H^1_{loc}(\Omega)$ weak, and $grad(w_n) - grad(w_\infty)$ defines an H-measure $\mu \in \mathcal{M}(\Omega \times \mathbb{S}^{N-1}; \mathbb{C}^{N\times N})$, then*

$$\mu^{j,k} = \xi_j \xi_k \pi \textit{ for } j,k = 1,\ldots,N, \textit{ with } \pi \in \mathcal{M}(\Omega \times \mathbb{S}^{N-1}), \pi \geq 0. \tag{28.41}$$

Proof: Here $U^n = grad(w_n)$ is curl free, but the same result is true if $U^n \rightharpoonup U^\infty$ in $L^2_{loc}(\Omega)$ weak, $U^n - U^\infty$ defines an H-measure μ, and $\frac{\partial U_j^n}{\partial x_k} - \frac{\partial U_k^n}{\partial x_j}$ stays in a compact of $H^{-1}_{loc}(\Omega)$ strong for all j,k. Indeed, by Theorem 28.7

$$\begin{aligned} &\xi_k \mu^{j,\ell} - \xi_j \mu^{k,\ell} = 0 \text{ for } j,k,\ell = 1,\ldots,N, \text{ so that} \\ &\mu^{j,\ell} = \xi_j \nu^\ell \text{ (and } \nu^\ell = \textstyle\sum_k \xi_k \mu^{k,\ell}) \text{ for } j,\ell = 1,\ldots,N, \end{aligned} \tag{28.42}$$

and then the Hermitian symmetry of μ gives

$$\begin{aligned} &\xi_j \nu^k = \xi_k \overline{\nu^j} \text{ for } j,k = 1,\ldots,N, \text{ so that} \\ &\nu^k = \xi_k \pi \text{ (and } \pi = \overline{\textstyle\sum_j \xi_j \nu^j}) \text{ for } k = 1,\ldots,N, \end{aligned} \tag{28.43}$$

and then $\pi = \sum_k \xi_k \nu^k = \sum_k \mu^{k,k} \geq 0$ by the nonnegativity of μ. □

Corollary 28.10. *If $w_n \rightharpoonup w_\infty$ in $H^1_{loc}(\Omega)$ weak, $\Omega \subset \mathbb{R}^N \times \mathbb{R}$, $grad_{x,t}(w_n) - grad_{x,t}(w_\infty)$ defines an H-measure $\mu \in \mathcal{M}(\Omega \times \mathbb{S}^N; \mathbb{C}^{(N+1)\times(N+1)})$, and*

$$\frac{\partial}{\partial t}\Big(\varrho\frac{\partial w_n}{\partial t}\Big) - \sum_{i,j=1}^N \frac{\partial}{\partial x_i}\Big(a_{i,j}\frac{\partial w_n}{\partial x_j}\Big) \textit{ stays in a compact of } H^{-1}_{loc}(\Omega) \textit{ strong,} \tag{28.44}$$

with ϱ, $a_{i,j} \in C(\Omega)$ *for all* i,j, *then, using* x_{N+1} *for* t, *one has*

$$\begin{aligned} &\mu^{j,k} = \xi_j\xi_k\pi \textit{ for } j,k = 1,\ldots,N+1, \textit{ with } \pi \in \mathcal{M}(\Omega\times\mathbb{S}^N), \pi \geq 0,\\ &Q\,\pi = 0, \textit{ with } Q(x,\xi) = \varrho(x)\,\xi_{N+1}^2 - \textstyle\sum_{i,j=1}^N a_{i,j}(x)\,\xi_i\xi_j. \end{aligned} \tag{28.45}$$

Proof: That $\mu^{j,k} = \xi_j\xi_k\pi$ with a nonnegative $\pi \in \mathcal{M}(\Omega\times\mathbb{S}^N)$, for all j,k, follows from Corollary 28.9. One extends the vector $U^n = grad_{x,t}(w_n)$ by adding $N+1$ new components, with

$$\begin{aligned} &U^n_{N+1+i} = -\textstyle\sum_{j=1}^N a_{i,j}U^n_j \text{ for } i = 1,\ldots,N,\\ &U^n_{2N+2} = \varrho\,U^n_{N+1}, \end{aligned} \tag{28.46}$$

and the extended sequence U^n converges in $L^2_{loc}(\Omega)$ weak to an extended U^∞ satisfying (28.46) with n replaced by ∞, and the extended $U^n - U^\infty$ defines an extended H-measure μ which is now a $(2N+2)\times(2N+2)$ Hermitian symmetric matrix, such that

$$\begin{aligned} &\mu^{N+1+i,\ell} = -\sum_{j=1}^N a_{i,j}\mu^{j,\ell} = -\Big(\sum_{j=1}^N a_{i,j}\xi_j\Big)\xi_\ell\pi, i = 1,\ldots,N, \text{ for all } \ell,\\ &\mu^{2N+2,\ell} = \varrho\,\mu^{N+1,\ell} = \varrho\,\xi_{N+1}\xi_\ell\pi \text{ for all } \ell, \end{aligned} \tag{28.47}$$

but (28.44) and Theorem 28.7 imply

$$\begin{aligned} &\textstyle\sum_{i=1}^{N+1}\frac{\partial U^n_{N+1+i}}{\partial x_i} \text{ stays in a compact of } H^{-1}_{loc}(\Omega) \text{ strong, so that}\\ &\textstyle\sum_{i=1}^{N+1}\xi_i\mu^{N+1+i,\ell} = 0 \text{ for all } \ell, \text{ i.e., } Q\,\xi_\ell\pi = 0 \text{ for all } \ell, \end{aligned} \tag{28.48}$$

by using (28) and the definition of Q, and it means $Q\,\pi = 0$. □

One should notice that it is not assumed that (28.44) is a wave equation for Corollary 28.10 to hold, and the coefficients could all take complex values, for example.

Corollary 28.11. *If* $U^n \rightharpoonup U^\infty$ *in* $L^2_{loc}(\Omega;\mathbb{C}^p)$ *weak and*

$$\sum_{j=1}^N\sum_{k=1}^p \frac{\partial(A_{i,j,k}U^n_k)}{\partial x_j} \in \textit{ compact of } H^{-1}_{loc}(\Omega) \textit{ strong}, i = 1,\ldots,q, \tag{28.49}$$

with $A_{i,j,k} \in C(\Omega)$ *for* $i = 1,\ldots,q$, $j = 1,\ldots,N$, $k = 1,\ldots,p$ *and*

$$Q(x;U^n) = \sum_{k,\ell=1}^p q_{k,\ell}U^n_k\overline{U^n_\ell} \rightharpoonup \nu \textit{ in } \mathcal{M}(\Omega) \textit{ weak} \star, \tag{28.50}$$

with $q_{k,\ell} \in C(\Omega)$ *and* $q_{\ell,k} = \overline{q_{k,\ell}}$ *for* $k, \ell = 1, \ldots, p$, *then*

$$\begin{aligned}&\nu \geq Q(x; U^\infty) \text{ in } \Omega \text{ if for all } (x,\xi) \in \Omega \times \mathbb{S}^{N-1}, \lambda \in \Lambda_{x,\xi}, Q(x;\lambda) \geq 0,\\ &\nu = Q(x; U^\infty) \text{ in } \Omega \text{ if for all } (x,\xi) \in \Omega \times \mathbb{S}^{N-1}, \lambda \in \Lambda_{x,\xi}, Q(x;\lambda) = 0,\\ &\Lambda_{x,\xi} = \{\lambda \in \mathbb{C}^p \mid \textstyle\sum_{j=1}^N \sum_{k=1}^p \xi_j A_{i,j,k}\lambda_k = 0, \text{ for } i = 1, \ldots, q\}.\end{aligned} \tag{28.51}$$

Proof: Using (28.30) for $\psi = 1$ and the Plancherel formula, one finds that for a sequence converging weakly to 0 and defining an H-measure μ, the weak $\star$ limit of $U_k^n \overline{U_\ell^n}$ is the projection of $\mu^{k,\ell}$ in x, formally written as $\int_{\mathbb{S}^{N-1}} d\mu^{k,\ell}$. Here, $U^n \rightharpoonup U^\infty$, a subsequence $U^m - U^\infty$ defines an H-measure μ, and

$$Q(x; U^m) \rightharpoonup Q(x; U^\infty) + \int_{\mathbb{S}^{N-1}} \sum_{k,l} q_{k,l} d\mu^{k,\ell} \text{ in } \mathcal{M}(\Omega) \text{ weak } \star, \tag{28.52}$$

so that (28.51) is proven by showing that $\sum_{k,\ell} q_{k,\ell} d\mu^{k,\ell} \geq 0$ or $\sum_{k,\ell} q_{k,\ell} d\mu^{k,\ell} = 0$ in $\Omega \times \mathbb{S}^{N-1}$, since by Corollary 28.6 one can always extract subsequences defining an H-measure. Using the Radon–Nikodym theorem, one writes $\mu = M(x,\xi)\,\pi$ with $\pi = \sum_k \mu^{k,k}$, and π a.e. $M(x,\xi)$ is a nonnegative Hermitian symmetric matrix, and one must show that π a.e. one has $\sum_{k,\ell} q_{k,\ell} M_{k,\ell}(x,\xi) \geq 0$ or $\sum_{k,\ell} q_{k,\ell} M_{k,\ell}(x,\xi) = 0$. Then, Theorem 28.7 and (28.49) imply

$$\sum_{j=1}^N \sum_{k=1}^p \xi_j A_{i,j,k} \mu^{k,\ell} = 0, i = 1, \ldots, q, \tag{28.53}$$

so that the range of $M^T(x,\xi)$ is included in $\Lambda_{x,\xi}$, but $M(x,\xi)$ being Hermitian nonnegative it is then a sum of $\lambda \otimes \lambda$ with $\lambda \in \Lambda_{x,\xi}$, showing (28.51). □

It suggested that I develop a calculus with a class of symbols, and in [105], I used functions in $C_0(\mathbb{R}^N)$ in order to use the first commutation lemma (Lemma 28.2 with Corollary 28.3), but as $M_1 = I$ commutes with P_a, one may consider continuous functions having a limit at ∞, which I denote $C(\overline{\mathbb{R}^N})$.[48] Definition 28.12 is adapted to sequences converging weakly in $L^2(\mathbb{R}^N; \mathbb{C}^p)$.

Definition 28.12. *A function* s *on* $\mathbb{R}^N \times \mathbb{S}^{N-1}$ *is a* symbol *if*

$$\begin{aligned}&s(x,\xi) = \textstyle\sum_k a_k\left(\frac{\xi}{|\xi|}\right) b_k(x) \text{ with } a_k \in C(\mathbb{S}^{N-1}), b_k \in C(\overline{\mathbb{R}^N})\\ &\text{and } \textstyle\sum_k ||a_k||_{C(\mathbb{S}^{N-1})} ||b_k||_{C(\overline{\mathbb{R}^N})} < +\infty,\end{aligned} \tag{28.54}$$

and the standard operator *with symbol* s *is*

$$S_s = \sum_k P_{a_k} M_{b_k}, \tag{28.55}$$

[48] I use $\overline{\mathbb{R}^N}$ for the Aleksandrov one point compactification of $\mathbb{R}^N$.

An operator $T \in \mathcal{L}\big(L^2(\mathbb{R}^N); L^2(\mathbb{R}^N)\big)$ is said to have symbol s if $T - S_s$ is a compact operator from $L^2(\mathbb{R}^N)$ into itself.

For $v \in L^2(\mathbb{R}^N)$, one has $(\mathcal{F} S_s v)(\xi) = \sum_k a_k\big(\frac{\xi}{|\xi|}\big)(\mathcal{F} M_{b_k} v)(\xi)$, and as $(\mathcal{F} M_{b_k} v)(\xi) = \int_{\mathbb{R}^N} b_k(x) v(x)\, e^{-2i\,\pi\,(x,\xi)}\, dx$ one deduces that

$$(\mathcal{F} S_s v)(\xi) = \int_{\mathbb{R}^N} s\Big(x, \frac{\xi}{|\xi|}\Big)\, v(x)\, e^{-2i\,\pi\,(x,\xi)}\, dx \text{ in } \mathbb{R}^N, \tag{28.56}$$

for all $v \in L^2(\mathbb{R}^N)$, so that S_s only depends upon the symbol s (and not upon its decomposition). An example of an operator with symbol s is

$$L_s = \sum_k M_{b_k} P_{a_k}, \tag{28.57}$$

because the sum of the norms of the commutators $[M_{b_k}, P_{a_k}]$ is finite, and as each of them is compact, the sum is compact. Using $P_{a_k} v = \mathcal{F}^{-1}(a_k \mathcal{F} v) = \int_{\mathbb{R}^N} a_k\big(\frac{\xi}{|\xi|}\big)\, (\mathcal{F} v)(\xi)\; e^{+2i\,\pi\,(x,\xi)}\, d\xi$, one deduces that

$$L_s v(x) = \int_{\mathbb{R}^N} s\Big(x, \frac{\xi}{|\xi|}\Big)\, (\mathcal{F} v)(\xi)\, e^{+2i\,\pi\,(x,\xi)}\, d\xi \text{ in } \mathbb{R}^N, \tag{28.58}$$

so that L_s only depends upon the symbol s. The classical theory of pseudo-differential operators uses L_s, with smooth symbols, and it corresponds to $L_s(e^{+2i\,\pi\,(\cdot,\eta)}) = s(\cdot,\eta)(e^{+2i\,\pi\,(\cdot,\eta)})$ for all $\eta \in \mathbb{R}^N$, explaining what the symbols mean. I introduced the standard operators because they have a meaning on $\Omega \times \mathbb{S}^{N-1}$ with $\Omega \neq \mathbb{R}^N$ if the b_k have their support in a fixed compact K of Ω, although I only described the symbols corresponding to using sequences converging in $L^2(\mathbb{R}^N; \mathbb{C}^p)$ weak, and this explains the use of φ_1, φ_2 in Lemma 28.13, which tells one that an H-measure is adapted to computing some weak $\star$ limits, of sesqui-linear forms using operators with various symbols.

Lemma 28.13. *If T_1 and T_2 have symbols s_1 and s_2, then $T_1 T_2$ (as well as $T_2 T_1$) has the symbol $s_1 s_2$, and the adjoint T_1^* has the symbol $\overline{s_1}$.*

Proof: If $T_j = S_j + K_j$ for $j = 1, 2$, where S_j is the standard operator of symbol s_j and K_j is compact, then $T_1 T_2 = S_1 S_2 + (K_1 S_2 + S_1 K_2 + K_1 K_2)$, and the operators $K_1 S_2$, $S_1 K_2$, and $K_1 K_2$ are compact. Notice that $S_1 S_2$ is not in general the standard operator of symbol $s_1 s_2$: if $S_1 = \sum_m P_{a_m} M_{b_m}$ with $\sum_m ||a_m||\,||b_m|| < +\infty$, and $S_2 = \sum_n P_{c_m} M_{d_m}$ with $\sum_n ||c_n||\,||d_n|| < +\infty$, then $S_1 S_2 = \sum_{m,n} P_{a_m} M_{b_m} P_{c_m} M_{d_m}$ and for each m, n one has

$$\begin{aligned}
&(P_{a_m} M_{b_m})\,(P_{c_m} M_{d_m}) = P_{a_m c_n} M_{b_m d_n} + K_{m,n},\\
&\text{with } K_{m,n} = P_{a_m} [M_{b_m}, P_{c_n}]\, M_{d_m} \text{ compact,}\\
&\text{and } ||K_{m,n}|| \le 2||a_m||\,||b_m||\,||c_n||\,||d_n||,
\end{aligned} \tag{28.59}$$

so that the sum of the $K_{m,n}$ is compact as the sum of their norms is finite, and $\sum_{m,n} ||a_m c_n||\,||b_m d_n|| < +\infty$, as $||a_m c_n|| \leq ||a_m||\,||c_n||$, and $||b_m d_n|| \leq ||b_m||\,||d_n||$.

Similarly, $(S_1 + K_1)^* = S_1^* + K_1^*$, and K_1^* is compact, and $S_1^* = \sum_m M_{\overline{b_m}}^* P_{a_m}^*$. ☐

Lemma 28.14. *If $U^m \rightharpoonup U^\infty$ in $L^2_{loc}(\Omega;\mathbb{C}^p)$ weak, and $U^m - U^\infty$ defines an H-measure μ, then for all $\varphi_1, \varphi_2 \in C_c(\Omega)$, T_1, T_2 having symbols s_1, s_2, and $k, \ell = 1, \ldots, p$, one has*

$$T_1(\varphi_1 U_k^m)\,\overline{T_2(\varphi_2 U_\ell^m)} \rightharpoonup T_1(\varphi_1 U_k^\infty)\,\overline{T_2(\varphi_2 U_\ell^\infty)} + \nu \text{ in } \mathcal{M}(\Omega) \text{ weak} \star, \\ \text{with } \langle \nu, \varphi\rangle = \langle \mu^{k,\ell}, \varphi\,\varphi_1 s_1 \overline{\varphi_2 s_2}\rangle \text{ for all } \varphi \in C_c(\Omega), \tag{28.60}$$

or formally $\nu = \int_{\mathbb{S}^{N-1}} \varphi_1 s_1 \overline{\varphi_2 s_2}\, d\mu^{k,\ell}$.

Proof: By replacing U^n by $U^n - U^\infty$, one may assume that $U^\infty = 0$. One then needs to compute the limit of $\int_{\mathbb{R}^N} T_1(\varphi_1 U_k^m)\,\overline{T_2(\varphi_2 U_\ell^m)}\,\varphi\, dx$, and one may replace T_1 by S_{s_1} and T_2 by S_{s_2} without changing the limits. It is then enough to consider one element in each sum, i.e., identify the limit of $\int_{\mathbb{R}^N} P_{a_1} M_{b_1}(\varphi_1 U_k^m)\,\overline{P_{a_2} M_{b_3}(\varphi_2 U_\ell^m)}\,\varphi\, dx$, which is done using Corollary 28.3 and Corollary 28.6. ☐

As for examples, I first checked the periodic case,[49] and then the modulated periodic case, choosing the unit cube $Q = (0,1)^N$ as the period in order to simplify the Fourier transform, which makes the dual lattice appear, so that I considered

$$u_n(x) = v\big(x, \tfrac{x}{\varepsilon_n}\big) \text{ with } v(x,y) \text{ being } Q\text{-periodic in } y, \\ v(x,y) = \textstyle\sum_{m\in\mathbb{Z}^N} v_m(x) e^{2i\,\pi\,(m,y)}, \tag{28.61}$$

and I assumed v smooth enough and $v_0 = 0$, so that $u_n \rightharpoonup 0$ in $L^2_{loc}(\Omega)$; a sufficient regularity hypothesis is to assume v continuous in x with values in $L^\infty(Q)$, which I thought too restrictive, and I only needed v continuous in x with values in $L^2(Q)$ by defining the sequence in a slightly different way, which I find more natural,[50] and choosing $z \in \mathbb{R}^N$ as origin, I used

$$u_n(x) = v\big(x_p, \tfrac{x}{\varepsilon_n}\big) \text{ for } x \in z + p\,\varepsilon_n Q, p \in \mathbb{Z}^N, \\ \text{with } x_p \text{ chosen in } z + p\,\varepsilon_n Q, p \in \mathbb{Z}^N. \tag{28.62}$$

[49] In the small-amplitude homogenization problem that I looked, which I describe in Chap. 29, the periodic case was simple, but its knowledge was not of much help for showing how to invent *H-measures*!

[50] I do not like the idea of comparing a general sequence to a modulated periodic one, and I heard a few years ago that Jacques-Louis LIONS found a counter-example to the original result of two-scale convergence, so that, unless he made a mistake or his statement was misinterpreted, it seems that no new "proof" of that result could be correct!

One transforms the case (28.61) into the case (28.62) by using a choice function π giving $\pi(x) = x_p$ in the cube $z + p\,\varepsilon_n Q$, and on any compact set K the difference between $v\big(x, \frac{x}{\varepsilon_n}\big)$ and $v\big(\pi(x), \frac{x}{\varepsilon_n}\big)$ is uniformly small in the L^∞ norm, and it does not change the H-measure.

Lemma 28.15. *Under (28.62), the whole sequence defines the H-measure $\mu = \sum_{m\neq 0} |v_m|^2 \otimes \delta_{\frac{m}{|m|}}$, i.e., for all $\Phi \in C_c(\mathbb{R}^N \times \mathbb{S}^{N-1})$ one has*

$$\langle \mu, \Phi\rangle = \sum_{m\in\mathbb{Z}^N\setminus\{0\}} \int_{\mathbb{R}^N} |v_m(x)|^2 \Phi\Big(x, \frac{m}{|m|}\Big)\,dx. \tag{28.63}$$

Proof: One approaches π by piecewise constant functions on disjoint small open sets G_α which do not shrink with ε_n, any compact K being the union of a finite number of G_α plus a set of Lebesgue measure zero. As v is continuous in x with values in $L^2(Q)$, one makes a small error in the $L^2(K)$ norm, which creates a small change for the H-measure (in the norm of $\mathcal{M}(K \times \mathbb{S}^{N-1})$). The new sequence w_n obtained is such that $|w_n|^2$ converges weakly in $L^1(K)$, so that its associated H-measure cannot charge the boundaries of the open sets G_α, and one only needs to identify ν on each $G_\alpha \times \mathbb{S}^{N-1}$, which amounts to finding what the H-measure μ is for the periodic case

$$\begin{aligned} &w(y) = \textstyle\sum_{m\in\mathbb{Z}^N\setminus\{0\}} w_m e^{2i\,\pi\,(m,y)},\\ &\mathcal{F}\big[\varphi\, w\big(\tfrac{\cdot}{\varepsilon_n}\big)\big] = \textstyle\sum_{m\in\mathbb{Z}^N\setminus\{0\}} w_m \mathcal{F}\varphi\big(\cdot - \tfrac{m}{\varepsilon_n}\big), \end{aligned} \tag{28.64}$$

for $\varphi \in \mathcal{S}(\mathbb{R}^N)$ with $\mathcal{F}\varphi \in C_c^\infty(\mathbb{R}^N)$, and for ε_n small enough

$$\begin{aligned} &\big|\textstyle\sum_{m\in\mathbb{Z}^N\setminus\{0\}} w_m \mathcal{F}\varphi\big(\xi - \tfrac{m}{\varepsilon_n}\big)\big|^2 = \textstyle\sum_{m\in\mathbb{Z}^N\setminus\{0\}} |w_m|^2 \big|\mathcal{F}\varphi\big(\xi - \tfrac{m}{\varepsilon_n}\big)\big|^2,\\ &\lim \textstyle\int_{\mathbb{R}^N} \big|\mathcal{F}\varphi\big(\xi - \tfrac{m}{\varepsilon_n}\big)\big|^2 \psi\big(\tfrac{\xi}{|\xi|}\big)\,d\xi = \psi\big(\tfrac{m}{|m|}\big) \int_{\mathbb{R}^N} |\mathcal{F}\varphi(\xi)|^2\,d\xi \text{ for } m \neq 0, \end{aligned} \tag{28.65}$$

for $\psi \in C(\mathbb{S}^{N-1})$, so that one finds

$$\begin{aligned} &\langle \mu, |\varphi|^2 \otimes \psi\rangle = \big(\textstyle\int_{\mathbb{R}^N} |\varphi(x)|^2\,dx\big)\,\big[\textstyle\sum_{m\in\mathbb{Z}^N\setminus\{0\}} |w_m|^2 \psi\big(\tfrac{m}{|m|}\big)\big],\\ &\quad\text{i.e., } \mu = 1 \otimes \big(\textstyle\sum_{m\in\mathbb{Z}^N\setminus\{0\}} |w_m|^2 \delta_{\frac{m}{|m|}}\big) \end{aligned} \tag{28.66}$$

by density of the functions of the form $\Phi = |\varphi|^2 \otimes \psi$. □

Lemma 28.16. *If u_n (showing a concentration effect at $z \in \mathbb{R}^N$) is given by*

$$u_n(x) = \varepsilon_n^{-N/2} f\Big(\frac{x-z}{\varepsilon_n}\Big) \text{ with } f \in L^2(\mathbb{R}^N), \tag{28.67}$$

it defines an H-measure $\mu = \delta_z \otimes \nu$ *where* ν *has a surface density*

$$\begin{aligned} &\nu(\xi) = \int_0^\infty |\mathcal{F}f(t\,\xi)|^2\,dt \text{ for } \xi \in \mathbb{S}^{N-1}, \text{ i.e.,} \\ &\langle \mu, \Phi\rangle = \int_{\mathbb{R}^N} |\mathcal{F}f(\xi)|^2 \Phi\big(z, \tfrac{\xi}{|\xi|}\big)\,d\xi \text{ for all } \Phi \in C_c(\mathbb{R}^N \times \mathbb{S}^{N-1}). \end{aligned} \tag{28.68}$$

Proof: For $\varphi \in C_c(\Omega)$, $\varphi\, u_n - \varphi(z)\, u_n \to 0$ in $L^2(R^N)$ and one notices that

$$\int_{\mathbb{R}^N} |\mathcal{F}u_n(\xi)|^2 \psi\Big(\frac{\xi}{|\xi|}\Big)\,d\xi = \int_{\mathbb{R}^N} |\mathcal{F}f(\xi)|^2 \psi\Big(\frac{\xi}{|\xi|}\Big)\,d\xi, \tag{28.69}$$

and it gives the value of $\langle \mu, |\varphi|^2 \otimes \psi\rangle$. □

Corollary 28.17. *If* $\mu \in \mathcal{M}_b(\Omega \times \mathbb{S}^{N-1})$ *has norm* A^2, *then for* $B > A$ *there exists a sequence* u_n *defining the H-measure* μ *satisfying* $||u_n||_{L^2(\mathbb{R}^N)} \le B$.

Proof: One easily constructs a sequence $\mu_n \rightharpoonup \mu$ in $\mathcal{M}(\Omega \times \mathbb{S}^{N-1})$ weak $\star$, with total mass $\le M < B^2$, which can be obtained as an H-measure, either using Lemma 28.15 or Lemma 28.16. □

Finally, it is useful to notice that Theorem 28.7 has a converse.

Lemma 28.18. *Let* $U^n \rightharpoonup U^\infty$ *in* $L^2_{loc}(\Omega; \mathbb{C}^p)$ *weak, such that* $U^n - U^\infty$ *defines an H-measure* μ, *and* $A_{j,k} \in C(\Omega)$ *for* $j = 1, \ldots, N$, *and* $k = 1, \ldots, p$. *If (28.34) holds, then (28.33) must hold.*

Proof: (28.33) is equivalent to the condition that for all $\varphi \in C^1_c(\Omega)$ the sequence $V^n = \varphi\,(U^n - U^\infty)$ satisfies

$$\sum_{j=1}^N \sum_{k=1}^p \frac{\partial(A_{j,k} V^n_k)}{\partial x_j} \to 0 \text{ in } H^{-1}(\mathbb{R}^N) \text{ strong}, \tag{28.70}$$

itself equivalent to

$$w_n = \sum_{j=1}^N \sum_{k=1}^p R_j(A_{j,k} V^n_k) \to 0 \text{ in } L^2(\mathbb{R}^N) \text{ strong}, \tag{28.71}$$

and if one defines $S_k(x,\xi) = \sum_{j=1}^N \xi_j A_{j,k}$, then

$$\text{the H-measure of } w_n \text{ is } \pi = \sum_{k,\ell=1}^p S_k(x,\xi)\overline{S_\ell(x,\xi)}\mu^{k,\ell}. \tag{28.72}$$

(28.34) means

$$\sum_{k=1}^p S_k(x,\xi)\mu^{k,\ell} = 0 \text{ for } \ell = 1, \ldots, p, \tag{28.73}$$

and multiplying by $\overline{S_\ell(x,\xi)}$ and summing in ℓ gives $\pi = 0$, equivalent to $w_n \to 0$ in $L^2_{loc}(\mathbb{R}^N)$ strong; w_n having its support in a fixed compact set, one deduces that $w_n \to 0$ in $L^2(\mathbb{R}^N)$ strong, and it implies (28.33). □

Lemma 28.18 shows that conditions of belonging to a compact of $H^{-1}_{loc}(\Omega)$ strong, which I started using in the late 1970s, are equivalent to the statement about *H-measures*, which I only introduced in the late 1980s. Because I did not see such a condition used before I introduced it myself, it suggests that anyone using it follows one of my ideas, even for those who carefully avoid mentioning my name for most of my ideas!

Additional footnotes: Nachman ARONSZAJN,[51] Michael ATIYAH,[52] Erik BALDER,[53] Raoul BOTT,[54] Clement XII,[55] Pierre DELIGNE,[56] DE PAUL,[57]

[51] Nachman ARONSZAJN, Polish-born mathematician, 1907–1980. He worked at The University of Kansas, Lawrence, KS, where I met him during my first visit to United States, in 1971.

[52] Sir Michael Francis ATIYAH, British mathematician, born in 1929. He received the Fields Medal in 1966 primarily for his work in topology. He received the Abel Prize in 2004 with Isadore M. SINGER for their discovery and proof of the index theorem, bringing together topology, geometry and analysis, and their outstanding role in building new bridges between mathematics and theoretical physics. He worked in Oxford, holding the Savilian chair of geometry, in Cambridge, England, and in Edinburgh, Scotland.

[53] Erik Jan BALDER, Dutch mathematician, born in 1949. He works in Utrecht, The Netherlands.

[54] Raoul BOTT, Hungarian-born mathematician, 1923–2005. He received the Wolf Prize in 2000 for his deep discoveries in topology and differential geometry and their applications to Lie groups, differential operators and mathematical physics, jointly with Jean-Pierre SERRE. He worked at University of Michigan, Ann Arbor, MI, and at Harvard University, Cambridge, MA. In 1948, Raoul BOTT was one of the first students in the graduate program at Carnegie Tech (Carnegie Institute of Technology), Pittsburgh, PA, and he was a PhD student of my late colleague Richard DUFFIN.

[55] Clement XII (Lorenzo CORSINI), Italian Pope, 1652–1740. He was elected Pope in 1730.

[56] Pierre DELIGNE, Belgian-born mathematician, born in 1944. He received the Fields Medal in 1978 for his work in algebraic geometry. He received the Crafoord Prize in 1988, jointly with Alexandre GROTHENDIECK, who declined it. He worked at IHES (Institut des Hautes Études Scientifiques), Bures sur Yvette, France, and at IAS (Institut for Advanced Study), Princeton, NJ.

[57] Saint Vincent DE PAUL, French Catholic priest, 1581–1660. He was canonized by Clement XII in 1737. He founded many charitable organizations. DePaul University in Chicago, IL, is named after him.

Emilio GAGLIARDO,[58] Lars GÅRDING,[59] Pippo (Giuseppe) GEYMONAT,[60] Alexandre GROTHENDIECK,[61] Martin LAZAR,[62] RAMANUJAN,[63] SATO,[64] SAVILE,[65] Jean-Pierre SERRE,[66] Isadore SINGER,[67] TATE,[68] WASHINGTON,[69] WAYNE.[70]

---

[58] Emilio GAGLIARDO, Italian mathematician, 1930–2008. He worked in Genova (Genoa), and in Pavia, Italy.

[59] Lars GÅRDING, Swedish mathematician, born in 1919. He worked at Lund University, Lund, Sweden.

[60] Giuseppe GEYMONAT, Italian-born mathematician, born in 1939. He worked in Torino (Turin), Italy, at ENS (Ecole Normale Supérieure) Cachan, France, and he works now at Université des Sciences et Techniques de Languedoc (Montpellier II), Montpellier, France.

[61] Alexander GROTHENDIECK, German-born mathematician, born in 1928. He received the Fields Medal in 1966 for his work in algebraic geometry. He received the Crafoord Prize in 1988, jointly with Pierre DELIGNE, but he declined it. He worked at CNRS (Centre National de la Recherche Scientifique) in Paris, at IHES (Institut des Hautes Études Scientifiques) in Bures sur Yvette, at Collège de France (visiting for two years), Paris, and in Montpellier, France.

[62] Martin LAZAR, Croatian mathematician, born in 1975. He works in Zagreb, Croatia.

[63] Srinivasa Aiyangar RAMANUJAN, Indian mathematician, 1887–1920.

[64] Mikio SATO, Japanese mathematician, born in 1928. He received the Wolf Prize for 2002/2003, for his creation of "algebraic analysis," including hyperfunction and microfunction theory, holonomic quantum field theory, and a unified theory of soliton equations, jointly with John T. TATE. He worked at RIMS (Research Institute for Mathematical Sciences) at Kyoto University, Kyoto, Japan.

[65] Sir Henry SAVILE, English mathematician, 1549–1622. He endowed in 1619 a chair at Oxford, England, named after him, the Savilian chair of geometry.

[66] Jean-Pierre SERRE, French mathematician, born in 1926. He received the Fields Medal in 1954 for his work in algebraic topology. He received the Wolf Prize in 2000 for his many fundamental contributions to topology, algebraic geometry, algebra, and number theory and his inspirational lectures and writing, jointly with Raoul BOTT. He received the Abel Prize in 2003 for playing a key role in shaping the modern form of many parts of mathematics, including topology, algebraic geometry and number theory. He held a chair at Collège de France (algebra and geometry, 1956–1994), Paris.

[67] Isadore Manual SINGER, American mathematician, born in 1924. He received the Abel Prize in 2004 with Sir Michael Francis ATIYAH for their discovery and proof of the index theorem, bringing together topology, geometry and analysis, and their outstanding role in building new bridges between mathematics and theoretical physics. He worked at MIT (Massachusetts Institute of Technology), Cambridge, MA.

[68] John Torrence TATE, American mathematician, born in 1925. He received the Wolf Prize for 2002/2003, for his creation of fundamental concepts in algebraic number theory, jointly with Mikio SATO. He worked at Harvard University, Cambridge, MA, and at University of Texas, Austin, TX.

[69] George WASHINGTON, American general, 1732–1799. He was the first President of the United States.

[70] Anthony WAYNE, American general, 1745–1796. Wayne State University, Detroit, MI, is named after him.

Chapter 29
Small-Amplitude Homogenization

In June 1980, after finding the correct interval for the conductivity of an effective isotropic mixture of two isotropic conductors, I did not think about checking the (inexact) formula that LANDAU and LIFSHITZ wrote for the conductivity of a mixture in [47]. It was only in the fall of 1986 that I thought about it: their formula is one of the two Hashin–Shtrikman bounds![1] At that time, I understood a little better than in 1974 about the way physicists think, and I continued reading what they wrote after:[2] for the case when the conductivity $a(x)$ does not vary much, they proposed the formula

$$a_{eff} \approx \overline{a} - \frac{\overline{(a-\overline{a})^2}}{3\overline{a}}, \text{ in dimension } N = 3, \tag{29.1}$$

where a bar denotes an average value, and they deduced

$$a_{eff}^{1/3} \approx \overline{a^{1/3}}, \text{ in dimension } N = 3, \tag{29.2}$$

and using the sign $\approx$ instead of $=$ is important.[3] The functionals which I used for proving bounds on effective coefficients in 1980 are not limited to mixing only two materials,[4] and after carrying the computations for the case of several materials with Gilles FRANCFORT and François MURAT, we were surprised to find that (29.1) is a good approximation for the effective isotropic case (when the variations in conductivity are small, of course). It is always

---

[1] Although they did not mention any difference between the two materials, they obviously considered one as inclusions in a matrix made of the other.

[2] I realize now that in 1986 I looked at a different edition than the one I had read in the early 1970s.

[3] Unfortunately, physicists often use $=$ between nearby quantities, instead of $\approx$, and then they are surprised that taking derivatives of their "equality" creates problems, but if they had used $\approx$ it would be more easy to explain that nearby quantities may have very different derivatives!

[4] In the effective isotropic case first, and then in the effective anisotropic case with François MURAT, with the results shown in Lemma 21.6 and Lemma 21.7.

L. Tartar, *The General Theory of Homogenization*, Lecture Notes of the Unione Matematica Italiana 7, DOI 10.1007/978-3-642-05195-1_29,

amazing for mathematicians to find that physicists' arguments defying elementary logic give good results, and it often happens because one puts in the hypothesis what one wants in the conclusion, but it was not the case there, and I needed to find a rational explanation for the efficiency of the result.

My interpretation of (29.1) was that it comes from a question of small-amplitude homogenization

$$A^n = A + \gamma\, B^n \text{ with } |\gamma| \text{ small, and } B^n \rightharpoonup 0 \text{ in } L^\infty\big(\Omega; \mathcal{L}(\mathbb{R}^N;\mathbb{R}^N)\big) \text{ weak}\star, \tag{29.3}$$

where after extraction of a subsequence A^m which H-converges to $A^{\mathit{eff}}(\cdot;\gamma)$ for all γ small, one finds an effective limit which is analytic in γ,

$$A^{\mathit{eff}}(\cdot;\gamma) = A + \gamma^2 M^2 + \gamma^3 M^3 + \ldots, \tag{29.4}$$

and that (29.1) is about expressing M^2 in a particular situation:

$$\begin{aligned}&\text{if } A = a\,I, \text{ if } B^n = b_n\,I, \text{ with } b_n \rightharpoonup 0, b_n^2 \rightharpoonup \sigma \text{ in } L^\infty(\Omega) \text{ weak}\star,\\ &\text{and if } M^2 = m_2 I, \text{ then } m_2 = -\tfrac{\sigma}{N\,a}.\end{aligned} \tag{29.5}$$

For the general case, a new tool is necessary, for which I first had the intuition in 1984 for the problem described in Chap. 19, and I introduced *H-measures* for proving more general results than (29.5), like

$$\begin{aligned}&\text{if } B^n = b_n\,I, b_n \rightharpoonup 0 \text{ and defines an H-measure } \mu \in \mathcal{M}(\Omega \times \mathbb{S}^{N-1}),\\ &\text{then } M^2(x) = -\int_{\mathbb{S}^{N-1}} \tfrac{\xi\otimes\xi}{(A\,\xi,\xi)}\, d\mu(x,\xi) \text{ a.e. in } \Omega,\end{aligned} \tag{29.6}$$

from which it can be deduced, since if $A = a\,I$ one has $trace(M^2) = -\frac{\sigma}{a}$, since σ is the projection (in x) of μ.

My interpretation of (29.2) was that it is about finding Φ such that

$$\Phi\big(A^{\mathit{eff}}(\cdot;\gamma)\big) = \text{weak} \star \text{ limit of } \Phi(A^m) + O(\gamma^3), \tag{29.7}$$

and if one assumes that $A^{\mathit{eff}}(\cdot;\gamma)$ is isotropic, at least at order 2, i.e.

$$A^{\mathit{eff}}(\cdot;\gamma) = a\,I - \frac{\gamma^2\sigma}{N\,a}\,I + O(\gamma^3), \tag{29.8}$$

by (29.5), and if $\Phi(\kappa\,I) = \Phi_0(\kappa)\,I$ for all $\kappa > 0$, one uses the Taylor expansion of Φ_0 at order 2 for identifying the coefficients of σ, and one obtains

$$\frac{\Phi_0''(a)}{2} = -\frac{\Phi_0'(a)}{N\,a}, \tag{29.9}$$

and it is true for all $a > 0$ if one takes

$$\begin{aligned} \Phi_0(t) &= t^{(N-2)/N} \text{ for all } t > 0 \text{ if } N \geq 3 \\ \Phi_0(t) &= \log t \text{ for all } t > 0 \text{ if } N = 2. \end{aligned} \tag{29.10}$$

Of course, although (29.1) or (29.2) are only shown to be accurate for small variations of the conductivity, one may be optimistic and use the average of $\Phi_0(a)$ as a guess for $\Phi_0(a_{\mathit{eff}})$, even for large variations of the conductivity.

Georges MATHERON used the case $N = 2$ in the 1960s, for estimating the coefficient in the Darcy law, and since the coefficient of porosity does not contain enough information, he introduced other geometrical quantities,[5] calling his approach stereometry.[6]

The general theory of *H-measures* applies to all problems of small-amplitude homogenization, with sometimes lengthy computations of linear algebra to perform for systems,[7] but here I only study a general second-order equation, without assuming A^n symmetric,[8] and I consider a sequence

$$A^n(x;\gamma) = A(x) + \gamma\, B^n(x) + \gamma^2 C^n(x), x \in \Omega, |\gamma| \text{ small} \tag{29.11}$$

where all tensors involved have entries in a bounded set of $L^\infty(\Omega)$. For simplification, I assume that A is continuous,[9] and I assume that there exists $\alpha > 0$ such that $(A(x)\, v, v) \geq \alpha\, |v|^2$ for $x \in \Omega$ and $v \in \mathbb{R}^N$, that B^n converges weakly $\star$ to B^∞, that C^n converges weakly $\star$ to C^∞, and that $B^n - B^\infty$ corresponds to an H-measure μ. After extracting a subsequence, there is an effective limit $A^{\mathit{eff}}(x;\gamma)$ defined for small γ (eventually complex) and analytic in γ, by an argument similar to that used in Chap. 22, which I recall quickly. Since Ω is a countable union of bounded open sets, one may assume

---

[5] Because of surface tension, the pressure needed to push a liquid in an empty pipe with small diameter is large, and one considers the volume of void attainable from the outside by using passages of a minimum diameter, a function of the pressure used; if in an open set Ω the solid part is $\Omega \setminus \omega$ for an open set ω, the global porosity is the ratio $\mathit{measure}(\omega)/\mathit{measure}(\Omega)$ of the volume of void to the total volume, but if for $s > 0$ one defines the open sets $A_s = \{x \in \omega \mid B(x,s) \subset \omega\}$, $B_s = \bigcup_{x \in A_s} B(x,s)$, and ω_s is the connected component of B_s which touches $\partial\omega$, then the usable porosity is $\lim_{s\to 0} \mathit{measure}(\omega_s)/\mathit{measure}(\Omega)$.

[6] I do not know if Georges MATHERON pointed out that for a cracked medium it is imperative to consider elastic deformations (in nonlinear elasticity, of course, and not linearized elasticity!), which may open some cracks, and create a dramatic increase in the effective porosity.

[7] In my computations for linearized elasticity, I made a mistake which Gilles FRANCFORT pointed out to me, and a numerical constant is then wrong in [106].

[8] Although symmetry holds for most applications in continuum mechanics or physics, Graeme MILTON has studied a question of (classical) Hall effect which requires a nonsymmetric conductivity tensor.

[9] The continuity hypothesis can be avoided by using the Meyers regularity theorem.

Ω bounded, let $V = H_0^1(\Omega)$ and $\mathcal{A}^n(\gamma)$ be the operator from V into V' defined by $\langle \mathcal{A}^n(\gamma)\, u, v\rangle = \int_\Omega \big(A^n(\cdot;\gamma)\, grad(u), grad(v)\big)\, dx$ for $u, v \in V$, and notice that for $|\gamma|$ small the Lax–Milgram lemma applies; for f in a countable dense set of V', and for γ in a countable dense set of a small disc $|z| < \gamma_c$, subsequences of $u_n = \big(\mathcal{A}^n(\gamma)\big)^{-1} f$ and $A^n(\cdot;\gamma)\, grad(u_n)$ converge weakly in $L^2(\Omega;\mathbb{C}^N)$, so that by a Cantor diagonal argument and equicontinuity in γ a subsequence exists for which the convergence holds for all $f \in V'$ and all γ in the small disc, and the limits are analytic in γ.

Theorem 29.1. *The effective conductivity tensor $A^{\mathit{eff}}(\cdot;\gamma)$ satisfies*

$$A^{\mathit{eff}}(\cdot;\gamma) = A + \gamma\, B^\infty + \gamma^2 (C^\infty + M) + O(\gamma^3), \tag{29.12}$$

and the second-order H-correction $M \in L^\infty\big(\Omega; \mathcal{L}(\mathbb{R}^N;\mathbb{R}^N)\big)$ satisfies

$$\int_\Omega M_{i,j}\varphi\, dx = -\sum_{k,\ell=1}^N \Big\langle \mu^{i,k;\ell,j}, \frac{\varphi\, \xi_k \xi_\ell}{(A\,\xi,\xi)} \Big\rangle \text{ for all } \varphi \in C_c(\Omega), i,j = 1,\ldots,N. \tag{29.13}$$

Proof. Remark that, since $B^n - B^\infty$ is indexed by a pair of numbers, the H-measure μ is indexed by two pairs, or four indices. For every u_* in V, one takes $f = \mathcal{A}^{\mathit{eff}}(\gamma)\, u_*$ and one constructs the (sub)sequences $E^m(\cdot;\gamma) = grad(u_m)$ and $D^m(\cdot;\gamma) = A^m(\cdot;\gamma)\, E^m(\cdot;\gamma)$, which satisfy

$$\begin{aligned} &\tfrac{\partial E^m_j}{\partial x_k} - \tfrac{\partial E^m_k}{\partial x_j} = 0 \text{ in } \Omega, j,k = 1,\ldots,N,\\ &\textstyle\sum_{j=1}^N \tfrac{\partial D^m_j}{\partial x_j} = -\mathcal{A}^{\mathit{eff}}(\gamma)\, u_* \text{ in } \Omega,\\ &E^m_j \rightharpoonup \tfrac{\partial u_*}{\partial x_j} \text{ in } L^2(\Omega) \text{ weak}, j = 1,\ldots,N,\\ &D^m_j \rightharpoonup \textstyle\sum_{k=1}^N A^{\mathit{eff}}_{j,k}(\cdot;\gamma)\, \tfrac{\partial u_*}{\partial x_k} \text{ in } L^2(\Omega) \text{ weak}, j = 1,\ldots,N. \end{aligned} \tag{29.14}$$

Using the definition of u_m and the analyticity in γ, one has

$$\begin{aligned} &E^m(\cdot;\gamma) = grad(u_*) + \gamma\, E^{m,1} + \gamma^2 E^{m,2} + O(\gamma^3),\\ &E^{m,1} \rightharpoonup 0, E^{m,2} \rightharpoonup 0 \text{ in } L^2(\Omega;\mathbb{R}^N) \text{ weak},\\ &D^m(\cdot;\gamma) = A\, grad(u_*) + \gamma\, D^{m,1} + \gamma^2 D^{m,2} + O(\gamma^3), \end{aligned} \tag{29.15}$$

and, using the definition (29.11) of $A^m(\cdot;\gamma)$, one has

$$\begin{aligned} &D^{m,1} = A\, E^{m,1} + B^m grad(u_*) \rightharpoonup B^\infty grad(u_*) \text{ in } L^2(\Omega;\mathbb{R}^N) \text{ weak},\\ &D^{m,2} = A\, E^{m,2} + B^m E^{m,1} + C^m grad(u_*). \end{aligned} \tag{29.16}$$

Since $A^{\mathit{eff}}(\cdot;\gamma)$ is analytic in γ, it has an expansion

$$A^{\mathit{eff}}(\cdot;\gamma) = A + \gamma\, B + \gamma^2 C + O(\gamma^3) \tag{29.17}$$

and since u_* is arbitrary in V, (29.16) implies $B = B^\infty$, and

$$C\,grad(u_*) = C^\infty grad(u_*) + \text{ weak limit of } B^m E^{m,1}. \tag{29.18}$$

Identifying the corrector of order 2 in γ then requires the computation of the weak limit of $(B^m - B^\infty)\,E^{m,1}$ (since $E^{m,1} \rightharpoonup 0$), computed from the H-measure π associated to (a subsequence of) $(B^m - B^\infty, E^{m,1})$, which one can deduce from the H-measure μ corresponding to $B^m - B^\infty$ alone. One has

$$\begin{aligned} &\frac{\partial E_j^{m,1}}{\partial x_k} - \frac{\partial E_k^{m,1}}{\partial x_j} = 0 \text{ in } \Omega, j, k = 1, \ldots, N,\\ &\textstyle\sum_{j=1}^N \frac{\partial D_j^{m,1}}{\partial x_j} = div\big(B^\infty grad(u_*)\big) \text{ in } \Omega, \end{aligned} \tag{29.19}$$

and, using (29.16) for expressing $D^{m,1}$ in terms of $E^{m,1}$, one finds

$$div\big(A\,E^{m,1} + (B^m - B^\infty)\,grad(u_*)\big) = 0. \tag{29.20}$$

If one was on the whole space $\mathbb{R}^N$, and if $grad(u_*)$ was continuous (and using the continuity of A), (29.19)–(29.20) would mean that $E^{m,1}$ is obtained from $B^m - B^\infty$ by an operator S of order zero, whose symbol transforms a matrix b into the vector $\frac{(b\,grad(u_*),\xi)}{(A\,\xi,\xi)}\,\xi$,[10] and one would deduce the weak limit of $G^m = (B^m - B^\infty)\,E^{m,1}$ in terms of μ; however, there are some technical details to check, because if A is not constant, it is not clear if $\frac{1}{(A\,\xi,\xi)}$ belongs to my class of symbols.

The result follows from applying the localization principle (Theorem 28.7), as I did in [105], but here I want to show the algebraic manipulations in a different way, by working with operators and their symbols, like the (M.) Riesz operator R_j, $j = 1, \ldots, N$ (whose symbol is $\frac{i\,\xi_j}{|\xi|}$); one needs to localize (in x) before using them, so for $\psi \in C_c^1(\Omega)$, one observes that $e^m = \psi\,E^{m,1}$, and $f^m = \psi\,(B^m - B^\infty)\,grad(u_*)$ satisfy

$$\begin{aligned} &e^m, f^m \rightharpoonup 0 \text{ in } L^2(\mathbb{R}^N;\mathbb{R}^N) \text{ weak},\\ &\frac{\partial e_j^m}{\partial x_k} - \frac{\partial e_k^m}{\partial x_j} \to 0 \text{ in } H^{-1}(\mathbb{R}^N) \text{ strong}, j, k = 1, \ldots, N,\\ &div(A\,e^m + f^m) \to 0 \text{ in } H^{-1}(\mathbb{R}^N) \text{ strong}, \end{aligned} \tag{29.21}$$

[10] Essentially, the computations are done by freezing the values of A and $grad(u_*)$, and using the Fourier transform. It is one advantage of *H-measures* that they explain why computing with constant coefficients or with periodic data gives results which can be used in the case of (continuous) variable coefficients; physicists may say that they "knew" that for a long time, but the truth is that they were just doing it and they expected that it was valid: as far as I know, no one showed that it was a valid procedure before I did with my theory of *H-measures*!

and one rewrites the differential information on e^m, f^m as

$$\begin{aligned} &R_k e_j^m - R_j e_k^m \to 0 \text{ in } L^2(\mathbb{R}^N) \text{ strong}, j,k = 1,\ldots,N, \\ &\textstyle\sum_{j,k} R_j \widetilde{A}_{j,k} e_k^m + \sum_j R_j f_j^m \to 0 \text{ in } L^2(\mathbb{R}^N) \text{ strong}, \end{aligned} \tag{29.22}$$

where $\widetilde{A}_{j,k}$ stands for the operator of multiplication by a continuous function $\widetilde{A}_{j,k}$, which coincides with $A_{j,k}$ on the support of ψ, and tends to a constant at ∞.[11] Using $\sum_k R_k^2 = -I$, the information on e^m can be summarized as

$$\begin{aligned} &w_m = -\textstyle\sum_k R_k e_k^m \rightharpoonup 0 \text{ in } L^2(\mathbb{R}^N) \text{ weak}, \\ &e_j^m - R_j w_m \to 0 \text{ in } L^2(\mathbb{R}^N) \text{ strong}, j = 1,\ldots,N, \end{aligned} \tag{29.23}$$

by writing $e_j^m = -(\sum_k R_k^2)\, e_j^m = -\sum_k R_k(R_k e_j^m - R_j e_k^m) + R_j w_m$. Using the information on f^m then gives

$$\Big(\sum_{j,k=1}^N R_j \widetilde{A}_{j,k} R_k\Big) w_m + \sum_\ell R_\ell f_\ell^m \to 0 \text{ in } L^2(\mathbb{R}^N) \text{ strong}, \tag{29.24}$$

so that if S is the standard operator of symbol $s(x,\xi) = \sum_{j,k=1}^N \widetilde{A}_{j,k}(x)\, \xi_j \xi_k$, and recalling that $f^m = \psi\,(B^m - B^\infty)\, grad(u_*)$, one has

$$S\, w_m - \sum_{\ell,j=1}^N R_\ell \psi\, (B_{\ell,j}^m - B_{\ell,j}^\infty)\, \frac{\partial u_*}{\partial x_j} \to 0 \text{ in } L^2(\mathbb{R}^N) \text{ strong}. \tag{29.25}$$

One wants to compute the weak limit of $G^m = (B^m - B^\infty)\, E^{m,1}$ in terms of μ, and write that it is $M\, grad(u_*)$, and thus identify M; $M_{i,j}$ is then the coefficient of $\frac{\partial u_*}{\partial x_j}$ in the weak limit of $G_i^m = \sum_k (B_{i,k}^m - B_{i,k}^\infty)\, E_k^{m,1}$, which is the projection (in x) of $\sum_k \pi^{i,k;k}$, where π is the H-measure associated to (a subsequence of) $(B^m - B^\infty, E^{m,1})$. Replacing $E_k^{m,1}$ by $S\,\psi\, E_k^{m,1}$ multiplies $\pi^{i,k;k}$ by $s\,\psi$, but since $e_k^m = \psi\, E_k^{m,1}$ behaves like $R_k w_m$, (29.25) implies

$$s\,\psi\,\pi^{i,k;k} = -\sum_{\ell,j=1}^N \xi_k \xi_\ell \psi\, \mu^{i,k;\ell,j}\, \frac{\partial u_*}{\partial x_j}, \tag{29.26}$$

from which one deduces (29.13). □

---

[11] One chooses $\theta \in C_c(\Omega)$ with $0 \le \theta \le 1$ in Ω and $\theta = 1$ on the support of ψ, and $\widetilde{A} = \theta\, A + (1-\theta)\, I$, and the reason for doing this is to have A defined on $\mathbb{R}^N$ and having a limit at ∞, since I used $C(\overline{\mathbb{R}^N})$ in defining my class of symbols.

For a sequence B^n, the knowledge of all the H-correction terms for various matrices A almost characterizes the H-measure μ, since the contributions at ξ and $-\xi$ are always added and cannot be separated, so that one can only expect to characterize the even part of μ. More precisely, if a Radon measure ν on $\mathbb{S}^{N-1}$ is even in ξ and one defines the function g by $g(A) = \langle \nu, \frac{1}{(A\,\xi,\xi)} \rangle$ for positive matrices A, then the mapping $\nu \mapsto g$ is injective. Indeed, by differentiating successively in directions $C_1, \ldots, C_m$ at $A = I$, one obtains the integral of $(C_1 x, x) \cdots (C_m x, x)$, and all these even moments of ν characterize it.

It would be useful to avoid the Fourier transform in defining *H-measures*, and this property seems interesting in that respect, but a defect is that applying the Lax–Milgram lemma to a perturbation requires B^n to be bounded in $L^\infty(\Omega; \mathcal{L}(\mathbb{R}^N; \mathbb{R}^N))$, and defining *H-measures* for sequences in $L^2(\Omega)$ is crucial for many applications, where "energy" is quadratic.

I had an intuitive idea of *H-measures* in 1984, when I revisited the model problem of hydrodynamics described in Chap. 19, which I introduced in 1976 with the purpose of understanding some questions related to turbulence effects.[12] In my model problem, the oscillating coefficients do not appear in the highest-order terms, and *H-measures* grasp the exact effective behaviour, not just the beginning of a Taylor expansion as in Theorem 29.1. In Lemma 19.1, choosing $\lambda = 1$, I considered a stationary velocity field u^n satisfying

$$-\nu\,\Delta\,u^n + \big(u^n \times curl(v^\infty + w^n)\big) + grad(p_n) = f^n, div(u^n) = 0 \text{ in } \Omega \subset \mathbb{R}^3, \tag{29.27}$$

with $v^\infty \in L^3_{loc}(\Omega; \mathbb{R}^3)$, and I assumed that $u^n \rightharpoonup u^\infty$ in $H^1_{loc}(\Omega; \mathbb{R}^3)$ weak, $p_n \rightharpoonup p_\infty$ in $L^2_{loc}(\Omega)$ weak, $f^n \to f^\infty$ in $H^{-1}_{loc}(\Omega; \mathbb{R}^3)$ strong; for the forcing oscillating field w^n, I assumed that $w^n \rightharpoonup 0$ in $L^3_{loc}(\Omega; \mathbb{R}^3)$ weak, and under these hypotheses, I showed (as in [102]) that u^∞ satisfies

$$-\nu\,\Delta\,u^\infty + \big(u^\infty \times curl(v^\infty)\big) + M^{eff} u^\infty + grad(p_\infty) = f^\infty, div(u^\infty) = 0 \text{ in } \Omega, \tag{29.28}$$

where M^{eff} is a symmetric nonnegative tensor with coefficients in $L^{3/2}_{loc}(\Omega)$, and I noticed a quadratic dependence of M^{eff} with respect to w^n. I now express M^{eff} in terms of the H-measure of a subsequence w^m, as in [105].

Lemma 29.2. *If w^m defines an H-measure μ, then*

$$\int_\Omega M^{eff}_{i,j}\,\varphi\,dx = \tfrac{1}{\nu}\Big(\textstyle\sum_{k=1}^3 \langle \mu^{k,k}, \varphi \otimes \xi_i \xi_j\rangle - \sum_{k,\ell=1}^3 \langle \mu^{k,\ell}, \varphi \otimes \xi_i \xi_j \xi_k \xi_\ell\rangle\Big)$$
$$\text{for all } \varphi \in C_c(\Omega), i, j = 1, \ldots, N, \tag{29.29}$$

[12] The appearance of supplementary terms like $M^{eff} u$ in the equation should be completed with the appearance of nonlocal effects, like those which I described in Chap. 23, more particularly the example studied by Youcef AMIRAT, Kamel HAMDACHE, and Hamid ZIANI [1,2].

i.e. $M^{eff}_{i,j}$ *is the projection (in* x*) of* $\frac{\xi_i\xi_j}{\nu}\left(\sum_{k=1}^3 \mu^{k,k} - \sum_{k,\ell=1}^3 \xi_k\xi_\ell\mu^{k,\ell}\right)$.

Proof. For identifying M^{eff} on a bounded open set ω with $\overline{\omega} \subset \Omega$, the proof of Lemma 19.1 showed that for $k \in \mathbb{R}^3$

$$z^m \times curl(w^m) \rightharpoonup M^{eff} k \text{ in } H^{-1}(\omega;\mathbb{R}^3) \text{ weak}, \tag{29.30}$$

with $z^n \rightharpoonup 0$ in $H^1_0(\omega;\mathbb{R}^3)$ weak, $q_n \rightharpoonup 0$ in $L^2(\omega)$ weak, such that

$$-\nu\,\Delta\, z^n + \big(k \times curl(w^n)\big) + grad(q_n) = 0, div(z^n) = 0 \text{ in } \omega. \tag{29.31}$$

Changing w^m to be 0 outside ω, one defines Z^m and Q_m in $\mathbb{R}^3$ by

$$-\nu\,\Delta\, Z^m + \big(k \times curl(w^m)\big) + grad(Q_m) = 0, div(Z^m) = 0 \text{ in } \mathbb{R}^3, \tag{29.32}$$

which by regularity theory (using the Calderón–Zygmund theorem) has solutions with Q_m and all $\frac{\partial Z^m_j}{\partial x_i}$ converging to 0 in $L^3(\mathbb{R}^3)$ weak, and all $\frac{\partial (Z^m_j - z^m_j)}{\partial x_i}$ converging to 0 in $L^3_{loc}(\omega)$ strong, so that one may use Z^m instead of z^m in (29.30). Using the Fourier transform, (29.32) gives

$$4\pi^2\nu\,|\xi|^2\mathcal{F}\,Z^m = -P_{\xi^\perp}\big(k \times (2i\,\pi\,\xi \times \mathcal{F}w^m)\big), \tag{29.33}$$

where $P_{\xi^\perp}$ is the orthogonal projection onto $\xi^\perp$, so that for $v \in \mathbb{R}^3$

$$-P_{\xi^\perp}\big(k \times (2i\,\pi\,\xi \times v)\big) = -P_{\xi^\perp}\big((k,v)\,2i\,\pi\,\xi - (k,2i\,\pi\,\xi)\,v\big) = (k,2i\,\pi\,\xi)\,P_{\xi^\perp}v, \tag{29.34}$$

and since $P_{\xi^\perp}v = v - \frac{(v,\xi)\,\xi}{|\xi|^2}$, one deduces that

$$(P_{\xi^\perp}\mathcal{F}w^m)_s = \mathcal{F}\Big(w^m_s + R_s\sum_{\ell=1}^3 R_\ell w^m_\ell\Big), s = 1,2,3, \tag{29.35}$$

$$\frac{\partial Z^m_s}{\partial x_i} = \frac{1}{\nu}R_i\Big(\sum_{j=1}^3 k_j R_j\Big)\Big(w^m_s + R_s\sum_{\ell=1}^3 R_\ell w^m_\ell\Big), i,s = 1,2,3. \tag{29.36}$$

Using $z^m \to 0$ in $L^p(\omega;\mathbb{R}^3)$ strong for $2 \le p < 6$, and $div(z^m) = 0$, (29.30) is

$$-\sum_s \frac{\partial z^m_s}{\partial x_i}w^m_s \rightharpoonup (M^{eff}k)_i \text{ in } H^{-1}(\omega) \text{ weak}, i = 1,2,3, \tag{29.37}$$

and one may replace $\frac{\partial z^m_s}{\partial x_i}$ by $\frac{\partial Z^m_s}{\partial x_i}$ without changing the limit, so that $M^{eff}_{i,j}$ is the limit of

$$-\frac{1}{\nu}\sum_s R_i R_j\Big(w^m_s + R_s\sum_{\ell=1}^3 R_\ell w^m_\ell\Big)w^m_s, \tag{29.38}$$

which gives (29.29). □

The fact that Lemma 19.1 also had the information that $M_{i,j}^{\mathit{eff}} \in L^{3/2}_{loc}(\Omega)$ is related to the following general result.

Lemma 29.3. *For an open set $\omega \subset \mathbb{R}^N$, if for some $q > 2$, $u_n \rightharpoonup 0$ in $L^q(\omega)$ weak and corresponds to an H-measure μ, then the projection (in x) of μ belongs to $L^{q/2}(\omega)$. Let $U^n \rightharpoonup 0$ in $L^2(\omega;\mathbb{R}^p)$ weak, corresponding to an H-measure ν; if U_i^n is bounded in $L^{q_i}(\omega)$ and U_j^n is bounded in $L^{q_j}(\omega)$ with either $q_i > 2$ or $q_j > 2$, then the projection (in x) of $|\nu^{i,j}|$ belongs to $L^r(\omega)$ with $\frac{1}{r} = \frac{1}{q_i} + \frac{1}{q_j}$.*

Proof. Since u_n^2 is bounded in $L^{q/2}(\omega)$, and $\frac{q}{2} > 1$, a subsequence u_m^2 converges to σ in $L^{q/2}(\omega)$ weak ($L^\infty(\omega)$ weak $\star$ if $q = \infty$), and σ is the projection of μ.

For $k, \ell = 1, \ldots, p$, let $\pi^{k,\ell} \geq 0$ be the projection of $|\nu^{k,\ell}|$, so that if $q_i = 2$ and $q_j > 2$,[13] one has $\pi^{i,i} = f_i\,dx + \sigma$ with $f_i \in L^1(\omega)$ and σ singular with respect to the Lebesgue measure, and $\pi^{j,j} = f_j\,dx$ with $f_j \in L^{q_j/2}(\omega)$. Since ν is Hermitian nonnegative, one has

$$\langle |\nu^{i,j}|, \varphi\rangle^2 \leq \langle \nu^{i,i}, \varphi\rangle\,\langle \nu^{j,j}, \varphi\rangle \text{ for all } \varphi \in C_c(\omega \times \mathbb{S}^{N-1}), \varphi \geq 0, \tag{29.39}$$

and taking φ independent of ξ gives

$$\langle \pi^{i,j}, \varphi\rangle^2 \leq \langle \pi^{i,i}, \varphi\rangle\,\langle \pi^{j,j}, \varphi\rangle \text{ for all } \varphi \in C_c(\omega), \varphi \geq 0. \tag{29.40}$$

Since σ lives on a Borel set of Lebesgue measure 0, one deduces from (29.40) that $\pi^{i,j} = g\,dx$ and $g^2 \leq f_i f_j$ a.e. in ω, so that $g \in L^r(\omega)$. □

Another application of *H-measures* related to the small-amplitude homogenization approach is to deduce the Taylor expansion at order 2 on the diagonal for the functions attached to geometries which I described in Chap. 22.

Lemma 29.4. *Let a sequence of characteristic functions $\chi_n \rightharpoonup \theta$ in $L^\infty(\Omega)$ weak $\star$, with $\chi_n - \theta$ defining an H-measure μ and $A^m = \chi_m M^1 + (1-\chi_m)M^2$ H-converging to $A^{\mathit{eff}} = F(\cdot; M^1, M^2)$ for all $M^1, M^2 \in \mathcal{L}_+(\mathbb{R}^N;\mathbb{R}^N)$. Then, if $A \in \mathcal{L}_+(\mathbb{R}^N;\mathbb{R}^N)$ and for $Q \in \mathcal{L}(\mathbb{R}^N;\mathbb{R}^N)$ with $||Q||$ small enough*

$$F(\cdot; A+Q, A) = A + \theta\,Q - Q\left(\int_{\mathbb{S}^{N-1}} \frac{\xi\otimes\xi}{(A\,\xi,\xi)}\,d\mu\right)Q + o(||Q||^2) \text{ in } \Omega. \tag{29.41}$$

Proof. One applies Theorem 29.1 to $A + \gamma\,\chi_m Q$, so that $B^m = \chi_m Q$, $B^\infty = \theta\,Q$ and $\mu^{i,k;\ell,j} = Q_{i,k}Q_{\ell,j}\mu$ for all i,j,k,ℓ, and one obtains $A^{\mathit{eff}}(\cdot;\gamma) = A + \gamma\,\theta\,Q + \gamma^2 M + O(\gamma^3)$ with

[13] If $i = j$, the first part of the argument applies.

$$M_{i,j} = -\int_{\mathbb{S}^{N-1}} \Big(\sum_{k,\ell=1}^N Q_{i,k} Q_{\ell,j} \frac{\xi_k \xi_\ell}{(A\,\xi,\xi)} \Big)\, d\mu \text{ in } \Omega, i,j = 1,\ldots,N, \tag{29.42}$$

and then $\sum_{k,\ell=1}^N Q_{i,k}Q_{\ell,j}\xi_k\xi_\ell = (Q\,\xi)_i(Q^T\xi)_j = (Q(\xi\otimes\xi)Q)_{i,j}$. □

In the late 1980s, while François MURAT was visiting me in Pittsburgh, we used the same kind of argument for characterizing the *H-measures* associated to characteristic functions, and I mentioned the result in [105].

Lemma 29.5. *If a sequence of characteristic functions $\chi_n \rightharpoonup \theta$ in $L^\infty(\Omega)$ weak $\star$, with $\chi_n - \theta$ defining an H-measure μ, then $\mu \geq 0$ is even in ξ, and with projection (in x) $\theta\,(1-\theta)\,dx$. Conversely, if $0 \leq \theta \leq 1$ a.e. in Ω and $\mu \in \mathcal{M}(\Omega\times\mathbb{S}^{N-1})$ is nonnegative, even in ξ, and with projection $\theta\,(1-\theta)\,dx$, then there exists a sequence of characteristic functions χ_n converging to θ in $L^\infty(\Omega)$ weak $\star$, with $\chi_n - \theta$ defining the H-measure μ.*

Proof. For a real scalar sequence $u_n \rightharpoonup u_\infty$ in $L^2_{loc}(\Omega)$ weak with $u_n - u_\infty$ defining an H-measure μ, $\mu \in \mathcal{M}(\Omega\times\mathbb{S}^{N-1})$ is automatically nonnegative and even in ξ, and its projection is the limit of $(u_n-u_\infty)^2$ in $L^1_{loc}(\Omega)$ weak $\star$; here $(\chi_n-\theta)^2 = \chi_n(1-2\theta)+\theta^2$ converges to $\theta\,(1-\theta)$ in $L^\infty(\Omega)$ weak $\star$.

Then, we interpreted the formulas which I described in Chap. 27 for repeated laminations, for example (27.21), which says that laminating materials with tensors M^1 and M^2 in proportions η and $1-\eta$ orthogonally to $e\in\mathbb{S}^{N-1}$ gives an effective tensor M given by

$$\begin{aligned} &M = \eta\,M^1 + (1-\eta)\,M^2 - \eta\,(1-\eta)(M^2-M^1)\,R\,(M^2-M^1),\\ &\text{with } R = \tfrac{e\otimes e}{(1-\eta)\,(M^1\,e,e)+\eta\,(M^2\,e,e)}, \end{aligned} \tag{29.43}$$

so that if one has two mixtures of $A+\gamma\,Q$ and A giving $M^1 = A+\theta_1\gamma\,Q - \gamma^2 Q\,N^1 Q + o(\gamma^2)$ and $M^2 = A+\theta_2\gamma\,Q-\gamma^2 Q\,N^2Q+o(\gamma^2)$, then M^2-M^1 is $O(\gamma)$ and one needs R at order 0 in γ, which is $\frac{e\otimes e}{(A\,e,e)}$, and one obtains

$$\begin{aligned} &M = A + (\eta\,\theta_1+(1-\eta)\,\theta_2)\,\gamma\,Q - \gamma^2 Q\,N\,Q + o(\gamma^2)\\ &\text{with } N = \eta_1 N^1 + (1-\eta)\,N^2 + \eta\,(1-\eta)\,(\theta_2-\theta_1)^2 \tfrac{e\otimes e}{(A\,e,e)}. \end{aligned} \tag{29.44}$$

The coefficient of $\gamma\,Q$ gives the proportion used of $A+\gamma\,Q$, and formula (29.44) tells one that if a first mixture corresponds to a proportion θ_1 and an H-measure μ_1, and a second mixture corresponds to a proportion θ_2 and an H-measure μ_2, then one can create a mixture corresponding to proportion $\eta\,\theta_1+(1-\eta)\,\theta_2$ and H-measure $\eta\,\mu_1+(1-\eta)\,\mu_2+\eta\,(1-\eta)\,(\theta_2-\theta_1)^2\delta_e$.[14]

[14] This uses (29.41) and the fact already mentioned that if a Radon measure ν on $\mathbb{S}^{N-1}$ is even in ξ and one defines the function g by $g(A) = \Big\langle \nu, \frac{1}{(A\,\xi,\xi)}\Big\rangle$ for positive matrices A, then the mapping $\nu\mapsto g$ is injective.

One notices that if $\theta_1 = \theta_2$, one does not add a Dirac mass at $e \in \mathbb{S}^{N-1}$ by laminating orthogonally to e, and one performs a convex combination for μ, so that mixing various materials obtained with the same proportion θ and showing a term $\theta\,(1-\theta)\,\delta_{e_j}$ gives a term $\theta\,(1-\theta)\,\nu$ where ν is any finite probability on $\mathbb{S}^{N-1}$.

In the preceding argument, all the mixtures considered use a constant proportion, on an open set ω, and after having obtained all piecewise constant proportions and *H-measures* which are finite probabilities, one uses a closure argument based on the metrizability of the topology used. □

I shall describe this result again as Lemma 33.2, in particular the last step for θ not constant.

Chapter 30
H-Measures and Bounds on Effective Coefficients

The method for obtaining bounds on effective coefficients, which I devised in the fall of 1977 while I was visiting MRC in Madison, WI, used the *compensated compactness method* and correctors, generalizing the method which I used earlier with François MURAT, based on the *div–curl lemma.* Since *H-measures* represent a better way to deal with the *compensated compactness* ideas which I introduced with François MURAT,[1] it was natural that I try to use them for studying bounds in homogenization.

As usual with H-convergence, $E^n = grad u_n$ and $D^n = A^n E^n$.[2] For F a homogeneous polynomial of degree 2 in E, D and A, some bounds are derived by computing the weak limit of $F(E^n, D^n, A^n)$ in two different ways.[3] The first way uses the classical idea of Young measures and bounds the quantity $F(E^n, A^n E^n, A^n)$ using the Young measure corresponding to the sequence A^n and the constraints that $E^n \rightharpoonup E^\infty$ and $A^n E^n \rightharpoonup A^{\mathit{eff}} E^\infty$ in $L^2(\Omega; \mathbb{R}^N)$ weak. The second way uses *H-measures* and computes the limit using the weak limits E^∞, D^∞ and A^∞ together with the H-measure μ' corresponding to the sequence $(E^n - E^\infty, D^n - D^\infty, A^n - A^\infty)$; then one bounds the result using only the H-measure μ corresponding to $A^n - A^\infty$ and taking into account the fact that $curl(E^n)$ and $div(D^n)$ are bounded. Comparing the two results may give some interesting inequalities satisfied by A^{eff}.

One defines the polynomial G of degree 2 in E and A by

$$G(E, A) = F(E, A\,E, A), \tag{30.1}$$

and the first question is to find the optimal information on the weak limit of $G(E^n, A^n)$ knowing that the Young measure associated to the sequence A^n is ν, and that the weak limits (in $L^2(\Omega; \mathbb{R}^N)$ weak) of E^n and $A^n E^n$ are

---

[1] However, H-measures have not made obsolete my *compensated compactness method,* based on using "entropies," since not much is understood yet about relations between the H-measures corresponding to U^n and to $F(U^n)$ for various nonlinear mappings F.

[2] One may consider more than one test field E^n, and use the correctors P^n.

[3] Adding terms of order $\leq$1 to F is not useful, since the weak limit of these terms is obvious.

L. Tartar, *The General Theory of Homogenization,* Lecture Notes of the Unione Matematica Italiana 7, DOI 10.1007/978-3-642-05195-1_30,

respectively E^∞ and $D^\infty = A^{\mathit{eff}} E^\infty$. For that, one uses the convex conjugate function G^* of G in the variable E, i.e.,

$$G^*(E^*, A) = \sup_{E \in \mathbb{R}^N} \big((E, E^*) - G(E, A)\big) \tag{30.2}$$

and for $e, e' \in \mathbb{R}^N$ one takes the limit of the inequality

$$G^*(e + A^T e', A) \geq (E, e) + (A\,E, e') - G(E, A). \tag{30.3}$$

One obtains

$$\begin{aligned}\langle \nu, G^*(e + A^T e', A)\rangle &= \lim_{n\to\infty} G^*(e + (A^n)^T e', A^n) \\ &\geq (E^\infty, e) + (A^{\mathit{eff}} E^\infty, e') - \lim_{n\to\infty} G(E^n, A^n),\end{aligned} \tag{30.4}$$

which implies

$$\lim_{n\to\infty} G(E^n, A^n) \geq \sup_{e,e' \in \mathbb{R}^N} \big((E^\infty, e) + (A^{\mathit{eff}} E^\infty, e') - \langle \nu, G^*(e + A^T e', A)\rangle\big). \tag{30.5}$$

In the case where G is a polynomial of degree 2 in E, one needs G convex in E in order to have $G^* \neq +\infty$, and G^* is then also a convex polynomial of degree 2 in E^*; more precisely, if

$$G(E, A) = \frac{1}{2}\,(\alpha(A)E, E) - (\beta(A), E), \tag{30.6}$$

with $\alpha(A) > 0$, then G^* has the form

$$G^*(E^*, A) = \frac{1}{2}\,\big(\alpha(A)^{-1}[E^* + \beta(A)], E^* + \beta(A)\big), \tag{30.7}$$

and this gives

$$\langle \nu, G^*(e + A^T e', A)\rangle = \frac{1}{2}\,\big\langle \nu, \big(\alpha(A)^{-1}[e + A^T e' + \beta(A)], e + A^T e' + \beta(A)\big)\big\rangle. \tag{30.8}$$

In order to find the best $e, e' \in \mathbb{R}^N$, one must solve the system

$$\begin{aligned}&\langle \nu, \alpha(A)^{-1}[e_0 + A^T e_0' + \beta(A)]\rangle = E^\infty \\ &\langle \nu, A\,\alpha(A)^{-1}[e_0 + A^T e_0' + \beta(A)]\rangle = A^{\mathit{eff}} E^\infty,\end{aligned} \tag{30.9}$$

and this gives a lower bound

$$\begin{aligned}\lim_{n\to\infty} F(E^n, A^n E^n, A^n) &\geq \tfrac{1}{2}\langle \nu, \alpha(A)^{-1}(e_0 + A^T e_0'), e_0 + A^T e_0'\rangle \\ &\quad - \tfrac{1}{2}\big\langle \nu, \big(\alpha(A)^{-1}\beta(A), \beta(A)\big)\big\rangle.\end{aligned} \tag{30.10}$$

Note that the condition $\alpha(A) > 0$ need only hold on the support of ν.

Using the H-measure μ' of the complete sequence $(E^n - E^\infty, D^n - D^\infty, A^n - A^\infty)$ the weak limit is $F(E^\infty, A^{\mathit{eff}} E^\infty, A^\infty)$ + an integral term involving μ'. One important fact to take into account is that, at the point $\xi \in \mathbb{S}^{N-1}$, the H-measure only sees E parallel to ξ and D perpendicular to ξ and that the part of μ' corresponding to $A^n - A^\infty$ alone is the given H-measure μ; however, there is one more constraint, which comes from the fact that $A^n E^n$ converges to $A^{\mathit{eff}} E^\infty$.

One assumes that F is concave in (E, D) when restricted to the subspace of vectors E parallel to ξ and D perpendicular to ξ, and one defines Φ by

$$\Phi(A, e, \xi) = \inf_{E \| \xi, D \perp \xi} \big((A\,E, e) - F(E, D, A)\big), \tag{30.11}$$

which gives Φ as a polynomial of degree 2 in A, since F is assumed to be a homogeneous polynomial of degree 2.

$$\begin{aligned}\lim\nolimits_{n\to\infty}\big(F(E^n, A^n E^n, A^n) - (A^n E^n, e)\big) &= F(E^\infty, D^\infty, A^\infty) - (A^\infty E^\infty, e)\\ &+ \quad \langle\!\langle \mu', F(E, D, A) - (A\,E, e)\rangle\!\rangle,\end{aligned} \tag{30.12}$$

where the notation ⟪H-measure, quadratic function⟫, which I introduced in [105], is defined as follows.

Definition 30.1. *If $U^n \rightharpoonup U^\infty$ in $L^2_{loc}(\Omega; \mathbb{R}^p)$ weak, and $U^n - U^\infty$ defines an H-measure μ, and $Q(x, \xi, U) = \sum_{i,j} q_{i,j}(x, \xi) U_i \overline{U_j}$, then $\langle\!\langle \mu, Q(x, \xi, U)\rangle\!\rangle$ means $\sum_{i,j} \int_{\mathbb{S}^{N-1}} q_{i,j}\, d\mu^{i,j} \in \mathcal{M}(\Omega)$.*

Using

$$F(E, D, A) - (A\,E, e) \le -\Phi(A, e, \xi) \text{ when } E \| \xi \text{ and } D \perp \xi, \tag{30.13}$$

and

$$\langle\!\langle \mu', Q(E, D, A)\rangle\!\rangle = 0 \text{ when } Q \text{ is a multiple of } E \times \xi \text{ or } D \cdot \xi, \tag{30.14}$$

one deduces that

$$\langle\!\langle \mu', F(E, D, A) - (A\,E, e)\rangle\!\rangle \le -\langle\!\langle \mu', \Phi(A, e, \xi)\rangle\!\rangle = -\langle\!\langle \mu, \Phi(A, e, \xi)\rangle\!\rangle, \tag{30.15}$$

and this gives

$$\begin{aligned}\lim\nolimits_{n\to\infty} F(E^n, A^n E^n, A^n) - (D^\infty, e) &\le F(E^\infty, D^\infty, A^\infty) - (A^\infty E^\infty, e)\\ &- \quad \langle\!\langle \mu, \Phi(A, e, \xi)\rangle\!\rangle.\end{aligned} \tag{30.16}$$

Taking the infimum in e finally gives an upper bound

$$\begin{aligned}&\lim_{n\to\infty} F(E^n, A^n E^n, A^n) \le F(E^\infty, A^{\mathit{eff}} E^\infty, A^\infty)\\ &+ \quad \inf_{e\in\mathbb{R}^N}\big((A^{\mathit{eff}} - A^\infty)E^\infty, e) - \langle\!\langle \mu, \Phi(A, e, \xi)\rangle\!\rangle\big).\end{aligned} \tag{30.17}$$

The comparison of the lower and upper bounds for the weak $\star$ limit of $F(E^n, A^n E^n, A^n)$ gives some bounds for the effective tensor A^{eff} in terms of the Young measure ν and the H-measure associated to subsequences of A^n and $A^n - A^\infty$.

This method, like all the preceding ones concerned with obtaining bounds for effective coefficients, faces an enormous amount of technical computation if one does not start with a good function $F(E, D, A)$. I shall now show a special case, which has the advantage of giving a known optimal lower bound for the case of mixing two isotropic materials; I shall also recover the optimal upper bound in that case, but from a computation using the H-measure associated to $(A^n)^{-1}$: since not much is known yet about the relations between the *H-measures* corresponding to different functions of A^n (except for two-component mixtures), this method should certainly be improved.

Lemma 30.2. *Assuming that A^n are symmetric positive matrices,*[4] *then for every symmetric matrix M satisfying $0 < M \le A^n$ for all n, one has the following inequality, implying a lower bound for A^{eff}:*

$$\begin{pmatrix} A^{\mathit{eff}} - M & A^\infty & I \\ A^\infty & R(M) + M^{1,1} & M^{0,1} \\ I & M^{1,0} & M^{0,0} \end{pmatrix} \ge 0 \tag{30.18}$$

where $R(M)$ and the moments $M^{i,j}$ are defined from the H-measure μ associated to $A^m - A^\infty$ and the Young measure ν associated to A^m by

$$(R(M)\, v, v) = \left\langle\!\!\left\langle \mu, \frac{(A\,\xi, v)^2}{(M\,\xi,\xi)} \right\rangle\!\!\right\rangle \text{ for every } v \in \mathbb{R}^N, \tag{30.19}$$

$$M^{i,j} = \langle \nu, A^j (A - M)^{-1} A^i\rangle. \tag{30.20}$$

Proof: One first notices that, as a special case of the general argument

$$(M\,E^n, E^n) - 2(A^n E^n, v) \rightharpoonup (M\,E^\infty, E^\infty) - 2(A^\infty E^\infty, v) + X, \tag{30.21}$$

with the H-correction X satisfying

$$X \ge -\left\langle\!\!\left\langle \mu, \frac{(A\,\xi, v)^2}{(M\,\xi,\xi)} \right\rangle\!\!\right\rangle = -(R(M)\, v, v). \tag{30.22}$$

[4] The symmetry of all the matrices used is taken as a simplification.

Then one notices that

$$\begin{aligned}&(A^n E^n, E^n) - (M\,E^n, E^n) + 2(A^n E^n, v) + 2(E^n, w) \geq \\ &\quad -\big((A^n - M)^{-1}(A^n v + w), (A^n v + w)\big) \to \\ &\quad\quad -\big\langle \nu, \big((A - M)^{-1}(A\,v + w), (A\,v + w)\big)\big\rangle.\end{aligned} \tag{30.23}$$

Then, one adds the inequalities obtained and this gives

$$\begin{aligned}&(A^{\mathit{eff}} E^\infty, E^\infty) \geq (M\,E^\infty, E^\infty) - (R(M)\,v, v) \\ &- \quad \big\langle \nu, \big((A - M)^{-1}(A\,v + w), (A\,v + w)\big)\big\rangle - 2(A^\infty E^\infty, v) - 2(E^\infty, w),\end{aligned} \tag{30.24}$$

an inequality valid for all vectors $E^\infty, v, w \in \mathbb{R}^N$ (and any matrix M satisfying the constraints $0 < M \leq A_n$ for all n), and this is what (30.18) says. □

The preceding result contains more information than bounds which use only Young measures (which corresponds to using statistics on the components of a mixture), since the matrix $R(M)$ contains some information on the micro-geometry of the mixture. For example, in the case of a mixture made with homogeneous fibres of arbitrary cross-section, but with axes parallel to x_N, one finds that $R(M)$ has a zero eigenvector in this direction, since the H-measure will be supported by $\xi_N = 0$.

It is through this kind of result that one can, in principle, obtain relations between the effective coefficients corresponding to different physical properties. However, at the moment, I do not know an efficient way to perform all the necessary technical computations.

Corollary 30.3. *Assuming that all A^n are symmetric positive matrices, then for every symmetric matrix M satisfying $0 < M \leq A_n$ for all n, one has the following lower bound for A^{eff}:*

$$(A^{\mathit{eff}} - M)^{-1} \leq (A^\infty - M)^{-1} + (A^\infty - M)^{-1} R(M) (A^\infty - M)^{-1}. \tag{30.25}$$

Proof: Indeed, this simpler inequality results from the special case $w = -M\,v$, which corresponds to taking the limit of the inequality

$$\big((A^n - M)E^n, E^n\big) + 2\big((A^n - M)\,E^\infty, v\big) + \big((A^n - M)\,v, v\big) \geq 0, \tag{30.26}$$

which gives

$$\begin{aligned}&\big((A^{\mathit{eff}} - M)\,E^\infty, E^\infty\big) + 2\big((A^\infty - M)\,E^\infty, v\big) + \big((A^\infty - M)\,v, v\big) \geq \\ &\quad -\big\langle\!\big\langle \mu, \tfrac{(A\,\xi, v)^2}{(M\,\xi, \xi)} \big\rangle\!\big\rangle = -(R(M)\,v, v),\end{aligned} \tag{30.27}$$

which is equivalent to

$$(A^\infty - M) - (A^\infty - M)(A^{\rm eff} - M)^{-1}(A^\infty - M) + R(M) \geq 0, \qquad (30.28)$$

itself equivalent to (30.25). □

In the case of mixing two isotropic materials with conductivity $\alpha < \beta$ and proportions θ and $(1-\theta)$ one can take $M = \gamma\, I$, $\gamma < \alpha$ and then let γ tend to α; this amounts to taking $w = -\alpha\, v$ and using the preceding inequalities. In that case, the Young measure ν is $\theta\, \delta_{\alpha\, I} + (1-\theta)\, \delta_{\beta\, I}$ and the H-measure μ is a nonnegative measure on $\mathbb{S}^{N-1}$ with total mass $\theta\,(1-\theta)\,(\beta-\alpha)^2$; $R(\alpha\, I)$ is a nonnegative symmetric matrix of trace $\theta\,(1-\theta)\,\alpha^{-1}(\beta-\alpha)^2$. The preceding computation gives the inequality

$$(A^{\rm eff} - \alpha\, I)^{-1} \leq \frac{R(\alpha\, I)}{(1-\theta)^2(\beta-\alpha)^2} + \frac{I}{(1-\theta)\,(\beta-\alpha)}, \qquad (30.29)$$

which implies one of the optimal bounds found in 1981 with François MURAT:

$$Trace(A^{\rm eff} - \alpha\, I)^{-1} \leq \frac{\theta}{(1-\theta)\,\alpha} + \frac{N}{(1-\theta)\,(\beta-\alpha)}. \qquad (30.30)$$

Lemma 30.4. *Assuming that A_n are symmetric positive matrices and that $(B^\infty)^{-1}$ is the weak $\star$ limit of $(A^n)^{-1}$, then for any symmetric matrix N satisfying $0 < N^{-1} \leq (A^n)^{-1}$, i.e., $N \geq A^n$, for all n, one has the following inequality, implying an upper bound for $A^{\rm eff}$:*

$$\begin{pmatrix} (A^{\rm eff})^{-1} - N^{-1} & (B^\infty)^{-1} & I \\ (B^\infty)^{-1} & S(N) + N^{1,1} & N^{0,1} \\ I & N^{1,0} & N^{0,0} \end{pmatrix} \geq 0, \qquad (30.31)$$

where $S(N)$ and the moments $N^{i,j}$ are defined from the H-measure π associated to $(A^m)^{-1} - (B^\infty)^{-1}$ and the Young measure ν associated to A^m by

$$(S(N)v, v) = \langle\langle \pi, (N\, A^{-1}v, A^{-1}v)\rangle\rangle - \left\langle\left\langle \pi, \frac{(N\, A^{-1}v, \xi)^2}{(N\,\xi, \xi)}\right\rangle\right\rangle \text{ for every } v \in \mathbb{R}^N, \qquad (30.32)$$

$$N^{i,j} = \langle \nu, A^{-j}(A^{-1} - N^{-1})^{-1}A^{-i}\rangle. \qquad (30.33)$$

Proof: One first notices that

$$(N^{-1}D^n, D^n) - 2\big((A^n)^{-1}D^n, v\big) \to (N^{-1}D^\infty, D^\infty) - 2\big((B^\infty)^{-1}D^\infty, v\big) + Y, \qquad (30.34)$$

with the H-correction Y given by

$$Y = \langle\!\langle \pi', (N^{-1}D, D) - 2(A^{-1}D, v) \rangle\!\rangle, \tag{30.35}$$

where π' is the H-measure corresponding to $\big((A^n)^{-1} - (B^\infty)^{-1}, D^n - D^\infty\big)$. Following the general argument, one defines

$$\Psi(A^{-1}, v, \xi) = \inf_{D \perp \xi} \big((N^{-1}D, D) - 2(A^{-1}D, v)\big), \tag{30.36}$$

which one computes by making the changes of variables

$$D = N^{\frac{1}{2}} f \text{ and } v = A\, N^{-\frac{1}{2}} g, \tag{30.37}$$

so that

$$\Psi(A^{-1}, v, \xi) = \inf_{f \perp N^{\frac{1}{2}}\xi} \big(|f|^2 - 2(f, g)\big), \tag{30.38}$$

and the best f is the projection of g on the orthogonal of $N^{1/2}\xi$. This gives

$$\Psi(A^{-1}, v, \xi) = -|g|^2 + \frac{(g, N^{1/2}\xi)^2}{(N\,\xi, \xi)} = -(N\,A^{-1}v, A^{-1}v) + \frac{(N\,A^{-1}v, \xi)^2}{(N\,\xi, \xi)}, \tag{30.39}$$

which implies that

$$\begin{aligned} Y = \langle\!\langle \pi', (N^{-1}D, D) - 2(A^{-1}D, v) \rangle\!\rangle &\geq -\langle\!\langle \pi, (N\,A^{-1}v, A^{-1}v) \rangle\!\rangle \\ + \quad \langle\!\langle \pi, \tfrac{(N\,A^{-1}v,\xi)^2}{(N\,\xi,\xi)} \rangle\!\rangle. \end{aligned} \tag{30.40}$$

Then, one notices that

$$\begin{aligned} \big((A^n)^{-1}D^n, D^n\big) - (N^{-1}D^n, D^n) + 2\big((A^n)^{-1}D^n, v\big) + 2(D^n, w) \geq \\ - \quad \big(\big[(A^n)^{-1} - N^{-1}\big]\big((A^n)^{-1}v + w\big), (A^n)^{-1}v + w\big), \end{aligned} \tag{30.41}$$

$$\begin{aligned} \lim_{n\to\infty} \big(\big[(A^n)^{-1} - N^{-1}\big]\big((A^n)^{-1}v + w\big), (A^n)^{-1}v + w\big) \\ = \big\langle \nu, \big((A^{-1} - N^{-1})(A^{-1}v + w), A^{-1}v + w\big) \big\rangle. \end{aligned} \tag{30.42}$$

Then, one adds these inequalities, and one takes the limit; this gives

$$\begin{aligned} \big((A^{\mathit{eff}})^{-1}D^\infty, D^\infty\big) + 2(D^\infty, w) &\geq (N^{-1}D^\infty, D^\infty) - 2\big((B^\infty)^{-1}D^\infty, v\big) \\ - \quad \langle\!\langle \pi, (N\,A^{-1}v, A^{-1}v) \rangle\!\rangle &+ \langle\!\langle \pi, \tfrac{N\,A^{-1}v,\xi)^2}{(N\,\xi,\xi)} \rangle\!\rangle \\ - \quad \big\langle \nu, \big((A^{-1} - N^{-1})^{-1}&(A^{-1}v + w), (A^{-1}v + w)\big) \big\rangle, \end{aligned} \tag{30.43}$$

an inequality valid for all vectors $D^\infty, v, w \in \mathbb{R}^N$ (and any matrix N satisfying $0 < N^{-1} \leq (A^n)^{-1}$ for all n), and this corresponds to (30.31). □

Corollary 30.5. *Assuming that A^n are symmetric positive matrices, then for any symmetric matrix N satisfying $0 < N^{-1} \le (A^n)^{-1}$ for all n, one has the following upper bound for A_{eff}:*

$$\begin{aligned}&\big[(A^{\mathit{eff}})^{-1} - N^{-1}\big]^{-1} \le \big[(B^\infty)^{-1} - N^{-1}\big]^{-1}\\ &+ \quad \big[(B^\infty)^{-1} - N^{-1}\big]^{-1} S(N)\big[(B^\infty)^{-1} - N^{-1}\big]^{-1}.\end{aligned} \tag{30.44}$$

Proof: Indeed, one can obtain this simpler inequality from the special case $w = -N^{-1}v$, which corresponds to taking the limit of the inequality

$$\begin{aligned}&\big(\big[(A^n)^{-1} - N^{-1}\big]D^n, D^n\big) + 2\big(\big[(A^n)^{-1} - N^{-1}\big]D^n, v\big)\\ &+ \quad \big(\big[(A^n)^{-1} - N^{-1}\big]v, v\big) \ge 0.\end{aligned} \tag{30.45}$$

One obtains in this case

$$\begin{aligned}&\big(\big[(A^{\mathit{eff}})^{-1} - N^{-1}\big]D^\infty, D^\infty\big) + 2\big(\big[(B^\infty)^{-1} - N^{-1}\big]D^\infty, v\big)\\ &+ \quad \big(\big[(B^\infty)^{-1} - N^{-1}\big]v, v\big) \ge -\langle\!\langle \pi, (N\,A^{-1}v, A^{-1}v)\rangle\!\rangle\\ &+ \quad \langle\!\langle \pi, \tfrac{(N\,A^{-1}v,\xi)^2}{(N\,\xi,\xi)}\rangle\!\rangle = -(S(N)v, v),\end{aligned} \tag{30.46}$$

which is equivalent to

$$\begin{aligned}&\big[(B^\infty)^{-1} - N^{-1}\big] - \big[(B^\infty)^{-1} - N^{-1}\big]\big[(A^{\mathit{eff}})^{-1} - N^{-1}\big]\big[(B^\infty)^{-1} - N^{-1}\big]^{-1}\\ &+ \quad S(N) \ge 0,\end{aligned} \tag{30.47}$$

an equivalent form of the corollary. □

In the case of mixing two isotropic materials with conductivity $\alpha < \beta$ and proportions θ and $(1-\theta)$, one can take $N = \gamma\,I$ with $\gamma > \beta$ and then let γ tend to β; this amounts to taking $w = -\beta^{-1}v$ and using the preceding inequalities. In that case, the Young measure ν is $\theta\,\delta_{\alpha\,I} + (1-\theta)\,\delta_{\beta\,I}$ and the H-measure π is a nonnegative measure on $\mathbb{S}^{N-1}$ with total mass $\theta\,(1-\theta)\,(\frac{\beta-\alpha}{\alpha\,\beta})^2$; $S(\beta\,I)$ is a nonnegative matrix of trace $\theta\,(1-\theta)\left(\frac{\beta-\alpha}{\alpha\,\beta}\right)^2(N-1)\,\beta$. The preceding computation gives the inequality

$$\left((A^{\mathit{eff}})^{-1} - \frac{I}{\beta}\right)^{-1} \le \frac{\alpha\,\beta\,I}{\theta\,(\beta-\alpha)} + \left(\frac{\alpha\,\beta}{\theta\,(\beta-\alpha)}\right)^2 S(\beta\,I), \tag{30.48}$$

which implies the other optimal bound found in 1981 with François Murat,

$$\begin{aligned}&Trace\big[(\beta\,I - A^{\mathit{eff}})^{-1}\big] + \tfrac{N}{\beta} + \tfrac{1}{\beta^2}Trace\Big[\Big(A^{\mathit{eff}} - \tfrac{I}{\beta}\Big)^{-1}\Big]\\ &\le \tfrac{N\,\alpha}{\theta\,\beta\,(\beta-\alpha)} + \tfrac{N-1}{\beta\,\theta} + \tfrac{1}{\beta}.\end{aligned} \tag{30.49}$$

Chapter 31
H-Measures and Propagation Effects

It was an early observation that *the knowledge of the proportions of materials used in a mixture is not sufficient for determining some of its effective properties*,[1] and I felt that developing a mathematical theory for such questions is of crucial importance for correcting the defects of the models which physicists proposed for explaining how the world functions.

Of course, some properties are additive, and if it is the case for the density of mass ϱ and the density of linear momentum q, linked by the equation of conservation of mass $\frac{\partial \varrho}{\partial t} + div(q) = 0$, it leads to the observation (which I learned from Joel ROBBIN) that some quantities are coefficients of differential forms, for which weak topologies are natural. However, besides the need to introduce other topologies for other quantities, for example for the velocity of transport of mass u which results from writing $q = \varrho\, u$, one also observes that conserved quantities may be hiding at various mesoscopic levels under the form of oscillations and concentration effects, and one needs to introduce more general mathematical tools for following their movement, i.e. develop a mathematical theory that could follow the evolution of microstructures, and discover which type of microstructures are bound to be observed as a consequence of various mathematical models, and assert which types are seen in nature.

One important defect of a theory like thermodynamics is that it implicitly asserts that internal energy has only one form, and that it interprets an observed irreversibility by postulating a local dissipation.[2] It is a consequence of POINCARÉ's ideas concerning the principle of relativity, which he may not have perceived well enough, and which EINSTEIN did not seem to understand well after him, that at a basic level information is carried by waves, and one should not be lured by the fact that physicists gave names of "particles" to some of these waves. The main reason for irreversibility is then that

[1] Except in dimension $N = 1$, but we seem to be living in a three-dimensional world, with a fourth dimension of time showing a curious type of irreversibility.

[2] One may consider that the Fourier heat equation is an approximation obtained by letting the speed of light c tend to ∞ in a more realistic physical model, probably a semi-linear hyperbolic system.

L. Tartar, *The General Theory of Homogenization*, Lecture Notes of the Unione Matematica Italiana 7, DOI 10.1007/978-3-642-05195-1_31,

energy stored at mesoscopic levels is carried away by waves, so that without understanding which kind of waves exist it is hardly possible to recuperate this energy at a macroscopic level.

After developing *H-measures*, it was then natural that I checked if these new mathematical objects are able to describe the transport of oscillations and concentration effects, and I first studied the case of a first-order scalar equation in order to see if propagation occurs along bicharacteristic rays.[3] After that, I observed that my method of proof extends to more interesting physical models, like the wave equation, the Maxwell–Heaviside system, the Dirac system, or even to systems which are not so good physical models, like the Lamé system of linearized elasticity.[4]

My method relies on what I called the second commutation lemma.

Lemma 31.1. *Let $a \in C^1(\mathbb{S}^{N-1})$, extended to be homogeneous of degree 0 in $\mathbb{R}^N \setminus \{0\}$, and $b \in \mathcal{F}L^1(\mathbb{R}^N)$ with $\frac{\partial b}{\partial x_i} \in \mathcal{F}L^1(\mathbb{R}^N)$ for $i = 1, \ldots, N$. Then,*

$$\begin{aligned} &C = [P_a, M_b] = P_a M_b - M_b P_a \in \mathcal{L}\big(L^2(\mathbb{R}^N); H^1(\mathbb{R}^N)\big)\\ &\tfrac{\partial}{\partial x_j} C \text{ has the symbol } \tfrac{\xi_j}{|\xi|} \textstyle\sum_{k=1}^N \tfrac{\partial a}{\partial \xi_k}\big(\tfrac{\xi}{|\xi|}\big) \tfrac{\partial b}{\partial x_k}, j = 1, \ldots, N. \end{aligned} \tag{31.1}$$

Proof. Since $b \in C_0(\mathbb{R}^N)$, the first commutation lemma (Lemma 28.2) applies, and C is a compact operator from $L^2(\mathbb{R}^N)$ into itself. For $u \in \mathcal{S}(\mathbb{R}^N)$

$$\mathcal{F}\Big(\frac{\partial(C\,u)}{\partial x_j}\Big)(\xi) = 2i\,\pi\,\xi_j \int_{\mathbb{R}^N} \Big(a\Big(\frac{\xi}{|\xi|}\Big) - a\Big(\frac{\eta}{|\eta|}\Big)\Big)\mathcal{F}b(\xi-\eta)\mathcal{F}u(\eta)\,d\eta, \tag{31.2}$$

and one must bound it in $L^2(\mathbb{R}^N)$ using only the norm of u in $L^2(\mathbb{R}^N)$. Using

$$\Big|\frac{\xi}{|\xi|} - \frac{\eta}{|\eta|}\Big| \le 2\frac{|\xi-\eta|}{|\xi|} \text{ for } \xi \text{ and } \eta \ne 0, \tag{31.3}$$

and denoting K the Lipschitz constant of a on $\mathbb{S}^{N-1}$, one deduces that

$$\Big|\mathcal{F}\Big(\frac{\partial(C\,u)}{\partial x_j}\Big)(\xi)\Big| \le 4\pi\,K \int_{\mathbb{R}^N} |\xi-\eta|\,|\mathcal{F}b(\xi-\eta)|\,|\mathcal{F}u(\eta)|\,d\eta, \tag{31.4}$$

$$\Big|\Big|\frac{\partial(C\,u)}{\partial x_j}\Big|\Big|_{L^2(\mathbb{R}^N)} \le 4\pi\,K\,||u||_{L^2(\mathbb{R}^N)} \int_{\mathbb{R}^N} |\xi|\,|\mathcal{F}b|\,d\xi. \tag{31.5}$$

---

[3] Lars HÖRMANDER's theory concerns the propagation of microlocal regularity, which happens along bicharacteristic rays. Since microlocal regularity has no physical interest, it is presented as propagation of singularities for reasons of propaganda, but although physicists seem to like singularities (probably because they believe in "particles"), I find this notion of little physical interest.

[4] Gabriel LAMÉ, French mathematician, 1795–1870. He worked in St. Petersburg, Russia, and in Paris, France.

In order to compute the symbol of the operator $\frac{\partial}{\partial x_j} C$ one first approaches b by a sequence $b_n \in S(\mathbb{R}^N)$ with $\mathcal{F}b_n \in C_c^\infty(\mathbb{R}^N)$, in such a way that $(1+|\xi|)\mathcal{F}(b_n - b)$ tends to 0 in $L^1(\mathbb{R}^N)$ strong,[5] and then $C_n = [P_a, M_{b_n}]$ converges to C in norm in $\mathcal{L}\big(L^2(\mathbb{R}^N); L^2(\mathbb{R}^N)\big)$, since $b_n - b$ tends to 0 in $C_0(\mathbb{R}^N)$ strong, and $\frac{\partial}{\partial x_j} C_n$ tends to $\frac{\partial}{\partial x_j} C$ in norm in $\mathcal{L}\big(L^2(\mathbb{R}^N); L^2(\mathbb{R}^N)\big)$, by an estimate like (31.5). If one shows that $\frac{\partial}{\partial x_j} C_n$ has the symbol $s_n = \frac{\xi_j}{|\xi|} \sum_{k=1}^N \frac{\partial a}{\partial \xi_k}\big(\frac{\xi}{|\xi|}\big) \frac{\partial b_n}{\partial x_k}$,[6] i.e. that it is $S_{s_n} + K_n$ with K_n compact, then S_{s_n} converges in norm to S_s with $s = \frac{\xi_j}{|\xi|} \sum_{k=1}^N \frac{\partial a}{\partial \xi_k}\big(\frac{\xi}{|\xi|}\big) \frac{\partial b}{\partial x_k}$ since each $\frac{\partial b_n}{\partial x_k}$ converges to $\frac{\partial b}{\partial x_k}$ in $C_0(\mathbb{R}^N)$ strong, hence K_n converges in norm, to a limit which must be a compact operator, showing that $\frac{\partial}{\partial x_j} C$ has the symbol s.

In (31.2) written for C_n, the integral is taken for $|\eta - \xi| \le \rho_n$, so that if $|\xi| \ge \frac{1}{\varepsilon}$ and $\varepsilon > 0$ is small enough, one has $\big|\frac{\eta}{|\eta|} - \frac{\xi}{|\xi|}\big| \le 2\varepsilon\, \rho_n$ by (31.3), and $\big|\frac{\eta}{|\eta|} - \frac{\eta}{|\xi|}\big| \le \varepsilon\, \rho_n$; then, since a is of class C^1, one deduces from its Taylor expansion at $\frac{\xi}{|\xi|}$ that

$$\begin{aligned} 2i\,\pi\,\xi_j\big(a\big(\tfrac{\xi}{|\xi|}\big) - a\big(\tfrac{\eta}{|\eta|}\big)\big) &= 2i\,\pi\,\xi_j \textstyle\sum_{k=1}^N \frac{\partial a}{\partial \xi_k}\big(\frac{\xi}{|\xi|}\big)\big(\frac{\xi_k}{|\xi|} - \frac{\eta_k}{|\eta|}\big) + o(\varepsilon\,\rho_n) \\ &= \tfrac{\xi_j}{|\xi|} \textstyle\sum_{k=1}^N \frac{\partial a}{\partial \xi_k}\big(\frac{\xi}{|\xi|}\big) 2i\,\pi\,(\xi_k - \eta_k) + O(\varepsilon\,\rho_n). \end{aligned} \tag{31.6}$$

Replacing $2i\,\pi\,\xi_j\big(a\big(\frac{\xi}{|\xi|}\big) - a\big(\frac{\eta}{|\eta|}\big)\big)$ by $\frac{\xi_j}{|\xi|} \sum_{k=1}^N \frac{\partial a}{\partial \xi_k}\big(\frac{\xi}{|\xi|}\big) 2i\,\pi\,(\xi_k - \eta_k)$ for $|\xi| \ge \frac{1}{\varepsilon}$ gives an error bounded by

$$O(\varepsilon\,\rho_n) \int_{\mathbb{R}^N} |\mathcal{F}b_n(\xi-\eta)|\,|\mathcal{F}u(\eta)|\,d\eta = O(\varepsilon\,\rho_n)\,|\mathcal{F}b_n| \star |\mathcal{F}u|, \tag{31.7}$$

i.e. an operator of norm $O(\varepsilon\,\rho_n)\,||\mathcal{F}b_n||_{L^1(\mathbb{R}^N)}$ in $\mathcal{L}\big(L^2(\mathbb{R}^N); L^2(\mathbb{R}^N)\big)$. Making the same replacement for $|\xi| \le \frac{1}{\varepsilon}$, and necessarily $|\eta| \le \rho_n + \frac{1}{\varepsilon}$, gives a Hilbert–Schmidt operator,[7] so that $\frac{\partial}{\partial x_j} C$ differs by a compact operator from the operator which to $u \in L^2(\mathbb{R}^N)$ associates

$$\begin{aligned} &\textstyle\int_{\mathbb{R}^N} \frac{\xi_j}{|\xi|} \sum_{k=1}^N \frac{\partial a}{\partial \xi_k}\big(\frac{\xi}{|\xi|}\big) 2i\,\pi(\xi_k - \eta_k)\mathcal{F}b_n(\xi-\eta)\,\mathcal{F}u(\eta)\,d\eta \\ &\quad = \tfrac{\xi_j}{|\xi|} \textstyle\sum_{k=1}^N \frac{\partial a}{\partial \xi_k}\big(\frac{\xi}{|\xi|}\big)\mathcal{F}\big(\frac{\partial b_n}{\partial x_k}\,u\big)(\xi), \end{aligned} \tag{31.8}$$

i.e. the standard operator with symbol $\frac{\xi_j}{|\xi|} \sum_{k=1}^N \frac{\partial a}{\partial \xi_k}\big(\frac{\xi}{|\xi|}\big) \frac{\partial b_n}{\partial x_k}$. □

---

[5] Defining $\beta_n = b\,\theta\big(\frac{\xi}{n}\big)$ with $\theta \in C_c^\infty(\mathbb{R}^N)$ equal to 1 on the unit ball, $(1+|\xi|)\mathcal{F}(\beta_n - b)$ tends to 0 in $L^1(\mathbb{R}^N)$ strong by the Lebesgue dominated convergence theorem; one then regularizes β_n by convolution in order to create the sequence b_n.

[6] This symbol is admissible, since $\frac{\xi_j}{|\xi|} \frac{\partial a}{\partial \xi_k}\big(\frac{\xi}{|\xi|}\big)$ is homogeneous of degree 0 and is continuous on $\mathbb{S}^{N-1}$, and $\frac{\partial b_n}{\partial x_k} \in C_0(\mathbb{R}^N)$.

[7] For this, it would be enough to have $\mathcal{F}b_n \in L^2_{loc}(\mathbb{R}^N)$.

Definition 31.2. *The Poisson bracket* $\{g,h\}$ *of two functions* g,h *on* $\mathbb{R}^N \times \mathbb{R}^N$ *is defined by*

$$\{g,h\} = \sum_{k=1}^{N} \Big(\frac{\partial g}{\partial \xi_k} \frac{\partial h}{\partial x_k} - \frac{\partial g}{\partial x_k} \frac{\partial h}{d\xi_k} \Big). \tag{31.9}$$

Since a is independent of x and b is independent of ξ, (31.1) then means that the symbol of $\frac{\partial}{\partial x_j}\,[P_a, M_b]$ is $\xi_j\{a,b\}$; this result extends to standard operators by Lemma 31.3.

Lemma 31.3. *If* S_1, S_2 *are the standard operators with symbols* $s_k(x,\xi) = a_k(\xi) b_k(x)$, $k = 1,2$, *and* a_k, b_k *satisfy the hypotheses of Lemma 31.3 for* $k = 1,2$, *then the operator* $\frac{\partial}{\partial x_j}\,[S_1, S_2]$ *has the symbol* $\xi_j\{s_1, s_2\}$. *The result extends to general sums, if the series of products of corresponding norms converge.*

Proof. One must find the symbol of $\frac{\partial}{\partial x_j}\,(P_{a_1} M_{b_1} P_{a_2} M_{b_2} - P_{a_2} M_{b_2} P_{a_1} M_{b_1})$, and using the fact that $\frac{\partial}{\partial x_j}$ commutes with P_{a_k}, $k = 1,2$, one finds that

$$\begin{aligned} \tfrac{\partial}{\partial x_j}\, P_{a_1} M_{b_1} P_{a_2} M_{b_2} &= P_{a_1} \big(\tfrac{\partial}{\partial x_j}\,(M_{b_1} P_{a_2} - P_{a_2} M_{b_1}) \big)\, M_{b_2} \\ &+ \; P_{a_1} P_{a_2} \tfrac{\partial}{\partial x_j}\, M_{b_1} M_{b_2} \\ \tfrac{\partial}{\partial x_j}\, P_{a_2} M_{b_2} P_{a_1} M_{b_1} &= P_{a_2} \big(\tfrac{\partial}{\partial x_j}\,(M_{b_2} P_{a_1} - P_{a_1} M_{b_2}) \big)\, M_{b_1} \\ &+ \; P_{a_2} P_{a_1} \tfrac{\partial}{\partial x_j}\, M_{b_2} M_{b_1}, \end{aligned} \tag{31.10}$$

and using the fact that P_{a_1} and P_{a_2} commute, and that M_{b_1} and M_{b_2} commute, one deduces that the desired symbol is $a_1[-\xi_j\{a_2,b_1\}]b_2 + a_2[\xi_j\{a_1, b_2\}]b_1$, and one checks that it is equal to $\xi_j\{a_1 b_1, a_2 b_2\}$. □

The hypothesis that b belongs to the space which I denoted $X^1(\mathbb{R}^N)$ in [105], i.e. $b, \frac{\partial b}{\partial x_1}, \ldots, \frac{\partial b}{\partial x_N} \in \mathcal{F}L^1(\mathbb{R}^N)$, is a little restrictive because in applications b is a coefficient of a partial differential equation that one studies, and it is useful to avoid asking too much regularity for these coefficients. Since $\mathcal{F}L^1(\mathbb{R}^N) \subset C_0(\mathbb{R}^N)$, one has $X^1(\mathbb{R}^N) \subset C_0^1(\mathbb{R}^N)$, and since $\mathcal{F}L^1(\mathbb{R}^N)$ is a multiplicative algebra, so is $X^1(\mathbb{R}^N)$, but if one wants to compare to Sobolev spaces, one has $H^s(\mathbb{R}^N) \subset \mathcal{F}L^1(\mathbb{R}^N)$ if (and only if) $s > \frac{N}{2}$,[8] so that $H^\sigma(\mathbb{R}^N) \subset X^1(\mathbb{R}^N)$ if (and only if) $\sigma > \frac{N}{2} + 1$.

When I mentioned my result to Pierre-Louis LIONS, he told me about a result of Guy DAVID and JOURNÉ which applied to my situation [21],[9,10]

[8] More generally, if $1 \le p \le 2$, $W^{s,p}(\mathbb{R}^N) \subset \mathcal{F}L^1(\mathbb{R}^N)$ if (and only if) $s > \frac{N}{p}$.

[9] Guy DAVID, French mathematician. He works at Université de Paris Sud, Orsay, France.

[10] Jean-Lin JOURNÉ, French mathematician.

and when I checked it I found that they used a result of Raphaël COIFMAN and Yves MEYER [18], who themselves mentioned an article by Alberto CALDERÓN [12], and I used his result in [105], which permits one to have $b \in C_0^1(\mathbb{R}^N)$ and a a little smoother $(X_{loc}^1(\mathbb{R}^N \setminus 0))$.[11] However, I shall rely here on Lemma 31.1, despite the fact that the smoothness hypothesis can be improved, because I feel uneasy when I use results of others which I have not studied well enough for explaining the reason for all the hypotheses, since Alberto CALDERÓN based his proofs on complex methods which I do not know well (and he also considered the case of functions b with derivatives in L^p), while Guy DAVID and JOURNÉ used real methods which I understand better, with BMO spaces and the Cotlar lemma,[12,13] but they also used constructions of Raphaël COIFMAN and Yves MEYER which I did not read.

The difference between the static property of the localization principle (Theorem 28.7) based on the first commutation lemma (Lemma 28.2), and the dynamic property of transport of *H-measures* along bicharacteristic rays based on the second commutation lemma (Lemma 31.1) can be seen on a scalar first-order linear equation

$$\sum_{j=1}^{N} b_j \frac{\partial u_n}{\partial x_j} + c\, u_n = f_n \text{ in } \Omega \subset \mathbb{R}^N, \tag{31.11}$$

with $b_j \in C^1(\Omega)$, $j = 1, \ldots, N$, and $c \in C(\Omega)$. Assuming that

$$\begin{aligned} &u_m \rightharpoonup 0 \text{ in } L^2_{loc}(\Omega) \text{ weak},\\ &u_m \text{ defines an H-measure } \mu \in \mathcal{M}(\Omega \times \mathbb{S}^{N-1}),\\ &f_m \to 0 \text{ in } H^{-1}_{loc}(\Omega) \text{ strong}, \end{aligned} \tag{31.12}$$

the localization principle applies to (31.11) and gives (Corollary 28.8)

$$P\,\mu = 0 \text{ in } \Omega \times \mathbb{S}^{N-1}, \text{ with } P(x,\xi) = \sum_{j=1}^{N} b_j(x)\xi_j \text{ in } \Omega \times \mathbb{S}^{N-1}, \tag{31.13}$$

so that the coefficient c plays no role, and the term $c_n u_n$ can be absorbed into the term f_n. The support of μ is included in the zero set of P, which

---

[11] It is not so restrictive, because in applications a is a test function. $X_{loc}^1(\Omega)$ is the subspace of $u \in C^1(\Omega)$ such that $\varphi u \in X^1(\mathbb{R}^N)$ for all $\varphi \in C_c^\infty(\Omega)$.

[12] Mischa COTLAR, Ukrainian-born mathematician, 1913–2007. He worked in Buenos Aires and in La Plata, Argentina, at Rutgers University, Piscataway, NJ, and in Caracas, Venezuela.

[13] Although I heard Mischa COTLAR speak at the Lions–Schwartz seminar at IHP in Paris in the late 1960s, I only read about his lemma in an article by Charles FEFFERMAN [27], who points out that Mischa COTLAR considered the commutative case, and that the non-commutative case is due to STEIN and KNAPP.

can be decomposed into bicharacteristic rays along which the oscillations and concentration effects described by the H-measure propagate, but this requires some supplementary hypotheses. The bicharacteristic rays are defined in $\Omega \times (\mathbb{R}^N \setminus \{0\})$ by the equations

$$\frac{dx_j}{d\tau} = \frac{\partial P}{\partial \xi_j} \text{ and } \frac{d\xi_j}{d\tau} = -\frac{\partial P}{\partial x_j}, \tag{31.14}$$

which imply $\frac{d[P(x,\xi)]}{d\tau} = 0$; the second part of (31.14) is homogeneous of degree 1 in ξ, and induces an equation for half lines, showing one defect of having chosen $\mathbb{S}^{N-1}$ as a way to pick one point in each equivalent class.

In order to prove a propagation property, one needs more hypotheses. One hypothesis is that $b_j \in X^1_{loc}(\Omega)$ for $j = 1, \ldots, N$, but it is not so important, and it is due to my choice of using only Lemma 31.1 and not the improvement possible by using a theorem of Alberto CALDERÓN [12]. More important is the hypothesis that the b_j are real, since no propagation should be expected if (31.11) is not hyperbolic. The hypothesis that $f_n \rightharpoonup 0$ in $L^2_{loc}(\Omega)$ weak intuitively means that one puts some control on the oscillations or concentration effects that the source term f_n may create, but then it allows the possibility that f_n looks like $T\,u_n$ for an operator T with a symbol, and this explains that the H-measure ν defined by a subsequence (u_m, f_m) plays a role.

Lemma 31.4. *If b_j is real and belongs to $X^1_{loc}(\Omega)$ for $j = 1, \ldots, N$, if $f_n \rightharpoonup 0$ in $L^2_{loc}(\Omega)$ weak, and (u_m, f_m) defines an H-measure ν, so that $\mu = \nu^{1,1}$, then μ satisfies a first-order partial differential equation in (x, ξ):*

$$\langle \mu, \{\Phi, P\}\rangle + \big\langle \big(-div(b) + 2\Re(c)\big)\mu, \Phi\big\rangle = \langle 2\Re\nu^{1,2}, \Phi\rangle \tag{31.15}$$

for all test functions $\Phi \in C^1_c(\Omega \times \mathbb{S}^{N-1})$.[14]

Proof. One multiplies (31.11) by $\varphi \in C^1_c(\Omega)$, so that

$$\sum_{j=1}^N b_j \frac{\partial(\varphi\, u_n)}{\partial x_j} = g_n = \varphi\,(f_n - c\, u_n) + \Big(\sum_{j=1}^N b_j \frac{\partial \varphi}{\partial x_j}\Big)\, u_n, \tag{31.16}$$

and the sequence $(\varphi\, u_n, g_n)$ corresponds to the H-measure π, with

$$\pi^{1,1} = |\varphi|^2\mu \text{ and } \pi^{2,1} = |\varphi|^2\nu^{2,1} + \Big(-c\,\varphi + \sum_{j=1}^N b_j \frac{\partial\varphi}{\partial x_j}\Big)\overline{\varphi}\,\mu. \tag{31.17}$$

Using $v_n = \varphi\, u_n$, (31.16) is now valid in $\mathbb{R}^N$, and one multiplies it by P_a, with $a \in C^\infty(\mathbb{S}^{N-1})$; since P_a commutes with $\frac{\partial}{\partial x_j}$, $j = 1, \ldots, N$, one obtains

---

[14] Φ is extended to be homogeneous of degree zero in ξ.

$$\sum_{j=1}^N \frac{\partial[(P_a b_j - b_j P_a)v_n]}{\partial x_j} + \sum_{j=1}^N \frac{\partial(b_j P_a v_n)}{\partial x_j} - P_a div(b)\, v_n = P_a\, g_n \tag{31.18}$$

and one applies Lemma 31.1 to the first term, and Lemma 28.2 to the third term, i.e. $div(b)P_a - P_a div(b)$ is compact, and one obtains

$$\sum_{j=1}^N b_j \frac{\partial(P_a v_n)}{\partial x_j} + K\, v_n = P_a\, g_n, \tag{31.19}$$

$$K \text{ has the symbol } \sum_{j=1}^N \xi_j \{a, b_j\} = \{a, P\}. \tag{31.20}$$

Then, one uses both equations

$$\sum_{j=1}^N b_j \frac{\partial v_n}{\partial x_j} = g_n \text{ and } \sum_{j=1}^N b_j \frac{\partial(P_a v_n)}{\partial x_j} + K\, v_n = P_a g_n \tag{31.21}$$

in order to obtain a sesqui-linear conservation form

$$\sum_{j=1}^N b_j \frac{\partial(P_a v_n \overline{v_n})}{\partial x_j} = (P_a g_n - K\, v_n)\overline{v_n} + P_a v_n \overline{g_n}, \tag{31.22}$$

and it is here that one uses the hypothesis that the coefficients b_j are real. One applies (31.22) to a test function $w \in C_c^1(\Omega)$, and one takes the limit

$$-\Big\langle \pi^{1,1}, a \sum_{j=1}^N \frac{\partial(b_j w)}{\partial x_j} \Big\rangle = \langle \pi^{2,1}, w\, a\rangle - \langle \pi^{1,1}, w\,\{a, P\}\rangle + \langle \pi^{1,2}, w\, a\rangle, \tag{31.23}$$

and one notices that

$$w\{a, P\} - a \sum_{j=1}^N \frac{\partial(b_j w)}{\partial x_j} = \{w\, a, P\} - a\, w\, div(b), \tag{31.24}$$

so that

$$\langle \pi^{1,1}, \{w\, a, P\} - w\, a\, div(b)\rangle = \langle 2\Re \pi^{1,2}, w\, a\rangle. \tag{31.25}$$

Using (31.17) one obtains

$$\begin{aligned} &\langle \mu, |\varphi|^2\{w\, a, P\} - |\varphi|^2 w\, a\, div(b)\rangle \\ &= 2\Big\langle \Re\Big(-|\varphi|^2 c + \textstyle\sum_{j=1}^N b_j \frac{\partial \varphi}{\partial x_j}\overline{\varphi}\Big)\mu + |\varphi|^2 \Re \nu^{1,2}, w\, a\Big\rangle, \end{aligned} \tag{31.26}$$

and since $|\varphi|^2\{w\,a,P\} - 2\Re\big((\sum_j b_j \frac{\partial \varphi}{\partial x_j})\overline{\varphi}\big) w\,a = \{|\varphi|^2 w\,a, P\}$ one has

$$\langle \mu, |\varphi|^2\{w\,a,P\} - |\varphi|^2 w\,a\,div(b) + 2|\varphi|^2\Re(c) w\,a\rangle = \langle 2\Re\nu^{1,2}, |\varphi|^2 w\,a\rangle, \quad (31.27)$$

which is (31.15) for $\Phi = |\varphi|^2 w\,a$. Since linear combinations of these functions Φ are dense in $C_c^1(\Omega \times \mathbb{S}^{N-1})$, one deduces (31.15). □

One should notice that the Radon measure $\nu^{1,2}$ is not arbitrary with respect to μ, because of the Hermitian character of *H-measures*, so that for every Borel set $E \subset \Omega \times \mathbb{S}^{N-1}$ one has $|\nu^{1,2}(E)|^2 \leq \mu(E)\nu^{2,2}(E)$, and if $E \subset K \times \mathbb{S}^{N-1}$ for a compact $K \subset \Omega$, one deduces that $|\nu^{1,2}(E)|^2 \leq F\,\mu(E)$, where F is an upper bound for $\int_K |f_n|^2\,dx$.

In the case where $f_n = T\,u_n$ for an operator T having an admissible symbol s, one has $\nu^{1,2} = \overline{s}\,\mu$, and (31.15) becomes a homogeneous equation for μ.

The important step in the proof of Lemma 31.4 is that

$$\overline{D\,v_n}\,P_a v_n + \overline{v_n} D\,P_a v_n = D(P_a v_n \overline{v_n}), \quad (31.28)$$

since $\overline{D\,v_n} = D\overline{v_n}$, because the coefficients b_j are real.

The same scheme can be applied to most linear (or semi-linear) systems of continuum mechanics or physics: one first localizes in x (which could be (x,t)), so that instead of working in Ω one works in $\mathbb{R}^N$, and then one applies the localization principle, which makes some characteristic polynomial appear; after that, one applies an operator in standard form and some commutator shows up, which uses the Poisson bracket of the characteristic polynomial and a symbol, and finally comes the crucial step which needs a sesqui-linear conservation law valid for complex solutions, but physical systems are usually endowed with a conservation law for a quadratic quantity, the energy, and one must only check that this conservation extends to complex solutions, if one replaces the quadratic quantity by a sesqui-linear quantity.

A particular reason for checking what happens for a scalar wave equation is to make a comparison with what the formal theory of geometrical optics says,[15] and what it means for the energy to propagate along bicharacteristic rays, and why my method using *H-measures* is not bothered by the phase.

I consider a scalar wave equation in *a medium whose properties vary smoothly*, with *coefficients independent of* t, and for convenience, I sometimes replace t by x_0 and denote the dual variable by ξ_0:

$$\varrho\,\frac{\partial^2 u_n}{\partial t^2} - \sum_{i,j=1}^N \frac{\partial}{\partial x_i}\Big(a_{i,j}\frac{\partial u_n}{\partial x_j}\Big) = f_n \text{ in } \Omega \times (0,T), \quad (31.29)$$

[15] Of course, real light is not described by a scalar wave equation! One needs to use the Maxwell–Heaviside system, which explains what polarization is.

and I assume that

$$\begin{aligned}&u_n \rightharpoonup 0 \text{ in } H^1_{loc}\big(\Omega\times(0,T)\big) \text{ weak},\\ &\tfrac{\partial u_n}{\partial t} \rightharpoonup 0 \text{ in } L^2_{loc}\big(\Omega\times(0,T)\big) \text{ weak},\\ &U^n_j = \tfrac{\partial u_n}{\partial x_j}, j=0,\ldots,N \text{ defines an H-measure } \mu,\end{aligned} \tag{31.30}$$

and the localization principle applies to the information

$$\frac{\partial U^n_j}{\partial x_k} - \frac{\partial U^n_k}{\partial x_j} = 0, j,k=0,\ldots,N, \tag{31.31}$$

and gives (Corollary 28.9)

$$\mu^{j,k} = \xi_j\xi_k\nu, j,k=0,\ldots,N, \text{ with } \nu\in\mathcal{M}_+\big(\Omega\times(0,T)\big). \tag{31.32}$$

For applying the localization principle to (31.29), like in Corollary 28.10, one assumes that ϱ and $a_{i,j}$ are continuous for $j=1,\ldots,N$, and that $f_n\to 0$ in $H^{-1}_{loc}\big(\Omega\times(0,T)\big)$ strong, and one obtains

$$Q\,\nu = 0 \text{ in } \Omega\times(0,T)\times\mathbb{S}^N, \text{ with } Q(x,\xi) = \varrho(x)\xi_0^2 - \sum_{i,j=1}^N a_{i,j}(x)\xi_i\xi_j. \tag{31.33}$$

In order to obtain a transport property for the H-measure μ, or equivalently for the measure ν defined by (31.32), one assumes that $f_n \rightharpoonup 0$ in $L^2_{loc}\big(\Omega\times(0,T)\big)$ weak, that the coefficients ϱ and $a_{i,j}$ are real and belong to $X^1_{loc}(\Omega)$,[16] and that $a_{j,i}=a_{i,j}$ for $i,j=1,\ldots,N$.[17] If one makes the natural assumptions that (31.29) is indeed a wave equation, i.e. $\varrho\ge\varrho_0>0$ in Ω and that there exists $\alpha>0$ such that $\sum_{i,j=1}^N a_{i,j}\lambda_i\lambda_j \ge \alpha\,|\lambda|^2$ for all $\lambda\in\mathbb{R}^N$ in Ω, then (31.33) shows that ξ_0 cannot be zero on the support of ν. Also, it is under such conditions that, using adequate boundary conditions (either on $\partial\,\Omega$ or further away if one only observes in Ω functions which are defined on a larger set), one can prove the existence of such solutions, together with the conservation of energy for real solutions

$$\frac{\partial}{\partial t}\Big(\frac{\varrho}{2}\Big(\frac{\partial u_n}{\partial t}\Big)^2 + \sum_{i,j=1}^N \frac{a_{i,j}}{2}\frac{\partial u_n}{\partial x_i}\frac{\partial u_n}{\partial x_j}\Big) - \sum_{i,j=1}^N \frac{\partial}{\partial x_i}\Big(a_{i,j}\frac{\partial u_n}{\partial x_j}\frac{\partial u_n}{\partial t}\Big) = f_n\frac{\partial u_n}{\partial t}, \tag{31.34}$$

[16] One could have coefficients in $C^1(\Omega)$ by a result of Alberto CALDERÓN [12].

[17] One could have $a_{i,j}$ complex, satisfying $a_{j,i}=\overline{a_{i,j}}$ for $i,j=1,\ldots,N$.

and for proving transport properties, it is crucial that a similar result holds for complex solutions,[18] namely

$$\frac{\partial}{\partial t}\Big(\frac{\varrho}{2}\Big|\frac{\partial u_n}{\partial t}\Big|^2 + \sum_{i,j=1}^N \frac{a_{i,j}}{2}\frac{\partial u_n}{\partial x_i}\overline{\frac{\partial u_n}{\partial x_j}}\Big) - \sum_{i,j=1}^N \frac{\partial}{\partial x_i}\Big(a_{i,j}\frac{\partial u_n}{\partial x_j}\overline{\frac{\partial u_n}{\partial t}}\Big) = f_n\overline{\frac{\partial u_n}{\partial t}}. \tag{31.35}$$

Under these hypotheses, a subsequence of (U^m, f_m) defines an H-measure π, with indices running from 0 to $N+1$, and $\pi^{i,j} = \mu^{i,j} = \xi_i\xi_j\nu$ for $i,j = 0,\ldots,N$, but the localization principle applied to (31.31) also implies

$$\xi_k\pi^{j,N+1} - \xi_j\pi^{k,N+1} = 0, j,k = 0,\ldots,N, \text{ so that } \pi^{j,N+1} = \xi_j\sigma, j = 0,\ldots,N, \tag{31.36}$$

and $\sigma = \sum_{k=0}^N \xi_k\pi^{k,N+1} \in \mathcal{M}(\Omega\times(0,T)\times\mathbb{S}^N)$ may be complex.

Lemma 31.5. *Assuming that $\varrho \in X^1_{loc}(\Omega)$ is real with $\varrho \ge \varrho_0 > 0$ in Ω, that $a_{i,j} \in X^1_{loc}(\Omega)$ is real with $a_{j,i} = a_{i,j}$ for $i,j = 1,\ldots,N$, and there exists $\alpha > 0$ such that $\sum_{i,j=1}^N a_{i,j}\lambda_i\lambda_j \ge \alpha\,|\lambda|^2$ for all $\lambda \in \mathbb{R}^N$ in Ω, that (31.30) holds, and that (U^m, f_m) defines an H-measure π with (31.36), ν satisfies*

$$\langle \nu, \{\Phi, Q\}\rangle = \langle 2\Re\sigma, \Phi\rangle, \tag{31.37}$$

for all test functions $\Phi \in C^1_c(\Omega\times(0,T)\times\mathbb{S}^N)$ (extended to be homogeneous of degree zero in ξ).

Proof. For simplicity, I assume that one has replaced u_n by $\psi\, u_n$ with $\psi \in C^\infty_c(\Omega\times(0,T))$, so that the equation already holds in $\mathbb{R}^{N+1}$, with u_n and f_n having their support in a fixed compact set. Multiplying the equation by P_a with $a \in C^1(\mathbb{S}^N)$, one obtains a wave equation for $P_a u_n$

$$\begin{aligned}&\varrho\tfrac{\partial^2 P_a u_n}{\partial t^2} - \textstyle\sum_{i,j=1}^N \tfrac{\partial}{\partial x_i}\Big(a_{i,j}\tfrac{\partial P_a u_n}{\partial x_j}\Big) + \tfrac{\partial}{\partial t}\Big((P_a\varrho - \varrho P_a)\tfrac{\partial u_n}{\partial t}\Big)\\ &\quad - \textstyle\sum_{i,j=1}^N \tfrac{\partial}{\partial x_i}\Big((P_a a_{i,j} - a_{i,j}\,P_a)\tfrac{\partial u_n}{\partial x_j}\Big) = P_a f_n.\end{aligned} \tag{31.38}$$

One defines $K_{0,0}, K_{i,j} \in \mathcal{L}\big(L^2(\mathbb{R}^{N+1}); L^2(\mathbb{R}^{N+1})\big)$, and $g_n \in L^2(\mathbb{R}^{N+1})$ by

$$\begin{aligned}&K_{0,0} = \tfrac{\partial}{\partial t}(P_a M_\varrho - M_\varrho P_a)\\ &K_{i,j} = \tfrac{\partial}{\partial x_i}(P_a M_{a_{i,j}} - M_{a_{i,j}} P_a), i,j = 1,\ldots,N\\ &g_n = P_a f_n - K_{0,0}U^n_0 + \textstyle\sum_{i,j=1}^N K_{i,j}U^n_j,\end{aligned} \tag{31.39}$$

[18] Applying "pseudo-differential operators" to real functions may give complex-valued functions, so that one must consider complex solutions.

and then, multiplying by $\overline{P_a \frac{\partial u_n}{\partial t}}$ and taking the real part, one obtains

$$\begin{aligned}&\tfrac{\partial}{\partial t}\Big(\tfrac{\varrho}{2}\Big|\tfrac{\partial P_a u_n}{\partial t}\Big|^2+\textstyle\sum_{i,j=1}^N \tfrac{a_{i,j}}{2}\tfrac{\partial P_a u_n}{\partial x_i}\overline{\tfrac{\partial P_a u_n}{\partial x_j}}\Big)\\&-\Re\textstyle\sum_{i,j=1}^N \tfrac{\partial}{\partial x_i}\Big(a_{i,j}\tfrac{\partial P_a u_n}{\partial x_i}\overline{\tfrac{\partial P_a u_n}{\partial t}}\Big)=\Re\Big(g_n\overline{\tfrac{\partial P_a u_n}{\partial t}}\Big).\end{aligned} \tag{31.40}$$

Then, one applies this equation to a test function $\varphi \in C_c^\infty(\mathbb{R}^{N+1})$, and one takes the limit, and the different terms are:

$$\begin{aligned}&\lim_{n\to\infty}\tfrac{1}{2}\Big\langle\tfrac{\partial}{\partial t}\Big(\varrho\Big|\tfrac{\partial P_a u_n}{\partial t}\Big|^2+\textstyle\sum_{i,j=1}^N a_{i,j}\tfrac{\partial P_a u_n}{\partial x_i}\overline{\tfrac{\partial P_a u_n}{\partial x_j}}\Big),\varphi\Big\rangle\\&\quad=-\tfrac{1}{2}\Big\langle\Big(\varrho|a|^2\mu^{0,0}+\textstyle\sum_{i,j=1}^N a_{i,j}|a|^2\mu^{i,j}\Big),\tfrac{\partial\varphi}{\partial t}\Big\rangle\\&\quad=-\tfrac{1}{2}\Big\langle\nu,\Big(\varrho\,\xi_0^2+\textstyle\sum_{i,j=1}^N a_{i,j}\xi_i\xi_j\Big)|a|^2\tfrac{\partial\varphi}{\partial t}\Big\rangle,\end{aligned} \tag{31.41}$$

$$\begin{aligned}&\lim_{n\to\infty}\Big\langle-\Re\textstyle\sum_{i,j=1}^N\tfrac{\partial}{\partial x_i}\Big(a_{ij}\tfrac{\partial P_a u_n}{\partial x_j}\overline{\tfrac{\partial P_a u_n}{\partial t}}\Big),\varphi\Big\rangle\\&\quad=\textstyle\sum_{i,j=1}^N\Big\langle\Re(a_{ij}|a|^2\mu^{j,0}),\tfrac{\partial\varphi}{\partial x_i}\Big\rangle\\&\quad=\Big\langle\nu,\textstyle\sum_{i,j=1}^N a_{ij}\xi_j\xi_0|a|^2\tfrac{\partial\varphi}{\partial x_i}\Big\rangle,\end{aligned} \tag{31.42}$$

$$\begin{aligned}&\lim_{n\to\infty}\Big\langle\Re\Big(g_n\overline{\tfrac{\partial P_a u_n}{\partial t}}\Big),\varphi\Big\rangle\\&\quad=\lim_{n\to\infty}\Big\langle\Re\Big(\Big(P_a f_n-K_{0,0}U_0^n+\textstyle\sum_{i,j=1}^N K_{ij}U_j^n\Big)\overline{\tfrac{\partial P_a u_n}{\partial t}}\Big),\varphi\Big\rangle\\&\quad=\Big\langle\Re\Big(|a|^2\mu^{N+1,0}-s_{0,0}\overline{a}\mu^{0,0}+\textstyle\sum_{i,j=1}^N s_{i,j}\overline{a}\mu^{j,0}\Big),\varphi\Big\rangle\\&\quad=\langle\Re\sigma,\xi_0|a|^2\varphi\rangle+\Big\langle\nu,\Big(-s_{0,0}\overline{a}\xi_0^2+\textstyle\sum_{i,j=1}^N s_{i,j}\overline{a}\xi_j\xi_0\Big)\varphi\Big\rangle,\end{aligned} \tag{31.43}$$

where $s_{0,0}$ and $s_{i,j}$ are the symbols of $K_{0,0}$ and $K_{i,j}$, given by Lemma 31.1:

$$\begin{aligned}s_{0,0}(x,\xi)&=\xi_0\textstyle\sum_{k=1}^N\tfrac{\partial a}{\partial\xi_k}\tfrac{\partial\varrho}{\partial x_k}\\s_{i,j}(x,\xi)&=\xi_i\textstyle\sum_{k=1}^N\tfrac{\partial a}{\partial\xi_k}\tfrac{\partial a_{i,j}}{\partial x_k}\text{ for }i,j=1,\ldots,N.\end{aligned} \tag{31.44}$$

All the terms in (31.41)–(31.43) involve the test function Φ given by

$$\Phi(x,\xi)=\varphi(x)\,\xi_0|a(\xi)|^2, \tag{31.45}$$

and one uses $Q\,\nu = 0$ in (31.41), so that the different terms are then

$$\begin{aligned}&-\Big\langle\nu,\varrho\,\xi_0\tfrac{\partial\Phi}{\partial t}\Big\rangle,\quad\Big\langle\nu,\textstyle\sum_{i,j=1}^N a_{i,j}\xi_j\tfrac{\partial\Phi}{\partial x_i}\Big\rangle\text{ and}\\&\langle\Re\sigma,\Phi\rangle+\tfrac{1}{2}\textstyle\sum_{k=1}^N\Big\langle\nu,\tfrac{\partial\Phi}{\partial\xi_k}\Big(-\xi_0^2\tfrac{\partial\varrho}{\partial x_k}+\textstyle\sum_{i,j=1}^N\xi_i\xi_j\tfrac{\partial a_{i,j}}{\partial x_k}\Big)\Big\rangle.\end{aligned} \tag{31.46}$$

Summing these terms makes the Poisson bracket of Φ and Q appear, so that one obtains the transport equation (31.37) with Φ given by (31.45), and since $\xi_0 \neq 0$ on the support of ν, (31.37) is valid for every smooth Φ.

If one starts from a sequence u_n which does not have a compact support, then choosing a smooth test function ψ with compact support and writing the wave equation for $\psi\,u_n$,

$$\varrho\frac{\partial^2(\psi\,u_n)}{\partial t^2} - \sum_{i,j=1}^N \frac{\partial}{\partial x_i}\Big(a_{i,j}\frac{\partial(\psi\,u_n)}{\partial x_j}\Big) = \psi\,f_n + 2\varrho\frac{\partial\psi}{\partial t}U_0^n - 2\sum_{i,j=1}^N a_{i,j}\frac{\partial\psi}{\partial x_i}U_j^n$$
$$+ \quad \Big(\varrho\frac{\partial^2\psi}{\partial t^2} - \sum_{i,j=1}^N \frac{\partial}{\partial x_i}\Big(a_{i,j}\frac{\partial\psi}{\partial x_j}\Big)\Big)u_n, \tag{31.47}$$

so that the preceding analysis can be applied with ν replaced by $|\psi|^2\nu$ and $2\Re\sigma$ replaced by

$$2\Re|\psi|^2\sigma+2\Re\Big(2\varrho\frac{\partial\psi}{\partial t}\overline{\psi}\xi_0\Big)\nu-2\Re\Big(2\sum_{i,j=1}^N a_{i,j}\frac{\partial\psi}{\partial x_i}\overline{\psi}\xi_j\Big)\nu = 2\Re|\psi|^2\sigma+\{Q,|\psi|^2\}\nu, \tag{31.48}$$

and one obtains (31.37) for ν with the test function Φ replaced by $|\psi|^2\Phi$. □

Once the tedious work is done for a typical equation like the preceding one, it appears advantageous to avoid doing useless computations in future examples, but it is by doing such computations at least once that one has some chance to guess about an intrinsic setting valid for more general examples.

It is important to notice the difference between Lemma 31.5 and the formal theory of geometrical optics. In that theory, one constructs particular asymptotic solutions of the wave equation of the form $A(x,t)\,e^{i\,\Phi(x,t)}$ for an amplitude A and a phase Φ, and with a frequency ν tending to ∞, one looks for $\Phi = \nu\,\Phi_1 + \Phi_0 + \nu^{-1}\Phi_{-1} + \ldots$ and $A = A_0 + \nu^{-1}A_{-1} + \ldots$; one finds that the phase Φ_1 must satisfy an eikonal equation $\varrho\,\big|\frac{\partial\Phi_1}{\partial t}\big|^2 = \sum_{i,j=1}^N a_{i,j}\frac{\partial\Phi_1}{\partial x_i}\frac{\partial\Phi_1}{\partial x_j}$, which is a Hamilton–Jacobi equation,[19] which has singularities on caustics; then one finds that, outside caustics, A_0 satisfies a transport equation which uses the gradient of Φ_1. The defect of geometrical optics is that it pretends that solutions of the wave equation look like distorted plane waves, and it is certainly not true near caustics. At best, geometrical optics says that there are solutions of the wave equation which show transport of energy along bicharacteristic rays, so that one conjectures that it is also true in other situations, but one does not say explicitly that it is only a conjecture, of course!

H-measures do not care about a phase, since they use no characteristic lengths,[20] and they cannot be bothered by caustics, unless one wants to know if the *H-measures* which appear have a smooth density with respect to the Lebesgue measure on $\mathbb{S}^N$, which is not a question of physical interest. *H-measures* are not bothered by situations where countably many plane waves pass through a point x at time t, since they pass at different points

---

[19] Some mathematicians seem to like geometrical optics because it involves geometry, but this approach is a dead end if one wants to prove that in the limit of infinite frequency all oscillating solutions of the wave equation or Maxwell–Heaviside equation or some other system have their energy propagating along some curves. However, it seems that one reason why many advocate this approach is precisely because it seems a dead end for understanding more physics.

[20] Since too many were stuck in studying only periodically modulated materials, I systematically avoided using characteristic lengths in homogenization!

in the $\big((t,x),(\tau,\xi)\big)$ space. The variable $\xi \in \mathbb{S}^N$ corresponds to the direction of the gradient of Φ_1 in geometrical optics, but Lemma 31.5 says that for all ways of preparing initial data and boundary conditions in putting a finite amount of energy in high frequencies, in the limit all solutions satisfy (31.37), which says that the energy hidden at a meso-scopic level will propagate along bicharacteristic rays! However, as pointed out by Patrick GÉRARD, the equation for bicharacteristic rays

$$\begin{aligned} \tfrac{dx_0}{d\tau} &= \tfrac{\partial Q}{\partial \xi_0} = 2\varrho\,\xi_0\\ \tfrac{dx_k}{d\tau} &= \tfrac{\partial Q}{\partial \xi_k} = -2\textstyle\sum_{j=1}^N a_{k,j}\xi_j, k = 1,\ldots,N\\ \tfrac{d\xi_0}{d\tau} &= -\tfrac{\partial Q}{\partial x_0} = 0\\ \tfrac{d\xi_k}{d\tau} &= -\tfrac{\partial Q}{\partial x_k} = \tfrac{\partial \varrho}{\partial x_k}\xi_0^2 - \textstyle\sum_{i,j}\tfrac{\partial a_{i,j}}{\partial x_k}\xi_i\xi_j, k = 1,\ldots,N \end{aligned} \tag{31.49}$$

is not exactly like (31.14), which decouples into a differential equation for x ($\frac{dx_j}{d\tau} = \frac{\partial P}{\partial \xi_j} = b_j(x)$ for $j = 1,\ldots,N$) followed by a linear equation for ξ ($\frac{d\xi_j}{d\tau} = -\frac{\partial P}{\partial x_j} = \sum_{k=1}^N \frac{\partial b_k}{\partial x_j}\xi_k$ for $j = 1,\ldots,N$) which have unique solutions for b_k Lipschitz continuous for $k = 1,\ldots,N$. Having the coefficients ϱ and $a_{i,j}$ of class C^1 does not seem enough for obtaining uniqueness of solutions of (31.49), so that the weak solutions of (31.37) may not be unique, and it is better to assume that the coefficients are of class C^2 for avoiding such problems. Notice that multiplying the initial data for ξ by λ, the solution replaces $\big(x(\tau),\xi(\tau)\big)$ by $\big(x(\lambda\,\tau),\lambda\,\xi(\lambda\,\tau)\big)$, showing (31.49) is actually an equation for rays in ξ.

I did not create *H-measures* for proving Lemma 31.5, and my intuition came from homogenization questions, as recalled by the prefix H-, and once the tool existed I naturally considered the question of transport of oscillations and concentration effects for a first-order scalar hyperbolic equation, and it is only because my proof showed how to extend the result to systems that I looked at the wave equation. If I had wanted to improve geometrical optics, it would have been difficult for me to choose to avoid using a phase, and this problem is general in research, that one should stay alert to new ideas, and think about a few different problems in order to be able to imagine completely different approaches to old problems.

Although *H-measures* permit one to explain why some computations done in a periodic setting can sometimes be used in a general framework without any periodicity or long-range order, it is not clear for example how to extend the concept of Bloch waves to a non-periodic setting.

After deriving the transport equations for a scalar first-order equation in Lemma 31.4 and for the wave equation in Lemma 31.5, it was natural that I study the question of initial and boundary conditions for these partial differential equations in (x,ξ) that I derived. In [105], I discussed the case of a homogeneous scalar equation, i.e. (31.11) with $c = 0$ and $f_n = 0$, and I chose to give "initial data" v_n on the hyperplane $H = \{x \in \mathbb{R}^N \mid x_N = 0\}$, assuming that it is not characteristic, i.e. $b_N \neq 0$ on H. I wondered how to use the H-measure μ_0 defined by a subsequence v_m for describing the initial data for μ, and an obvious problem is that μ_0 lives in $H \times \mathbb{S}^{N-2}$, considering $\mathbb{S}^{N-2} \subset H \subset \mathbb{R}^N$, and that as initial data for μ one needs an object in $H \times \mathbb{S}^{N-1}$, considering $\mathbb{S}^{N-1} \subset \mathbb{R}^N$, and one must lift the information from μ_0 at a point $(x, \xi') \in H \times \mathbb{S}^{N-2}$ to a suitable point $(x, \xi) \in H \times \mathbb{S}^{N-1}$; it is natural to observe that $\xi' = (\xi_1, \ldots, \xi_{N-1})$ must be associated with $\xi = (\xi', \xi_N)$ with $\sum_{j=1}^N b_j(x)\xi_j = 0$, since μ lives on the zero set of P, and ξ_N is well defined since $b_N(x) \neq 0$, and once more one must observe that my choice of the sphere as a way to pick one point in each equivalent class is not good, since the mapping $\xi' \mapsto \xi$ does not map S^{N-2} into S^{N-1}. In other words, if one prepares oscillating data in the direction ξ', the question is that μ_0 does not know at what "velocity" these oscillations will start moving, but since the support of μ is in the zero set of P, there is only one point (ξ', ξ_N) in the zero set of P for a given ξ'.

I did not think about checking the same question for the wave equation, and in that case, there are two opposite values of ξ_0 which can be associated with each ξ, so that if one prepares oscillating data in the direction ξ one must determine how much energy is sent away at each of the two velocities; however, the initial data at $t = 0$ contain the value of u_n (or $grad_x(u_n)$) and the value of $\frac{\partial u_n}{\partial t}$, and there are some algebraic computations to perform for deciding which information moves one way and which information moves the other way. These computations were done by Gilles FRANCFORT and François MURAT, with the technical help of Patrick GÉRARD, but they used C^∞ coefficients and the classical theory of pseudo-differential operators, using ideas which do not seem easy to extend to general systems, so that the general question of taking into account the initial data is far from being settled.

For the question of boundary conditions, I only made an observation about specular reflection for the classical wave equation (isotropic with constant coefficients) with Dirichlet condition, by extending the solution to be odd, and applying the propagation result to the whole space. For more general cases, like curved boundaries, I studied in [105] what I thought to be a first step, and I considered how *H-measures* change in a change of variable: if $u_n \rightharpoonup 0$ in $L^2(\mathbb{R}^N)$ weak and defines an H-measure μ_0, and one defines v_n by

$$v_n(x) = u_n\big(F(x)\big) \text{ in } \mathbb{R}^N, \tag{31.50}$$

where F is a local diffeomorphism of $\mathbb{R}^N$ into itself, and if v_n defines an H-measure μ_1, find the way to compute μ_1 from μ_0. The definition of *H-measures* through the Fourier transform is not adapted to this question, but the case where F is affine suggests that one has

$$\begin{aligned}&\langle \mu_1, \Phi\rangle = \langle \mu_0, \Psi\rangle \text{ for all } \Phi \text{ of class } C^1 \text{ with small support, with}\\ &\Psi(x,\xi) = \tfrac{1}{det[\nabla F(F^{-1}(x))]}\Phi\big(F^{-1}(x), (\nabla F)^T(F^{-1}(x))\,\xi\big),\end{aligned} \tag{31.51}$$

and I proved this formula in [105] by considering a differential equation

$$\frac{\partial w_n}{\partial t} + \sum_{j=1}^{N} b_j \frac{\partial w_n}{\partial x_j} = 0 \tag{31.52}$$

with $b_j \in C^1(\mathbb{R}^N)$, $j = 1, \ldots, N$, and such that

$$w_n \,|_{t=0} = u_n \text{ near } x_0 \text{ implies } w_n \,|_{t=1} = v_n \text{ near } F(x_0), \tag{31.53}$$

and then used my results of transport of *H-measures*, and of taking into account the initial (and final) condition.

Additional footnotes: EÖTVÖS,[21] Charles FEFFERMAN,[22] KNAPP,[23] LOVASZ,[24] STEIN.[25]

[21] Baron Loránd EÖTVÖS, Hungarian physicist, 1848–1919. Eötvös University, Budapest, Hungary, is named after him.

[22] Charles Louis FEFFERMAN, American mathematician, born in 1949. He received the Fields Medal in 1978 for his work in classical analysis. He worked at The University of Chicago, Chicago, IL, and he works now at Princeton University, Princeton, NJ.

[23] Anthony William KNAPP, American mathematician, born in 1941. He worked at Cornell University, Itaca, NY, and at SUNY (State University of New York) at Stony Brook, NY.

[24] Laszlo LOVASZ, Hungarian-born mathematician, born in 1948. He received the Wolf Prize in 1999, for his outstanding contributions to combinatorics, theoretical computer science and combinatorial optimization, jointly with Elias M. STEIN. He works at Yale University, New Haven, CT, and at Eötvös University, Budapest, Hungary.

[25] Elias M. STEIN, Belgian-born mathematician, born in 1931. He received the Wolf Prize in 1999, for his contributions to classical and "Euclidean" Fourier analysis and for his exceptional impact on a new generation of analysts through his eloquent teaching and writing, jointly with Laszlo LOVASZ. He worked at The University of Chicago, Chicago, IL, and he works now at Princeton University, Princeton, NJ.

Chapter 32
Variants of H-Measures

H-measures are defined without using characteristic lengths, and I described a variant using one characteristic length in a talk at Collège de France in January 1990, and immediately after, Patrick GÉRARD sent me his work about the subject, and he called his variant semi-classical measures.

If $\Omega \subset \mathbb{R}^N$ and $U^n \rightharpoonup 0$ in $L^2_{loc}(\Omega; \mathbb{R}^p)$ weak, my idea for introducing a variant of *H-measures* using a characteristic length ε_n tending to 0 was to introduce the sequence V^n defined on $\Omega \times \mathbb{R}$ by

$$V^n(x, x_{N+1}) = U^n(x) \cos \frac{x_{N+1}}{\varepsilon_n}, x \in \Omega, x_{N+1} \in \mathbb{R}, \tag{32.1}$$

so that $V^n \rightharpoonup 0$ in $L^2_{loc}(\Omega \times \mathbb{R}; \mathbb{R}^p)$ weak, and to extract a subsequence V^m associated to an H-measure π. Actually, I could have multiplied it by $\sqrt{2}$, so that the weak $\star$ limit of $(V^n)^2$ is that of $(U^n)^2$ extended to be independent of x_{N+1}, but a better choice is to consider

$$V^n(x, x_{N+1}) = U^n(x)\, e^{\frac{2i\,\pi\, x_{N+1}}{\varepsilon_n}}, x \in \Omega, x_{N+1} \in \mathbb{R}. \tag{32.2}$$

Lemma 32.1. *If V^n is given by (32.2) and defines an H-measure π, then π is independent of x_{N+1}.*

Proof. For $h \in \mathbb{R}$ one writes τ_h for the operator of translation of h in the direction x_{N+1}. For any $h \in \mathbb{R}$ there is a multiple h_n of ε_n such that $|h - h_n| \le \varepsilon_n$, and since $h_n \to h$, $\tau_{h_n} V^n$ and $\tau_h V^n$ define the same H-measure: indeed, when one localizes by multiplying by φ, it amounts to multiplying V^n respectively by $\tau_{-h_n}\varphi$ and $\tau_{-h}\varphi$, and $\tau_{-h_n}\varphi - \tau_{-h}\varphi$ tends to 0 uniformly, because of the uniform continuity of φ (due to its support being compact). Since $\tau_{h_n} V^n = V^n$, and $\tau_h V^n$ defines the H-measure $\tau_h \pi$, one deduces that $\tau_h \pi = \pi$. □

Of course, the same result holds if V^n is given by (32.1). My idea then adds a variable ξ_{N+1}, without really adding a variable x_{N+1}.

L. Tartar, *The General Theory of Homogenization*, Lecture Notes of the Unione Matematica Italiana 7, DOI 10.1007/978-3-642-05195-1_32,

Patrick GÉRARD's idea was to define a Hermitian nonnegative matrix of Radon measures in $\Omega \times \mathbb{R}^N$, denoted μ_{sc} and which he called a semi-classical measure,[1] by extracting a subsequence for which, for all $j, k = 1, \ldots, p$

$$\begin{aligned}&\lim_{m\to\infty} \int_{\mathbb{R}^N} \mathcal{F}(\varphi\, U_j^m)\overline{\mathcal{F}(\varphi\, U_k^m)}\, \psi(\varepsilon_m \xi)\, d\xi = \langle \mu_{sc}^{j,k}, |\varphi|^2 \otimes \psi\rangle,\\ &\quad\text{for all } \varphi \in C_c^\infty(\Omega), \psi \in \mathcal{S}(\mathbb{R}^N), \text{ and with } \mu_{sc}^{j,k} \in \mathcal{M}(\Omega \times \mathbb{R}^N).\end{aligned} \tag{32.3}$$

His intuition was that for oscillating sequences using a characteristic length ε_n, like periodically modulated sequences $v\big(x, \frac{x}{\varepsilon_n}\big)$, the Fourier transform essentially lives at a distance of order $\frac{1}{\varepsilon_n}$ in ξ,[2] so that the rescaling $\psi(\varepsilon_n \xi)$ is a way to focus on the places where the Fourier transform of $\varphi\, U^n$ is expected to be important.

In a computation done with Patrick GÉRARD for a sequence using two characteristic lengths, in a situation where the oscillations and concentration effects at the two different scales interact, we observed that the preceding intuition may be wrong, and I shall discuss this point in a moment. However, the following scalar example explains the intuition when only one scale is present, and already shows some defects of our initial approaches.

Lemma 32.2. *If $\eta_n \to 0$, then for any unit vector $e \in \mathbb{S}^{N-1}$*

$$u_n(x) = e^{\frac{2i\pi\,(x,e)}{\eta_n}}, \textit{ for } x \in \mathbb{R}^N, \textit{ defines a semi-classical measure } \mu_{sc} \tag{32.4}$$

if $\frac{\varepsilon_n}{\eta_n} \to \infty$, with $\mu_{sc} = 0$, if $\frac{\varepsilon_n}{\eta_n} \to 0$, with $\mu_{sc} = 1 \otimes \delta_0$, or more generally if $\frac{\varepsilon_n}{\eta_n} \to \kappa \in (0, \infty)$, with $\mu_{sc} = 1 \otimes \delta_{\kappa\, e}$.

Proof. One has

$$\mathcal{F} u_n = \delta_{\frac{e}{\eta_n}}, \text{ and } \mathcal{F}(\varphi\, u_n)(\xi) = \mathcal{F}\varphi\Big(\xi - \frac{e}{\eta_n}\Big), \text{ for } \varphi \in C_c^\infty(\mathbb{R}^N), \tag{32.5}$$

so that one needs to find the limit as $n \to \infty$ of

$$\int_{\mathbb{R}^N} \Big|\mathcal{F}\varphi\Big(\xi - \frac{e}{\eta_n}\Big)\Big|^2 \psi(\varepsilon_n \xi)\, d\xi = \int_{\mathbb{R}^N} |\mathcal{F}\varphi(\xi)|^2 \psi\Big(\varepsilon_n \xi + \frac{\varepsilon_n e}{\eta_n}\Big)\, d\xi. \tag{32.6}$$

[1] Because physicists use the term semi-classical for a game that they invented, of letting the Planck constant h tend to 0 in the postulated Schrödinger equations, and recovering the classical mechanics framework for the Hamiltonians used in generating the Schrödinger equations.

[2] Since the Fourier transform uses $e^{\pm 2i\pi\,(x,\xi)}$, the quantity (x, ξ) should have no dimension, and the characteristic length ε_n in x forces to use the characteristic length $\frac{1}{\varepsilon_n}$ in ξ.

Using the Lebesgue dominated convergence theorem, one sees easily that if $\mathcal{F}\varphi \in L^2(\mathbb{R}^N)$ and $\psi \in C_0(\mathbb{R}^N)$ the limit is 0 if $\frac{\varepsilon_n}{\eta_n} \to \infty$, since ψ is 0 at ∞, but for $\psi \in C_b(\mathbb{R}^N)$ and $\frac{\varepsilon_n}{\eta_n} \to \kappa$ the limit is $\int_{\mathbb{R}^N} |\mathcal{F}\varphi(\xi)|^2 \psi(\kappa\, e)\, d\xi$. □

Lemma 32.3. *If* $\eta_n \to 0$, $e \in \mathbb{S}^{N-1}$, $u_n(x) = e^{\frac{2i\,\pi\,(x,e)}{\eta_n}}$ *for* $x \in \mathbb{R}^N$,

$$v_n(x, x_{N+1}) = u_n(x)\, e^{2i\,\pi\,\frac{x_{N+1}}{\varepsilon_n}} \text{ defines an H-measure } \pi \in \mathcal{M}(\mathbb{R}^{N+1} \times \mathbb{S}^N) \tag{32.7}$$

if $\frac{\varepsilon_n}{\eta_n} \to \infty$, *with* $\pi = 1 \otimes \delta_e$, *if* $\frac{\varepsilon_n}{\eta_n} \to 0$, *with* $\pi = 1 \otimes \delta_{e_{N+1}}$, *or more generally if* $\frac{\varepsilon_n}{\eta_n} \to \kappa \in (0,\infty)$, *with* $\pi = 1 \otimes \delta_{M_\kappa}$, *with* $M_\kappa = \frac{\kappa\, e + e_{N+1}}{\sqrt{\kappa^2+1}}$.

Proof. One has

$$\mathcal{F}v_n = \delta_{P_n}, \text{ with } P_n = \frac{e}{\eta_n} + \frac{e_{N+1}}{\varepsilon_n}, \tag{32.8}$$

so that the H-measure creates a Dirac mass at the limit of $\frac{P_n}{|P_n|}$, which is M_κ if $0 \le \kappa \le \infty$. □

For sequences using one characteristic length η_n, the semi-classical measures of Patrick GÉRARD then lose information at ∞ if η_n tends to 0 much faster than ε_n does, while my *H-measures* see that information on the equator of $\mathbb{S}^N$, which is $\mathbb{S}^{N-1}$. However, both our approaches have the defect that if ε_n tends to 0 much faster than η_n does, the information is found at 0 or at e_{N+1}, but without remembering which direction of oscillations e was used, so that one mixes information from different directions. One sees that our approaches are related, by considering that the space $\mathbb{R}^N$ in ξ in the definition of semi-classical measures is like the tangent hyperplane to $\mathbb{S}^N$ at e_{N+1}.

If in (32.3) one uses $\psi \in C(\mathbb{S}^{N-1})$, extended to $\mathbb{R}^N \setminus \{0\}$ as a homogeneous function of degree 0, then one is in the situation of the definition of *H-measures*, and the desired subsequence exists, but for $\psi \in \mathcal{S}(\mathbb{R}^N)$ one must prove a different lemma than the first commutation lemma (Lemma 28.2). For the sake of generality, I avoid assuming ψ to be smooth.[3]

Lemma 32.4. *If* $\varepsilon_n \to 0$, $b \in C_0(\mathbb{R}^N)$, $\psi \in BUC(\mathbb{R}^N)$,[4] *and* ψ_n *is defined by* $\psi_n(\xi) = \psi(\varepsilon_n \xi)$ *for* $\xi \in \mathbb{R}^N$, *then the commutator* $C_n = M_b P_{\psi_n} - P_{\psi_n} M_b$ *tends to 0 in norm in* $\mathcal{L}\big(L^2(\mathbb{R}^N); L^2(\mathbb{R}^N)\big)$.

Proof. Like for the proof of Lemma 28.2, one constructs $b_m \in \mathcal{S}(\mathbb{R}^N)$ with $||b - b_m||_{L^\infty(\mathbb{R}^N)} \le \frac{1}{m}$ and $\mathcal{F}b_m$ having compact support, inside $|\xi| \le \rho_m$, and

$$\begin{aligned} \text{if } C_{m,n} = M_{b_m} P_{\psi_n} - P_{\psi_n} M_{b_m}, \text{ one has } & ||C_n - C_{m,n}||_{\mathcal{L}(L^2(\mathbb{R}^N);L^2(\mathbb{R}^N))} \\ \le \quad 2||\psi_n||_{L^\infty(\mathbb{R}^N)} ||b - b_m||_{L^\infty(\mathbb{R}^N)} \le & \frac{2||\psi||_{L^\infty(\mathbb{R}^N)}}{m}. \end{aligned} \tag{32.9}$$

[3] Patrick GÉRARD needed ψ smooth for proving a localization principle which is not restricted to the class of first-order equations, as I did for H-measures.

[4] $BUC(\mathbb{R}^N)$ is the space of bounded uniformly continuous functions on $\mathbb{R}^N$.

Then, for $v \in L^2(\mathbb{R}^N)$, one has

$$\mathcal{F}(C_{m,n}v)(\xi) = \int_{\mathbb{R}^N} \big(\psi(\varepsilon_n\eta) - \psi(\varepsilon_n\xi)\big)\, \mathcal{F}b_m(\xi-\eta)\, \mathcal{F}v(\eta)\, d\eta, \tag{32.10}$$

so that, if ω is the modulus of uniform continuity of ψ, one has

$$\begin{aligned} |\mathcal{F}(C_{m,n}v)(\xi)| &\le \omega(\varepsilon_n\rho_m) \textstyle\int_{\mathbb{R}^N} |\mathcal{F}b_m(\xi-\eta)|\, |\mathcal{F}v(\eta)|\, d\eta \\ ||\mathcal{F}(C_{m,n}v)||_{L^2(\mathbb{R}^N)} &\le \omega(\varepsilon_n\rho_m)\, ||\mathcal{F}b_m||_{L^1(\mathbb{R}^N)} ||\mathcal{F}v||_{L^2(\mathbb{R}^N)} \\ ||C_{m,n}||_{\mathcal{L}(L^2(\mathbb{R}^N);L^2(\mathbb{R}^N))} &\le \omega(\varepsilon_n\rho_m)\, ||\mathcal{F}b_m||_{L^1(\mathbb{R}^N)}, \end{aligned} \tag{32.11}$$

showing that for m fixed $C_{m,n}$ tends to 0 in norm as $n \to \infty$, and with (32.9) it implies that C_n tends to 0 in norm as $n \to \infty$. □

The space $BUC(\mathbb{R}^N)$ equipped with the sup norm is a Banach space, but it is not separable,[5] and it is simpler to restrict attention to a separable subspace of $BUC(\mathbb{R}^N)$, closed for the sup norm.[6] Instead of the choice $\mathcal{S}(\mathbb{R}^N)$ of Patrick GÉRARD, one may choose $C_0(\mathbb{R}^N)$, but in order to avoid losing information at ∞ it is better to compactify $\mathbb{R}^N$ with a sphere Σ_∞ at ∞, and in order not to mix information from different directions at 0, it is better to open a hole at 0, i.e. consider $\mathbb{R}^N \setminus \{0\}$, and compactify it by also adding a sphere Σ_0 at 0; this leads to the following definition.

Definition 32.5. *$K_\infty(\mathbb{R}^N)$ is the compactification of $\mathbb{R}^N$ obtained by adding a sphere Σ_∞ at ∞: $C\big(K_\infty(\mathbb{R}^N)\big)$ is the space of continuous functions f on $\mathbb{R}^N$ such that there exists $f_\infty \in C(\mathbb{S}^{N-1})$, with $f(\xi) - f_\infty\big(\frac{\xi}{|\xi|}\big) \to 0$ as $|\xi| \to \infty$.*

$K_{0,\infty}(\mathbb{R}^N)$ is the compactification of $\mathbb{R}^N \setminus \{0\}$ obtained by adding a sphere Σ_0 at 0 and a sphere Σ_∞ at ∞: $C\big(K_{0,\infty}(\mathbb{R}^N)\big)$ is the space of continuous functions g on $\mathbb{R}^N \setminus \{0\}$ such that there exists $g_0, g_\infty \in C(\mathbb{S}^{N-1})$, with $g(\xi) - g_0\big(\frac{\xi}{|\xi|}\big) \to 0$ as $|\xi| \to 0$, and $g(\xi) - g_\infty\big(\frac{\xi}{|\xi|}\big) \to 0$ as $|\xi| \to \infty$.

Lemma 32.6. *If $\varepsilon_n \to 0$ and $U^n \rightharpoonup 0$ in $L^2(\Omega;\mathbb{R}^p)$ weak, there exists a subsequence U^m and a $p \times p$ Hermitian symmetric matrix $\mu_{K_{0,\infty}}$ of Radon measures on $\Omega \times K_{0,\infty}(\mathbb{R}^N)$ such that for all $\varphi_1, \varphi_2 \in C_c(\Omega)$, all $\psi \in C\big(K_{0,\infty}(\mathbb{R}^N)\big)$, and all $j,k \in \{1,\ldots,p\}$, one has*

---

[5] The continuous functions f affine in each interval $(n, n+1)$ with $f(n) \in \{-1,+1\}$ for all $n \in \mathbb{Z}$ are Lipschitz continuous with constant ≤ 2 and at distance 2 apart in the sup norm. This family is not countable, and cannot be covered by countably many balls of radius <1, each ball containing at most one such function.

[6] If $f \in L^\infty(\mathbb{R}^N)$ and $\rho \in L^1(\mathbb{R}^N)$, the Young inequality gives $||f \star \rho||_{L^\infty(\mathbb{R}^N)} \le ||f||_{L^\infty(\mathbb{R}^N)} ||\rho||_{L^1(\mathbb{R}^N)}$, but one actually has $f \star \rho \in BUC(\mathbb{R}^N)$, since convolution commutes with translations and $||\tau_h\rho - \rho||_{L^1(\mathbb{R}^N)} \to 0$ as $|h| \to 0$. $BUC(\mathbb{R}^N)$ is actually the space of functions $f \in L^\infty(\mathbb{R}^N)$ for which $f \star \rho_n$ converges uniformly to f when ρ_n is a regularizing sequence.

$$\lim_{m\to\infty}\int_{\mathbb{R}^N}\mathcal{F}(\varphi_1 U_j^m)\overline{\mathcal{F}(\varphi_2 U_k^m)}\,\psi(\varepsilon_m\xi)\,d\xi=\langle\mu^{j,k}_{K_{0,\infty}},\varphi_1\overline{\varphi_2}\otimes\psi\rangle. \quad (32.12)$$

Proof. One first extracts a subsequence for which an H-measure μ exists, and if $\psi(\xi)=\psi_0\big(\frac{\xi}{|\xi|}\big)$ for all $\xi\in\mathbb{R}^N$ then $\psi(\varepsilon_n\xi)=\psi_0\big(\frac{\xi}{|\xi|}\big)$ for all $\xi\in\mathbb{R}^N$, and the limit exists. For $\psi\in C\big(K_{0,\infty}(\mathbb{R}^N)\big)$, one subtracts ψ_0 and one has $\psi-\psi_0\in C\big(K_\infty(\mathbb{R}^N)\big)$, which is included in $BUC(\mathbb{R}^N)$, so that Lemma 32.4 applies, and it serves in asserting that the limit only depends upon $\varphi_1\overline{\varphi_2}$. However, instead of repeating the steps in the proof of Theorem 28.5, I follow my approach, defining V^n on $\Omega\times\mathbb{R}$ by (32.2), and extracting a subsequence V^m which defines an H-measure π, independent of x_{N+1} by Lemma 32.1, so that π_0 denotes its projection on $\mathbb{R}^N\times\mathbb{S}^N$ obtained by forgetting x_{N+1}. Choosing a nonzero $\varphi\in\mathcal{S}(\mathbb{R})$ with $\mathcal{F}\varphi\in C_c^\infty(\mathbb{R})$, with *support*$(\varphi)\subset[-\rho,+\rho]$, one defines Φ_j on $\mathbb{R}^{N+1}$ for $j=1,2$, and $\Psi\in C(\mathbb{S}^N)$, already extended to $\mathbb{R}^{N+1}\setminus\{0\}$ to be homogeneous of degree 0, by

$$\begin{aligned}&\Phi_j(x,x_{N+1})=\varphi_j(x)\,\varphi(x_{N+1}), j=1,2,\\ &\Psi(\xi,\xi_{N+1})=\psi\big(\tfrac{\xi}{\xi_{N+1}}\big)\text{ if }\xi_{N+1}\neq0,\ \text{ and }\Psi(\xi,0)=\psi_\infty(\xi)\text{ if }\xi\neq0,\end{aligned} \quad (32.13)$$

and one writes

$$\lim_{m\to\infty}\int_{\mathbb{R}^{N+1}}\mathcal{F}(\Phi_1 V_j^m)\overline{\mathcal{F}(\Phi_2 V_k^m)}\Psi\,d\xi\,d\xi_{N+1}=\langle\pi^{j,k},\Phi_1\overline{\Phi_2}\otimes\Psi\rangle. \quad (32.14)$$

Then, $\mathcal{F}(\Phi_1 V_j^m)=\mathcal{F}(\varphi_1 U_j^m)\,\mathcal{F}(\varphi\, e^{\frac{2i\pi\cdot}{\varepsilon_m}})$, $\overline{\mathcal{F}(\Phi_2 V_k^m)}=\overline{\mathcal{F}(\varphi_2 U_k^m)}\,\overline{\mathcal{F}(\varphi\, e^{\frac{2i\pi\cdot}{\varepsilon_m}})}$, and $\mathcal{F}(\varphi\, e^{\frac{2i\pi\cdot}{\varepsilon_m}})=\tau_{1/\varepsilon_m}\mathcal{F}\varphi$, so that the integrand on the left of (32.14) has a term $|\tau_{1/\varepsilon_m}\mathcal{F}\varphi|^2$, whose support in ξ_{N+1} is in the interval $\big[\frac{1}{\varepsilon_m}-\rho,\frac{1}{\varepsilon_m}+\rho\big]$.

If one shows that $\frac{1}{\varepsilon_m}-\rho\leq\xi_{N+1}\leq\frac{1}{\varepsilon_m}+\rho$ implies $|\Psi(\xi,\xi_{N+1})-\psi(\varepsilon_m\xi)|\leq\alpha_m$ for all $\xi\in\mathbb{R}^N$, and that α_m tends to 0, then the limit of the left side of (32.14) is equal to the limit of the left side of (32.12) multiplied by $\int_{\mathbb{R}}|\tau_{1/\varepsilon_m}\mathcal{F}\varphi|^2\,d\xi_{N+1}$, which is $\int_{\mathbb{R}}|\varphi|^2\,dx_{N+1}$. Since the right side of (32.14) is $\langle\pi^{j,k},\Phi_1\overline{\Phi_2}\otimes\Psi\rangle=\langle\pi_0^{j,k},\varphi_1\overline{\varphi_2}\otimes\Psi\rangle\int_{\mathbb{R}}|\varphi|^2\,dx_{N+1}$, one deduces that the limit of the left side of (32.12) is $\langle\pi_0^{j,k},\varphi_1\overline{\varphi_2}\otimes\Psi\rangle$, and by using the explicit expression of Ψ in terms of ψ, one deduces what $\mu^{j,k}_{K_{0,\infty}}$ is in that case.

For computing α_m, one notices that (as soon as $\varepsilon_m\rho<1$) $\big|\frac{1}{\varepsilon_m}-\xi_{N+1}\big|\leq\rho$ implies $\big|\frac{1}{\xi_{N+1}}-\varepsilon_m\big|\leq\sigma_m$ with $\sigma_m\to0$, so that $\big|\frac{\xi}{\xi_{N+1}}-\varepsilon_m\xi\big|\leq\sigma_m|\xi|$ and $\big|\psi\big(\frac{\xi}{\xi_{N+1}}\big)-\psi(\varepsilon_m\xi)\big|\leq\omega(\sigma_m|\xi|)$, where ω is the modulus of uniform continuity of ψ; this gives an estimate for α_m when $|\xi|\leq r$, and for $|\xi|\geq r$ one uses $|\psi-\psi_\infty|\leq\beta(r)$, with $\beta(r)\to0$ as $r\to\infty$, and $\psi_\infty\big(\frac{\xi}{\xi_{N+1}}\big)=\psi_\infty(\varepsilon_m\xi)$, since ψ_∞ is homogeneous of degree 0. ☐

The semi-classical measures introduced by Patrick GÉRARD have the defect of forgetting the part of $\mu_{K_{0,\infty}}$ supported on the sphere Σ_0 at 0 and on the sphere Σ_∞ at ∞, while the improvement (32.2) of my initial proposal (32.1) only forgets the part of $\mu_{K_{0,\infty}}$ supported on the sphere Σ_0 at 0. The reason why I later introduced the variant $\mu_{K_{0,\infty}}$, from which one can deduce both the semi-classical measure and the H-measure associated to a subsequence, was to correct a mistake of Pierre-Louis LIONS and Thierry PAUL,[7] who wrote the false statement that one can deduce the H-measure of a sequence from its semi-classical measure, although Patrick GÉRARD explicitly mentioned the question of losing information at ∞ and at 0, which is why their statement is false. I was actually very puzzled, since I thought it impossible that they could believe in a world with only one characteristic length.[8] Even if he had not read or understood what Patrick GÉRARD wrote in his article, Pierre-Louis LIONS knew about the possibility of losing information at ∞, since he worked on such a question, calling his approach concentration-compactness, but as he told me that he chose this term in order to lure people with the similarity in names with *compensated compactness*, I must point out that *compensated compactness* is a microlocal theory, and that concentration-compactness has no microlocal character.[9] Actually, homogenization is a nonlinear microlocal theory!

Pierre-Louis LIONS and Thierry PAUL renamed to Wigner measures the semi-classical measures of Patrick GÉRARD, because they found a way to define them using the Wigner transform.[10] Since semi-classical measures are just a variant of *H-measures* using one characteristic length, and one can define other variants, some of them using many characteristic lengths, I do not find it wise to invent a new name for each variant. Being interested in physics problems, I do not understand either the interest of some mathematicians for physicists' problems, as if they did not understand what I explained about the defects of kinetic theory and of quantum mechanics, for example.

In the late 1970s, I mentioned once to George PAPANICOLAOU that I wanted to split Young measures by adding a variable ξ, and he mentioned

[7] Thierry PAUL, French mathematician. He works at Université Paris IX-Dauphine, Paris, France.

[8] Strangely enough, I actually heard such a silly statement in 2007 in the talk of a physicist, former Nobel laureate, who said that problems in biology are more difficult since they use many characteristic lengths, while problems in physics only use one! With physicists having no shame boasting about their ignorance of multiscale problems, which they must have heard about at least in questions of material sciences during their studies, it is time to train a new generation of mathematicians who hopefully will be less deluded about questions of scales.

[9] Pierre-Louis LIONS told me that Raghu VARADHAN mentioned to him that similar ideas to concentration-compactness were used earlier, by P. LÉVY.

[10] Jenõ Pál (Eugene Paul) WIGNER, Hungarian-born physicist, 1902–1995. He shared the Nobel Prize in Physics in 1963, for his contributions to the theory of the atomic nucleus and the elementary particles, particularly through the discovery and application of fundamental symmetry principles, jointly with Maria GOEPPERT-MAYER and J. Hans D. JENSEN. He worked at Princeton University, Princeton, NJ.

the Wigner transform, which consists in associating to a function $u \in L^2(\mathbb{R}^N)$ the wave function W defined on $\mathbb{R}^N \times \mathbb{R}^N$ by

$$W(x,\xi) = \int_{\mathbb{R}^N} u\Big(x+\frac{y}{2}\Big)\,\overline{u\Big(x-\frac{y}{2}\Big)}\,e^{-2i\,\pi\,(y,\xi)}\,dy, \tag{32.15}$$

giving $W \in C_b(\mathbb{R}^N \times \mathbb{R}^N)$, and under supplementary hypotheses, one has

$$\int_{\mathbb{R}^N} W(x,\xi)\,d\xi = |u(x)|^2, \tag{32.16}$$

$$\int_{\mathbb{R}^N} W(x,\xi)\,dx = |\mathcal{F}u(\xi)|^2. \tag{32.17}$$

A sufficient condition for having (32.16) is to have $u \in L^2(\mathbb{R}^N) \cap \mathcal{F}L^1(\mathbb{R}^N)$, so that both $u\big(x+\frac{\cdot}{2}\big)$ and $u\big(x-\frac{\cdot}{2}\big)$ belong to this space and their product f_x belongs to $L^1(\mathbb{R}^N) \cap \mathcal{F}L^1(\mathbb{R}^N)$,[11] so that $\mathcal{F}f_x \in C_0(\mathbb{R}^N) \cap L^1(\mathbb{R}^N)$ and $\int_{\mathbb{R}^N} \mathcal{F}f_x(\xi)\,d\xi = f_x(0)$, but $\mathcal{F}f_x(\xi) = W(x,\xi)$, and $f_x(0) = |u(x)|^2$. A sufficient condition for having (32.17) is to have $u \in L^2(\mathbb{R}^N) \cap L^1(\mathbb{R}^N)$ so that $u\big(x+\frac{y}{2}\big)\,\overline{u\big(x-\frac{y}{2}\big)}$ is integrable in (x,y), and one may use the Fubini theorem for the integrand, written as $\big[u\big(x+\frac{y}{2}\big)\,e^{-2i\,\pi\,(x+\frac{y}{2},\xi)}\big]\,\overline{\big[u\big(x-\frac{y}{2}\big)\,e^{-2i\,\pi\,(x-\frac{y}{2},\xi)}\big]}$, and notice that $d\big(x-\frac{y}{2}\big)\,d\big(x+\frac{y}{2}\big) = dx\,dy$. In the late 1970s, I had in mind to split Young measures in ξ, and I could not find a way to use the Wigner transform. Since I often tried to push researchers away from to the too particular case of periodically modulated materials, I wanted to avoid using a single characteristic length in my homogenization studies, and after introducing *H-measures* in the late 1980s, I only mentioned in passing my idea for using a variant with one characteristic length. Actually, what interests me is to understand situations with many characteristic lengths, and eventually a large number of them, since some reasonable specialists of hydrodynamics say that it is what one observes, in boundary layers,[12] and in turbulent flows.[13]

I could not have thought then of the idea of Pierre-Louis LIONS and Thierry PAUL, to rescale the Wigner transform by defining W_n by

$$W_n(x,\xi) = \int_{\mathbb{R}^N} u_n\Big(x+\frac{\varepsilon_n y}{2}\Big)\,\overline{u_n\Big(x-\frac{\varepsilon_n y}{2}\Big)}\,e^{-2i\,\pi\,(y,\xi)}\,dy, \tag{32.18}$$

[11] $\mathcal{F}L^1(\mathbb{R}^N)$ is a multiplicative algebra, since $L^1(\mathbb{R}^N)$ is a convolution algebra.

[12] The early approach of PRANDTL led to a one characteristic length model, but the later approach of STEWARTSON led to a three characteristic lengths model (the Stewartson triple deck), which I first heard about from Richard MEYER, and then from Jean-Pierre GUIRAUD, and from Edward FRAENKEL.

[13] As shown in Chap. 24, the class of first-order differential operators is not stable by homogenization. A consequence is that effective equations for describing turbulent flows could be quite different than what was proposed up to now. Of course, one should also avoid using a probabilistic language for this question!

for a sequence u_n converging to 0 in $L^2(\mathbb{R}^N)$ weak, with ε_n tending to 0, and showing that W_n converges to a nonnegative Radon measure on $\Omega \times \mathbb{R}^N$. Why did they propose to call them Wigner measures, without emphasizing that they are the same object that Patrick GÉRARD previously called semi-classical measures? They found a different way to introduce semi-classical measures, and it is sometimes useful to find different ways to introduce a known object, but giving many names to the same object can only make worse the chaotic situation which already exists in the academic world! After reading their false statement about deducing *H-measures* from semi-classical measures, I did not bother to look at the too lengthy technical details of their article, and when Patrick GÉRARD visited CMU (*Carnegie Mellon University*) afterward, I asked him what Pierre-Louis LIONS and Thierry PAUL were trying to do in their article; his explanation was that they wanted to show that the weak $\star$ limit of a subsequence W_m is ≥ 0,[14] and I immediately thought of a way to use two-point correlations and the Bochner theorem, which we checked; later, Patrick GÉRARD adopted this new way of using two-point correlations for introducing semi-classical measures.

I knew that one needs at least one characteristic length for defining correlations for a general sequence u_n, but before that discussion I did not find anything interesting to say about that. Assuming that $u_n \rightharpoonup 0$ in $L^2(\mathbb{R}^N)$ as a simplification, and if $\varepsilon_n \to 0$, one observes that for all $y, z \in \mathbb{R}^N$ the sequence $u_n(\cdot + \varepsilon_n y)\,\overline{u_n(\cdot + \varepsilon_n z)}$ is bounded in $L^1(\mathbb{R}^N)$, so that

$$\begin{aligned}&\text{for fixed } y, z \in \mathbb{R}^N, \text{ there is a subsequence such that}\\ &\quad u_m(\cdot + \varepsilon_m y)\,\overline{u_m(\cdot + \varepsilon_m z)} \rightharpoonup C_{y,z} \text{ in } \mathcal{M}_b(\mathbb{R}^N) \text{ weak } \star,\end{aligned} \tag{32.19}$$

and the basic observation is that

$$C_{y+h,z+h} = C_{y,z} \text{ for all } h \in \mathbb{R}^N, \tag{32.20}$$

using the same subsequence whatever $h \in \mathbb{R}^N$ is, since

$$\langle \tau_{-\varepsilon_m(y+h)} u_m \overline{\tau_{-\varepsilon_m(z+h)} u_m}, \varphi \rangle = \langle \tau_{-\varepsilon_m y} u_m \overline{\tau_{-\varepsilon_m z} u_m}, \tau_{\varepsilon_m h}\varphi \rangle, \tag{32.21}$$

[14] Marc FEIX told me afterward that WIGNER proved that the convolution of W with some Gaussian is ≥ 0, and I mistakenly thought in [119] that it was a Gaussian in ξ, but it must be a Gaussian in x, $e^{-\gamma\, x^2}$, and he characterized the best value of $\gamma > 0$. Marc FEIX said that he mentioned this fact to Pierre-Louis LIONS, but I am not sure if he mentioned it before or after he wrote the article with his coauthor, since they attribute the idea to someone else. I assume then that some technical details of their proof consist in noticing that the convolution of W_m with $e^{-\gamma\, \varepsilon_m^{-2} x^2}$ is ≥ 0, so that the limit of W_m is ≥ 0, since a multiple of that Gaussian converges to the Dirac mass at 0.

and $\tau_{\varepsilon_m h}\varphi - \varphi \to 0$ in the sup norm for all $\varphi \in C_0(\mathbb{R}^N)$, and even for all $\varphi \in BUC(\mathbb{R}^N)$, since it follows from uniform continuity. For a finite number of points $y^1, \ldots, y^q \in \mathbb{R}^N$ and arbitrary complex numbers $\lambda_1, \ldots, \lambda_q$, one has

$$\Big|\sum_{j=1}^{q} \lambda_j u_n(\cdot + \varepsilon_n y^j)\Big|^2 \rightharpoonup \sum_{j,k=1}^{q} \lambda_j \overline{\lambda_k} C_{y^j,y^k} \text{ in } \mathcal{M}(\mathbb{R}^N) \text{ weak } \star, \tag{32.22}$$

for any subsequence such that C_{y^j,y^k} exists for $j,k = 1,\ldots,q$; if $\varGamma \in \mathcal{M}(\mathbb{R}^N)$ is defined by $\varGamma(h) = C_{\frac{h}{2},\frac{-h}{2}}$, one has $C_{y,z} = \varGamma(y-z)$ and (32.22) implies

$$\sum_{j,k=1}^{q} \lambda_j \overline{\lambda_k} \varGamma(y^j - y^k) \geq 0. \tag{32.23}$$

If $\varGamma$ is a continuous function, and (32.23) is true for all choices of $q \in \mathbb{N}$, $y^1,\ldots,y^q \in \mathbb{R}^N$, and $\lambda_1,\ldots,\lambda_q \in \mathbb{C}$, then $\varGamma$ is by definition a function of positive type (in y), and by the Bochner theorem it is the Fourier transform of a nonnegative Radon measure, but without continuous dependence of $C_{y,z}$ in (y,z), one must use Laurent SCHWARTZ's extension of the Fourier transform to tempered distributions. One first defines $C_n \in L^1_{loc}(\mathbb{R}^N \times \mathbb{R}^N \times \mathbb{R}^N)$ by

$$C_n(x,y,z) = u_n(x+\varepsilon_n y)\,\overline{u_n(x+\varepsilon_n z)} \text{ for } x,y,z \in \mathbb{R}^N, \tag{32.24}$$

and one observes that

$$\text{a subsequence } C_m \rightharpoonup C \text{ in } \mathcal{M}(\mathbb{R}^N \times \mathbb{R}^N \times \mathbb{R}^N) \text{ weak } \star, \tag{32.25}$$

$$\text{for all } h \in \mathbb{R}^N, \tau_{0,h,h} C = C, \tag{32.26}$$

and (32.26) holds since one has

$$\langle \tau_{(0,h,h)} C_m, \varphi\rangle = \langle \tau_{(\varepsilon_m h,0,0)} C_m, \varphi\rangle = \langle C_m, \tau_{(-\varepsilon_m h,0,0)}\varphi\rangle \to \langle C, \varphi\rangle, \tag{32.27}$$

for all $\varphi \in C_c(\mathbb{R}^N \times \mathbb{R}^N \times \mathbb{R}^N)$, because $\tau_{(-\varepsilon_m h,0,0)}\varphi$ keeps its support in a compact, and converges uniformly to φ. If C is a function, $C(x,y,z) = D(x,y-z)$, but C being a Radon measure, it means that there exists $D \in \mathcal{M}(\mathbb{R}^N \times \mathbb{R}^N)$ such that

$$\begin{aligned} &\langle C,\varphi\rangle = \langle D,\psi\rangle, \text{ for all } \varphi \in C_c(\mathbb{R}^N \times \mathbb{R}^N \times \mathbb{R}^N),\\ &\quad \text{with } \psi \in C_c(\mathbb{R}^N \times \mathbb{R}^N) \text{ given by}\\ &\quad \psi(x,y) = \textstyle\int_{\mathbb{R}^N} \varphi(x,y+h,h)\,dh, \text{ for all } x,y \in \mathbb{R}^N. \end{aligned} \tag{32.28}$$

Then for $\varphi,\psi \in C_c(\mathbb{R}^N)$ one has

$$\begin{aligned} &\textstyle\int_{\mathbb{R}^N} \big|\int_{\mathbb{R}^N} u_m(x+\varepsilon_m y)\,\varphi(y)\,dy\big|^2 \psi(x)\,dx \to \langle C, \psi\otimes\varphi\otimes\overline{\varphi}\rangle\\ &\quad = \langle D, \psi\otimes\varPhi\rangle, \text{ with } \varPhi(y) = \textstyle\int_{\mathbb{R}^N} \varphi(y+h)\,\overline{\varphi(h)}\,dh, y \in \mathbb{R}^N, \end{aligned} \tag{32.29}$$

and the important property is that for $\varphi \in \mathcal{S}(\mathbb{R}^N)$ one has $\mathcal{F}\Phi = |\mathcal{F}\varphi|^2$. Since C_n is bounded in y, z with values in $L^1(\mathbb{R}^N)$, D is a tempered distribution and if $\mu = \overline{\mathcal{F}}_y D$ (so that $D = \mathcal{F}_\xi \mu$ if one let $\xi \in \mathbb{R}^N$ be the dual variable of $y \in \mathbb{R}^N$), then for $\psi \in \mathcal{S}(\mathbb{R}^N)$ satisfying $\psi \geq 0$, one has

$$0 \leq \langle D, \psi \otimes \Phi \rangle = \langle \mathcal{F}_\xi \mu, \psi \otimes \Phi \rangle = \langle \mu, \psi \otimes \mathcal{F}\Phi \rangle = \langle \mu, \psi \otimes |\mathcal{F}\varphi|^2 \rangle; \quad (32.30)$$

since $\mathcal{F}\varphi$ is arbitrary in $\mathcal{S}(\mathbb{R}^N)$, its square can approach (uniformly) any nonnegative function in $C_c(\mathbb{R}^N)$, and one deduces that $\mu \geq 0$.[15] Then,

$$\begin{aligned} &\text{if } D_m(x,y) = C_m\big(x, \tfrac{y}{2}, \tfrac{-y}{2}\big) \text{ on } \mathbb{R}^N \times \mathbb{R}^N, \\ &\text{then } D_m \rightharpoonup D \text{ in } \mathcal{M}(\mathbb{R}^N \times \mathbb{R}^N) \text{ weak } \star. \end{aligned} \quad (32.31)$$

Indeed, assuming that a subsequence converges weakly $\star$ to D_∞, one uses the change of variable $Y = y - z, h = \frac{y+z}{2}, X = x + \varepsilon_m h$, which gives $C_m(x,y,z) = D_m(X,Y)$, and $\varphi\big(X - \varepsilon_m h, \frac{Y}{2} + h, \frac{-Y}{2} + h\big)$ converges uniformly to $\varphi\big(X, \frac{Y}{2} + h, \frac{-Y}{2} + h\big)$, since h is bounded on the support of φ (as $\pm\frac{Y}{2} + h$ are bounded). Since $dx\,dy\,dz = dX\,dY\,dh$, the limit of the integral of $C_m \varphi$ is $\langle D, \psi \rangle$ with ψ given at (32.26), but it is also D_∞ (in the variables (X,Y)) applied to the function $\varphi\big(X, \frac{Y}{2} + h, \frac{-Y}{2} + h\big)$ integrated in $h \in \mathbb{R}^N$, and that is $\psi(X,Y)$.

The Fourier transform in y of D_m is well defined, and is W_m, but the natural bound in $\frac{1}{\varepsilon_m^N}$ for the $L^1(\mathbb{R}^N \times \mathbb{R}^N)$ norm of D_m creates a problem, so one uses Laurent SCHWARTZ's extension of the Fourier transform, by observing that D_m converges weakly to D in $\mathcal{S}'(\mathbb{R}^N \times \mathbb{R}^N)$, so that $W_m = \mathcal{F}_y D_m$ converges weakly in $\mathcal{S}'(\mathbb{R}^N \times \mathbb{R}^N)$ to $\mathcal{F}_y D = \mu \geq 0$.

Then, one must show that μ is the semi-classical measure μ_{sc} introduced by Patrick GÉRARD. Since $W_m \rightharpoonup \mu$ in $\mathcal{S}'(\mathbb{R}^N \times \mathbb{R}^N)$ weak, one has

$$\begin{aligned} \langle \mu, |\varphi|^2 \otimes \psi \rangle &= \lim_{m\to\infty} \textstyle\int_{\mathbb{R}^N \times \mathbb{R}^N} W_m(x,\xi)\, |\varphi(x)|^2 \psi(\xi)\, dx\, d\xi = \lim_{m\to\infty} \\ &\textstyle\int_{\mathbb{R}^N \times \mathbb{R}^N \times \mathbb{R}^N} u_m\big(x + \frac{\varepsilon_m y}{2}\big) \overline{u_m\big(x - \frac{\varepsilon_m y}{2}\big)}\, e^{-2i\pi\,(y,\eta)}\, |\varphi(x)|^2 \psi(\eta)\, dx\, dy\, d\eta, \end{aligned} \quad (32.32)$$

for all $\varphi, \psi \in \mathcal{S}(\mathbb{R}^N)$. For the semi-classical measure,

$$\begin{aligned} \langle \mu_{sc}, |\varphi|^2 \otimes \psi \rangle &= \lim_{m\to\infty} \textstyle\int_{\mathbb{R}^N} \mathcal{F}(\varphi\, u_m)(\xi)\, \overline{\mathcal{F}(\varphi\, u_m)}\, \psi(\varepsilon_m \xi)\, dx\, d\xi = \lim_{m\to\infty} \\ &\textstyle\int_{\mathbb{R}^N \times \mathbb{R}^N \times \mathbb{R}^N} \varphi(z^1) u_m(z^1) \overline{\varphi(z^2) u_m(z^2)}\, e^{-2i\pi\,(z^1 - z^2, \xi)}\, \psi(\varepsilon_m \xi)\, dz^1\, dz^2\, d\xi, \end{aligned} \quad (32.33)$$

[15] If $\mu \in \mathcal{M}(\Omega_1 \times \Omega_2)$ satisfies $\langle \mu, \varphi_1 \otimes \varphi_2 \rangle \geq 0$ whenever $\varphi_1 \in C_c(\Omega_1)$ and $\varphi_2 \in C_c(\Omega_2)$ are ≥ 0, then by a limiting process one deduces that $\mu(E) \geq 0$ if $E = E_1 \times E_2$ for particular Borel sets E_1, E_2 such that $\overline{E_1} \subset \Omega_1 \subset \mathbb{R}^{N_1}$, $\overline{E_2} \subset \Omega_2 \subset \mathbb{R}^{N_2}$; one then observes that any open set in $\Omega_1 \times \Omega_2$ is a countable disjoint union of such products, and one deduces that the μ-measure of any open set in $\Omega_1 \times \Omega_2$ is ≥ 0, so that $\mu \geq 0$.

for all $\varphi,\psi\in C_0(\mathbb{R}^N)$, and the change of variable $z^1=x+\frac{\varepsilon_m y}{2}$, $z^2=x-\frac{\varepsilon_m y}{2}$, $\varepsilon_m\xi=\eta$ gives $(z^1-z^2,\xi)=(y,\eta)$ and $dz^1\,dz^2\,d\xi=dx\,dy\,d\eta$, so that (32.32) and (32.33) look quite similar, except that $|\varphi(x)|^2$ in (32.32) is replaced by $\varphi(x+\frac{\varepsilon_m y}{2})\overline{\varphi(x-\frac{\varepsilon_m y}{2})}$ in (32.33), and the difference is important, since there is no bound on y. Of course, the two quantities are equal if one can take $\varphi=1$, and so for $\varphi_0\in C_c^\infty(\mathbb{R}^N)$, one considers $\varphi_0 u_m$, so that C is replaced by $|\varphi_0|^2C$ in (32.24)–(32.25),[16] D is replaced by $|\varphi_0|^2D$ in (32.28), and μ is replaced by $|\varphi_0|^2\mu$ in (32.30); of course, μ_{sc} is also replaced by $|\varphi_0|^2\mu_{sc}$, and then one uses any $\varphi\in\mathcal{S}(\mathbb{R}^N)$ which is equal to 1 on the support of φ_0, so that it amounts to taking $\varphi=1$, and one deduces that $\langle\mu,|\varphi_0|^2\otimes\psi\rangle=\langle\mu_{sc},|\varphi_0|^2\otimes\psi\rangle$ for all $\varphi_0\in C_c^\infty(\mathbb{R}^N)$ and all $\psi\in\mathcal{S}(\mathbb{R}^N)$, i.e. $\mu=\mu_{sc}$.

My idea (32.1), or the improvement (32.2), leads to using *H-measures*, so that the localization principle (Theorem 28.7) applies, since a first-order partial differential equation for U^n gives a first-order partial differential equation for V^n. However, Patrick GÉRARD thought of using partial differential equations of order >1 if the higher-order derivatives come with a corresponding power of ε_n; actually, Patrick GÉRARD defined semi-classical measures without extracting a subsequence converging weakly and subtracting the limit, so that I am adapting his idea to my general approach. Using the notation of WHITNEY,[17] I consider relations in conservative form

$$\sum_{1\le|\alpha|\le r,\alpha\ge 0}\sum_{j=1}^{p}\varepsilon_n^{|\alpha|-1}D^\alpha(\varphi_{\alpha,j}U_j^n)=f_n\text{ in }\Omega,\tag{32.34}$$

with all $\varphi_{\alpha,j}\in C(\Omega)$, and f_n converging to 0 in a suitable way.

Lemma 32.7. *Let $\varepsilon_n\to 0$, $U^n\rightharpoonup 0$ in $L^2(\Omega;\mathbb{R}^p)$ weak, and a $p\times p$ Hermitian symmetric matrix $\mu_{K_{0,\infty}}$ of Radon measures on $\Omega\times K_{0,\infty}(\mathbb{R}^N)$ be associated to a subsequence U^m like in Lemma 32.6. If (32.34) holds, with f_n satisfying*

$$\text{for all }\varphi\in C_c^\infty(\Omega),\ \frac{\mathcal{F}(\varphi f_n)}{1+\sum_{s=1}^r\varepsilon_n^{s-1}|\xi|^s}\to 0\text{ in }L^2(\mathbb{R}^N)\text{ strong},\tag{32.35}$$

[16] Since test functions have compact support in y and z.

[17] Laurent SCHWARTZ told me in the late 1990s that he was often wrongly credited for the simplifying notation introduced by WHITNEY, but I must say that he never mentioned this name when he used the notation in his course at École Polytechnique in 1965–1966 [86], and I do not remember seeing the name of WHITNEY in his book on distributions either [85]. For multi-indices $\alpha=(\alpha_1,\ldots,\alpha_N)$, $\beta=(\beta_1,\ldots,\beta_N)$, one uses x^α to mean $x_1^{\alpha_1}\cdots x_N^{\alpha_N}$, $|\alpha|$ to mean $|\alpha_1|+\ldots+|\alpha_N|$, $\alpha\ge\beta$ to mean $\alpha_1\ge\beta_1,\ldots,\alpha_N\ge\beta_N$; for $\alpha\ge 0$ one uses $\alpha!$ to mean $\alpha_1!\cdots\alpha_N!$, and for $\alpha\ge\beta\ge 0$ one uses $\binom{\alpha}{\beta}$ to mean $\binom{\alpha_1}{\beta_1}\cdots\binom{\alpha_N}{\beta_N}=\frac{\alpha!}{\beta!\,(\alpha-\beta)!}$.

then $\mu_{K_{0,\infty}}$ *satisfies*

$$\sum_{1\le|\alpha|\le r,\alpha\ge 0}\sum_{j=1}^{p}\Big\langle\varphi_{\alpha,j}\,(2i\,\pi)^{|\alpha|}\frac{\xi^\alpha}{|\xi|+|\xi|^r}\mu_{K_{0,\infty}}^{j,\ell},\varphi\otimes\psi\Big\rangle=0 \text{ for } \ell=1,\ldots,p,$$
$$\text{for all } \varphi\in C_c(\Omega) \text{ and all } \psi\in C\big(K_{0,\infty}(\mathbb{R}^N)\big) \text{ equal to 0 near } \Sigma_0. \tag{32.36}$$

Proof. Notice that for $\alpha\ge 0$ and $1\le|\alpha|\le r$ one has $\frac{\xi^\alpha}{|\xi|+|\xi|^r}\in C\big(K_{0,\infty}(\mathbb{R}^N)\big)$. One uses the Leibniz formula,[18] which says that for every multi-index $\alpha\ge 0$

$$D^\alpha(\varphi\,S)=\sum_{0\le\beta\le\alpha}\binom{\alpha}{\beta}D^\beta\varphi\,D^{\alpha-\beta}S, \text{ for all } \varphi\in C^\infty(\Omega), S\in\mathcal{D}'(\Omega); \tag{32.37}$$

one chooses $S\in C_c^\infty(\Omega)$ and one applies (32.37) to $T\in\mathcal{D}'(\Omega)$, so that

$$\varphi\,D^\alpha T=\sum_{0\le\beta\le\alpha}(-1)^{|\beta|}\binom{\alpha}{\beta}D^{\alpha-\beta}\big((D^\beta\varphi)\,T\big), \text{ for all } \varphi\in C^\infty(\Omega), T\in\mathcal{D}'(\Omega); \tag{32.38}$$

one chooses $\varphi\in C_c^\infty(\Omega)$ and one applies (32.38) to T being any of the terms $\varphi_{\alpha,j}U_j^n$ appearing in (32.34), so that

$$\begin{aligned}&\varphi\,f_n=\textstyle\sum(-1)^{|\beta|}\binom{\alpha}{\beta}\varepsilon_n^{|\alpha|-1}D^{\alpha-\beta}\big((D^\beta\varphi)\,\varphi_{\alpha,j}U_j^n\big),\\&\mathcal{F}(\varphi\,f_n)=\textstyle\sum(-1)^{|\beta|}\binom{\alpha}{\beta}\varepsilon_n^{|\alpha|-1}(2i\,\pi)^{|\alpha|-|\beta|}\xi^{\alpha-\beta}\mathcal{F}\big((D^\beta\varphi)\,\varphi_{\alpha,j}U_j^n\big),\end{aligned} \tag{32.39}$$

where the sums are taken over α,β,j satisfying $0\le\beta\le\alpha$, $1\le|\alpha|\le r$, and $1\le j\le p$. For $|\xi|\ge 1$ the ratio $\frac{|\xi|+\varepsilon_n^{r-1}|\xi|^r}{1+\sum_{s=1}^r\varepsilon_n^{s-1}|\xi|^s}$ is bounded above and below by positive constants so that by (32.35) $\frac{\mathcal{F}(\varphi\,f_n)}{|\xi|+\varepsilon_n^{r-1}|\xi|^r}$ tends to 0 in $L^2\big(\mathbb{R}^N\setminus B(0,1)\big)$ strong. For $\psi\in C\big(K_{0,\infty}(\mathbb{R}^N)\big)$ which is 0 in a neighbourhood of 0, say $|\xi|\le\eta$ with $\eta>0$, for $\varphi_1\in C_c(\Omega)$ and for $\ell=1,\ldots,p$, one multiplies the second equation of (32.39) by $\frac{\psi(\varepsilon_n\xi)}{|\xi|+\varepsilon_n^{r-1}|\xi|^r}\overline{\mathcal{F}(\varphi_1U_\ell^n)}$, one integrates over $\mathbb{R}^N$, and one identifies the limit of each term. One notices that no problem arises near 0, since $\psi(\varepsilon_n\xi)=0$ for $|\xi|\le\frac{\eta}{\varepsilon_n}$, so that

$$\int_{\mathbb{R}^N}\frac{\mathcal{F}(\varphi\,f_n)}{|\xi|+\varepsilon_n^{r-1}|\xi|^r}\psi(\varepsilon_n\xi)\,\overline{\mathcal{F}(\varphi_1U_\ell^n)}\,d\xi\to 0, \tag{32.40}$$

[18] Gottfried Wilhelm VON LEIBNIZ, German mathematician, 1646–1716. He worked in Frankfurt, in Mainz, Germany, in Paris, France, and in Hanover, Germany, but never in an academic position.

and for $|\beta|\geq 1$

$$\int_{\mathbb{R}^N}\varepsilon_n^{|\alpha|-1}\mathcal{F}\big((D^\beta\varphi)\,\varphi_{\alpha,j}U_j^n\big)\frac{\xi^{\alpha-\beta}\psi(\varepsilon_n\xi)}{|\xi|+\varepsilon_n^{r-1}|\xi|^r}\,\overline{\mathcal{F}(\varphi_1 U_\ell^n)}\,d\xi\to 0, \qquad (32.41)$$

since $\frac{\varepsilon_n^{|\alpha|-1}\xi^{\alpha-\beta}\psi(\varepsilon_n\xi)}{|\xi|+\varepsilon_n^{r-1}|\xi|^r}=\varepsilon_n^{|\beta|}\widetilde{\psi}(\varepsilon_n\xi)$ with $\widetilde{\psi}\in C\big(K_{0,\infty}(\mathbb{R}^N)\big)$, due to the fact that ψ is 0 in a neighbourhood of 0 for the case where $\beta=\alpha$; then for $\beta=0$

$$\begin{aligned}&\textstyle\int_{\mathbb{R}^N}\varepsilon_n^{|\alpha|-1}\mathcal{F}\big(\varphi\,\varphi_{\alpha,j}U_j^n\big)\frac{\xi^\alpha\psi(\varepsilon_n\xi)}{|\xi|+\varepsilon_n^{r-1}|\xi|^r}\,\overline{\mathcal{F}(\varphi_1 U_\ell^n)}\,d\xi\to\\&\big\langle\mu_{K_{0,\infty}}^{j,\ell},\varphi\,\varphi_{\alpha,j}\overline{\varphi_1}\otimes\tfrac{\xi^\alpha\psi}{|\xi|+|\xi|^r}\big\rangle,\end{aligned} \qquad (32.42)$$

showing (32.36) with $\varphi\overline{\varphi_1}$ instead of $\varphi\in C_c(\Omega)$, but one may take φ_1 equal to 1 on the support of φ, and then let $\varphi\in C_c^\infty(\Omega)$ approach uniformly any function in $C_c(\Omega)$. □

Corollary 32.8. *Under the hypotheses of Lemma 32.7, the restriction μ_∞ of $\mu_{K_{0,\infty}}$ to the sphere Σ_∞ at ∞, whose entries belong to $\mathcal{M}(\Omega\times\Sigma_\infty)$, satisfies*

$$\sum_{|\alpha|=r,\alpha\geq 0}\sum_{j=1}^p\varphi_{\alpha,j}\frac{\xi^\alpha}{|\xi|^r}\mu_\infty^{j,\ell}=0 \text{ in } \Omega\times\Sigma_\infty \text{ for } \ell=1,\ldots,p. \qquad (32.43)$$

Proof. For $\lambda_1,\ldots,\lambda_p\in\mathbb{C}$, $\pi(\lambda)=\sum_{j,k=1}^p\mu_{K_{0,\infty}}^{j,k}\lambda_j\overline{\lambda_k}\in\mathcal{M}(\Omega\times K_{0,\infty})$ is nonnegative, and one may define its restriction $\pi_\infty(\lambda)=\sum_{j,k=1}^p\mu_\infty^{j,k}\lambda_j\overline{\lambda_k}\in\mathcal{M}(\Omega\times K_\infty)$ by the Lebesgue dominated convergence theorem,

$$\begin{aligned}&\langle\pi_\infty(\lambda),\varphi\otimes\psi\rangle=\lim_{n\to\infty}\langle\pi(\lambda),\varphi\otimes\psi_n\rangle \text{ for all } \varphi\in C_c(\Omega),\psi\in C(\Sigma_\infty),\\&\text{where } \psi_n\in C\big(K_{0,\infty}(\mathbb{R}^N)\big) \text{ stays uniformly bounded, and converges}\\&\text{pointwise everywhere, to 0 in } K_{0,\infty}(\mathbb{R}^N)\setminus\Sigma_\infty \text{ and to } \psi \text{ on } \Sigma_\infty,\end{aligned} \qquad (32.44)$$

for example by choosing $\psi_n(\xi)=0$ on Σ_0, $\psi_n=\psi$ on Σ_∞, and $\psi_n(\xi)=\psi\big(\frac{\xi}{|\xi|}\big)\tanh\frac{|\xi|}{n}$ on $\mathbb{R}^N\setminus\{0\}$; also, μ_∞ is well defined from all the $\pi_\infty(\lambda)$ since it is Hermitian symmetric. By using such a ψ_n in (32.36), one deduces (32.43), since $\frac{\xi^\alpha}{|\xi|+|\xi|^r}$ tends to 0 on Σ_∞ for $|\alpha|<r$, and to $\frac{\xi^\alpha}{|\xi|^r}$ for $|\alpha|=r$ (multiplied by a factor 2 if $r=1$). □

In order to express that no information is lost at ∞, Patrick GÉRARD used the hypothesis that $\varepsilon_n\frac{\partial v_n}{\partial x_j}$ be bounded in L^2 for all j; if $v_n\rightharpoonup v_\infty$ in $L^2_{loc}(\Omega)$ weak, and $\varphi\in C_c^1(\Omega)$, one finds $\varepsilon_n\frac{\partial[\varphi\,(v_n-v_\infty)]}{\partial x_j}=\varphi\,\varepsilon_n\frac{\partial v_n}{\partial x_j}+\varepsilon_n\frac{\partial\varphi}{\partial x_j}v_n-\varepsilon_n\frac{\partial(\varphi\,v_\infty)}{\partial x_j}$, so that there is a term bounded in $L^2(\Omega)$ and the others are ε_n times a term compact in $H^{-1}(\Omega)$, with supports in a fixed compact set of Ω. In my framework, this leads to the following generalization.

Lemma 32.9. *Let* $\varepsilon_n \to 0$, $U^n \rightharpoonup 0$ *in* $L^2(\Omega;\mathbb{R}^p)$ *weak, and a* $p\times p$ *Hermitian symmetric matrix* $\mu_{K_{0,\infty}}$ *of Radon measures on* $\Omega\times K_{0,\infty}(\mathbb{R}^N)$ *be associated to a subsequence* U^m *like in Lemma 32.6. If*

$$\sum_{k=1}^N\sum_{j=1}^p \varepsilon_n \frac{\partial(\varphi_{k,j}U_j^n)}{\partial x_k} = f_n + \varepsilon_n g_n \text{ in } \Omega, \text{ with all } \varphi_{k,j}\in C(\Omega), \\ \text{with } f_n \text{ bounded in } L^2_{loc}(\Omega), \text{ and } g_n \text{ compact in } H^{-1}_{loc}(\Omega) \text{ strong,} \tag{32.45}$$

then the restriction μ_∞ *of* $\mu_{K_{0,\infty}}$ *to the sphere* Σ_∞ *at* ∞ *satisfies*

$$\sum_{k=1}^N\sum_{j=1}^p \frac{\xi_k}{|\xi|}\varphi_{k,j}\mu_\infty^{j,\ell} = 0 \text{ in } \Omega\times\Sigma_\infty, \text{ for } \ell=1,\ldots,p. \tag{32.46}$$

Proof. For $\varphi\in C^1_c(\Omega)$,

$$\sum_{k=1}^N\sum_{j=1}^p \varepsilon_n \frac{\partial(\varphi\,\varphi_{k,j}U_j^n)}{\partial x_k} = \varphi\, f_n + \varepsilon_n\varphi\, g_n + \sum_{k=1}^N\sum_{j=1}^p \varepsilon_n \frac{\partial\varphi}{\partial x_k}\,\varphi_{k,j}U_j^n = F_n + \varepsilon_n G_n, \tag{32.47}$$

and since F_n and G_n have their supports in a fixed compact of Ω, one extracts a subsequence such that $F_m \rightharpoonup F_\infty$ in $L^2(\mathbb{R}^N)$ weak and $G_m \to G_\infty$ in $H^{-1}(\mathbb{R}^N)$ strong, i.e. $\frac{\mathcal{F}G_m}{1+|\xi|} \to H_\infty$ in $L^2(\mathbb{R}^N)$ strong. One uses (32.47) for the sequence U^m, and one multiplies by $\frac{\psi(\varepsilon_m\xi)}{\varepsilon_m|\xi|}\overline{\mathcal{F}(\varphi_1 U_\ell^m)}$ with $\varphi_1\in C_c(\Omega)$ and $\psi\in C\big(K_{0,\infty}(\mathbb{R}^N)\big)$ equal to 0 in a neighbourhood of Σ_0, so that $\psi(\xi)=0$ for $|\xi|\le\eta$ for some $\eta>0$, and one deduces that

$$\sum_{k=1}^N\sum_{j=1}^p \langle \mu_{K_{0,\infty}}^{k,\ell}, \varphi\,\varphi_{k,j}\overline{\varphi_1}\otimes \tfrac{2i\pi\,\xi_k\psi}{|\xi|}\rangle = \lim_{m\to\infty}(A_m+B_m), \\ \text{with } A_m = \int_{\mathbb{R}^N}\mathcal{F}F_m \tfrac{\psi(\varepsilon_m\xi)}{\varepsilon_m|\xi|}\overline{\mathcal{F}(\varphi_1 U_\ell^m)}\,d\xi, \\ B_m = \int_{\mathbb{R}^N}\mathcal{F}G_m \tfrac{\psi(\varepsilon_m\xi)}{|\xi|}\overline{\mathcal{F}(\varphi_1 U_\ell^m)}\,d\xi. \tag{32.48}$$

Since F_m and $\varphi_1 U_\ell^m$ are bounded in $L^2(\mathbb{R}^N)$, one deduces that

$$\limsup_{m\to\infty}|A_m| \le c\max_{\xi\ne 0}\frac{|\psi(\xi)|}{|\xi|}, \tag{32.49}$$

for a constant depending upon φ_1 but independent of ψ. One has

$$|B_m| \le \frac{\alpha}{2}\int_{\mathbb{R}^N}|\psi(\varepsilon_m\xi)|\,|\mathcal{F}(\varphi_1 U_\ell^m)|^2\,d\xi + \frac{1}{2\alpha}\int_{\mathbb{R}^N}|\psi(\varepsilon_m\xi)|\,\frac{|\mathcal{F}G_m|^2}{|\xi|^2}\,d\xi, \tag{32.50}$$

for $\alpha>0$, and the first term is $\le c\,\alpha\max_\xi|\psi(\xi)|$; since $\frac{|\mathcal{F}G_m|^2}{|\xi|^2}$ converges in L^1 strong to $\frac{|H_\infty|^2(1+|\xi|)^2}{|\xi|^2}$ for $|\xi|\ge 1$ for example, $|\psi|$ is uniformly bounded, and $|\psi(\varepsilon_m\xi)|$ is 0 on larger and larger balls as $m\to\infty$, one deduces that the second term tends to 0. This shows that $|\lim_{m\to\infty}(A_m + B_m)| \le c \max_{\xi\ne 0}\frac{|\psi(\xi)|}{|\xi|}$, and then one uses a sequence ψ_r taking a fixed value on Σ_∞, uniformly bounded independently of r, and equal to 0 for $|\xi| \le r$, and one deduces (32.46) by then letting r tend to ∞. □

I think that it was on the same occasion that I discussed with Patrick GÉRARD about using two-point correlations and the Bochner theorem, and about problems involving more than one characteristic length, a question which led us to study the following example with two characteristic lengths:

$$u_n(x) = \begin{cases} \sqrt{n} \text{ if } \frac{k}{n} < x < \frac{k}{n} + \frac{1}{n^2} \text{ for some } k \in \mathbb{Z}, \\ 0 \text{ if } \frac{k}{n} + \frac{1}{n^2} < x < \frac{k+1}{n} \text{ for some } k \in \mathbb{Z}, \end{cases} \tag{32.51}$$

which is such that $u_n^2 \rightharpoonup 1$ in $L^1(I)$ weak $\star$ (but not in $L^1(I)$ weak) for a bounded interval I,[19] and since $u_n \to 0$ in $L^1_{loc}(\mathbb{R})$ strong, one deduces that $u_n \rightharpoonup 0$ in $L^2_{loc}(\mathbb{R})$ weak. However, my reason for considering such a sequence was that it is a simple one-dimensional model showing a structure of walls and domains, which physicists like BLOCH, LANDAU, and NÉEL studied for three-dimensional questions in magnetism, and I wanted to get some intuition about the expected sizes of domains. Although u_n has period $\frac{1}{n}$ (and shows the other characteristic length $\frac{1}{n^2}$), we were surprised to find that the semi-classical measure for the choice $\varepsilon_n = \frac{1}{n}$ is 0! Indeed, $u_n(x) = f_n(n\,x)$ with f_n having period 1, and $f_n(y) = \sum_{m\in\mathbb{Z}} c(m,n)\, e^{2i\,\pi\, m\, y}$ gives

$$\begin{aligned} u_n(x) &= \textstyle\sum_{m\in\mathbb{Z}} c(m,n)\, e^{2i\,\pi\, m\, n\, x} \text{ in } L^2_{loc}(\mathbb{R}) \cap \mathcal{S}'(\mathbb{R}) \\ \mathcal{F}u_n &= \textstyle\sum_{m\in\mathbb{Z}} c(m,n)\, \delta_{m\,n} \text{ in } \mathcal{M}(\mathbb{R}) \cap \mathcal{S}'(\mathbb{R}), \\ &\text{with } \textstyle\sum_{m\in\mathbb{Z}} |c(m,n)|^2 = 1, \end{aligned} \tag{32.52}$$

$$\begin{aligned} c(m,n) &= \sqrt{n} \int_0^{\frac{1}{n}} e^{-2i\,\pi\, m\, y}\,dy \\ &= \frac{\sqrt{n}}{2i\,\pi\, m}(1 - e^{-2i\,\pi\, m/n}) \text{ if } m \ne 0, \frac{1}{\sqrt{n}} \text{ if } m = 0, \\ |c(m,n)| &= \frac{\sqrt{n}}{\pi\,|m|} \sin\Big(\frac{\pi\,|m|}{n}\Big) \le \min\Big\{\frac{\sqrt{n}}{\pi\,|m|}, \frac{1}{\sqrt{n}}\Big\} \text{ if } m \ne 0, = \frac{1}{\sqrt{n}} \text{ if } m = 0. \end{aligned} \tag{32.53}$$

For $\varphi \in \mathcal{S}(\mathbb{R})$ one has

$$\mathcal{F}(\varphi\, u_n) = \mathcal{F}\varphi \star \mathcal{F}u_n = \sum_{m\in\mathbb{Z}} c(m,n)\, \tau_{m\,n}\mathcal{F}\varphi \text{ in } C_b(\mathbb{R}), \tag{32.54}$$

and one then assumes also that $\mathcal{F}\varphi \in C_c^\infty(\mathbb{R})$, and the supports of the translated $\mathcal{F}\varphi$ do not overlap if *support*$(\mathcal{F}\varphi) \subset [-\rho, +\rho]$ and $n \ge \rho$, so that

[19] It means that $\int_\mathbb{R} u_n^2 \varphi\,dx \to \int_\mathbb{R} \varphi\,dx$ for all $\varphi \in C_c(\mathbb{R})$, but this fails for some $\varphi \in L^\infty(\mathbb{R})$ with compact support. Indeed, if $I = (0,1)$ the support of u_n has measure $\frac{1}{n}$, if A is the union of the supports of all u_m for a subsequence with $\sum \frac{1}{m} < 1$, and if χ is the characteristic function of $(0,1) \setminus A$, one has $\int_\mathbb{R} u_m^2 \chi\,dx = 0$.

$$|\mathcal{F}(\varphi\, u_n)|^2 = \sum_{m\in\mathbb{Z}} |c(m,n)|^2\, |\tau_{m\,n}\mathcal{F}\varphi|^2 \text{ in } C_b(\mathbb{R}). \tag{32.55}$$

If $\psi \in C_c(\mathbb{R})$ with $support(\psi) \subset [-M,+M]$ for a positive integer M, one has

$$\Big|\int_{\mathbb{R}} |\mathcal{F}(\varphi\, u_n)|^2 \psi\Big(\frac{\xi}{n}\Big)\, d\xi\Big| \le ||\psi||_{L^\infty} ||\varphi||_{L^2}^2 \sum_{m=-M}^{m=+M} |c(m,n)|^2, \tag{32.56}$$

and by (32.53) the sum is $\le \frac{2M+1}{n}$, and tends to 0 as n tends to ∞.

For positive integers k and K one has

$$\begin{aligned} &\textstyle\sum_{|m|\le k} |c(m,n)|^2 \le \frac{2k+1}{n},\\ &\textstyle\sum_{|m|\ge K} |c(m,n)|^2 \le \frac{2n}{\pi^2} \sum_{m=K}^{+\infty} \frac{1}{m^2} \le \frac{2n}{\pi^2(K-1)}, \end{aligned} \tag{32.57}$$

which shows that most of the relevant values for m are of order n, and the correct characteristic length to choose is then $\varepsilon_n = \frac{1}{n^2}$,[20] and one must identify

$$\lim_{n\to\infty} \int_{\mathbb{R}} |\mathcal{F}(\varphi\, u_n)|^2 \psi\Big(\frac{\xi}{n^2}\Big)\, d\xi, \text{ for } \psi \in C_0(\mathbb{R}). \tag{32.58}$$

By uniform continuity of ψ, $\psi(\frac{\xi}{n^2})$ does not vary much if ξ changes of ρ, and by (32.55) the limit is then the same as

$$\lim_{n\to\infty} ||\varphi||_{L^2(\mathbb{R})}^2 \sum_{m\in\mathbb{Z}} |c(m,n)|^2 \psi\Big(\frac{m\,n}{n^2}\Big), \tag{32.59}$$

and the sum being $\frac{1}{n}\,\psi(0) + \sum_{m\in\mathbb{Z}\setminus 0} \frac{n}{\pi^2 m^2} \sin^2\Big(\frac{\pi\, m}{n}\Big)\psi\big(\frac{m}{n}\big)$ is a Riemann sum for $\int_{\mathbb{R}} \frac{\sin^2(\pi\,\xi)}{\pi^2\xi^2}\, \psi(\xi)\, d\xi$, so that the semi-classical measure (for $\varepsilon_n = \frac{1}{n^2}$) is

$$\mu_{sc} = \frac{\sin^2(\pi\,\xi)}{\pi^2\xi^2}\, dx\, d\xi. \tag{32.60}$$

The initial intuition that a characteristic length ε_n in a sequence u_n implies that the Fourier transform of u_n mostly lives at distance $\frac{1}{\varepsilon_n}$ in ξ is then wrong, since for the example (32.51) and $\varepsilon_n = \frac{1}{n}$, the Fourier transform lives at distance of order n^2, and shows fluctuations at a characteristic length n. Actually, this corresponds to a classical remark that we should have thought about, concerning beats:[21] if f and g are periodic with period 1,

---

[20] (32.57) shows that the measure $\mu_{K_{0,\infty}}$ does not charge the spheres Σ_0 at 0 and Σ_∞ at ∞, so that the semi-classical measure does not lose any information.

[21] It is the phenomenon used by a piano tuner, who hits the A key in the middle of the keyboard, together with a tuning fork vibrating at the correct frequency for

then $f(\frac{x}{\varepsilon_n}) + g(\frac{x}{\delta_n})$ has Fourier transform $\varepsilon_n \mathcal{F}f(\varepsilon_n\xi) + \delta_n \mathcal{F}g(\delta_n\xi)$, but the product $f(\frac{x}{\varepsilon_n})g(\frac{x}{\delta_n})$ has a more complicated Fourier transform, where frequencies are added or subtracted, like in classical trigonometric formulas such as $2\cos(a\,x)\cos(b\,x) = \cos\big((a+b)\,x\big) + \cos\big((a-b)\,x\big)$.

This example also made me understand something puzzling which physicists say in their explanations of the rays of absorption observed in spectroscopy for light passing through hydrogen gas. Of course, the rules invented by physicists for explaining experiments in spectroscopy are silly, since it is a problem of homogenization for some hyperbolic system, and according to the examples described in Chaps. 23 and 24 a kernel appearing in an effective equation is related to the fluctuation of some parameters in the gas, which by the way is supposed to contain the Avogadro number (6.023×10^{23}) of molecules in each mole,[22] occupying a volume of 22.4 L, so that a game played with "one" electron attached to "one" proton seems quite irrelevant, and what happens in the gas might be like a complex organization of walls between domains, like for the phase boundaries between the various crystalline phases of a poly-crystal in the case of a solid.[23] What puzzled me in those silly rules of the physicists' games was that the rays of absorption are all attributed to "the" electron, and none to "the" proton, but I then guessed that the frequencies of absorption correspond to small wavelengths comparable to the size attributed to an electron, and that is like the $\frac{1}{n^2}$ in the preceding example, while no absorption occurs for the longer wavelengths comparable to the size attributed to a proton, and that is like the $\frac{1}{n}$ in the preceding example; however, it is the corresponding positions of the various rays of absorption (and their finite size showing a Lorentzian density of absorption) which then tell one about how "the" electron and "the" proton interact.[24]

The example (32.51) suggests defining new variants of *H-measures* able to deal with many characteristic lengths in a hierarchical way: after localizing in x by multiplication by φ and considering the Fourier transform $\mathcal{F}(\varphi\, u_n)$, one may need to rescale in various ways in order to create

---

this key, and putting it in contact with the piano couples the two frequencies and one hears then a modulation at the difference of the frequencies; one makes such a modulation disappear by changing the tension of the corresponding wire. One also hears a modulation at the sum of the frequencies, but this one resembles the harmonic tone and is not detected so well by the ear.

[22] Lorenzo Romano Amedeo Carlo AVOGADRO, count of Quaregna and Cerreto, Italian physicist, 1776–1856. He worked in Torino (Turin), Italy.

[23] This possible analogy between gases and solids only occurred to me while I was writing this chapter.

[24] The silly rules of quantum mechanics were invented in part because of a formula found by BALMER and extended by RYDBERG, giving the position of the rays for the hydrogen gas (proportional to $\frac{1}{n^2} - \frac{1}{m^2}$ for distinct positive integers m and n), and one should explain instead the finite size and the Lorentzian shape of the "rays" of absorption, and find what it tells one about the state of the gas.

subsequences $|\mathcal{F}(\varphi\, u_m)|^2$ which converge weakly $\star$ to nonzero Radon measures, and then, for each of these subsequences one may need to repeat an analysis with *H-measures* or variants on the rescaled sequence from $\mathcal{F}(\varphi\, u_m)$, because of oscillations or concentration effects.

However, inventing new variants is not really so difficult, and the priority seems to determine which class of variants are adapted to the important goal of confirming or correcting some guesses made by physicists or engineers, on questions like boundary layers or turbulent flows, for example, and one important difficulty for these problems is that they are related to nonlinear partial differential equations, of course.

It was during my CBMS–NSF course at UCSC in the summer of 1993 that I mentioned to Patrick GÉRARD my way of handling the heat equation $\frac{\partial u_n}{\partial t} - \varepsilon_n^2 \kappa\, \Delta\, u_n = 0$, using a bound in L^2 for $\varepsilon_n grad(u_n)$, and he showed me how to obtain the same result using only a bound in L^2 for u_n.

My observation was that if $u_n \rightharpoonup 0$ in $L^2_{loc}(\Omega)$ weak and corresponds to a semi-classical measure μ_{sc} and if $v_n = \varepsilon_n \frac{\partial u_n}{\partial x_j}$ is bounded in $L^2_{loc}(\Omega)$ and corresponds to a semi-classical measure ν_{sc}, then $\varphi\, v_n - \varepsilon_n \frac{\partial(\varphi\, u_n)}{\partial x_j} \to 0$ in $L^2(\Omega)$ strong for $\varphi \in C^1_c(\Omega)$, so that $\mathcal{F}(\varphi\, v_n)$ behaves like $2i\,\pi\, \varepsilon_n \xi_j \mathcal{F}(\varphi\, u_n)$, and this implies $\nu_{sc} = -4\pi^2 \xi_j^2 \mu_{sc}$, from which one deduces a relation between the two-point correlation measure of v_n and that of u_n.

The idea which I learned from Patrick GÉRARD may be similar to observations of WIGNER, who showed that his function W satisfies a transport equation when u solves a free Schrödinger equation,[25] and this only has a historical interest for me, since I understood in 1983 what is wrong with quantum mechanics and why the Schrödinger equation is not good physics.[26] Although most physicists still want to interpret the world in terms of particles,[27] I understood that there are only waves and no particles at all: it is a question of mathematics to understand the behaviour of highly oscillatory waves in a different way than the non-oscillatory ones,[28] by finding some effective equations for them, but there is no need for those effective equations to have an interpretation in terms of "particles" (or "anti-particles")!

---

[25] When George PAPANICOLAOU told me about the Wigner transform in the late 1970s, he did not mention how WIGNER used it.

[26] I proposed instead to study the oscillations and concentration effects of the Dirac equation coupled with the Maxwell–Heaviside equation, but without mass term in the Dirac equation, since I consider that this term should be left to appear by a homogenization effect.

[27] It shows that physicists have not understood yet that particles correspond to an eighteenth century point of view about mechanics!

[28] What puzzles physicists is that "particles sometimes behave like waves", while it is only the oscillatory waves, i.e. those using high frequencies, which are described by a different effective equation, which may or may not have a simple interpretation in terms of "particles".

If $u_n \in H^1_{loc}(\Omega \times (0,T))$,[29] $u_n \rightharpoonup 0$ in $L^2_{loc}(\Omega \times (0,T))$ weak, and

$$\frac{\partial u_n}{\partial t} + i\,\kappa\,\varepsilon_n \Delta\, u_n = f_n \to 0 \text{ in } L^2_{loc}\big(\Omega \times (0,T)\big) \text{ strong,} \tag{32.61}$$

then, for an arbitrary $y \in \mathbb{R}^N$

$$\begin{aligned}&\tfrac{\partial[u_n(x+\varepsilon_n y,t)\overline{u_n(x,t)}]}{\partial t} + i\,\kappa\,\varepsilon_n \Delta\, u_n(x+\varepsilon_n y,t)\cdot\overline{u_n(x,t)}\\ &-\quad i\,\kappa\,\varepsilon_n \Delta\,\overline{u_n(x,t)}\cdot u_n(x+\varepsilon_n y,t) \to 0 \text{ in } L^1_{loc}\big(\Omega \times (0,T)\big) \text{ strong.}\end{aligned} \tag{32.62}$$

After that, one uses $y \in \mathbb{R}^N$ as a variable and for $j = 1,\ldots,N$,

$$\frac{\partial[u_n(x+\varepsilon_n y,t)]}{\partial y_j} = \varepsilon_n \frac{\partial[u_n(x+\varepsilon_n y,t)]}{\partial x_j}, \text{ and } \frac{\partial u_n(x,t)}{\partial y_j} = 0, \tag{32.63}$$

in the sense of distributions in $\Omega \times \mathbb{R}^N \times (0,T)$, so that

$$i\,\kappa\,\varepsilon_n \Delta\, u_n(x+\varepsilon_n y,t)\cdot\overline{u_n(x,t)} = i\,\kappa \textstyle\sum_{j=1}^N \frac{\partial}{\partial y_j}\Big(\frac{\partial[u_n(x+\varepsilon_n y,t)]}{\partial x_j}\,\overline{u_n(x,t)}\Big), \tag{32.64}$$

$$\begin{aligned}-i\,\kappa\,\varepsilon_n \Delta\,\overline{u_n(x,t)}\cdot u_n(x+\varepsilon_n y,t) &= -i\,\kappa\,\varepsilon_n \Delta\,[u_n(x+\varepsilon_n y,t)\,\overline{u_n(x,t)}]\\ +\quad i\,\kappa\,\varepsilon_n \Delta\, u_n(x+\varepsilon_n y,t)\cdot\overline{u_n(x,t)} &+ 2i\,\kappa\,\varepsilon_n \textstyle\sum_{j=1}^N \frac{\partial[u_n(x+\varepsilon_n y,t)]}{\partial x_j}\,\frac{\partial\overline{u_n(x,t)}}{\partial x_j},\end{aligned} \tag{32.65}$$

$$\begin{aligned}&2i\,\kappa\,\varepsilon_n \textstyle\sum_{j=1}^N \frac{\partial[u_n(x+\varepsilon_n y,t)]}{\partial x_j}\,\frac{\partial\overline{u_n(x,t)}}{\partial x_j}\\ &= 2i\,\kappa \textstyle\sum_{j=1}^N \frac{\partial}{\partial y_j}\big(u_n(x+\varepsilon_n y,t)\,\frac{\partial\overline{u_n(x,t)}}{\partial x_j}\big).\end{aligned} \tag{32.66}$$

Using (32.64)–(32.66), (32.62) becomes

$$\begin{aligned}&\big(\tfrac{\partial}{\partial t} - i\,\kappa\,\varepsilon_n \Delta + 2i\,\kappa \textstyle\sum_{j=1}^N \frac{\partial^2}{\partial x_j \partial y_j}\big)\,[u_n(x+\varepsilon_n y,t)\overline{u_n(x,t)}]\\ &\to 0 \text{ in } L^1_{loc}\big(\Omega \times \mathbb{R}^N \times (0,T)\big) \text{ strong.}\end{aligned} \tag{32.67}$$

Then, if u_n defines a semi-classical measure μ_{sc}, $u_n(x+\varepsilon_n y,t)\overline{u_n(x,t)}$ converges weakly $\star$ to $\mathcal{F}_\xi \mu_{sc}$, and (32.67) gives

$$\begin{aligned}&\big(\tfrac{\partial}{\partial t} + 2i\,\kappa \textstyle\sum_{j=1}^N \frac{\partial^2}{\partial x_j \partial y_j}\big)\mathcal{F}_\xi \mu_{sc} = 0,\\ &\big(\tfrac{\partial}{\partial t} + 4\pi\,\kappa \textstyle\sum_{j=1}^N \xi_j \frac{\partial}{\partial x_j}\big)\mu_{sc} = 0.\end{aligned} \tag{32.68}$$

[29] One does not assume a bound independent of n in this space, since this hypothesis only serves in having all terms defined in the sense of distributions.

If now one replaces the Schrödinger equation (32.61) by the heat equation

$$\frac{\partial u_n}{\partial t} - \kappa\,\varepsilon_n^2 \Delta\, u_n = f_n \to 0 \text{ in } L^2_{loc}\big(\Omega \times (0,T)\big) \text{ strong}, \tag{32.69}$$

noticing the different power of ε_n used, then for $y \in \mathbb{R}^N$

$$\begin{aligned}&\tfrac{\partial[u_n(x+\varepsilon_n y,t)\overline{u_n(x,t)}]}{\partial t} - \kappa\,\varepsilon_n^2 \Delta\, u_n(x+\varepsilon_n y,t)\cdot\overline{u_n(x,t)}\\ &- \quad \kappa\,\varepsilon_n^2 \Delta\,\overline{u_n(x,t)}\cdot u_n(x+\varepsilon_n y,t) \to 0 \text{ in } L^1_{loc}\big(\Omega\times(0,T)\big) \text{ strong},\end{aligned} \tag{32.70}$$

$$\begin{aligned}&-\kappa\,\varepsilon_n^2 \Delta\, u_n(x+\varepsilon_n y,t)\cdot\overline{u_n(x,t)} - \kappa\,\varepsilon_n^2 \Delta\,\overline{u_n(x,t)}\cdot u_n(x+\varepsilon_n y,t)\\ &= -\kappa\,\varepsilon_n^2 \Delta\,[u_n(x+\varepsilon_n y,t)\overline{u_n(x,t)}] + 2\kappa\,\varepsilon_n^2 \textstyle\sum_{j=1}^N \frac{\partial[u_n(x+\varepsilon_n y,t)]}{\partial x_j}\,\frac{\partial\overline{u_n(x,t)}}{\partial x_j},\end{aligned} \tag{32.71}$$

$$\begin{aligned}&2\kappa\,\varepsilon_n^2 \textstyle\sum_{j=1}^N \frac{\partial[u_n(x+\varepsilon_n y,t)]}{\partial x_j}\,\frac{\partial\overline{u_n(x,t)}}{\partial x_j} = 2\kappa\,\varepsilon_n \sum_{j=1}^N \frac{\partial}{\partial y_j}\big(u_n(x+\varepsilon_n y,t)\,\frac{\partial\overline{u_n(x,t)}}{\partial x_j}\big)\\ &= 2\kappa\,\varepsilon_n \textstyle\sum_{j=1}^N \frac{\partial^2[u_n(x+\varepsilon_n y,t)\,\overline{u_n(x,t)}]}{\partial x_j \partial y_j} - 2\kappa \sum_{j=1}^N \frac{\partial^2[u_n(x+\varepsilon_n y,t)\,\overline{u_n(x,t)}]}{\partial y_j^2},\end{aligned} \tag{32.72}$$

so that, using (32.71)–(32.72), (32.70) becomes

$$\begin{aligned}&\big(\tfrac{\partial}{\partial t} - \kappa\,\varepsilon_n^2\Delta + 2\kappa\,\varepsilon_n \textstyle\sum_{j=1}^N \frac{\partial^2}{\partial x_j\partial y_j} - 2\kappa \sum_{j=1}^N \frac{\partial^2}{\partial y_j^2}\big)[u_n(x+\varepsilon_n y,t)\,\overline{u_n(x,t)}]\\ &\to 0 \text{ in } L^1_{loc}\big(\Omega\times(0,T)\big) \text{ strong},\end{aligned} \tag{32.73}$$

and if u_n defines a semi-classical measure μ_{sc}, it satisfies

$$\begin{aligned}&\big(\tfrac{\partial}{\partial t} - 2\kappa\,\Delta_y\big)\mathcal{F}_\xi \mu_{sc} = 0,\\ &\big(\tfrac{\partial}{\partial t} + 8\,\kappa\,|\xi|^2\big)\mu_{sc} = 0.\end{aligned} \tag{32.74}$$

With Patrick GÉRARD, we then applied this idea to k-point correlations, with $k \ge 3$. Assuming that $u_n \rightharpoonup 0$ in $L^k_{loc}\big(\Omega\times(0,T)\big)$ weak, one may extract a subsequence u_m such that

$$u_m(x+\varepsilon_m y^1)\cdots u_m(x+\varepsilon_m y^k) \rightharpoonup C_k \text{ in } \mathcal{M}\big(\Omega\times\ldots\times\Omega\times(0,T)\big) \text{ weak } \star, \tag{32.75}$$

noticing that for $k \ge 3$ one does not use complex conjugates. One has

$$\tau_{(0,h,\ldots,h)} C_k = C_k \text{ for all } h \in \mathbb{R}^N, \tag{32.76}$$

but there is no analogue of the Bochner theorem. If u_n satisfies (32.69), one is led to assume that $f_n \to 0$ in $L^k_{loc}\big(\Omega\times(0,T)\big)$ strong, and one deduces that

$$\Big(\frac{\partial}{\partial t} - \kappa\,\Delta_{y^1} - \ldots - \kappa\,\Delta_{y^k}\Big)\,C_k = 0 \text{ in } \Omega \times \ldots \times \Omega \times (0,T). \tag{32.77}$$

Without an analogue for the Bochner theorem, one has only one form of the equation, but with (32.76) one may eliminate a variable like y^k.

The idea of Pierre-Louis LIONS and Thierry PAUL then became more useful than just providing a new way to define the semi-classical measures of Patrick GÉRARD, since it helped us to think in terms of correlations instead of using the Fourier transform, and the previous approaches were not adapted to using k-point correlations. However, I think that there should be a more geometric approach behind the idea of using correlations, and for k-point correlations with $k \geq 3$ one should also find a way to use different estimates, since L^k estimates with $k \neq 2$ are not natural for hyperbolic systems,[30] and realistic physical models should use hyperbolic systems.[31] Also, but that seems less important, it would be useful to find a replacement for the Bochner theorem when $k \geq 3$.

Handling variable coefficients requires a little more care, and a term in $\varepsilon_n^2 \frac{\partial}{\partial x_i}\big(a_{i,j}\frac{\partial u_n}{\partial x_j}\big)$ with $a_{i,j} \in C^1\big(\Omega \times (0,T)\big)$ creates the terms

$$\begin{aligned}&\varepsilon_n^2 \tfrac{\partial}{\partial x_i}\big(a_{i,j}(x+\varepsilon_n y,t)\tfrac{\partial[u_n(x+\varepsilon_n y,t)]}{\partial x_j}\big)\,\overline{u_n(x,t)}\\ &= \tfrac{\partial}{\partial y_i}\big(a_{i,j}(x+\varepsilon_n y,t)\tfrac{\partial[u_n(x+\varepsilon_n y,t)\,\overline{u_n(x,t)}]}{\partial y_j}\big),\end{aligned} \tag{32.78}$$

and, omitting (x,t) for simplification,

$$\begin{aligned}&\varepsilon_n^2 \tfrac{\partial}{\partial x_i}\big(\overline{a_{i,j}(x,t)}\,\tfrac{\partial \overline{u_n(x,t)}}{\partial x_j}\big)\,u_n(x+\varepsilon_n y,t)\\ &= \varepsilon_n^2 \tfrac{\partial}{\partial x_i}\big(\overline{a_{i,j}}\,u_n(x+\varepsilon_n y,t)\,\tfrac{\partial \overline{u_n}}{\partial x_j}\big) - \varepsilon_n \tfrac{\partial}{\partial y_i}\big(\overline{a_{i,j}}\,u_n(x+\varepsilon_n y,t)\,\tfrac{\partial \overline{u_n}}{\partial x_j}\big),\end{aligned} \tag{32.79}$$

whose two terms are taken care of by the formulas

$$\begin{aligned}\varepsilon_n^2 \tfrac{\partial}{\partial x_i}\big(\overline{a_{i,j}}\,u_n(x+\varepsilon_n y,t)\,\tfrac{\partial \overline{u_n}}{\partial x_j}\big) &= \varepsilon_n^2 \tfrac{\partial}{\partial x_i}\big(\overline{a_{i,j}}\,\tfrac{\partial[u_n(x+\varepsilon_n y,t)\,\overline{u_n}]}{\partial x_j}\big)\\ &- \quad \varepsilon_n \tfrac{\partial^2 \overline{a_{i,j}}\,[u_n(x+\varepsilon_n y,t)\,\overline{u_n}]}{\partial x_i \partial y_j},\end{aligned} \tag{32.80}$$

---

[30] One may need to use ideas like *compensated integrability* and *compensated regularity*, for which I refer to [115].

[31] The Schrödinger equation should be considered as coming from a hyperbolic system like the Dirac equation (without mass term) in which one lets the speed of light c tend to ∞; also, the potential used in the Schrödinger equation is but the electrostatic part of the potential in the Maxwell–Heaviside equation, and it is only a first approximation that one may impose it. The heat equation and other diffusion equations are often introduced by using the unphysical "Brownian" motion, where unrealistic jumps in position appear instead of more realistic jumps in velocity resulting from collisions, which lead to the Fokker–Planck equation (but advocates of fake mechanics often use this name for equations without a velocity variable), and one finds a diffusion equation by letting c tend to ∞, like in the Rosseland approximation.

$$
\begin{aligned}
-\varepsilon_n \tfrac{\partial}{\partial y_i}\big(\overline{a_{i,j}}\, u_n(x+\varepsilon_n y,t)\, \tfrac{\partial \overline{u_n}}{\partial x_j}\big) &= -\varepsilon_n \tfrac{\partial}{\partial y_i}\big(\overline{a_{i,j}}\, \tfrac{\partial [u_n(x+\varepsilon_n y,t)\, \overline{u_n}]}{\partial x_j}\big) \\
+ \quad \tfrac{\partial^2 \overline{a_{i,j}}\, [u_n(x+\varepsilon_n y,t)\, \overline{u_n}]}{\partial y_i \partial y_j}, &
\end{aligned}
\tag{32.81}
$$

and since $u_n(x+\varepsilon_n y,t)\,\overline{u_n}$ is bounded in $L^1_{loc}(\Omega)$, all the terms converge in the sense of distributions, because $a_{i,j} \in C^1\big(\Omega \times (0,T)\big)$.

A term in $b_j \frac{\partial u_n}{\partial x_j}$ with b_j real and $b_j \in C^1\big(\Omega \times (0,T)\big)$ creates the terms

$$
\begin{aligned}
& b_j(x+\varepsilon_n y,t)\, \overline{u_n(x,t)}\, \tfrac{\partial [u_n(x+\varepsilon_n y,t)]}{\partial x_j} + b_j(x,t)\, u_n(x+\varepsilon_n y,t)\, \tfrac{\partial [\overline{u_n(x,t)}]}{\partial x_j} \\
& = b_j(x,t)\, \tfrac{\partial [u_n(x+\varepsilon_n y,t)\overline{u_n(x,t)}]}{\partial x_j} + X \\
& X = c_j^n(x,y,t)\, \tfrac{\partial [u_n(x+\varepsilon_n y,t)\overline{u_n(x,t)}]}{\partial y_j}, \text{ with } = c_j^n(x,y,t)\, \tfrac{b_j(x+\varepsilon_n y,t) - b_j(x,t)}{\varepsilon_n} \\
& = \tfrac{\partial [c_j^n(x,y,t)\, u_n(x+\varepsilon_n y,t)\overline{u_n(x,t)}]}{\partial y_j} - \tfrac{\partial c_j^n(x,y,t)}{\partial y_j}\, [u_n(x+\varepsilon_n y,t)\overline{u_n(x,t)}],
\end{aligned}
\tag{32.82}
$$

and as n tends to ∞ one has

$$
\begin{aligned}
& c_j^n(x,y,t) \to \textstyle\sum_{k=1}^N y_k \tfrac{\partial b_j(x,t)}{\partial x_k} \text{ in } C\big(\Omega \times (0,T)\big), \\
& \tfrac{\partial c_j^n(x,y,t)}{\partial y_j} = \tfrac{\partial b_j(x+\varepsilon_n y,t)}{\partial x_j} \to \tfrac{\partial b_j(x,t)}{\partial x_j} \text{ in } C\big(\Omega \times (0,T)\big).
\end{aligned}
\tag{32.83}
$$

Of course, the same results hold for a first-order equation, but one should remember that the computations are performed under the hypothesis that $u_n \in H^1_{loc}(\Omega)$, so that all the terms introduced have a meaning in the sense of distributions. One should notice that the details of the proof are quite different from those used in Lemma 31.4 for obtaining the transport equation for *H-measures*.

Patrick GÉRARD also did an interesting computation for the Dirac equation, considering that the scalar potential V and the vector potential A are given C^1 functions,[32] so that the equation for ψ becomes a linear hyperbolic system, having only the speed of light c as characteristic speed, and he called $\frac{m_0}{\varepsilon_n^2}\,\psi$ the mass term in the equation for ψ,[33] and he used ε_n for defining a semi-classical measure, and he then looked for the transport equation that it satisfies. Since this question does not correspond to my intuition about what one should do with the semi-linear coupled Dirac/Maxwell–Heaviside system, I did not read the detail of his computations, but what he found is that the effective equation can be interpreted as describing two kinds of particles, having relativistic mass $\frac{m_0}{\sqrt{1-|v|^2/c^2}}$ when moving at velocity v (with $|v| < c$, of course), having elementary charge $+e$ for "positrons" or $-e$ for "electrons", and feeling the Lorentz force $\pm e\,(E + v \times B)$, with $E = -grad(V) + \frac{\partial A}{\partial t}$

[32] Patrick GÉRARD did not use the Maxwell–Heaviside equation, where the density of charge ϱ and the density of current j are given by sesqui-linear quantities in ψ.

[33] With m_0 being the "rest mass" of the electron.

and $B = -curl(A)$. Of course, it shows that DIRAC had done a superb job, cheating exactly in the right way with the rules of quantum mechanics (as he did not feel bound to act according to its silly dogmas) for keeping the symmetries of relativity.

One interesting consequence is that the Lorentz force only exists at a mesoscopic level, since there are only waves and no "particles" at a microscopic level!

My research programme is to develop better mathematical tools for analysing semi-linear hyperbolic systems, with the goal of starting with the coupled Maxwell–Heaviside/Dirac system without mass term, and show that the atomic level corresponds to concentrations effects creating the "particles" which physicists talk about (extending the idea of BOSTICK concerning "electrons"). After that, similar computations to those of Patrick GÉRARD could be needed for describing what happens at mesoscopic levels.

Additional footnotes: BALMER,[34] Marc FEIX,[35] FOKKER,[36] GOEPPERT-MAYER,[37] Jean-Pierre GUIRAUD,[38] JENSEN J.H.D.,[39] LÉVY P.,[40] Richard MEYER,[41] PRANDTL,[42] ROSSELAND,[43] RYDBERG,[44] STEWARTSON.[45]

---

[34] Johann Jakob BALMER, Swiss mathematician, 1825–1898. He worked in Basel, Switzerland.

[35] Marc R. FEIX, French physicist, 1928–2005. He worked in Orléans, France.

[36] Adriaan Daniël FOKKER, Dutch physicist and composer, 1887–1972. He worked in Leiden, The Netherlands. He wrote music under the pseudonym Arie DE KLEIN.

[37] Maria GOEPPERT-MAYER, German-born physicist, 1906–1972. She received the Nobel Prize in Physics in 1963, with J. Hans D. JENSEN, for their discoveries concerning nuclear shell structure, jointly with Eugene P. WIGNER. She worked in Chicago, IL, and at USCD (University of California at San Diego), La Jolla, CA.

[38] Jean-Pierre GUIRAUD, French mathematician. He worked at UPMC (Université Pierre et Marie Curie), Paris, and at ONERA (Office National d'Études et de Recherches Aéronautiques), Châtillon, France.

[39] J. Hans D. JENSEN, German physicist, 1907–1973. He received the Nobel Prize in Physics in 1963, with Maria GOEPPERT-MAYER, for their discoveries concerning nuclear shell structure, jointly with Eugene P. WIGNER. He worked in Hanover, and in Heidelberg, Germany.

[40] Paul Pierre LÉVY, French mathematician, 1886–1971. He worked in Paris, France.

[41] Richard Ernst MEYER, German-born mathematician, 1919–2008. He worked in Manchester, England, in Sidney, Australia, at Brown University, Providence, RI, and at UW (University of Wisconsin), Madison, WI.

[42] Ludwig PRANDTL, German physicist, 1875–1953. He worked in Hanover and in Göttingen, Germany.

[43] Svein ROSSELAND, Norwegian physicist, 1894–1985. He worked in Oslo, Norway.

[44] Johannes Robert RYDBERG, Swedish physicist, 1854–1919. He worked in Lund, Sweden.

[45] Keith STEWARTSON, English mathematician, 1925–1983. He worked in Bristol, in Durham, and in London, England.

Chapter 33
Relations Between Young Measures and H-Measures

By describing the $L^\infty(\Omega)$ weak $\star$ limits of all continuous functions of a bounded sequence U^n in $L^\infty(\Omega;\mathbb{R}^p)$, the Young measures give a natural mathematical way to interpret the local (one-point) statistics of the values taken by U^n without falling prey to the unrealistic fashion of imposing old probabilistic games which usually destroy the physical relevance, since nature behaves in a different way, which it is the role of scientists to discover. However, Young measures are limited and cannot discern any geometric pattern or take into account differential equations satisfied by U^n; actually, the theory has no need for Ω to be an open subset of $\mathbb{R}^N$ or a manifold, and it can be applied to a set endowed with a measure without atoms.

Starting from a scalar sequence $u_n \in L^\infty(\mathbb{R}^2)$ defining a Young measure ν, one creates another scalar sequence v_n by decomposing $\mathbb{R}^2$ into small squares of size 2^{-n},[1] and taking $v_n = u_n \circ \Phi_n$, where Φ_n rotates each little square of an arbitrary multiple of $\frac{\pi}{2}$, and one sees easily that any of these choices gives a sequence v_n which also defines the Young measure ν.[2] However, if u_n satisfies a differential equation it is unlikely that v_n will satisfy the same differential equation (or another given one), so that knowing the Young measure of a sequence does not tell us much about which differential equations are satisfied.

The *compensated compactness* theory, developed with François MURAT in the mid 1970s, can take into account differential relations, and it predicts inequalities for the weak $\star$ limits of some quadratic functions, so that it puts constraints on the Young measures. In the late 1970s I developed the *compensated compactness method* for trying to understand in a better way the interaction between the pointwise nonlinear constitutive relations and the linear differential balance equations [98], and for this I used the notion of entropies as defined by Peter LAX in the early 1970s [48], and it may be useful to repeat that entropies are not related to the hyperbolic character of the equations used (as some advocates of fake mechanics seem to believe, maybe

---

[1] More precisely, those squares whose vertices have coordinates which are integer multiples of 2^{-n}.

[2] If ω is a square whose coordinates are integer multiples of 2^{-k}, then for $n \geq k$ one has $\int_\omega F(v_n)\,dx = \int_\omega F(u_n)\,dx$ for all continuous functions F.

L. Tartar, *The General Theory of Homogenization*, Lecture Notes of the Unione Matematica Italiana 7, DOI 10.1007/978-3-642-05195-1_33,

in order to pretend that they were the first to introduce Young measures in the stationary partial differential equations of continuum mechanics, at least ten years after me!). My theory of *H-measures*, developed in the late 1980s [105] and discussed in Chap. 28, supersedes the *compensated compactness* theory, but not the *compensated compactness method*, and since *H-measures* are used for questions of small-amplitude homogenization, discussed in Chap. 29, and for bounds on effective coefficients, discussed in Chap. 30, it is important to understand the relations between Young measures and *H-measures*.

I worked on this question with François MURAT, and in the late 1980s we first characterized the case of characteristic functions, corresponding to mixing two materials [105]; then in 1991 we solved a more general question which I described in the fall at a conference in Ferrara, Italy, for the 600th anniversary of the foundation of the university [109],[3] and immediately after, I described our result to someone who later claimed our result as his at a conference in the fall of 1993 in Trieste, Italy; by that time we had a draft for an article, which we never worked on again afterward, but which contained improvements that I described at a conference in the summer of 1994 in Udine, Italy, [110]: I used the title "beyond Young measures" for emphasizing why Young measures do not characterize microstructures, but that it is still important to use Young measures, because of the erroneous belief that some engineers and physicists still have, that one can deduce effective coefficients from proportions, so that it then becomes important to characterize the sets of effective coefficients for different proportions of the materials used.

A first step in this direction is to perform the characterization in my simplified framework of small-amplitude homogenization, which is precisely the question to characterize which *H-measures* can be obtained for a given Young measure. The natural method of attack for studying this question is to couple the results of small-amplitude homogenization, Theorem 29.1 and Lemma 29.4, with the formulas for effective properties of composite multi-layered materials deduced from the formula for simple layering, Lemma 27.7.

Lemma 33.1. *Laminating orthogonally to* e *materials with tensors* $M^1, \ldots, M^r \in \mathcal{M}(\alpha,\beta)$ *(with* $0 < \alpha \le \beta < \infty$*), in proportions* $\eta_1, \ldots, \eta_r$ *(*≥ 0 *with* $\sum_{i=1}^r \eta_i = 1$*), with* $M^i = A^* + \gamma\, B^i + \gamma^2 C^i + O(\gamma^3)$ *(with* $|\gamma|$ *small enough), gives an effective tensor* M^{eff} *of the form*

$$\begin{aligned} M^{eff} &= A^* + \gamma \textstyle\sum_{i=1}^r \eta_i B^i + \gamma^2\big(\sum_{i=1}^r \eta_i C^i - Z\big) + O(\gamma^3),\\ Z &= \tfrac{1}{2}\textstyle\sum_{i,j=1}^r \eta_i\eta_j (B^i - B^j)\,\frac{e\otimes e}{(A^* e, e)}\,(B^i - B^j). \end{aligned} \tag{33.1}$$

[3] I did not know about the work of Antonio DE SIMONE at the time [23]. He did not look at the general question which François MURAT and myself considered, but for the problem of micromagnetism his constructions were simpler than ours.

Proof. Formula (27.30) is

$$\begin{aligned} &M^{\text{eff}} = \sum_{i=1}^r \eta_i\, M^i - \sum_{1\le i<j\le r} \eta_i\eta_j(M^i - M^j)R^{i,j}(M^i - M^j),\\ &R^{i,j} = \frac{1}{(M^i\, e,e)}\, \frac{e\otimes e}{H}\, \frac{1}{(M^j\, e,e)} \text{ for } i,j = 1,\ldots,r,\\ &H = \sum_{k=1}^r \frac{\eta_k}{(M^k\, e,e)}, \end{aligned} \tag{33.2}$$

and since $M^i - M^j = \gamma\,(B^i - B^j) + O(\gamma^2)$ it is enough to know $R^{i,j}$ at order 0 in γ, which requires one to know H at order 0 in γ, i.e., $H = \frac{1}{(A^* e,e)} + O(\gamma)$, so that $R^{i,j} = \frac{e\otimes e}{(A^* e,e)} + O(\gamma)$ for $i < j$, and (33.1) follows, noticing that $\sum_{1\le i<j\le r}$ and $\frac{1}{2}\sum_{i,j=1}^r$ give the same quantities, as one deals with quantities which are symmetric in i and j, and 0 for $i = j$. □

Lemma 33.2. *(= 29.5) Let a sequence of characteristic functions χ_n satisfy*

$$\chi_n \rightharpoonup \theta \text{ in } L^\infty(\Omega) \text{ weak } \star, \text{ and } \chi_n - \theta \text{ defines an H-measure } \mu. \tag{33.3}$$

Then

$$\begin{aligned} &\mu \in \mathcal{M}(\Omega \times \mathbb{S}^{N-1}) \text{ is nonnegative, even in } \xi \in \mathbb{S}^{N-1},\\ &\quad\text{and its projection on } \Omega \text{ is } \theta\,(1-\theta)\,dx, \end{aligned} \tag{33.4}$$

(written formally as $\int_{\mathbb{S}^{N-1}} d\mu = \theta\,(1-\theta)$ a.e. $x \in \Omega$). Conversely, if $\theta \in L^\infty(\Omega)$ satisfies $0 \le \theta \le 1$ a.e. in Ω and μ satisfies (33.4), then there exists a sequence of characteristic functions χ_n satisfying (33.3).

Proof. The H-measure corresponding to a sequence u_n converging in $L^2_{loc}(\Omega)$ weak to u_∞ is automatically nonnegative, and if u_n is real one has $\mathcal{F}u_n(-\xi) = \overline{\mathcal{F}u_n(\xi)}$ for all $\xi \in \mathbb{R}^N$, so that the H-measure is even in ξ, and then its projection on Ω is the weak $\star$ limit of $(u_n - u_\infty)^2$; since $\chi_n^2 = \chi_n$, one has $(\chi_n - \theta)^2 = \chi_n(1 - 2\theta) + \theta^2$, which converges in $L^\infty(\Omega)$ weak $\star$ to $\theta\,(1-\theta)$.

If for $A^* \in \mathcal{M}(\alpha,\beta)$, one extracts a subsequence such that Theorem 29.1 applies to $A^n = A^* + \gamma\,\chi_n I$ (for $|\gamma|$ small), then $A^{\text{eff}} = A^* + \gamma\,\theta\,I - \gamma^2 \int_{\mathbb{S}^{N-1}} \frac{\xi\otimes\xi}{(A^*\xi,\xi)}\, d\mu(\cdot,\xi) + O(\gamma^3)$. If θ is constant and $\chi_n \rightharpoonup \theta$ in $L^\infty(\mathbb{R})$ weak $\star$, then for $e \in \mathbb{S}^{N-1}$ the sequence $\chi_n\big((\cdot,e)\big)$ gives the H-measure $\mu = \theta\,(1-\theta)\,dx \otimes \frac{\delta_e + \delta_{-e}}{2}$. Then, using M^i corresponding to $e^i \in \mathbb{S}^{N-1}$, for $i = 1,\ldots,r$, one applies Lemma 33.1 for an arbitrary direction e and constant proportions $\eta_1,\ldots,\eta_r$ adding up to 1, and since all the B^i are equal to $\theta\,I$ and $C^i = \theta\,(1-\theta)\,\frac{e_i\otimes e_i}{(A^* e_i,e_i)}$, the correction Z is 0 and the coefficient of γ^2 is $\theta\,(1-\theta)\sum_{i=1}^r \eta_i \frac{e_i\otimes e_i}{(A^* e_i,e_i)}$, corresponding to the H-measure $\theta\,(1-\theta)\,dx \otimes \sum_{i=1}^r \eta_i \frac{\delta_{e_i} + \delta_{-e_i}}{2}$.[4]

---

[4] Since the (real analytic) mapping Φ_μ defined on $\mathcal{L}_{sym,+}(\mathbb{R}^N,\mathbb{R}^N)$ for $\mu \in \mathcal{M}(\mathbb{S}^{N-1})$, μ even, by $\Phi_\mu(A^*) = \int_{\mathbb{S}^{N-1}} \frac{\xi\otimes\xi}{(A^*\xi,\xi)}\, d\mu$ is injective.

One decomposes $\mathbb{R}^N$ in cubes of size 2^{-k} and for each cube Q^j intersecting Ω, one assumes that θ is a constant θ^j with $0 \leq \theta^j \leq 1$, and that μ is $dx \otimes \nu^j$ where ν^j is a finite combination of Dirac masses on $\mathbb{S}^{N-1}$ with total mass $\theta^j(1-\theta^j)$ which is ≥ 0 and even. One creates a sequence of characteristic functions in Ω by defining it on Q^j to coincide with a sequence of characteristic functions defined on $\mathbb{R}^N$ and creating at the limit θ^j and $dx \otimes \nu^j$; by gluing all these pieces together one obtains a sequence having the correct properties in Ω, thanks to Lemma 29.3 which implies that the H-measure cannot have any concentration effects (in x) on the boundaries of the cubes.

One concludes with an argument of density of such a piecewise constant (θ, μ) in those satisfying (33.4), and this part uses the metrizability of the $L^\infty(\Omega)$ and the $\mathcal{M}(\Omega \times \mathbb{S}^{N-1})$ weak $\star$ topologies on the bounded sets considered. For example, starting from $(\theta, \mu) \in L^\infty(\Omega) \times \mathcal{M}(\Omega \times \mathbb{S}^{N-1})$ satisfying (33.4), for each cube Q^j of size 2^{-k} with $meas(Q^j \cap \Omega) > 0$, one defines (θ_k, μ_k) on $Q^j \cap \Omega$ by averaging in x,[5] i.e.,

$$\begin{aligned}
&\theta_k(x) = \theta_k^j = \frac{\int_{Q^j\cap\Omega} \theta\,dx}{meas(Q^j\cap\Omega)} \text{ for } x \in Q^j \cap \Omega,\\
&\mu_k = dx \otimes \nu_k^j \text{ on } (Q^j \cap \Omega) \times \mathbb{S}^{N-1} \text{ with } \nu_k^j \in \mathcal{M}(\mathbb{S}^{N-1}) \text{ defined by}\\
&\langle \nu_k^j, \psi\rangle = \frac{\langle \mu, \chi_{Q^j\cap\Omega}\otimes\psi\rangle}{meas(Q^j\cap\Omega)} \text{ for all } \psi \in C(\mathbb{S}^{N-1}),
\end{aligned} \tag{33.5}$$

and one has $\theta_k \rightharpoonup \theta$ in $L^\infty(\Omega)$ weak $\star$ and $\mu_k \rightharpoonup \mu$ in $\mathcal{M}(\Omega \times \mathbb{S}^{N-1})$ weak $\star$ as k tends to ∞, as well as $\theta_k \to \theta$ in $L^p_{loc}(\Omega)$ strong for $1 \leq p < \infty$.[6] However, (θ_k, μ_k) does not in general satisfy (33.4), and for satisfying (33.4) one replaces μ_k by μ_k^* defined by

$$\begin{aligned}
&\mu_k^* = \mu_k + \alpha_k \otimes \frac{\delta_e + \delta_{-e}}{2} \text{ with } e \in \mathbb{S}^{N-1} \text{ and } \alpha_k \in L^\infty_{loc}(\Omega) \text{ defined by}\\
&\alpha_k(x) = \alpha_k^j \text{ for } x \in Q^j \cap \Omega, \text{ with } \alpha_k^j + \frac{\int_{Q^j\cap\Omega} \theta\,(1-\theta)\,dx}{meas(Q^j\cap\Omega)} = \theta_k^j(1-\theta_k^j),\\
&\quad\text{i.e., } \alpha_k^j = \frac{\int_{Q^j\cap\Omega} (\theta-\theta_k^j)^2\,dx}{meas(Q^j\cap\Omega)}
\end{aligned} \tag{33.6}$$

so that $\alpha_k \to 0$ in $L^1_{loc}(\Omega)$ strong since $\theta_k \to \theta$ in $L^2_{loc}(\Omega)$ strong. □

My motivation for considering more general constructions was that I heard strange remarks on a physicist's problem, micromagnetism [10], which is about stationary solutions of the Maxwell–Heaviside equation with $j = 0$, so that $H = grad(u)$ in $\mathbb{R}^3$, with the constitutive relation $B = H + m$ with $|m| = \kappa$ in Ω and $m = 0$ in $\mathbb{R}^3 \setminus \Omega$, explained by a statistical me-

[5] Since μ has a bounded density in x, it does not charge the boundaries of the sets $Q^j \cap \Omega$, and neither does μ_k as its definition by (33.5) makes sense.

[6] Because of the uniform continuity of continuous functions with compact support, and the density of $C_c(\Omega)$ in $L^p(\Omega)$ for $1 \leq p < \infty$.

chanics equilibrium argument (with κ depending upon temperature), and pretending that the "macroscopic spin" m manages to minimize a total energy obtained by adding the magnetostatic energy $\int_{\mathbb{R}^3} |grad(u)|^2\,dx$, a term $\int_\Omega (H_e, m)\,dx$ due to an applied magnetic field H_e, an "anisotropic term" $\int_\Omega \varphi(x,m)\,dx$ explained by a crystalline structure around x which selects preferred "easy directions" for m (and $\varphi(x,m)$ is invented so that its minima on the sphere $|m| = \kappa$ are the observed easy directions), and an "exchange energy" $\varepsilon \int_\Omega |\nabla m|^2\,dx$ explained by a quantum mechanics argument (which should be adapted to the local crystalline structure, I suppose). This game was chosen because of an observed structure of magnetic walls through which m "jumps," and the width of these walls is related to the small parameter ε, these walls separating magnetic domains, whose sizes are not so clear.

Although I could not confuse such an exotic minimization problem with the real physical problem (so that I call it a physicist's problem and not a physics problem), it led to an interesting mathematical question: when ε_n tends to 0 the limiting behavior of minimizing sequences m^n involves its Young measure for computing the limit of the anisotropic energy, and its H-measure for computing the limit of the magnetostatic energy, because of the equation $div(grad(u) + m) = 0$: there is then a natural relaxed problem involving the admissible pairs (ν, μ) of a Young measure and an H-measure associated to a sequence m^n taking its values in $\kappa\,\mathbb{S}^2$.[7]

One could consider that m is only defined in Ω, but it is important to know if *H-measures* do not charge $\partial\Omega$,[8] which can only be found by working in an open set containing $\overline{\Omega}$, or by generalizing *H-measures* to "smooth" manifolds with boundary;[9] also, the equation $div(grad(u) + m) = 0$ holds in $\mathbb{R}^3$, and if $m^n \rightharpoonup m^\infty$ in $L^\infty(\Omega;\mathbb{R}^3)$ (or $L^\infty(\mathbb{R}^3;\mathbb{R}^3)$) weak $\star$ and $m^n - m^\infty$ defines an H-measure μ, the limit of $\int_{\mathbb{R}^3} |grad(u_n)|^2\,dx$ is $\int_{\mathbb{R}^3} |grad(u_\infty)|^2\,dx$ plus a contribution of the H-measure μ, and since $\frac{\partial u_n}{\partial x_j} = R_j(\sum_k R_k w_k^n)$ this

---

[7] François MURAT and myself did not characterize the set of admissible pairs, but like for our homogenization method in optimal design [75], [111], the complete characterization is not needed, at least for computing the limit of the minimum "energy" $I(\varepsilon)$ as ε tends to 0; however, for understanding at which rate $I(\varepsilon)$ converges to that limiting value, I wanted to understand a question of characteristic lengths: the size of the walls must be of order ε, but the size of domains is not clear to me, and for that reason I investigated with Patrick GÉRARD a particular sequence with two characteristic lengths, described in Chap. 32.

[8] Here m^n is bounded in $L^\infty(\mathbb{R}^3;\mathbb{R}^3)$, and the H-measure inherits a bounded density in x, and cannot charge the boundary, if it has Lebesgue measure 0.

[9] When Patrick GÉRARD introduced *H-measures* independently for functions with values in a Hilbert space, under the name "microlocal defect measures" which I find misleading, he wrote that one cannot define them on a manifold but that it does not matter since only the support is important. Of course, he was thinking about a general topological manifold, but the formula for change of variables which I proved in [105] suggests that the correct hypothesis is to consider a manifold with a volume form. That he thought that only the support is important is a result of blindly following Lars HÖRMANDER, and I described in Chap. 28 the important differences between his wave front sets and my H-measures.

contribution is $\sum_{k,\ell}\langle\mu^{k,\ell},\xi_k\xi_\ell\rangle$, which makes sense if one assumes Ω bounded (or just of finite Lebesgue measure) since μ is 0 outside Ω.

For a given Young measure ν, it is not really necessary to characterize all the compatible *H-measures* μ, but to find those which minimize $\sum_{k,\ell}\langle\mu^{k,\ell},\xi_k\xi_\ell\rangle$, and the constructions of particular pairs (ν,μ) that I performed with François MURAT show that whatever ν is there is a compatible μ such that $\sum_{k,\ell}\langle\mu^{k,\ell},\xi_k\xi_\ell\rangle = 0$. In other terms, using Lemma 28.18, it means that given a sequence m^n defining a Young measure ν one can change it (in a way which our proof does not make explicit), in order to create another sequence m^n_* defining the same Young measure ν, but such that $div(m^n_*)$ stays in a compact of $H^{-1}_{loc}(\mathbb{R}^3)$ strong. More generally, if $U^n \rightharpoonup U^\infty$ in $L^\infty(\Omega;\mathbb{R}^p)$ weak $\star$ and defines a Young measure ν, then our constructions show that there exists another sequence U^n_* defining the same Young measure, and such that $\sum_{j=1}^N\sum_{k=1}^p A_{i,j,k}\frac{\partial(U^n_*)_j}{\partial x_k}$ stays in a compact of $H^{-1}_{loc}(\Omega)$ strong for $i = 1,\ldots,q$, under the important hypothesis that $\Lambda = \mathbb{R}^p$, using the definition (17.6) in our *compensated compactness* theory, described in Chap. 17.[10]

Lemma 33.1 permits one to study the effective properties of various classes of mixtures in the approximation of small-amplitude homogenization, from which one then identifies compatible pairs of a Young measure and an H-measure. François MURAT and myself considered simple and then multiple layering processes, rendered more complex by limiting steps.

Starting from a finite number of materials $M^1,\ldots,M^r \in \mathcal{M}(\alpha,\beta)$, one laminates them with various proportions $\eta_1,\ldots,\eta_r$, and various directions of layers $e \in \mathbb{S}^{N-1}$, Lemma 33.1 giving the effective tensor M^{eff} in the approximation of small-amplitude homogenization. Then, one repeats the process with a finite number of such first-generation materials, giving second-generation materials, and one repeats the process a finite number of times. This is done in the proof of Lemma 33.2 (starting with A^* and $A^* + \gamma I$), but replacing a finite combination of Dirac masses on $\mathbb{S}^{N-1}$ by an arbitrary nonnegative Radon measure on $\mathbb{S}^{N-1}$ then requires a limiting process. All the constructions used correspond to sequences $A^n \in \mathcal{M}(\alpha,\beta;\mathbb{R}^N)$ of the form

$$\begin{aligned} &A^n = \textstyle\sum_{k=1}^r \chi_{n,k} M^k,\\ &\chi_{n,k} \rightharpoonup \theta_k \text{ in } L^\infty(\mathbb{R}^N) \text{ weak } \star, k = 1,\ldots,r, \end{aligned} \tag{33.7}$$

where the $\chi_{n,k}$ are characteristic functions of disjoint sets, so that

$$0 \le \theta_k \le 1, k = 1,\ldots,r, \text{ and } \sum_{k=1}^r \theta_k = 1 \text{ a.e. in } \mathbb{R}^N; \tag{33.8}$$

---

[10] After returning from a conference [109] in Ferrara, Italy, I mentioned this statement to someone whom I heard talk about a similar result at a conference in Trieste, Italy, in the fall of 1993, and I told him publicly that the fact that he cannot learn *H-measures* is not a reason to avoid quoting my earlier results (unpublished, but which I described to him). He confirmed later this tendency to work on questions that I treated before, without properly attributing earlier results.

moreover, it is useful to assume that each θ_k is a constant, so that one may easily reiterate the process, and still know how much of each original material is used in the final mixture. Of course, using (27.30), each construction corresponds to a real analytic function Φ such that

$$M^{\text{eff}} = \Phi(M^1, \ldots, M^r), \tag{33.9}$$

but describing the exact Φ defined on $\big(\mathcal{M}(\alpha,\beta)\big)^r$ seems a daunting task. Instead, using $M^i = A^* + \gamma\, B^i + \gamma^2 C^i + O(\gamma^3)$ for $i = 1, \ldots, r$, with $|\gamma|$ small enough, one considers the Taylor expansion of Φ at order two in γ at A^*

$$\Phi(M^1, \ldots, M^r) = A^* + \gamma\Big(\sum_{i=1}^r \theta_i\, B^i\Big) + \gamma^2\Big(\sum_{i=1}^r \theta_i C^i - \Psi\Big) + O(\gamma^3), \tag{33.10}$$

and Ψ depends upon A^*, and $B^1, \ldots, B^r$, but not on $C^1, \ldots, C^r$.[11]

Lemma 33.3. *For $r \ge 2$, let $\pi^{i,j}$, $i,j = 1, \ldots, r$, be (even) probability measures on $\mathbb{S}^{N-1}$ with $\pi^{j,i} = \pi^{i,j}$, for $i,j = 1, \ldots, r$.*[12] *For $\theta_1, \ldots, \theta_r$ constant with (33.8), there exists a mixture using proportion θ_i of material M^i, $i = 1, \ldots, r$, having its effective tensor satisfying (33.10) with Ψ given by*

$$\begin{aligned} \Psi &= \tfrac{1}{2}\textstyle\sum_{i,j=1}^r \theta_i\theta_j (B^i - B^j) R^{i,j} (B^i - B^j),\\ R^{i,j} &= \textstyle\int_{\mathbb{S}^{N-1}} \frac{e\otimes e}{(A^* e, e)}\, d\pi^{i,j}(e), i,j = 1, \ldots, r. \end{aligned} \tag{33.11}$$

Proof. One uses an induction on r,[13] noticing that for $r = 2$, it is Theorem 29.1 (small-amplitude homogenization) together with Lemma 33.2. Assuming that Lemma 33.3 is true for $r-1$ ($\ge$2) materials, one starts by mixing materials $M^1, \ldots, M^{r-1}$ with proportions

$$\theta'_1 = \theta_1 + \theta_r, \text{ and } \theta'_j = \theta_j, j = 2, \ldots, r-1, \tag{33.12}$$

and one also sets $\theta'_r = 0$ so that the sums can be taken from $i = 1$ to $i = r$. Using Lemma 33.3 for $r-1$ materials, one can construct a mixture using proportion θ'_i of material M^i, $i = 1, \ldots, r$, having the effective conductivity

$$\begin{aligned} X &= A^* + \gamma\,(\textstyle\sum_{i=1}^r \theta'_i B^i) + \gamma^2(\textstyle\sum_{i=1}^r \theta'_i C^i - Z') + O(\gamma^3),\\ Z' &= \tfrac{1}{2}\textstyle\sum_{i,j=1}^r \theta'_i\theta'_j (B^i - B^j) R^{i,j} (B^i - B^j). \end{aligned} \tag{33.13}$$

Then one mixes materials $M^2, \ldots, M^r$ with proportions

---

[11] The various $O(\gamma^3)$ terms can be estimated by working with (27.30).

[12] In (33.11) one only integrates a function on $\mathbb{S}^{N-1}$ which is even, so that only the even part of $\pi^{i,j}$ (which is also a probability) is used.

[13] We first proved the case $r = 3$ using the result for $r = 2$, and then we noticed that the same argument applies to $r > 3$.

$$\theta''_j = \theta_j, j = 2, \ldots, r-1, \text{ and } \theta''_r = \theta_1 + \theta_r, \tag{33.14}$$

and one also sets $\theta''_1 = 0$ so that the sums can be taken from $i = 1$ to $i = r$. Using Lemma 33.3 for $r - 1$ materials, one can construct a mixture using proportion θ''_i of material M^i, $i = 1, \ldots, r$, having the effective conductivity

$$\begin{aligned} &Y = A^* + \gamma\left(\textstyle\sum_{i=1}^r \theta''_i B^i\right) + \gamma^2\left(\textstyle\sum_{i=1}^r \theta''_i C^i - Z''\right) + O(\gamma^3), \\ &Z'' = \tfrac{1}{2}\textstyle\sum_{i,j=1}^r \theta''_i \theta''_j (B^i - B^j) R^{i,j} (B^i - B^j). \end{aligned} \tag{33.15}$$

Finally one mixes materials X and Y with proportions $\frac{\theta_1}{\theta_1+\theta_r}$ and $\frac{\theta_r}{\theta_1+\theta_r}$; using Lemma 33.2 for the probability measure π^{1r} (i.e., Lemma 33.3 in the case $r = 2$), one obtains a material of effective conductivity M^{eff} given by

$$\begin{aligned} M^{\text{eff}} &= A^* + \gamma\left[\tfrac{\theta_1}{\theta_1+\theta_r}\left(\textstyle\sum_{i=1}^r \theta'_i B^i\right) + \tfrac{\theta_r}{\theta_1+\theta_r}\left(\textstyle\sum_{i=1}^r \theta''_i B^i\right)\right] \\ &+ \gamma^2\Big[\tfrac{\theta_1}{\theta_1+\theta_r}\left(\textstyle\sum_{i=1}^r \theta'_i C^i - Z'\right) + \tfrac{\theta_r}{\theta_1+\theta_r}\left(\textstyle\sum_{i=1}^r \theta''_i C^i - Z''\right) \\ &- \tfrac{\theta_1}{\theta_1+\theta_r}\,\tfrac{\theta_r}{\theta_1+\theta_r}\big((\theta_1+\theta_r)(B^1 - B^r)\big) R^{1,r} \big((\theta_1+\theta_r)(B^1 - B^r)\big)\Big] \\ &+ O(\gamma^3), \end{aligned} \tag{33.16}$$

showing the expected coefficients of B^i and C^i for $i = 1, \ldots, r$, since

$$\frac{\theta_1}{\theta_1+\theta_r}\theta'_i + \frac{\theta_r}{\theta_1+\theta_r}\theta''_i = \theta_i, i = 1, \ldots, r, \tag{33.17}$$

and a term Ψ is given by

$$\Psi = \frac{\theta_1}{\theta_1+\theta_r} Z' + \frac{\theta_r}{\theta_1+\theta_r} Z'' + \theta_1\theta_r (B^1 - B^r) R^{1,r} (B^1 - B^r), \tag{33.18}$$

and using (33.13) and (33.15), it is the expected formula for Ψ. □

For $r = 2$, the class described by Lemma 33.3 is stable by simple layering, since it corresponds to Lemma 33.2 for the H-measure of a sequence of characteristic functions, which is a characterization. However, the class described by Lemma 33.3 is not stable by simple layering for $r \geq 3$, and François MURAT and myself then looked for other constructions.

Lemma 33.4. *For $r \geq 3$, let π^i, $i = 1, \ldots, r$, be (even) probability measures on $\mathbb{S}^{N-1}$.*[14] *For $\theta_1, \ldots, \theta_r$ constant with (33.8), there exists a mixture using proportion θ_i of material M^i, $i = 1, \ldots, r$, having its effective tensor satisfying (33.10) with Ψ given by*

[14] In (33.19) one only integrates a function on $\mathbb{S}^{N-1}$ which is even, so that only the even part of π^i (which is also a probability) is used.

$$\begin{aligned}&\Psi = \sum_{i=1}^r \theta_i(B^i - G)R^i(B^i - G),\\ &R^i = \int_{\mathbb{S}^{N-1}} \frac{e\otimes e}{(A^* e,e)}\,d\pi^i(e), i = 1,\ldots,r,\\ &G = \sum_{i=1}^r \theta_i B^i.\end{aligned} \tag{33.19}$$

Proof. The proof is divided into several steps. One starts from any simple layered material $\mathcal{M}^0 = A^* + \gamma\, G + \gamma^2\varphi + O(\gamma^3)$, where the coefficient of γ is the desired $G = \sum_{i=1}^r \theta_i B^i$ of (33.19). In the first step, one constructs materials $\mathcal{M}^1,\ldots,\mathcal{M}^r$, by a layering of $\mathcal{M}^{j-1}$ with M^j, using a parameter $\alpha_j \in (0,1)$. In the second step, one shows that $\alpha_1,\ldots,\alpha_r$ can be chosen so that the coefficient of γ in $\mathcal{M}^r$ is G. In the third step, one observes that the coefficient φ of γ^2 in $\mathcal{M}^0$ is replaced in $\mathcal{M}^r$ by $\lambda\,\varphi + (1-\lambda)\left(\sum_{i=1}^r \theta_i C^i + \overline{\psi}\right)$ for a certain function $\overline{\psi}$ and some $\lambda \in (0,1)$. A countable reiteration of this process leads to the disappearance of φ,[15] and its replacement by $\sum_{i=1}^r \theta_i C^i + \overline{\psi}$. In the fourth step, one computes the function $\overline{\psi}$, which is found to converge to Ψ given by (33.19), when some parameter in the layering tends to 0.

First step: One chooses any initial layering giving an effective material

$$\mathcal{M}^0 = A^* + \gamma\, G + \gamma^2\varphi + O(\gamma^3) \tag{33.20}$$

where $G = \sum_{i=1}^r \theta_i B^i$. For a probability measure π^1 on $\mathbb{S}^{N-1}$ and $\alpha_1 \in (0,1)$ one uses Lemma 33.1 to construct a layering of M^1 and $\mathcal{M}^0$ with proportions α_1 and $1-\alpha_1$, giving an effective material $\mathcal{M}^1$, with

$$\begin{aligned}&\mathcal{M}^1 = A^* + \gamma\,\mathcal{B}^1 + \gamma^2\mathcal{C}^1 + O(\gamma^3),\\ &\mathcal{B}^1 = \alpha_1 B^1 + (1-\alpha_1)\,G,\\ &\mathcal{C}^1 = \alpha_1 C^1 + (1-\alpha_1)\,\varphi - \alpha_1(1-\alpha_1)\,(B^1 - G)R^1(B^1 - G),\end{aligned} \tag{33.21}$$

with R^1 given by (33.19). For $j = 2,\ldots,r$, for a probability measure π^j on $\mathbb{S}^{N-1}$ and $\alpha_j \in (0,1)$ one uses Lemma 33.1 to construct a layering of M^j and $\mathcal{M}^{j-1}$ with proportions α_j and $1-\alpha_j$, giving an effective material $\mathcal{M}^j$, with

$$\begin{aligned}&\mathcal{M}^j = A^* + \gamma\,\mathcal{B}^j + \gamma^2\mathcal{C}^j + O(\gamma^3), j = 2,\ldots,r,\\ &\mathcal{B}^j = \alpha_j B^j + (1-\alpha_j)\,\mathcal{B}^{j-1},\\ &\mathcal{C}^j = \alpha_j C^j + (1-\alpha_j)\,\mathcal{C}^{j-1} - \alpha_j(1-\alpha_j)\,(B^j - \mathcal{B}^{j-1})R^j(B^j - \mathcal{B}^{j-1}),\end{aligned} \tag{33.22}$$

with R^j given by (33.19); of course, the formula still holds for $j = 1$ if one sets $\mathcal{B}^0 = G$ and $\mathcal{C}^0 = \varphi$.

Second step: Since $\mathcal{B}^j = \alpha_j B^j + (1-\alpha_j)\mathcal{B}^{j-1}$ holds for $j = 1,\ldots,r$, and $\mathcal{B}_0 = G$, one wants to show that once $\alpha_1 \in (0,1)$ is chosen, one can choose $\alpha_2,\ldots,\alpha_r$ so that $\mathcal{B}^r = G$. Indeed, for $j = 1,\ldots,r-1$, one defines α_{j+1} by

---

[15] A similar process of partial disappearance of a coefficient was used by Graeme MILTON and by Enzo NESI.

$$\alpha_{j+1} = \frac{\alpha_j \theta_{j+1}}{\theta_j + \alpha_j \theta_{j+1}}, \text{ i.e., } 1 - \alpha_{j+1} = \frac{\theta_j}{\theta_j + \alpha_j \theta_{j+1}}, \tag{33.23}$$

which ensures that $0 < \alpha_j < 1$ for $j = 1, \ldots, r$, and

$$(1 - \alpha_{j+1})\frac{\alpha_j}{\theta_j} = \frac{\alpha_{j+1}}{\theta_{j+1}}, j = 1, \ldots, r-1. \tag{33.24}$$

By induction, (33.19) and (33.24) imply

$$\mathcal{B}^j = (1-\alpha_1)\cdots(1-\alpha_j)G + (\theta_1 B^1 + \ldots + \theta_j B^j)\frac{\alpha_j}{\theta_j} \text{ for } j = 1, \ldots, r, \tag{33.25}$$

and in order to have $\mathcal{B}^r = G$, one must have

$$(1 - \alpha_1)\cdots(1 - \alpha_r) + \frac{\alpha_r}{\theta_r} = 1, \tag{33.26}$$

which, using (33.24), amounts to

$$(1 - \alpha_1)\frac{\alpha_r}{\theta_r}\frac{\theta_1}{\alpha_1} + \frac{\alpha_r}{\theta_r} = 1. \tag{33.27}$$

For showing it, one rewrites (33.23) or (33.24) as

$$\frac{\theta_{j+1}}{\alpha_{j+1}} = \frac{\theta_j}{\alpha_j} + \theta_{j+1} \text{ for } j = 1, \ldots, r, \tag{33.28}$$

giving

$$\frac{\theta_k}{\alpha_k} = \frac{\theta_1}{\alpha_1} + \theta_2 + \ldots + \theta_k \text{ for } 2 \le k \le r, \tag{33.29}$$

so that for $k = r$ one has $\frac{\theta_r}{\alpha_r} = \frac{\theta_1}{\alpha_1} + 1 - \theta_1$, which is (33.27).

Third step: With any choice of $\alpha_1 \in (0,1)$, and $\alpha_2, \ldots, \alpha_r$, deduced from (33.23) like in (33.29), one constructed a mixture with $\mathcal{M}^r$ given by

$$\mathcal{M}^r = A^* + \gamma G + \gamma^2 \mathcal{C}^r + O(\gamma^3) \tag{33.30}$$

where $\mathcal{C}^r$ can be computed from (33.22). By an induction similar to that used for deriving (33.25), one may write $\mathcal{C}^r$ as

$$\mathcal{C}^r = (1-\alpha_1)\cdots(1-\alpha_r)\,\varphi + \frac{\alpha_r}{\theta_r}\sum_{i=1}^r \theta_j C^j + (1-(1-\alpha_1)\cdots(1-\alpha_r))\,\overline{\psi}, \tag{33.31}$$

which defines a function $\overline{\psi}$ depending upon $\alpha_j, \theta_j, B^j, R^j$, $j = 1, \ldots, r$, but not on C^j, $j = 1, \ldots, r$. One defines $\lambda \in (0,1)$ by

$$\lambda = (1 - \alpha_1)\cdots(1 - \alpha_r), \tag{33.32}$$

and since $\frac{\alpha_r}{\theta_r} = 1 - \lambda$ by (33.26), (33.31) is

$$\mathcal{C}^r = \lambda\,\varphi + (1-\lambda)\Big(\sum_{j=1}^{q} \theta_j C_j + \overline{\psi}\Big), \tag{33.33}$$

which explains the choice of the coefficient multiplying $\overline{\psi}$ in (33.31).

Starting from $\mathcal{M}^0$ defined by (33.20), one constructed $\mathcal{M}^r$ defined by (33.30), so that the coefficient of γ^2 changed from φ to $\mathcal{C}^r$ given by (33.33), and applying the same process to $\mathcal{M}^r$ gives $\mathcal{M}^{r,1}$, and repeating it gives

$$\mathcal{M}^{r,k} = A^* + \gamma\,G + \gamma^2 \mathcal{C}^{r,k} + O(\gamma^3), k \geq 1, \tag{33.34}$$

with

$$\mathcal{C}^{r,k} = (1-\lambda)\Big(\sum_{j=1}^{q} \theta_j C^j + \overline{\psi}\Big)(1+\lambda+\ldots+\lambda^k) + \lambda^{k+1}\varphi, k \geq 1, \tag{33.35}$$

and as k tends to ∞, the coefficient φ disappears progressively and is replaced by $\sum_{j=1}^{r} \theta_j C^j + \overline{\psi}$, so that one constructs a mixture with effective tensor[16]

$$\mathcal{M} = A^* + \gamma\,G + \gamma^2\Big(\sum_{j=1}^{r} \theta_j C^j + \overline{\psi}\Big) + O(\gamma^3). \tag{33.36}$$

Fourth step: The expression of $\overline{\psi}$ is intricate, but becomes simple if $\alpha_1 = \varepsilon\,\theta_1$ with ε small, since (33.23), (33.25), and (33.32) imply

$$\begin{aligned} &\alpha_j = \varepsilon\,\theta_j + O(\varepsilon^2), j = 1,\ldots,r,\\ &\mathcal{B}^j = G + O(\varepsilon), j = 1,\ldots,r,\\ &\lambda = 1 - \varepsilon + O(\varepsilon^2), \end{aligned} \tag{33.37}$$

[16] The validity of the limiting step comes from the small-amplitude homogenization Theorem 29.1. Any mixture considered here corresponds to a sequence $U^n \in L^\infty(\mathbb{R}^N;\mathbb{R}^r)$, where for $i = 1,\ldots,r$, one has $U_i^n = \chi_{i,n}$, the characteristic function of the set where one uses the material M^i, and $U^n \rightharpoonup U^\infty$ in $L^\infty(\mathbb{R}^N;\mathbb{R}^r)$ weak star, where $U_i^\infty = \theta_i$ for $i = 1,\ldots,r$; one may assume that $U^n - U^\infty$ defines an H-measure $\mu \in \mathcal{M}\big(\mathbb{R}^N \times \mathbb{S}^{N-1}; \mathcal{L}(\mathbb{C}^r;\mathbb{C}^r)\big)$, and Ψ depends linearly upon μ. Having a family indexed by $k \geq 1$ corresponding to the same U^∞ and various H-measures μ_k, if $\mu_k \rightharpoonup \mu_\infty$ in $\mathcal{M}\big(\mathbb{R}^N \times \mathbb{S}^{N-1}; \mathcal{L}(\mathbb{C}^r;\mathbb{C}^r)\big)$ weak $\star$, then there is a Cantor diagonal sequence V^n converging to U^∞ in $L^\infty(\mathbb{R}^N;\mathbb{R}^r)$ weak star, and such that $V^n - U^\infty$ defines the H-measure μ_∞, by an argument of metrizability of the corresponding weak $\star$ topologies. One deduces that the limit of Ψ_k corresponds to the correction Ψ_∞ associated to the mixture described by the sequence V^n.

and then (33.21)–(33.22) give

$$\begin{aligned}\mathcal{C}^1 &= \varepsilon\,\theta_1 C^1 + (1-\varepsilon\,\theta_1)\varphi - \varepsilon\,\theta_1(G-B^1)R^1(G-B^1) + O(\varepsilon^2),\\ \mathcal{C}^j &= \varepsilon\,\theta_j C^j + (1-\varepsilon\,\theta_j)\,\mathcal{C}^{j-1} - \varepsilon\,\theta_j(G-B^j)R^j(G-B^j) + O(\varepsilon^2).\end{aligned} \tag{33.38}$$

Using (33.37)–(33.38) in (33.31) gives

$$\overline{\psi} = -\sum_{k=1}^{r}\theta_k(B^k-G)R^k(B^k-G) + O(\varepsilon) = -\Psi + O(\varepsilon), \tag{33.39}$$

where Ψ given by (33.19), and one then lets ε tend to 0 in (33.39). □

The constructions which François MURAT and myself used for Lemma 33.3 and Lemma 33.4 reminded us of ways to compute a matrix of inertia for a discrete distribution of mass, extended to $\mathbb{R}^N$ with its usual Euclidean structure. If one uses weights w_i at points X_i, $i \in I$, the matrix of inertia at a point Y, which I write $J(w_i, X_i, i \in I \mid Y)$, is defined by

$$\begin{aligned}&\big(J(w_i, X_i, i \in I \mid Y)\,z, z\big) = \textstyle\sum_{i\in I} w_i(X_i - Y, z)^2\\ &\qquad\text{for all } z \in \mathbb{R}^N, \text{ i.e.,}\\ &J(w_i, X_i, i \in I \mid Y) = \textstyle\sum_{i\in I} w_i(X_i - Y)\otimes(X_i - Y);\end{aligned} \tag{33.40}$$

if $\sum_{i\in I} w_i \neq 0$, the center of mass is defined by $\big(\sum_{i\in I} w_i\big)\,X_* = \sum_{i\in I} w_i X_i$, and one has

$$\begin{aligned}J(w_i, X_i, i \in I \mid Y) &= J(w_i, X_i, i \in I \mid X_*)\\ &\quad+\big(\textstyle\sum_{i\in I} w_i\big)\,(Y - X_*)\otimes(Y - X_*).\end{aligned} \tag{33.41}$$

For nonnegative weights $\xi_i \geq 0$ not all 0, with $G = \frac{\sum_{i\in I}\xi_i Q^i}{\sum_{i\in I}\xi_i}$ being the center of mass of the family of points Q^i, $i \in I$, not necessarily distinct, I write the matrix of inertia at the center of mass

$$J_c(\xi_i, Q^i, i \in I) = \sum_{i\in I}\xi_i(Q^i - G)\otimes(Q^i - G); \tag{33.42}$$

if then each point Q^i is itself the center of mass of a distribution with weights $\xi_{i,j} \geq 0$ at points $Q^{i,j}$,[17] $j \in J_i$, with $\sum_{j\in J_i}\xi_{i,j} = \xi_i$ for all $i \in I$, i.e., $\xi_i Q^i = \sum_{j\in J_i}\xi_{i,j}Q^{i,j}$ for all $i \in I$, then the matrix of inertia of the distribution of mass $\xi_i\xi_{i,j}$ at points $Q^{i,j}$ is

[17] The same point may be used for two different values i_1 and i_2, and the decompositions be different for Q^{i_1} and Q^{i_2}. Also, the same point may be used as $Q^{i,j}$ for various values of i, or j.

$$\begin{aligned}J_c(\xi_i\xi_{i,j}, Q^{i,j}, i \in I, j \in J_i) = \textstyle\sum_{i\in I} \xi_i(Q^i - G) \otimes (Q^i - G)\\ + \textstyle\sum_{i\in I} \xi_i J_c(\xi_{i,j}, Q^{i,j}, j \in J_i),\end{aligned} \tag{33.43}$$

and the formula can be iterated by decomposing the points $Q^{i,j}$ as center of mass of other points. Lemma 33.2 is related to the case $I = \{1, 2\}$, where

$$J_c(\xi_i, Q^i, i = 1, 2) = \xi_1\xi_2\,(Q^1 - Q^2) \otimes (Q^1 - Q^2); \tag{33.44}$$

Lemma 33.3 is related to a first decomposition

$$\begin{aligned}G &= \tfrac{\theta_1}{\theta_1+\theta_r} X + \tfrac{\theta_r}{\theta_1+\theta_r} Y,\\ X &= (\theta_1 + \theta_r)B^1 + \textstyle\sum_{i=2}^{r-1} \theta_i B^i,\\ Y &= \textstyle\sum_{i=2}^{r-1} \theta_i B^i + (\theta_1 + \theta_r)B^r,\end{aligned} \tag{33.45}$$

and both X and Y are further decomposed, but not explicitly, since it is hidden inside the induction argument; Lemma 33.4 is related to the decomposition

$$G = \sum_i \theta_i B^i. \tag{33.46}$$

Lemma 33.6 will be a generalization extending these three examples, but it will not imply the characterization of Lemma 33.2. The idea is to describe in Definition 33.5 a family of admissible decompositions of the matrix of inertia T at G, which permits us to use various choices of (even) probabilities on $\mathbb{S}^{N-1}$, resulting in explicit functions Ψ in (33.10).

Definition 33.5. *Given points $B^i \in \mathbb{R}^N$, $i = 1, \ldots, r$, and $G = \sum_{i=1}^r \theta_i B^i$ (with $\theta_i \geq 0$ for $i = 1, \ldots, r$, and $\sum_{i=1}^r \theta_i = 1$), an admissible decomposition of the matrix of inertia $T = \sum_{i=1}^r \theta_i(B^i - G) \otimes (B^i - G)$ at G, is obtained in the following way: one writes*

$$\begin{aligned}G = \textstyle\sum_{j=1}^s \xi_j Q^j, 0 \leq \xi_j \leq 1, j = 1, \ldots, s, \sum_{j=1}^s \xi_j = 1,\\ Q^j \in \text{ convex hull of } \{B^1, \ldots, B^r\}, j = 1, \ldots, s,\\ T = \textstyle\sum_{j=1}^s \xi_j(Q^j - G) \otimes (Q^j - G) + \sum_{j=1}^s \xi_j T^j,\end{aligned} \tag{33.47}$$

and for $j = 1, \ldots, s$ each T^j is a matrix of inertia corresponding to a finite decomposition of the points Q^j, created in a similar way, but with no loops, and the ultimate points in this decomposition must be points from $\{B^1, \ldots, B^r\}$ (which have 0 as matrix of inertia); for $j = 1, \ldots, s$, an admissible decomposition of T_j is a list $\{\eta_k N^k \otimes N^k \mid k \in K_j\}$, and an admissible decomposition of T is a list

$$\begin{aligned}\{\xi_j(Q^j - G) \otimes (Q^j - G) \mid j = 1, \ldots, s\} \textstyle\bigcup \{\xi_1\eta_k N^k \otimes N^k \mid k \in K_1\}\\ \textstyle\bigcup \cdots \bigcup \{\xi_s\eta_k N^k \otimes N^k \mid k \in K_s\}.\end{aligned} \tag{33.48}$$

Definition 33.5 is recursive, since one needs to apply it to the terms T^j, $j = 1,\ldots,s$, mentioned in the definition, and the conditions that there are no loops and that the ultimate points in the decomposition belong to $\{B^1,\ldots,B^r\}$ are important.

For any B^i, $i = 1,\ldots,r$, the matrix of inertia is 0 and the list may be considered empty, since one may always remove from the list all the terms with $\eta_k = 0$, or with $N^k = 0$. Then if $Q = \sum_{i=1}^r \zeta^i B^i$, with $0 \le \zeta^i \le 1, i = 1,\ldots,r$, and $\sum_{i=1}^r \zeta^i = 1$, one has $T = \sum_{i=1}^r \zeta^i (B^i - Q)\otimes(B^i - Q)$ and the list is $\{\zeta^i(B^i - Q)\otimes(B^i - Q) \mid i = 1,\ldots,r\}$, from which one removes the zero terms. By induction on the number of steps needed to arrive at only combinations of $B^1,\ldots,B^r$, one sees that Definition 33.5 is not ambiguous.

There is no need to remember separately what η_k is and what N^k is, since only the rank-one operator $\eta_k N^k \otimes N^k$ plays a role.[18]

Lemma 33.6. *For $r \ge 3$,[19] for $\theta_1,\ldots,\theta_r$ constant with (33.8), let $G = \sum_{i=1}^r \theta_i B^i$ and let (33.48) be a list associated to an admissible decomposition of $T = \sum_{i=1}^r \theta_i (B^i - G)\otimes(B^i - G)$ satisfying (33.47). If for each index k in the list (33.48), π^k is an even probability measure on $\mathbb{S}^{N-1}$, then there exists a mixture using proportion θ_i of material M^i, $i = 1,\ldots,r$, and having its effective tensor satisfying (33.10) with Ψ given by*

$$\begin{aligned}\Psi &= \textstyle\sum_{i=1}^s \xi_j (Q^j - G)\,R^j (Q^j - G) + \sum_{k\in K_1} \xi_1 \eta_k N^k R^k N^k\\ &\quad + \ldots + \textstyle\sum_{k\in K_s} \xi_s \eta_k N^k R^k N^k,\\ R^k &= \textstyle\int_{\mathbb{S}^{N-1}} \frac{e\otimes e}{(A^* e,e)}\,d\pi^k(e) \text{ for all indices } k \text{ in the list (33.48).}\end{aligned} \tag{33.49}$$

Proof: What Definition 33.5 means is that there are points at level 0, which belong to $\{B^1,\ldots,B^r\}$, then points at level 1 which are not at level 0 and are convex combinations of only points at level 0, then points at level 2 which are not at level 1 and are convex combinations of only points at level ≤ 1, and so on for a finite number of levels.

The proof is by induction on the level. If G is at level 1, then each Q^j is at level 0, i.e., it belongs to $\{B^1,\ldots,B^r\}$ and $T^j = 0$, so that the result is that of Lemma 33.4.[20]

One assumes that the result is proven for points up to level $\ell \ge 1$, and that G is at level $\ell + 1$. By definition of level, for $j = 1,\ldots,s$, Q^j is at level $\le \ell$,

[18] If $0 \ne \eta_k N^k \otimes N^k = \zeta\, P\otimes P$, there exists $\lambda \ne 0$ with $P = \lambda\, N^k$ and $\zeta = \frac{\eta_k}{\lambda^2}$.

[19] For $r = 2$, $G = \theta\, B^1 + (1-\theta)\, B^2$, one has $\Psi = \theta\,(1-\theta)\,(B^2 - B^1)\,R\,(B^2 - B^1)$ with $R = \int_{\mathbb{S}^{N-1}} \frac{e\otimes e}{(A^* e,e)}\,d\pi(e)$ for an even probability π on $\mathbb{S}^{N-1}$, either by Lemma 33.1 or by Lemma 33.2, which shows that it is a characterization.

[20] It is not necessary that $B^1,\ldots,B^r$ be distinct, so if $B = B^i = B^j$ with $i \ne j$, the two terms $\theta_i (B^i - G)\,R^i (B^i - G)$ and $\theta_j (B^j - G)\,R^j (B^j - G)$ may be lumped together as $(\theta_i + \theta_j)\,(B - G)\,R\,(B - G)$ with $R = \frac{\theta_i R^i + \theta_j R^j}{\theta_i + \theta_j}$ and R comes from using the (even) probability $\pi = \frac{\theta_i \pi^1 + \theta_j \pi^j}{\theta_i + \theta_j}$ on $\mathbb{S}^{N-1}$.

and the induction argument applies to Q^j with its admissible decomposition of T^j. If the admissible decomposition of T^j, $j = 1, \ldots, s$, corresponds to

$$Q^j = \sum_{i=1}^{r} \zeta_{j,i} B^i, j = 1, \ldots, s, \tag{33.50}$$

then[21]

$$\theta_i = \sum_{j=1}^{s} \xi_j \zeta_{j,i}, i = 1, \ldots, r. \tag{33.51}$$

Applying the induction argument to Q^j with its admissible decomposition of T^j, there exists a mixture using proportion $\zeta_{j,i}$ of material M^i, $i = 1, \ldots, r$, and having its effective tensor P^j such that

$$\begin{aligned} P^j &= A^* + \gamma \widetilde{B}^j + \gamma^2 \widetilde{C}^j + O(\gamma^3), \\ \widetilde{B}^j &= \textstyle\sum_{i=1}^{r} \zeta_{j,i} B^i, \\ \widetilde{C}^j &= \textstyle\sum_{i=1}^{r} \zeta_{j,i} C^i - \Psi^j, \\ \Psi^j &= \textstyle\sum_{k \in K_j} \eta_k N^k R^k N^k. \end{aligned} \tag{33.52}$$

By Lemma 33.4, there exists a mixture using proportion ξ_j of material P^j, $j = 1, \ldots, s$, and having its effective tensor M^{eff} such that

$$\begin{aligned} &M^{\mathit{eff}} = A^* + \gamma \textstyle\sum_{j=1}^{s} \xi_j \widetilde{B}^j + \gamma^2 \Big(\textstyle\sum_{j=1}^{s} \xi_j \widetilde{C}^j - \Psi_P \Big) + O(\gamma^3), \\ &\Psi_P = \textstyle\sum_{j=1}^{s} \xi_j (Q^j - G) R^j (Q^j - G). \end{aligned} \tag{33.53}$$

Using (33.51)–(33.52), one has

$$\begin{aligned} &\textstyle\sum_{j=1}^{s} \xi_j \widetilde{B}^j = \textstyle\sum_{i=1}^{r} \theta_i B^i, \\ &\textstyle\sum_{j=1}^{s} \xi_j \widetilde{C}^j - \Psi_P = \textstyle\sum_{i=1}^{r} \theta_i C^i - \Psi, \\ &\quad \text{with } \Psi \text{ given by (33.49)}, \end{aligned} \tag{33.54}$$

showing that the result holds at level $\ell + 1$. □

In my work with François MURAT, we could not obtain a characterization, in part since we were not able to analyze loops in a simple way. I only thought of using triangular loops while listening to a conference in Pont à Mousson, France (in the summer of 1993, I believe), but Graeme MILTON and Enzo NESI may have used them before in this context. I first used loops ten years earlier, in the spring of 1983 while I was visiting MSRI in Berkeley, CA, but they were rectangular and adapted to a question about separate convexity,[22]

---

[21] If the B^i are not all distinct, (33.51) is the definition of the coefficients θ_i.

[22] It was in the summer of 1997, at a conference in Oberwolfach, Germany, that I first heard about an earlier rectangular construction for separate convexity, by AUMANN

and I made more general constructions in 1983, which I only described at a conference in Minneapolis, MN, in the fall of 1990 [108], but triangular loops are useless in that context.

For obtaining information on *H-measures* when one uses a Young measure which is a finite combination of Dirac masses at points $F^1, \ldots, F^r \in \mathbb{R}^p$, i.e., $\nu = \sum_{i=1}^r \theta_i \delta_{F^i}$ with $0 \leq \theta_i \leq 1$ for $i = 1, \ldots, r$, and $\sum_{i=1}^r \theta_i = 1$, one may interpret the preceding results for Ψ by using the small-amplitude homogenization Theorem 29.1 (as we did initially), but it is better to follow the proofs of the preceding results and to adapt them to the construction of *H-measures* for simple Young measures.

Lemma 33.7. *Let $\alpha_1, \ldots, \alpha_r, \beta_1, \ldots, \beta_r \in [0,1]$ with $\sum_{i=1}^r \alpha_i = \sum_{i=1}^r \beta_i = 1$. Let $U^{\alpha,n} = \sum_{i=1}^r \chi_i^{\alpha,n} F^i$, where for $i = 1, \ldots, r$, $\chi_i^{\alpha,n}$ are characteristic functions of disjoint Lebesgue-measurable sets, with $\chi_i^{\alpha,n} \rightharpoonup \alpha_i$ in $L^\infty(\mathbb{R}^N)$ weak $\star$ as $n \to \infty$ for $i = 1, \ldots, r$; writing $U^{\alpha,\infty} = \sum_{i=1}^r \alpha_i F^i$, assume moreover that $U^{\alpha,n} - U^{\alpha,\infty}$ defines the H-measure $dx \otimes \mu_\alpha$. Let $U^{\beta,n} = \sum_{i=1}^r \chi_i^{\beta,n} F^i$ be a similar sequence, with $\chi_i^{\beta,n} \rightharpoonup \beta_i$ in $L^\infty(\mathbb{R}^N)$ weak $\star$ as $n \to \infty$ for $i = 1, \ldots, r$; writing $U^{\beta,\infty} = \sum_{i=1}^r \beta_i F^i$, assume moreover that $U^{\beta,n} - U^{\beta,\infty}$ defines the H-measure $dx \otimes \mu_\beta$. Then, for $\theta \in (0,1)$ and $e \in \mathbb{S}^{N-1}$, there exists a sequence $U^n = \sum_{i=1}^r \chi_i^n F^i$, where for $i = 1, \ldots, r$, χ_i^n are characteristic functions of disjoint Lebesgue-measurable sets, and $\chi_i^n \rightharpoonup \theta_i = \theta\,\alpha_i + (1-\theta)\,\beta_i$ in $L^\infty(\mathbb{R}^N)$ weak $\star$ as $n \to \infty$ for $i = 1, \ldots, r$; writing $U^\infty = \sum_{i=1}^r \theta_i F^i$, $U^n - U^\infty$ defines the H-measure μ given by*[23]

$$\begin{aligned}\mu = {} & \theta\, dx \otimes \mu_\alpha + (1-\theta)\, dx \otimes \mu_\beta \\ & + \theta(1-\theta)\, dx \otimes \Big((U^{\beta,\infty} - U^{\alpha,\infty}) \otimes (U^{\beta,\infty} - U^{\alpha,\infty}) \frac{\delta_e + \delta_{-e}}{2} \Big).\end{aligned} \tag{33.55}$$

Proof. For a positive integer m, let

$$\begin{aligned} X_\alpha^m &= \big\{ x \in \mathbb{R}^N \mid \tfrac{\ell}{m} \leq (x,e) < \tfrac{\ell+\theta}{m} \text{ for an integer } \ell \in \mathbb{Z} \big\}, \\ X_\beta^m &= \big\{ x \in \mathbb{R}^N \mid \tfrac{\ell+\theta}{m} \leq (x,e) < \tfrac{\ell+1}{m} \text{ for an integer } \ell \in \mathbb{Z} \big\}, \end{aligned} \tag{33.56}$$

and define $V^{m,n}$ by

$$V^{m,n}(x) = \begin{cases} U^{n,\alpha}(x) \text{ if } x \in X_\alpha^m \\ U^{n,\beta}(x) \text{ if } x \in X_\beta^m \end{cases}, \tag{33.57}$$

---

and HART. For a few years after that, my name was not mentioned anymore for my construction, but then some people started calling it "Tartar's squares," although I used rectangles! Maybe it was an answer to my political opponents who systematically attribute all my ideas to others, but why not call the method "Aumann–Hart/Tartar rectangles"?

[23] Of course, U^n defines the Young measure $\sum_{i=1}^r \theta_i \delta_{F^i}$ as $n \to \infty$.

so that

$$
\begin{aligned}
& V^{m,n} \rightharpoonup V^{m,\infty} = \chi_{X^m_\alpha} U^{\alpha,\infty} + \chi_{X^m_\beta} U^{\beta,\infty} \text{ as } n \to \infty \\
& \quad \text{in } L^\infty(\mathbb{R}^N;\mathbb{R}^p) \text{ weak } \star, \\
& \chi_{X^m_\alpha} U^{\alpha,\infty} + \chi_{X^m_\beta} U^{\beta,\infty} \rightharpoonup \theta \textstyle\sum_{i=1}^r \alpha_i F^i + (1-\theta) \sum_{i=1}^r \beta_i F^i \\
& \quad = U^\infty = \textstyle\sum_{i=1}^r \theta_i F^i \text{ as } m \to \infty \text{ in } L^\infty(\mathbb{R}^N;\mathbb{R}^p) \text{ weak } \star.
\end{aligned}
\tag{33.58}
$$

For $\varphi \in C_c(\mathbb{R}^N)$, $\psi \in C(\mathbb{S}^{N-1})$ (extended to $\mathbb{R}^N\setminus\{0\}$ as a homogeneous function of degree 0), for $j,k \in \{1,\ldots,r\}$, then by definition of the *H-measures* $dx \otimes \mu_\alpha$ and $dx \otimes \mu_\beta$, and since $\chi_{X^m_\alpha}\chi_{X^m_\beta} = 0$ a.e. in $\mathbb{R}^N$, one has

$$
\begin{aligned}
& \textstyle\int_{\mathbb{R}^N} \mathcal{F}(\varphi\,\chi_{X^m_\alpha}(V_j^{m,n} - V_j^{m,\infty}))\,\overline{\mathcal{F}(\varphi\,\chi_{X^m_\alpha}(V_k^{m,n} - V_k^{m,\infty}))}\,\psi\big(\tfrac{\xi}{|\xi|}\big)\,d\xi \\
& \quad \to \langle dx \otimes \mu_\alpha^{j,k}, |\varphi|^2\,\chi_{X^m_\alpha} \otimes \psi\rangle \text{ as } n \to \infty, \\
& \textstyle\int_{\mathbb{R}^N} \mathcal{F}(\varphi\,\chi_{X^m_\alpha}(V_j^{m,n} - V_j^{m,\infty}))\,\overline{\mathcal{F}(\varphi\,\chi_{X^m_\beta}(V_k^{m,n} - V_k^{m,\infty}))}\,\psi\big(\tfrac{\xi}{|\xi|}\big)\,d\xi \\
& \quad \to 0 \text{ as } n \to \infty, \\
& \textstyle\int_{\mathbb{R}^N} \mathcal{F}(\varphi\,\chi_{X^m_\beta}(V_j^{m,n} - V_j^{m,\infty}))\,\overline{\mathcal{F}(\varphi\,\chi_{X^m_\alpha}(V_k^{m,n} - V_k^{m,\infty}))}\,\psi\big(\tfrac{\xi}{|\xi|}\big)\,d\xi \\
& \quad \to 0 \text{ as } n \to \infty, \\
& \textstyle\int_{\mathbb{R}^N} \mathcal{F}(\varphi\,\chi_{X^m_\beta}(V_j^{m,n} - V_j^{m,\infty}))\,\overline{\mathcal{F}(\varphi\,\chi_{X^m_\beta}(V_k^{m,n} - V_k^{m,\infty}))}\,\psi\big(\tfrac{\xi}{|\xi|}\big)\,d\xi \\
& \quad \to \langle dx \otimes \mu_\beta^{j,k}, |\varphi|^2\,\chi_{X^m_\beta} \otimes \psi\rangle \text{ as } n \to \infty,
\end{aligned}
\tag{33.59}
$$

which implies

$$
\begin{aligned}
& \textstyle\lim_{n\to\infty} \int_{\mathbb{R}^N} \mathcal{F}(\varphi\,V_j^{m,n})\,\overline{\mathcal{F}(\varphi\,V_k^{m,n})}\,\psi\big(\tfrac{\xi}{|\xi|}\big)\,d\xi \\
= \; & \textstyle\int_{\mathbb{R}^N} \mathcal{F}(\varphi\,V_j^{m,\infty})\,\overline{\mathcal{F}(\varphi\,V_k^{m,\infty})}\,\psi\big(\tfrac{\xi}{|\xi|}\big)\,d\xi \\
+ \; & \langle dx \otimes \mu_\alpha^{j,k}, |\varphi|^2\,\chi_{X^m_\alpha} \otimes \psi\rangle + \langle dx \otimes \mu_\beta^{j,k}, |\varphi|^2\,\chi_{X^m_\beta} \otimes \psi\rangle.
\end{aligned}
\tag{33.60}
$$

Since

$$
\begin{aligned}
& V^{m,\infty} = \chi_{X^m_\alpha}(U^{\alpha,\infty} - U^{\beta,\infty}) + U^{\beta,\infty}, \\
& \chi_{X^m_\alpha} \rightharpoonup \theta \text{ in } L^\infty(\mathbb{R}) \text{ weak } \star,
\end{aligned}
\tag{33.61}
$$

and since $\chi_{X^m_\alpha}$ only depends upon (x,e)

$$
\begin{aligned}
& \chi_{X^m_\alpha} - \theta \text{ defines the H-measure } \theta\,(1-\theta)\,dx \otimes \tfrac{\delta_e + \delta_{-e}}{2}, \\
& V^{m,\infty} - U^\infty \text{ defines the H-measure} \\
& \quad \theta\,(1-\theta)\,dx \otimes \big((U^{\alpha,\infty} - U^{\beta,\infty}) \otimes (U^{\alpha,\infty} - U^{\beta,\infty})\tfrac{\delta_e + \delta_{-e}}{2}\big),
\end{aligned}
\tag{33.62}
$$

$$
\begin{aligned}
& \textstyle\lim_{m\to\infty}\big[\lim_{n\to\infty} \int_{\mathbb{R}^N} \mathcal{F}(\varphi\,V_j^{m,n})\,\overline{\mathcal{F}(\varphi\,V_k^{m,n})}\,\psi\big(\tfrac{\xi}{|\xi|}\big)\,d\xi\big] \\
& = \textstyle\int_{\mathbb{R}^N} \mathcal{F}(\varphi\,U_j^\infty)\,\overline{\mathcal{F}(\varphi\,U_k^\infty)}\,\psi\big(\tfrac{\xi}{|\xi|}\big)\,d\xi \\
& \quad + \; \langle \theta\,(1-\theta)\,(U^{\alpha,\infty} - U^{\beta,\infty})_j\overline{(U^{\alpha,\infty} - U^{\beta,\infty})_k}\,dx \otimes \tfrac{\delta_e + \delta_{-e}}{2}, |\varphi|^2 \otimes \psi\rangle \\
& \quad + \; \langle dx \otimes \mu_\alpha^{j,k}, \theta\,|\varphi|^2 \otimes \psi\rangle + \langle dx \otimes \mu_\beta^{j,k}, (1-\theta)\,|\varphi|^2 \otimes \psi\rangle.
\end{aligned}
\tag{33.63}
$$

Choosing countable dense families of (φ, ψ), one extracts a subsequence $U^p = V^{f(p),g(p)}$ such that all $\lim_{p\to\infty}$ are equal to $\lim_{m\to\infty}[\lim_{n\to\infty}]$, and this formula corresponds to U^p defining the H-measure (33.55). □

Corollary 33.8. *For an even probability π on $\mathbb{S}^{N-1}$, there exists a sequence $U^n = \sum_{i=1}^r \chi_i^n F^i$ satisfying all the constraints in Lemma 33.8, such that $U^n - U^\infty$ defines the H-measure μ given by*

$$\begin{aligned}\mu &= \theta\, dx \otimes \mu_\alpha + (1-\theta)\, dx \otimes \mu_\beta \\ &+ \quad \theta\,(1-\theta)\, dx \otimes \big((U^{\beta,\infty} - U^{\alpha,\infty}) \otimes (U^{\beta,\infty} - U^{\alpha,\infty})\,\pi\big).\end{aligned} \tag{33.64}$$

Proof: If $U^{\beta,\infty} = U^{\alpha,\infty}$, then μ is a convex combination of μ_α and μ_β by (33.55). This applies to all the *H-measures* corresponding to various points $e \in \mathbb{S}^{N-1}$, since all the limits are $U^\infty = \sum_{i=1}^r \theta_i F^i$; by induction one obtains any finite even probability π, and by a limiting argument all the *H-measures* given by (33.64). □

Lemma 33.9. *For $r \geq 2$, let $\pi^{i,j}$, $i,j = 1,\ldots,r$, be (even) probability measures on $\mathbb{S}^{N-1}$ with $\pi^{j,i} = \pi^{i,j}$, for $i,j = 1,\ldots,r$. For $\theta_1,\ldots,\theta_r$ constant with (33.8), there exists a sequence defining the Young measure $\nu = \sum_{i=1}^r \theta_i \delta_{F^i}$ and the H-measure μ given by*

$$\mu = \frac{1}{2} \sum_{i,j=1}^r \theta_i \theta_j dx \otimes \big((F^i - F^j) \otimes (F^i - F^j)\,\pi^{i,j}\big). \tag{33.65}$$

Proof: Like in the proof of Lemma 33.3, one uses an induction on r, the case $r = 2$ being obtained by Corollary 33.8 (with $\alpha_1 = \beta_2 = 1$, $\alpha_2 = \beta_1 = 0$, and $\mu_\alpha = \mu_\beta = 0$). One applies the induction hypothesis for computing μ_α, with $\alpha_1 = \theta_1 + \theta_r$, $\alpha_i = \theta_i$ for $i = 2,\ldots,r-1$, and $\alpha_r = 0$, and for computing μ_β, with $\beta_1 = 0$, $\beta_i = \theta_i$ for $i = 2,\ldots,r-1$, and $\beta_r = \theta_1 + \theta_r$. Then, one applies Corollary 33.8 to μ_α and μ_β. □

Lemma 33.10. *For $r \geq 3$, let π^i, $i = 1,\ldots,r$, be (even) probability measures on $\mathbb{S}^{N-1}$. For $\theta_1,\ldots,\theta_r$ constant with (33.8), and $F_0 = \sum_{i=1}^r \theta_i F^i$, there exists a sequence defining the Young measure $\nu_0 = \sum_{i=1}^r \theta_i \delta_{F^i}$ and the H-measure μ given by*

$$\mu = \sum_{i=1}^r \theta_i dx \otimes \big((F^i - F_0) \otimes (F^i - F_0)\,\pi^i\big). \tag{33.66}$$

Proof. Like in the proof of Lemma 33.4, one starts with an initial H-measure μ_0 associated to the Young measure ν_0. For $\eta_1 \in (0,1)$, one uses Corollary 33.8 to construct an H-measure

$$\begin{aligned}&\mu_1 = (1-\eta_1)\,\mu_0 - \eta_1(1-\eta_1)dx \otimes \big((F^1 - F_0)\otimes(F^1-F_0)\,\pi^1\big),\\ &\quad\text{associated to the Young measure } \nu_1 = (1-\eta_1)\,\nu_0 + \eta_1\delta_{F^1},\\ &\quad\quad\text{whose centre of mass is } F_1,\end{aligned} \tag{33.67}$$

and then, for $j = 2,\ldots,r$, and $\eta_2,\ldots,\eta_r \in (0,1)$, one constructs *H-measures*

$$\begin{aligned}&\mu_j = (1-\eta_j)\,\mu_{j-1} - \eta_j(1-\eta_j)dx \otimes \big((F^j - F_{j-1})\otimes(F^j-F_{j-1})\,\pi^j\big),\\ &\quad\text{associated to the Young measure } \nu_j = (1-\eta_j)\,\nu_{j-1} + \eta_j\delta_{F^j},\\ &\quad\quad\text{whose centre of mass is } F_j, j = 2,\ldots,r.\end{aligned} \tag{33.68}$$

Choosing $\eta_1 \in (0,1)$ and then defining $\eta_2,\ldots,\eta_r$ by

$$\eta_j = \frac{\eta_{j-1}\theta_j}{\theta_{j-1} + \eta_{j-1}\theta_j}, j = 2,\ldots,r, \tag{33.69}$$

ensures that $0 < \eta_j < 1$ for $2 = 1,\ldots,r$, and that $\nu_r = \nu_0$; moreover

$$\mu_r = \lambda\,\mu_0 + (1-\lambda)\,\widetilde{\mu}, \text{ with } \lambda = (1-\eta_1)\cdots(1-\eta_r) < 1, \tag{33.70}$$

and $\widetilde{\mu}$ only depending upon $\theta_1,\ldots,\theta_r,F^1,\ldots,F^r,\pi^1,\ldots,\pi^r$, and η_1. Starting from an H-measure μ_0 associated to the Young measure ν_0, (33.70) gives another H-measure μ_r associated to ν_0, and iterating the process gives a sequence converging to $\widetilde{\mu}$, which is then also associated to ν_0. Choosing $\eta_1 = \varepsilon\,\theta_1$ with ε small gives $\eta_j = \varepsilon\,\theta_j + O(\varepsilon^2)$, for $j = 2,\ldots,r$, $\lambda = 1 - \varepsilon + O(\varepsilon^2)$ and $\widetilde{\mu} = \sum_{i=1}^r \theta_i dx \otimes \big((F^i - F_0)\otimes(F^i - F_0)\,\pi^i\big) + O(\varepsilon)$, which when ε tends to 0 gives (33.39). □

Lemma 33.11. *For $r \ge 3$, for $\theta_1,\ldots,\theta_r$ constant with (33.8), and $F_0 = \sum_{i=1}^r \theta_i F^i$, one considers an admissible decomposition of the matrix of inertia at F_0 like in Definition 33.5 and Lemma 33.6,*

$$\begin{aligned}&F_0 = \textstyle\sum_{j=1}^s \xi_j H^j,\\ &J = \textstyle\sum_{i=1}^r \theta_i(F^i - F_0)\otimes(F^i - F_0),\\ &\quad = \textstyle\sum_{j=1}^s \xi_j(H^j - F_0)\otimes(H^j - F_0) + \sum_{j=1}^s \xi_j U^j,\\ &\quad\quad 0 \le \xi_j \le 1 \textit{ for } j = 1,\ldots,s, \textstyle\sum_{j=1}^s \xi_j = 1,\end{aligned} \tag{33.71}$$

and for $j = 1,\ldots,s$, H^j belongs to the convex hull of $\{F^1,\ldots,F^r\}$, U^j is a matrix of inertia corresponding to a finite decomposition of H^j, created in a similar way, with no loops and the ultimate points in the decomposition belonging to $\{F^1,\ldots,F^r\}$, and an admissible decomposition of U^j is a list $\{\eta_k S^k \otimes S^k \mid k \in K_j\}$, so that an admissible decomposition of J is a list

$$\begin{aligned}&\{\xi_j(H^j - F_0)\otimes(H^j - F_0) \mid j = 1,\ldots,s\} \textstyle\bigcup \{\xi_1\eta_k S^k \otimes S^k \mid k \in K_1\}\\ &\quad \textstyle\bigcup \cdots \bigcup \{\xi_s\eta_k S^k \otimes S^k \mid k \in K_s\}.\end{aligned} \tag{33.72}$$

If for each index k in the list (33.72), π^k is an even probability measure on $\mathbb{S}^{N-1}$, then there exists a sequence $V^n = \sum_{i=1}^r \chi_{i,n} F^i$ with $\chi_{i,n}$ being characteristic functions of disjoint Lebesgue-measurable sets for $i = 1, \ldots, r$ (at n fixed), such that $\chi_{i,n} \rightharpoonup \theta_i$ in $L^\infty(\mathbb{R}^N)$ weak $\star$ for $i = 1, \ldots, r$ (as $n \to \infty$), so that V^n has the Young measure $\nu_0 = \sum_{i=1}^r \theta_i \delta_{F^i}$, and

$$\begin{aligned}&V^n - F_0 \text{ defines the H-measure } \mu, \text{ with}\\&\mu = \textstyle\sum_{j=1}^s \xi_j\, dx \otimes \big((H^j - F_0) \otimes (H^j - F_0)\, \pi^j\big)\\&\quad + \textstyle\sum_{j=1}^s \sum_{k \in K_j} \xi_j \eta_k\, dx \otimes \big((S^k \otimes S^k)\, \pi^k\big).\end{aligned} \tag{33.73}$$

Proof. Like in the proof of Lemma 33.6, the proof is by induction on the level: one assumes that the result is already proven at the level of the points $H^1, \ldots, H^s$, so that for $j = 1, \ldots, s$, there is a sequence $U^{n,j}$ using the correct proportions of $F^1, \ldots, F^r$ for H^j and such that

$$\begin{aligned}&U^{n,j} - H^j \text{ defines the H-measure}\\&\mu_{H^j} = \textstyle\sum_{k \in K_j} \eta_k\, dx \otimes \big((S^k \otimes S^k)\, \pi^k\big),\end{aligned} \tag{33.74}$$

and one wants to prove the result for F_0, i.e., show that (33.73) holds.

One needs a variant of Lemma 33.10, where each F^j was associated to the H-measure 0: now, each H^j is associated to the H-measure μ_{H^j}. Starting with any H-measure μ_0 associated to the Young measure $\nu_0^* = \sum_{j=1}^s \xi_j \delta_{H^j}$, using $\zeta_1, \ldots, \zeta_s \in (0,1)$, and writing $H_0 = F_0$, (33.67)–(33.68) is replaced by

$$\begin{aligned}&\mu_j = (1 - \zeta_j)\, \mu_{j-1} + \zeta_j \mu_{H^j}\\&\quad -\zeta_j (1 - \zeta_j) dx \otimes \big((H^j - H_{j-1}) \otimes (H^j - H_{j-1})\, \pi^j\big),\\&\text{associated to the Young measure } \nu_j^* = (1 - \zeta_j)\, \nu_{j-1}^* + \zeta_j \delta_{H^j},\\&\text{whose centre of mass is } H_j, j = 1, \ldots, s.\end{aligned} \tag{33.75}$$

After choosing $\zeta_1 \in (0,1)$, one defines $\zeta_2, \ldots, \zeta_s$ by

$$\zeta_j = \frac{\zeta_{j-1} \xi_j}{\xi_{j-1} + \zeta_{j-1} \xi_j}, j = 2, \ldots, s, \tag{33.76}$$

ensuring that $0 < \zeta_j < 1$ for $2 = 1, \ldots, s$, and that $\nu_s^* = \nu_0^*$; moreover

$$\mu_s = \lambda\, \mu_0 + (1 - \lambda)\, \widetilde{\mu}, \text{ with } \lambda = (1 - \zeta_1) \cdots (1 - \zeta_s) < 1, \tag{33.77}$$

and $\widetilde{\mu}$ depends upon $\xi_1, \ldots, \xi_s, H^1, \ldots, H^s, \mu_{H^1}, \ldots, \mu_{H^s}, \pi^1, \ldots, \pi^s$, and η_1; starting from an H-measure μ_0 associated to the Young measure ν_0^*, (33.77) gives another H-measure μ_s associated to ν_0^*, and iterating the process gives a sequence converging to $\widetilde{\mu}$, which is then also associated to ν_0^*. Choosing $\zeta_1 = \varepsilon\, \xi_1$ with ε small gives $\zeta_j = \varepsilon\, \xi_j + O(\varepsilon^2)$, for $j = 2, \ldots, r$, $\lambda = 1 - \varepsilon + O(\varepsilon^2)$

and $\widetilde{\mu} = \sum_{j=1}^{s} \xi_j \mu_{H^j} + \sum_{j=1}^{s} \xi_j dx \otimes \big((H^j - F_0) \otimes (H^j - F_0)\, \pi^j\big) + O(\varepsilon)$, which when ε tends to 0 shows that

$$\begin{gathered}\mu^* = \textstyle\sum_{j=1}^{s} \xi_j \mu_{H^j} + \sum_{j=1}^{s} \xi_j dx \otimes \big((H^j - F_0) \otimes (H^j - F_0)\, \pi^j\big) \\ \text{is an H-measure associated to the Young measure } \nu_0^*. \end{gathered} \tag{33.78}$$

Using (33.74) for $\mu_{H^1}, \ldots, \mu_{H^s}$ in (33.78) gives (33.73). □

The preceding results created sequences on $\mathbb{R}^N$, defining Young measures independent of x and finite combinations of Dirac masses, and *H-measures* of the form $dx \otimes \pi$ where π are real symmetric even matrices of Radon measures on $\mathbb{S}^{N-1}$. By localization, and a limiting process, one can deal with more general pairs of a Young measure and an associated H-measure, but I shall leave this exercise to the reader, and only show how the preceding construction permits one to create sequences satisfying some differential information.

Lemma 33.12. *For $\theta_1, \ldots, \theta_r$ constant with (33.8), assume that $A_{h,j,k}$ are real constants for $h = 1, \ldots, q$, $j = 1, \ldots, r$, $k = 1, \ldots, N$, and that for $i = 1, \ldots, r$, $F^i - F_0$ belongs to $\Lambda = \{\lambda \in \mathbb{R}^p \mid$ there exists $\xi \in \mathbb{S}^{N-1}$ with $\sum_{j=1}^{p} \sum_{k=1}^{N} A_{h,j,k} \lambda_j \xi_k = 0$, for $h = 1, \ldots, q\}$. Then, there exists a sequence $U^n = \sum_{i=1}^{r} \chi_i^n F^i$, where the functions χ_i^n are characteristic functions of disjoint Lebesgue-measurable sets with $\chi_i^n \rightharpoonup \theta_i$ in $L^\infty(\mathbb{R}^N)$ weak $\star$ for $i = 1, \ldots, r$, and such that $\sum_{j=1}^{p} \sum_{k=1}^{N} A_{h,j,k} \frac{\partial U_j^n}{\partial x_k}$ belongs to a compact of $H^{-1}_{loc}(\Omega)$ for $h = 1, \ldots, q$.*

Proof. By Lemma 28.18, if μ is the H-measure of the sequence U^n, the differential condition is equivalent to $\sum_{j=1}^{p} \sum_{k=1}^{N} A_{h,j,k} \xi_k \mu^{j,\ell} = 0$ for $\ell = 1, \ldots, p$. One then constructs a sequence according to Lemma 33.10, choosing π^i supported at $\pm\xi \in \mathbb{S}^{N-1}$ with ξ associated to $F^i - F_0$ in Λ, for $i = 1, \ldots, r$. □

Additional footnotes: AUMANN,[24] HART,[25] SCHELLING.[26]

[24] Robert John AUMANN, German-born mathematician, born in 1930. He received the Sveriges Riksbank Prize in Economic Sciences in Memory of Alfred Nobel in 2005, jointly with Thomas C. SCHELLING, for having enhanced our understanding of conflict and cooperation through game-theory analysis. He works at The Hebrew University, Jerusalem, Israel.

[25] Sergiu HART, Romanian-born mathematician, born in 1949. He works at The Hebrew University, Jerusalem, Israel.

[26] Thomas Crombie SCHELLING, American economist, born in 1921. He received the Sveriges Riksbank Prize in Economic Sciences in Memory of Alfred Nobel in 2005, jointly with Robert J. AUMANN, for having enhanced our understanding of conflict and cooperation through game-theory analysis. He worked at Yale University, New Haven, CT, at Harvard University, Cambridge, MA, and at UMD (University of Maryland) College Park, MD.

Chapter 34
Conclusion

My advisor, Jacques-Louis LIONS, once told me that when the plan of a book is made, then the book is almost written. He gave me the impression that he could write fifty pages in a weekend, give them to a secretary to type, and since he asked me to proofread the typed notes of his book [52] while he was teaching the course, I assume that he rarely read again himself what he had already written.

In contrast, I started my career not knowing how to write, and I thought that it was not important, since my choice was to be a researcher in mathematics (with an interest in applications). I was extremely shy, so my communication skills were rather limited, and if I could solve mathematical questions and explain my solutions at the blackboard, it was quite difficult for me to replace the chalk by a pen, and write these solutions on a blank sheet of paper. My advisor almost never insisted that I should publish something, and it created a problem, since I was not reading much and I had no way to compare what I was doing to the published articles,[1] and I interpreted his silence as meaning that the results which I showed him were not worth publishing.[2]

In the early 1970s, I gave a talk at the seminar of Jean LERAY at Collège de France, in Paris, about the second part of my thesis, whose details were never published, and I wrote a text for my talk without too much delay, and one reason may be that Jean LERAY was forty years older than me,

[1] My advisor gave me a list of journals that I should read regularly, but I never did, since I thought that research is about developing new ideas, and not about reading them in journals! Also, he added something like "even Jean-Pierre SERRE reads", which I interpreted as meaning that once short of personal ideas, even the best mathematicians read what others have done!

[2] In the early 1970s, I mostly replaced Jacques-Louis LIONS for advising one of his students, Jean-Pierre KERNEVEZ, to whom I explained how to adapt classical techniques based on the maximum principle for the problems which interested him (concerning diffusion and reactions in membranes), and my advisor told me to write an article on that, but I considered what I did as easy exercises.

L. Tartar, *The General Theory of Homogenization*, Lecture Notes of the Unione Matematica Italiana 7, DOI 10.1007/978-3-642-05195-1_34,

and since he always kept a distance from younger people,[3] it induced more respect towards him than towards other professors much older than me,[4] but another reason maybe that it was only later that I perceived some defects of the academic world, in particular the aggressions against me.

My difficulties about writing were known, so that students were asked to write the notes of my lectures, a few students of John NOHEL for the graduate course that I taught in 1974–1975 at UW, Madison, WI, and Bernard DACOROGNA for the notes of my lectures [98] in the summer of 1978 at Heriot–Watt University, Edinburgh, Scotland. Usually, I would not write unless I was put under some pressure, and for example Charles GOULAOUIC always insisted enough to make me write texts (like [99]) for the talks that I gave at the seminar that he organized with Laurent SCHWARTZ, and then also with Yves MEYER, at École Polytechnique in Palaiseau, France, but generally, I only wrote for the proceedings of the conferences where I gave talks.[5]

My difficulties increased in the late 1970s when I realized that some of those to whom I explained a mathematical result forgot to attribute the idea to me, and when I found the strength to ask some friends about that, they seemed not to understand why I was bothered. I usually answered all questions, since I found it to be my duty (learned from the parable of talents), even questions from those who already showed that they were dishonest, so that I had no defense against stealing, and if everyone found it normal to publish my results, I was not just bothered, I was deeply hurt!

Then, from the fall of 1979 to the summer of 1982, I did my duty of French citizen to oppose the invention of results of votes sent by my university to the minister in charge of the French universities, and when I wrote in 1986 a text (still unpublished) about these events,[6] "Moscou sur Yvette, souvenirs d'un mathématicien exclu" (Moscou upon Yvette, recollections of

[3] In November 1986, Jean LERAY did not leave immediately after the seminar at Collège de France, which was unusual, and he came to the back of the room where I sat with a few others, and he talked about himself: it was his 80th birthday. He said "do you know that I was born on the same day as the October revolution?" (which happened in November because of the difference between the Gregorian calendar and the Julian calendar, still used in Russia in 1917), and after a pause he added "it is not exact: I was already 11 years old when it happened!".

[4] My relations were more friendly with my advisor, Jacques-Louis LIONS, who was eighteen years older than me, and with his own advisor, Laurent SCHWARTZ, who was thirty-two years older than me.

[5] This helped those who wanted to steal my results, since most others would only read journals, and not know that they were reading about my ideas.

[6] After learning about word processing, the writing process was enormously simplified for me. Although I read the book [45] of KNUTH about TeX in the summer of 1986, it still took me a few years before I started writing in (bad) TeX.

an excluded mathematician),[7] I used many of the letters that I wrote to Laurent SCHWARTZ in order to make him react to the inadmissible acts of his political friends, but I could not foresee that later the friends of my political opponents would organize on a larger scale the attribution of my ideas to others,[8] and even use slander and threat against me, and this delayed for many years describing my ideas in writing.

I am unable to make a precise plan of a course or of a book, and my first three lecture notes [116, 117, 119] correspond to graduate courses that I taught at CMU (*Carnegie Mellon University*) in 1999, 2000, and 2001; at the beginning of each of these courses, I only had a vague idea about what I would teach, and it evolved during the semester, but I had no trouble writing after each lecture the few pages describing what I had just covered in class.

For this book, I chose to follow a loose chronological order concerning my ideas, since I thought that it was sufficient material for a book; I excluded from the project most of my ideas concerning the compensation effects in partial differential equations, for which I refer to [115], since I wrote the basic results in this survey article, but also since I realized that the book would already be thick enough without the description of this important aspect; for the same reason, I did not look for what others wrote, and I only mentioned what I already heard about in conferences or in discussions.

While writing, I often realized that what I already wrote for one chapter was too long, so that it was better to split the chapter in two, and this happened since the beginning, when I decided to make Chap. 1 out of material which I first planned for the Preface, and to split the overview account into Chaps. 2 and 3. I found natural to split the questions of correctors into Chaps. 13 and 14, and the questions of holes into Chaps. 15 and 16, although keeping them in one chapter would not have given something very long. Then, it was natural to split the questions of nonlocal effects into Chaps. 23 and 24, and the constructions for optimal bounds into Chaps. 25 and 26, and even 27. Finally, I had not anticipated that *H-measures* would take six chapters (Chaps. 28–33) and almost one hundred pages, but I saw no natural way to split Chaps. 28, 32, or 33 into smaller parts.

From the start, I decided that I would not describe my work with François MURAT on homogenization in optimal design, since I already wrote a detailed

---

[7] Yvette is a small river, going through Gif sur Yvette, where a few laboratories of CNRS (Centre National de la Recherche Scientifique) are located, Bures sur Yvette, where IHES (Institut des Hautes Études Scientifiques) is located, and then Orsay, where one campus of Université de Paris-Sud is located. I was the only mathematician who opposed the invention of results of votes, and among my colleagues only Jacques DENY once mentioned that he agreed with me, but he looked quite afraid to make it known.

[8] It was not only due to my opposition to the invention of results of votes in Orsay: I started explaining the defects of many theories in continuum mechanics and physics a long time before understanding why those who faked escapes from the east (and even those who were allowed to leave) usually advertised fake mechanics/physics.

account [111] for a CIME/CIM course held in June 1998 in Tróia, Portugal. However, this type of question is not only important for finding interesting designs in engineering applications, but also since nature uses some intricate designs, and it suggests looking at more general questions of identifying the sets in which some effective coefficients lie, in order to understand what one observes.

Of course, nature does not minimize any criterion in doing that, since it uses hyperbolic systems and waves for moving information around, and it is only by letting the speed of light c tend to ∞ that one obtains simplified equations, which may still be conservative like the Schrödinger equation, or dissipative and of parabolic type like the heat equation. Often, the effective parameters observed belong to the boundary of the set of effective coefficients, and it is important to study which geometries for microstructures give points near the boundary of such an admissible set, but the goal should be to understand which one nature chooses, and by what evolution process it finds them, and not to make a mistake of pseudo-logic type by using a function which is minimum at the boundary of this set and pretending that nature minimizes this function:[9] the knowledge of the whole set of effective coefficients is important, even though for a given application only some boundary points may be used, since it may fail to be true for other applications.

I discussed in [114] two problems involving microstructures adapted to evacuating heat and creating some elastic isotropy, snow flakes and quasicrystals, but although my text was not accepted for publication,[10] I decided

[9] The advocates of fake mechanics seem unable to understand that even if they want to restrict their attention to linearized elasticity, the system of equations to consider is hyperbolic and conservative, and no potential energy is minimized, and I wrote [113] for pushing them to acquire a little common sense. For those who know how solutions of hyperbolic systems behave, it is easy to understand that a part of the conserved energy may use high frequencies and be "trapped" at a mesoscopic level, and it is the internal energy that the first principle of thermodynamics talks about. Since the internal energy is a sum of different modes travelling in different ways, its evolution cannot be given by a simple differential equation, so that the second principle of thermodynamics is flawed; however, one should acknowledge that it was difficult in the nineteenth century to perceive a larger class of equations, involving pseudo-differential operators, for example, since the nonlinear analogue is still not understood (Constantine DAFERMOS told me that HEAVISIDE perceived that the relation between E and D should use integral operators, a way to express the dependence in the frequency of the dielectric permittivity ε in the linear homogeneous case).

[10] I wrote [114] in French, in memory of my friend Hamid (Abdelhamid ZIANI), and since he never wanted to describe much about the racism which he encountered in the French academic world, I described as an example that which I encountered myself. I also compared some dogmas in physics with those in some religions, since similar mistakes were made, but maybe my text was more upsetting because I repeated that one should also attribute the principle of relativity to POINCARÉ (as I first read in FEYNMAN's course [28]), the Maxwell equation also to HEAVISIDE, that light is not described by the wave equation, that EINSTEIN said a few silly things, that the

not to translate my conjectures into English for explaining them in this book, which I find long enough. One of my ideas in [114] is that which I repeated many times, that physicists invented games which are not compatible with what one understands now about effective properties of mixtures, and that the geometry used by microstructures is important, but more precisely, the notion of latent heat is ill defined, and the energy released in freezing a material should depend upon which path is followed in the space of binary mixtures, made of the liquid phase and "the" solid phase. In other words, it is silly to imagine that there is only one type of solid phase, in the same way that it is silly to claim that there is only one type of poly-crystal, and another one of my ideas in [114] is that games using surface energies depending upon the normal to an interface are quite unphysical, and not so helpful for understanding what nature produces.

One should develop improved physical models using both Young measures and *H-measures*, as described in Chap. 33, and since some problems use variants of *H-measures*, a similar study should be done for such variants, but an important step is to develop a general approach using information like three-point correlations, only hinted at in Chap. 32.

Graeme MILTON once pointed out to me that scattering phenomena involve three-point correlations, and I deduced that it might be difficult to understand what happens in an experiment in spectroscopy before one develops more general mathematical objects than *H-measures*; since an important step in my research programme is to develop tools for semi-linear hyperbolic systems, a new tool seeing three-point correlations should probably be useful for that question too, but probably not sufficient.

In the late 1980s, with François MURAT, we computed corrections in γ^3 in small-amplitude homogenization in the periodic case, i.e. $A^n(x) = A^* + \gamma\, b\big(\frac{x}{\varepsilon_n}\big)\, I$, and b being Y-periodic, and for simplification we took for Y the unit cube.[11] The quantity which appears is a sum of terms $C(m,n,p)\, b_m b_n b_p$, where for $j \in \mathbb{Z}^N$ one writes b_j for the Fourier coefficient $\int_Y b(y)\, e^{-2i\,\pi\,(j,y)}\, dy$ (and one assumes that $b_0 = 0$), the sum being extended over the multi-indices m,n,p satisfying $m+n+p=0$, and $C(m,n,p)$ is a matrix, not symmetric in m,n,p, but homogeneous of degree 0 in (m,n,p), so that one may group all the terms proportional to (m,n,p). We were not able to deduce from this technical computation how to define a general mathematical object for the nonperiodic case!

For the term in γ^2, the sum uses $\frac{m\otimes m}{(A^* m,m)}\, b_m b_{-m}$, and if b is real then $b_m = \overline{b_m}$, and the sum has the form $\int_{\mathbb{S}^{N-1}} \frac{\xi\otimes\xi}{(A^*\xi,\xi)}\, d\mu(\cdot,\xi)$ for the H-measure

tilings of Roger PENROSE have nothing to do with quasi-crystals (one more example of pseudo-logic), and so on!

[11] It is the dual lattice of Y which plays a role in the extension of the Fourier transform of Laurent SCHWARTZ, and by taking Y to be the unit cube the dual lattice uses then points with integer coordinates in an orthonormal basis.

$\mu = \sum_{m \in \mathbb{Z}^N \setminus 0} |b_m|^2 \delta_{\frac{m}{|m|}}$, but it is difficult to invent a correct definition for *H-measures* after seeing this formula, although it seems to be one reason why some people think that *H-measures* were used before, a confusion common in some circles.[12] There is a difference between defining the $L^\infty(\Omega)$ weak $\star$ topology (as was probably first done by BANACH), defining the weak $\star$ convergence (as was probably first done by F. RIESZ), and replacing $f\big(x, \frac{x}{\varepsilon_n}\big)$ by $\overline{f}(x) = \int_0^1 f(x,y)\,dy$ if f has period 1 in its second variable and ε_n is small, which is probably a quite old practice. Non-mathematicians rarely understand the differences and how difficult it is to start from an example and to invent an interesting general theory, or more than one, since there may be different ways to generalize a given result, not always with one being more interesting than the others.[13] Mathematicians from my generation were usually trained to perceive such differences, and to appreciate the creativity needed for inventing generalizations, but I wonder if it is still the case for younger generations!

In this book, I avoided saying much about examples using periodicity assumptions, since they are not so helpful for finding the generalizations that I am thinking about; another defect of the periodic framework is that it orients towards models with only one characteristic length. Apart from engineering situations, where the designers may choose to repeat a few patterns, I consider periodic homogenization only as a possible first step for understanding realistic situations, and this type of hypothesis should be put to scrutiny.

In the initial work [26] of Horia ENÉ and Évariste SANCHEZ-PALENCIA who derived the Darcy law from the Stokes equation in the early 1970s, the periodicity assumption was useful since it gave a way to use asymptotic expansions. The next step was to give a mathematical proof of the result, and after Jacques-Louis LIONS asked me to check if I could construct an extension of the "pressure" in the solid part with useful bounds, I solved the problem in the late 1970s for an over-academic geometry (appendix of [83]), and Grégoire ALLAIRE later extended the proof to more general geometries. I am not aware that anyone has found a way to attack more realistic questions of flows in porous media, and after being in contact with people at IFP in the mid 1980s, I learned some facts about real porous media, and besides real

[12] If one does not tell a physicist that one has a mathematical answer to a given question which interested physicists in the past, and one asks him/her which formula he/she considers right among those published, he/she might pick one, but he/she might avoid doing that if instead one says that one knows the mathematical answer to the question; however, if one first shows the mathematical answer, the usual answer is that it is "well known" and one may be shown a similar published argument (probably containing a part defying logic), but one rarely hears a mention of all the other wrong results which are published!

[13] The result that every continuous function on $[0,1]$ attains its maximum and its minimum and takes all the values between them uses the two distinct notions of compactness and of connectedness, not one being better than the other!

rocks showing intricate geometries using many length scales, one must take into account elasticity questions since small cracks may open under adapted stress fields,[14] and one then conjectures that the effective equation is an evolution equation containing nonlinear nonlocal effects, on which very little is known.

The reason why I developed *H-measures*, which use no characteristic length, is that they are natural for the question of small-amplitude homogenization that I first looked at, which was motivated by questions about real mixtures. After that, I tried to use my *H-measures* for propagation effects in some linear hyperbolic systems, and because of linearity the effects were independent of frequency, and *H-measures* were adapted to that question too, but some smoothness hypotheses are necessary for my proof to be valid; in particular, no discontinuities are allowed, so that my results do not say anything concerning refraction effects at interfaces.

Long before I learned in high school about what an index of refraction is, I saw that sunlight is decomposed into the colours of the spectrum by a prism made of crystal, and I later learned that this type of crystal contains lead and has a higher index of refraction than glass.[15] According to some computations made by physicists, like those of my physics courses at École Polytechnique in 1965–1967, refraction effects are frequency-dependent, but the computations used a crystalline structure, probably with cubic symmetry, so that the index of refraction is that of an isotropic material, given by a positive scalar (instead of a symmetric positive definite tensor), and also a modelling of what atoms are. This suggests that real materials are not well described by the Maxwell–Heaviside equation with a pointwise constitutive relation $D = \varepsilon E$, but that one needs a nonlocal relation (already studied by HEAVISIDE in some cases), involving a class of "pseudo-differential" operators, where the dielectric permittivity tensor ε depends upon frequency.[16] One should be careful then, not to deduce that the Maxwell–Heaviside equation shows frequency-dependent effects because of refraction effects when the coefficients are discontinuous!

GTD, the geometrical theory of diffraction which Joe KELLER developed in the 1950s, shows important frequency-dependent effects, even for constant coefficients (the Maxwell–Heaviside equation, the wave equation), and it goes beyond the approximation of geometrical optics, which is the limit when the

[14] Imagining a periodic repartition of cracks can only be a simplified first assumption, of course, but those who use such hypotheses in double porosity models do not seem to think of a more realistic next step in their research.

[15] Later, I also learned that glass has no crystalline structure.

[16] It is not clear what the exact class is, except for the case of the whole space without dependence in x, where one deals with a convolution operator compatible with the causality principle, i.e. the solution at time t only uses times $s \leq t$.

frequency tends to ∞.[17] It started with T. YOUNG in 1802,[18] who revived the work of HUYGENS,[19] on the wave nature of light,[20] and there were contributions by FRESNEL,[21] FRAUNHOFER,[22] AIRY,[23] SOMMERFELD, RAMAN,[24] FOCK,[25] LEONTOVICH,[26] and Joe KELLER, who summarized many computations made before him. Besides proposing a law of reflection for rays hitting an edge or a vertex, GTD considers grazing rays and makes them follow geodesics of the boundary and lose their energy exponentially fast with a coefficient in $|k|^{1/3}$ for the wave "number" k, in the convex parts of the boundary, and that is where the frequency dependence is. The curvature of the boundary

---

[17] The limit when the frequency tends to 0 is given by the Mie approximation.

[18] Thomas YOUNG, English scientist, 1773–1829. He worked at the Royal Institution in London, England. He then practised as a physician, and he did some deciphering from the Rosetta stone, not as decisive as he thought, as the final deciphering of Egyptian hieroglyphs by CHAMPOLLION showed. The Young modulus in elasticity is named after him.

[19] Christiaan HUYGENS, Dutch mathematician, astronomer and physicist, 1629–1695. He worked in Paris, France, and in The Hague, The Netherlands. The Huygens principle is named after him.

[20] The cult of personality toward NEWTON in England must have made it difficult to defend a position opposed to his ideas about the particle nature of light!

[21] Augustin-Jean FRESNEL, French engineer, 1788–1827. He worked in Paris, France. He invented the Fresnel lens for lighthouses, with many applications today. The Fresnel number is $F = \frac{a^2}{L\,\lambda}$, with a a characteristic size of the aperture, L the distance from the aperture to the screen, and λ the wavelength of the light. The Fresnel diffraction, or near-field diffraction, corresponds to $F \geq 1$.

[22] Joseph VON FRAUNHOFER, German optician, 1787–1826. He worked in Benediktbeuern, Germany. He invented the spectroscope in 1814, and discovered 574 dark lines appearing in the solar spectrum, named Fraunhofer lines after him, although he was not the first to observe them. The Fraunhofer diffraction, or far-field diffraction, corresponds to a Fresnel number $F << 1$.

[23] George Biddell AIRY, English mathematician and astronomer, 1801–1892. He was Lucasian professor of mathematics (1826–1828), and then Plumian professor of astronomy (1828–1835) at University of Cambridge, Cambridge, England, before becoming (7th) Astronomer Royal (1835–1881). The Airy stress function in elasticity is named after him. The Airy function, named after him, is defined by $Ai(x) = \frac{1}{2\pi} \int_{-\infty}^{+\infty} e^{i\,(x\,t+t^3/3)}\,dt$; it solves $\frac{d^2 Ai}{dx^2} + x\,Ai = 0$ on $\mathbb{R}$, and decays exponentially fast as $x \to +\infty$.

[24] Sir Chandrasekhara Venkata RAMAN, Indian physicist, 1888–1970. He received the Nobel Prize in Physics in 1930 for his work on the scattering of light and for the discovery of the effect named after him. He held the Palit chair of physics at Calcutta University, Calcutta, directed the Indian Institute of Science and the Raman Institute of Research, which he established and endowed by himself, in Bangalore, India. The Raman scattering, or the Raman effect, which he found, is the inelastic scattering of a "photon", resulting in a change in frequency.

[25] Vladimir Aleksandrovich FOCK, Russian physicist, 1898–1974. He worked in Petrograd/Leningrad (now St Petersburg), and at the Lebedev Physical Institute in Moscow, Russia. The Hartree–Fock method is partly named after him.

[26] Mikhail Aleksandrovich LEONTOVICH, Russian physicist, 1903–1981.

then plays a role, as well as which boundary condition is used (Dirichlet or Neumann condition for the wave equation, perfect conductor for the Maxwell–Heaviside equation), and using GTD for computing the backscattered energy of a plane wave hitting a smooth convex body, like a sphere of radius a,[27] one finds a good agreement with the exact result up to wavelengths of the order of a, although the asymptotic expansions used for guessing GTD need large (but not infinite) frequencies; however, GTD is not good near caustics, where it (wrongly) predicts an infinite amplitude.[28] Joe KELLER once told me an interesting observation, that the effect of light creeping in the shadow in GTD is like the tunnelling effect in quantum mechanics![29]

In May 2005 in Grenoble, France, after a discussion on the subject with Michael VOGELIUS, I thought that such results probably meant the existence of a boundary layer, which I (wrongly) thought to have thickness $O(|k|^{-1/3})$. As a consequence, I thought that an important goal would be to develop a variant of *H-measures*, with a few characteristic lengths, like for understanding the other problem of the same nature that I mentioned in Chap. 32, the Stewartson triple deck structure for boundary layers in hydrodynamics, which I first heard about from Richard MEYER, and for which I learned some explanations from Jean-Pierre GUIRAUD; hopefully, such a variant would catch the approximate propagation of energy along geodesics of the boundary in the boundary layer.

While writing this "conclusion", I found on the Internet the thesis [17] of John COATS,[30] obtained in 2002 in Oxford, England, under the supervision of Jon CHAPMAN and John OCKENDON,[31,32] with very precise estimates of the sizes, from which I understood that my initial guess for the thickness of the layer is probably wrong. In the case of an incident plane wave impinging tangentially on a convex surface, he describes a Fock–Leontovich region around the point where the ray is tangent, having length $O(|k|^{-1/3})$ and height $O(|k|^{-2/3})$ and at distance $O(1)$ along the boundary an Airy layer (along which the creeping rays occur) having height $O(|k|^{-2/3})$; in the case of a cylinder he describes a shadow boundary having height $O(|k|^{-1/2})$, with transition regions to the light and to the shadow of height $O(|k|^{-1/3})$.

[27] If a plane wave moves in direction $e \in \mathbb{S}^2$, the backscattered energy is the total energy that is sent in all directions ξ with $(\xi, e) < 0$.

[28] The conjecture is that the phase jumps of $\pm\frac{\pi}{2}$ when one crosses caustics.

[29] One should observe that light does not go through the body like if one opened a tunnel with some probability, but that it turns around the body.

[30] John COATS, English applied mathematician.

[31] Stephen Jonathan CHAPMAN, British applied mathematician. He works at OCIAM (Oxford Centre for Industrial Applied Mathematics) of University of Oxford, Oxford, England.

[32] John Richard OCKENDON, English applied mathematician, born in 1940. He works at OCIAM (Oxford Centre for Industrial Applied Mathematics) of University of Oxford, Oxford, England.

I then see a definite need for developing new variants of *H-measures* with enough variability about the scales involved so that an algorithm would permit one to determine which scales are present in a given problem, and where energy is located. Hopefully, such an improvement could then be useful for questions involving rough surfaces (without probabilistic ideas), or gratings, and for correcting the simplistic models using surface energies, like surface tension for liquids or surface energy density depending upon the normal for solids, which I do not consider good physics, since they come from unrealistic minimization principles, and one needs to understand evolution problems. Why not be optimistic, and hope that these questions will also lead to a good understanding of nonlinear problems.

It is not clear to me if questions concerning the phase, like the $\pm\frac{\pi}{2}$ conjectured jump in the phase across caustics could be seen by variants of *H-measures*; actually, it looks to me like a jump condition coming out of a weak formulation for a partial differential equation.

Although Bloch waves require a periodic medium in order to be defined, it is clear that physicists use some results from the theory in situations which are not exactly periodic, because of defects for example. It should be useful to understand if there is a natural topology for measuring how far a material is from a periodic medium, so that some results from the theory of Bloch waves still apply. Could a variant of *H-measures* using some characteristic lengths be useful for such a question?

A similar problem occurs for the question of X-ray diffraction, in relation with the classical W.L. Bragg law,[33] which is deduced from a crystalline structure, but X-rays create a diffraction pattern even though the material does not have a periodic structure! When experimentalists first used X-ray diffraction through a metallic ribbon and were surprised to see a 5-fold symmetry in the diffraction pattern, incompatible with all possible crystalline structures, they coined the term quasi-crystal, but as I explain in [114] a thickness of 0.1 mm corresponds to about a million atomic distances, and it is utopian to imagine that one deals with a two-dimensional structure, or that each atom sits just above another and that the same two-dimensional structure is repeated a million times, since such a microstructure would give quite bad macroscopic elastic properties to the ribbon, which would immediately change its microstructure to adapt to the imposed exterior forces. Since the ribbon was first heated above the Curie point for changing its magnetic structure, and then cooled down quickly in the hope of freezing its magnetic configuration, I argue in [114] that the microstructure evolved quickly in or-

[33] Sir William Lawrence BRAGG, Australian-born physicist, 1890–1971. He received the Nobel Prize in Physics in 1915, jointly with his father, Sir William Henry BRAGG, for their services in the analysis of crystal structure by means of X-rays. He was Langworthy professor of physics at Victoria University in Manchester, and Cavendish professor of experimental physics (1938–1953) at University of Cambridge, Cambridge, England. The Bragg law in X-ray diffraction is named after him.

der to change the macroscopic heat conductivity and elasticity properties of the ribbon, in order to evacuate heat and react to the imposed stresses, and my guess is that the result was a macroscopic transversally isotropic elastic material, but it remains to understand how the material manages to create an H-measure (or a variant) with very few Dirac masses. Actually, a similar property for an adapted variant should be sought, since *H-measures* have no characteristic length, so I doubt that they are the way to extend the Bragg law to more general materials than crystals.

Since my early work was concerned with homogenization and bounded coefficients, I neglected to study directly concentration effects, but when I was hearing about this question later I noticed that a common mistake is made by those who consider this question in continuum mechanics or physics, since they deal with measures in $x \in \Omega$ and not with measures in $(x, \xi) \in \Omega \times \mathbb{S}^{N-1}$, which I think are necessary for understanding how these concentration effects move around (considering t as x_0 or x_{N+1}).[34]

However, I think that new variants of *H-measures* should be developed in order to deal with concentration effects in semi-linear hyperbolic systems like the Maxwell–Heaviside/Dirac system, and it is important to study it without mass term, since I think that it is precisely the concentration effects which make similar terms appear, and that it is the explanation of what mass is!

Of course, nonlocal effects should be added and I think that hierarchies of systems will appear in order to explain how group invariance may be lost at the level of classical partial differential equations but recovered in that larger class. Once this is understood in a mathematical way, one might find similarities and differences with the guesses of FEYNMAN in his use of diagrams, which is not surprising because of an identical goal of understanding what nature does, but with the different perspectives of a physicist and of a mathematician, developing the mathematical theory which I perceived and called *beyond partial differential equations.* One may observe that a lot of what is often done under the name homogenization is not a part of what I described in this book, which is my plan, and which I propose to call GTH, the *General Theory of Homogenization.*

[34] I do not find surprising then that the advocates of fake mechanics are so interested in geometric measure theory!

Additional footnotes: BEYER,[35] BRAGG W.H.,[36] Caesar,[37] CHAMPOLLION,[38] Jacques DENY,[39] ERASMUS,[40] FROBEN,[41] Gregory XIII,[42] HARTREE,[43]

---

[35] Charles Frederick BEYER (Carl Friedrich BEYER) German-born engineer, 1813–1876. He endowed a chair of applied mathematics, named after him, at Owens College, predecessor of the actual University of Manchester, England.

[36] Sir William Henry BRAGG, English physicist and chemist, 1862–1942. He received the Nobel Prize in Physics in 1915, jointly with his son, William Lawrence BRAGG, for their services in the analysis of crystal structure by means of X-rays. He worked in Adelaide, Australia, was Cavendish professor of physics at Leeds, Quain professor of physics at UCL (University College London), and Fullerian professor of chemistry in the Royal Institution, London, England.

[37] Caesar (Gaius JULIUS Caesar), Roman military and political leader, 100 BCE–44 BCE. The term caesar for the Roman emperors, as well as kaiser in Germany, and czar in Russia come from his cognomen. The Julian calendar is named after him.

[38] Jean-François CHAMPOLLION, French classical scholar, philologist and orientalist, 1790–1832. He worked in Grenoble, and held a chair (Egyptology) at Collège de France, Paris, France. He deciphered the Egyptian hieroglyphs with the help of the work of his predecessors, among them Thomas YOUNG.

[39] Jacques DENY, French mathematician, born in 1918. He worked at Université Paris-Sud XI, Orsay, France, where he was my colleague from 1975 to 1982.

[40] Desiderius ERASMUS Roterodamus (i.e. of Rotterdam), Dutch humanist and theologian, 1466/1469–1536. His critical edition of the Greek New Testament included a Latin translation and annotations, and was published in 1516 by FROBEN in Basel, Switzerland. Erasmus University Rotterdam, in Rotterdam, The Netherlands, is named after him.

[41] Johann FROBEN (Johannes FROBENIUS), German-born printer and publisher, 1460–1527. He printed (in Basel, Switzerland) the works of his friend ERASMUS, who superintended his other editions.

[42] Gregory XIII (Ugo BONCOMPAGNI), Italian Pope, 1502–1585. He was elected Pope in 1572. The Gregorian calendar refers to him.

[43] Douglas Rayner HARTREE, English mathematical physicist, 1897–1958. He held the Beyer chair of applied mathematics in Manchester, and he was Plummer professor of mathematical physics at University of Cambridge, Cambridge, England. The Hartree–Fock method is partly named after him.

Jean-Pierre KERNEVEZ,[44] LANGWORTHY,[45] LEBEDEV,[46] LUTHER,[47] MIE,[48] OWENS,[49] PALIT,[50] PLUME,[51] PLUMMER,[52] QUAIN,[53] Queen Victoria.[54]

---

[44] Jean-Pierre KERNEVEZ, French mathematician, –2005. He worked at UTC (Université de Technologie de Compiègne), Compiègne, France.

[45] LANGWORTHY. I could not find much about this English philanthropist, who endowed a chair of physics, probably at Owens College, predecessor of the actual University of Manchester, England. Three recipients of the Langworthy chair became Nobel laureates!

[46] Pyotr Nikolaevich LEBEDEV, Russian physicist, 1866–1912. He worked in Moscow, Russia. The Lebedev Institute of Physics in Moscow is named after him.

[47] Martin LUTHER, German monk and theologian, 1483–1546. He worked in Wittenberg, Germany. He translated the New Testament into German in 1522, using the Latin translation (second edition, 1519) from the original Greek by ERASMUS, and he translated the Old Testament in 1534. MLU (Martin-Luther-Universität) at Halle-Wittenberg is named after him.

[48] Gustav Adolf Feodor Wilhelm Ludwig MIE, German physicist, 1869–1957. He worked in Greifswald, at MLU (Martin-Luther-Universität) of Halle-Wittenberg, and in Freiburg im Breisgau, Germany.

[49] John OWENS, English textile merchant and philanthropist, 1790–1846. He left money for the foundation of a college, Owens College, opened in 1851, which then became part of Victoria University in Manchester, England, which itself became The University of Manchester in 2004, after merging with UMIST (University of Manchester Institute of Science and Technology).

[50] Sir Taraknath PALIT, Indian lawyer and philanthropist, 1831–1914. He donated money to Calcutta University for science education, and he also donated money for establishing Calcutta Science College, Calcutta, India.

[51] Thomas PLUME, English churchman and philanthropist, 1630–1704. He founded the chair of Plumian professor of astronomy and experimental philosophy in 1704 at University of Cambridge, Cambridge, England.

[52] John Humphrey PLUMMER, English philanthropist, –1928. He endowed professorships in science at University of Cambridge, Cambridge, England.

[53] Richard QUAIN, Irish-born anatomist and surgeon, 1800–1887. He worked at University of London, England, now UCL (University College London), and he left funds to UCL that endowed professorships, named after him, in botany, English language and literature, jurisprudence, and physics.

[54] Alexandrina Victoria, 1819–1901. Queen of the United Kingdom of Great Britain and Ireland in 1837, empress of India in 1876. Several places were named after her, the Victoria University in Manchester, England being just one of them.

Chapter 35
Biographical Information

[In a reference a-*b*, a is the lecture number, 0 referring to the Preface, and *b* the footnote number in that lecture.]

L. Tartar, *The General Theory of Homogenization*, Lecture Notes of the Unione Matematica Italiana 7, DOI 10.1007/978-3-642-05195-1_35,

Chapter 36
Abbreviations and Mathematical Notation

Abbreviations for states:
For those not familiar with geography, I mentioned British Columbia, Ontario, and Québec, without saying that they are provinces of Canada, I mentioned England, Scotland, and Wales, without saying that they are part of the UK (United Kingdom), and for the United States of America I used DC = District of Columbia, and among the fifty states, CA = California, CT = Connecticut, GA = Georgia, IL = Illinois, IN = Indiana, KS = Kansas, KY = Kentucky, MA = Massachusetts, MD = Maryland, MI = Michigan, MN = Minnesota, MO = Missouri, NC = North Carolina, NJ = New Jersey, NM = New Mexico, NY = New York, OH = Ohio, PA = Pennsylvania, RI = Rhode Island, SC = South Carolina, TX = Texas, UT = Utah, VA = Virginia, WA = Washington, WI = Wisconsin.

Other abbreviations:
AIT = Asian Institute of Technology, Klongluang, Thailand
ALCOA = Aluminum Company of America, Alcoa Center, PA
AMS = American Mathematical Society, Providence, RI
ANU = Australian National University, Canberra, Australia
BAR = *Biblical Archaeology Review*, magazine published by BAS
BAS = Biblical Archaeology Society, Washington, DC
BCE = Before common era (instead of using BC = before Christ)
BR = *Bible Review*, magazine published by BAS
Caltech = California Institute of Technology, Pasadena, CA
Carnegie Tech = Carnegie Institute ot Technology, now part of CMU, Pittsburgh, PA
CBMS = Central Board of the Mathematical Sciences, Washington, DC
CE = Common era (instead of using AD = Anno Domini)
CEA = Commissariat à l'Énergie Atomique, France
CIM = Centro Internacional de Matemática, Portugal
CIME = Centro Internazionale Matematico Estivo, Italy
CMU = *Carnegie Mellon University*, Pittsburgh, PA
CNA = Center for Nonlinear Analysis, CMU, Pittsburgh, PA
CNRS = Centre National de la Recherche Scientifique, France

L. Tartar, *The General Theory of Homogenization*, Lecture Notes of the Unione Matematica Italiana 7, DOI 10.1007/978-3-642-05195-1_36,

DEA = Diplôme d'Études Approfondies
DoE = Department of Energy, Washington, DC
EDF = Électricité de France, France
ENS = École Normale Supérieure, France
ENSTA = École Normale Supérieure des Techniques Avancées, France
EPFL = École Polytechnique Fédérale de Lausanne, Lausanne, Switzerland
ERF = Église Réformée de France, France
ETH = Eidgenössische Technische Hochschule, Zürich, Switzerland
FermiLab = Fermi National Accelerator Laboratory, Batavia, IL
GTD = Geometrical Theory of Diffraction
GTH = General Theory of Homogenization
IAS = Institute for Advanced Study, Princeton, NJ
IBM = International Business Machines Corporation
ICM = International Congress of Mathematicians
IFP = Institut Français du Pétrole, Rueil-Malmaison, France
IHES = Institut des Hautes Études Scientifiques, Bures sur Yvette, France
IHP = Institut Henri Poincaré, Paris, France
IMA = Institute for Mathematics and its Applications, UMN, Minneapolis, MN
INRIA = Institut National de Recherche en Informatique et Automatique, France
IRIA = Institut de Recherche en Informatique et Automatique, Rocquencourt, France
IRCN = Institut de Recherches de la Construction Navale, France
LANL = Los Alamos National Laboratory, Los Alamos, NM
LJLL = Laboratoire Jacques-Louis Lions, UPMC, Paris, France
LCPC = Laboratoire Central des Ponts et Chaussées, Paris, France
MIT = Massachusetts Institute of Technology, Cambridge, MA
MLU = Martin-Luther-Universität, Halle and Wittenberg, Germany
MPI = Max Planck Institute, Germany
MRC = Mathematics Research Center, UW, Madison, WI
MSRI = Mathematical Sciences Research Institute, Berkeley, CA
NATO = North Atlantic Treaty Organization
NSF = National Science Foundation, Washington, DC
NYU = New York University, New York, NY
OCIAM = Oxford Centre for Industrial Applied Mathematics, Oxford, England
ONERA = Office National d'Études et de Recherches Aéronautiques, Châtillon, France
OSU = Ohio State University, Columbus, OH
Penn State = The Pennsylvania State University, State College, PA
RAND = Research ANd Development, Arlington, VA
RIMS = Research Institute for Mathematical Sciences, Kyoto, Japan
SIAM = Society for Industrial and Applied Mathematics, Philadelphia, PA
SISSA = Scuola Internazionale Superiore di Studi Avanzati, Trieste, Italy

SUNY = State University of New York, NY
UBC = University of British Columbia, Vancouver, British Columbia
UCB = University of California Berkeley, Berkeley, CA
UCL = University College London, London, England
UCLA = University of California Los Angeles, Los Angeles, CA
UCSB = University of California Santa Barbara, Santa Barbara, CA
UCSC = University of California Santa Cruz, Santa Cruz, CA
UCSD = University of California San Diego, La Jolla, CA
UMD = University of Maryland, College Park, MD
UMI = *Unione Matematica Italiana*, Italy
UMIST = University of Manchester Institute of Science and Technology, Manchester, England
UMN = University of Minnesota, Minneapolis, MN
UNC = University of North Carolina, Chapel Hill, NC
UPMC = Université Pierre et Marie Curie = Université Paris VI, Paris, France
USC = University of Southern California, Los Angeles, CA
USSR = Union of Socialist Sovietic Republics = Soviet Union
UTC = Université de Technologie de Compiègne, Compiègne, France
UW = University of Wisconsin, Madison, WI
VPISU = Virginia Polytechnic Institute and State University, Blacksburg, VA
WPI = Worcester Polytechnic Institute, Worcester, MA

Mathematical notation: First those beginning with a Latin letter, then those beginning with a Greek letter, then the other symbols.

- a.e.: Almost everywhere.
- $B(x,s)$: Open ball centered at x and radius $s > 0$, i.e., $\{y \in E \mid ||x-y||_E < s\}$ (in a normed space E).
- $BMO(\mathbb{R}^N)$: Space of functions of bounded mean oscillation on $\mathbb{R}^N$, i.e., semi-norm $||u||_{BMO} = \sup_{cubes\ Q} \frac{\int_Q |u-u_Q|\,dx}{|Q|} < \infty$ ($u_Q = \frac{\int_Q u\,dx}{|Q|}$, $|Q| = meas(Q)$).
- $BUC(\mathbb{R}^N)$: Banach space of bounded uniformly continuous functions on $\mathbb{R}^N$, with the sup norm.
- $\mathbb{C}$: Complex plane, i.e., $\mathbb{R} + i\,\mathbb{R}$.
- $\mathbb{C}^N$: Product of N copies of $\mathbb{C}$.
- $C(K)$: Banach space of scalar continuous (and bounded) functions on a compact K, equipped with the sup norm.
- $C(K;E)$: Banach space of scalar continuous (and bounded) functions on a compact K with values in a normed space E, equipped with the sup norm.
- $C(\Omega)$: Fréchet space of scalar continuous functions in an open set $\Omega \subset \mathbb{R}^N$ ($\mathcal{E}_0(\Omega)$ in the notation of Laurent SCHWARTZ).
- $C_0(\Omega)$: Banach space of scalar continuous (bounded) functions tending to 0 at ∞ and at the boundary of an open set $\Omega \subset \mathbb{R}^N$, equipped with the sup norm.

- $C_c(\Omega)$: Space of scalar continuous functions with compact support in an open set $\Omega \subset \mathbb{R}^N$.
- class C^k: Whose derivatives of order up to k are continuous.
- $C^k(\Omega)$: Fréchet space of scalar functions of class C^k in an open set $\Omega \subset \mathbb{R}^N$.
- $C^k(\Omega; E)$: Fréchet space of scalar functions of class C^k in an open set $\Omega \subset \mathbb{R}^N$, with values in a finite-dimensional space E.
- $C_c^k(\Omega)$: Space of scalar functions of class C^k with compact support in an open set $\Omega \subset \mathbb{R}^N$.
- $C^k(K)$: Banach space of restrictions to a compact $K \subset \mathbb{R}^N$ of functions in $C^k(\mathbb{R}^N)$.
- $conv(\cdot)$: Convex hull of $\cdot$.
- $\overline{conv}(\cdot)$: Closed convex hull of $\cdot$.
- $curl$: Rotational operator $\big(curl(u)\big)_i = \sum_{j,k} \varepsilon_{i,j,k} \frac{\partial u_j}{\partial x_k}$, used for open sets $\Omega \subset \mathbb{R}^3$ and functions u taking values in $\mathbb{R}^3$.
- D^α: $\frac{\partial^{\alpha_1}}{\partial x_1^{\alpha_1}} \cdots \frac{\partial^{\alpha_N}}{\partial x_N^{\alpha_N}}$ (for a multi-index α with α_j nonnegative integers, $j = 1, \ldots, N$).
- $\mathcal{D}'(\Omega)$: Space of distributions T in Ω, dual of $C_c^\infty(\Omega)$ ($\mathcal{D}(\Omega)$ in the notation of Laurent SCHWARTZ, equipped with its natural topology), i.e., for every compact $K \subset \Omega$ there exists $C(K)$ and an integer $m(K) \geq 0$ with $|\langle T, \varphi\rangle| \leq C(K) \sup_{|\alpha| \leq m(K)} ||D^\alpha \varphi||_\infty$ for all $\varphi \in C_c^\infty(\Omega)$ with support in K.
- det: Determinant.
- div: Divergence operator $div(u) = \sum_i \frac{\partial u_i}{\partial x_i}$.
- dx: Volume element $dx_1 \cdots dx_N$ when $x \in \mathbb{R}^N$.
- $\mathcal{F}$: Fourier transform, $\mathcal{F}f(\xi) = \int_{\mathbb{R}^N} f(x) e^{-2i\pi(x,\xi)}\, dx$ for $f \in L^1(\mathbb{R}^N)$, which Laurent SCHWARTZ extended to the space of tempered distributions $\mathcal{S}'(\mathbb{R}^N)$.
- $\overline{\mathcal{F}}$: Inverse Fourier transform, $\overline{\mathcal{F}}f(\xi) = \int_{\mathbb{R}^N} f(x) e^{+2i\pi(x,\xi)}\, dx$ for $f \in L^1(\mathbb{R}^N)$, which by Laurent SCHWARTZ extension is the inverse of $\mathcal{F}$. The notation is consistent, that $\overline{\mathcal{F}T} = \overline{\mathcal{F}}\,\overline{T}$ for all $T \in \mathcal{S}'(\mathbb{R}^N)$.
- $grad$: Gradient operator, $grad(u) = \big(\frac{\partial u}{\partial x_1}, \cdots \frac{\partial u}{\partial x_N}\big)$.
- $H^s(\mathbb{R}^N)$: (Sobolev space) Hilbert space of distributions $\in \mathcal{S}'(\mathbb{R}^N)$ (tempered), or functions in $L^2(\mathbb{R}^N)$ if $s \geq 0$, such that $(1 + |\xi|^2)^{s/2} \mathcal{F}u \in L^2(\mathbb{R}^N)$.
- $H^s(\Omega)$: For $s \geq 0$, Hilbert space of restrictions to Ω of functions from $H^s(\mathbb{R}^N)$, for an open set $\Omega \subset \mathbb{R}^N$.
- $H^s(\Omega; \mathbb{R}^N)$: Hilbert space of distributions u from Ω into $\mathbb{R}^N$ whose components belong to $H^s(\Omega)$, for an open set $\Omega \subset \mathbb{R}^N$.
- $H_0^s(\Omega)$: For $s \geq 0$, Hilbert space, closure of $C_c^\infty(\Omega)$ in $H^s(\Omega)$, for an open set $\Omega \subset \mathbb{R}^N$.
- $H^{-s}(\Omega)$: For $s \geq 0$, Hilbert space, dual of $H_0^s(\Omega)$, for an open set $\Omega \subset \mathbb{R}^N$.
- $H(div; \Omega)$: Hilbert space of functions $u \in L^2(\Omega; \mathbb{R}^N)$ with $div(u) \in L^2(\Omega)$, for an open set $\Omega \subset \mathbb{R}^N$.

- $\mathcal{H}^1(\mathbb{R}^N)$: (Hardy space) Banach space of functions $f \in L^1(\mathbb{R}^N)$ with $R_j f \in L^1(\mathbb{R}^N)$, $j = 1, \ldots, N$, where R_j, $j = 1, \ldots, N$ are the (M.) Riesz operators.
- $\Im$: Imaginary part of.
- $\mathcal{L}$: Laplace transform.
- $\mathcal{L}(E; F)$: Banach space of linear continuous operators M from the normed space E into the normed space F, with $||M||_{\mathcal{L}(E;F)} = \sup_{e \neq 0} \frac{||M\,e||_F}{||e||_E} < \infty$.
- $\mathcal{L}_{skew}(\mathbb{R}^N; \mathbb{R}^N)$: Finite-dimensional space of skew-symmetric N by N matrices.
- $\mathcal{L}_{sym}(E; E')$: Banach space of symmetric linear continuous operators M from the normed space E into its dual E'.
- $\mathcal{L}_{sym+}(\mathbb{R}^N; \mathbb{R}^N)$: Finite-dimensional open convex cone of symmetric positive definite N by N matrices.
- $\mathcal{L}_+(\mathbb{R}^N; \mathbb{R}^N)$: Finite-dimensional open convex cone of N by N matrices M satisfying $(M\,e, e) > 0$ for all nonzero $e \in \mathbb{R}^N$
- $L^p(A)$: (Lebesgue space) Banach space of (equivalence classes of a.e. equal) measurable functions u with $||u||_p = \left(\int_A |u(x)|^p\,dx\right)^{1/p} < \infty$ if $1 \leq p < \infty$, or $||u||_\infty = \inf\{M \mid |u(x)| \leq M \text{ a.e. in } A\} < \infty$, for a Lebesgue measurable set $A \subset \mathbb{R}^N$ (spaces also considered for the induced $(N-1)$-dimensional Hausdorff measure if $A = \partial\Omega$ for an open set $\Omega \subset \mathbb{R}^N$ with a smooth boundary).
- $L^p(A; \mathbb{R}^N)$: Banach space of (equivalence classes of a.e. equal) measurable functions from A into $\mathbb{R}^N$ whose components belong to $L^p(A)$.
- $L^p(A; E)$: Banach space of (equivalence classes of a.e. equal) measurable functions u from A into a separable Banach space E such that $||u||_E$ belongs to $L^p(A)$.
- $Lip(\Omega)$: Banach space of scalar Lipschitz continuous functions, also denoted $C^{0,1}(\Omega)$, i.e., bounded functions for which there exists M such that $|u(x) - u(y)| \leq M\,|x - y|$ for all $x, y \in \Omega \subset \mathbb{R}^N$; it is included in $C(\overline{\Omega})$.
- $_{loc}$: For any space Z of functions or distributions from an open set $\Omega \subset \mathbb{R}^N$ into a finite-dimensional space, Z_{loc} is the space of functions or distributions u such that $\varphi\,u \in Z$ for all $\varphi \in C_c^\infty(\Omega)$.
- M_b: Operator of multiplication by b.
- $\mathcal{M}(\alpha, \beta; \Omega)$: (Definition 6.3) the set of $A \in L^\infty(\Omega; \mathcal{L}(\mathbb{R}^N; \mathbb{R}^N))$ satisfying $(A(x)\xi, \xi) \geq \alpha\,|\xi|^2$, $(A(x)\xi, \xi) \geq \frac{1}{\beta}\,|A(x)\xi|^2$ for all $\xi \in \mathbb{R}^N$, a.e. $x \in \Omega$.
- $\mathcal{M}on(\alpha, \beta; \Omega)$: (Definition 11.1) the set of Carathéodory functions A defined on $\Omega \times \mathbb{R}^N$ satisfying $(A(x,a) - A(x,b), a - b) \geq \frac{1}{\beta}\,|A(x,a) - A(x,b)|^2$, $(A(x,a) - A(x,b), a - b) \geq \alpha\,|a - b|^2$ for all $a, b \in \mathbb{R}^N$, a.e. in Ω.
- $\mathcal{M}(\Omega)$: Space of Radon measures μ in an open set $\Omega \subset R^N$, dual of $C_c(\Omega)$ (equipped with its natural topology), i.e., for every compact $K \subset \Omega$ there exists $C(K)$ with $|\langle\mu, \varphi\rangle| \leq C(K)||\varphi||_\infty$ for all $\varphi \in C_c(\Omega)$ with support in K.

- $\mathcal{M}_b(\Omega)$: Banach space of Radon measures $\mu \in \mathcal{M}(\Omega)$ with finite total mass in an open set $\Omega \subset \mathbb{R}^N$, dual of $C_0(\Omega)$, i.e., there exists C with $|\langle \mu, \varphi \rangle| \le C\,||\varphi||_\infty$ for all $\varphi \in C_c(\Omega)$.
- $meas(\cdot)$: Lebesgue measure of $\cdot$, sometimes denoted $|\cdot|$.
- num: Numerical range (Definition 22.2).
- P_a: "Pseudo-differential" operator $\overline{\mathcal{F}} M_a \mathcal{F}$.
- p': Conjugate exponent of $p \in [1,\infty]$, i.e., $\frac{1}{p} + \frac{1}{p'} = 1$.
- p^*: Sobolev exponent of $p \in [1, N)$, i.e., $\frac{1}{p^*} = \frac{1}{p} - \frac{1}{N}$ or $p^* = \frac{N p}{N-p}$ for $\Omega \subset R^N$ and $N \ge 2$.
- $\cdot_{per}$: Space defined on a period cell (usually Y) with periodic conditions.
- $\mathbb{R}$: Real line, i.e., $(-\infty, \infty)$.
- $\mathbb{R}_+$: $(0, \infty)$.
- $\mathbb{R}^N$: Product of N copies of $\mathbb{R}$.
- $R(A)$: Range of a linear operator $A \in \mathcal{L}(E; F)$, i.e., $\{f \in F \mid f = A e$ for some $e \in E\}$.
- $\Re$: Real part of.
- R_j: (M.) Riesz operators, $j = 1, \ldots, N$, defined by $\mathcal{F}(R_j u)(\xi) = \frac{i\,\xi_j \mathcal{F}u(\xi)}{|\xi|}$ on $L^2(\mathbb{R}^N)$; natural extensions to $\mathbb{R}^N$ of the Hilbert transform, they map $L^p(\mathbb{R}^N)$ into itself for $1 < p < \infty$, and $L^\infty(\mathbb{R}^N)$ into $BMO(\mathbb{R}^N)$.
- $\mathcal{S}(\mathbb{R}^N)$: Fréchet space of functions $u \in C^\infty(\mathbb{R}^N)$ with $x^\alpha D^\beta u$ bounded for all multi-indices α, β with α_j, β_j nonnegative integers for $j = 1, \ldots, N$.
- $\mathcal{S}'(\mathbb{R}^N)$: Space of tempered distributions, dual of $\mathcal{S}(\mathbb{R}^N)$, i.e., $T \in \mathcal{D}'(\mathbb{R}^N)$ and there exists C and an integer $m \ge 0$ with $|\langle T, \psi \rangle| \le C \sup_{|\alpha|,|\beta| \le m} ||x^\alpha D^\beta \psi||_\infty$ for all $\psi \in \mathcal{S}(\mathbb{R}^N)$.
- $\star$: Convolution product $(f \star g)(x) = \int_{\mathbb{R}^N} f(x-y) g(y)\,dy$, or when used on a dual E' the weak $\star$ topology is that denoted $\sigma(E', E)$ in functional analysis, and not the weak topology $\sigma(E', E'')$.
- $\cdot^T$: Transpose of $\cdot$.
- $\mathcal{V}$: Full characteristic set ($\subset \mathbb{R}^p \times (\mathbb{R}^N \setminus 0)$) used in the *compensated compactness* theory (17.44).
- WF: Wave front set of.
- x: A point in $\mathbb{R}^N$.
- x': In $\mathbb{R}^N$, $x = (x', x_N)$, i.e., $x' = (x_1, \ldots, x_{N-1})$.
- x^α: $x_1^{\alpha_1} \ldots x_N^{\alpha_N}$ for a multi-index α with α_j nonnegative integers for $j = 1, \ldots, N$, for $x \in \mathbb{R}^N$.
- α: A positive scalar, or a multi-index $\alpha = (\alpha_1, \ldots, \alpha_N)$ with all α_i nonnegative integers, whose length is $|\alpha| = |\alpha_1| + \ldots + |\alpha_N|$, and $\alpha! = \alpha_1! \cdots \alpha_N!$.
- γ_0: Trace operator, defined for smooth functions by restriction to the boundary $\partial\Omega$, for an open set $\Omega \subset \mathbb{R}^N$ with a smooth boundary, and extended by density to functional spaces in which smooth functions are dense.
- Δ: Laplacian $\sum_{j=1}^N \frac{\partial^2}{\partial x_j^2}$, defined on any open set $\Omega \subset \mathbb{R}^N$.
- $\delta_{i,j}$: Kronecker symbol, equal to 1 if $i = j$ and equal to 0 if $i \ne j$ (for $i, j = 1, \ldots, N$).

- $\varepsilon_{i,j,k}$: For $i,j,k \in \{1,2,3\}$, completely antisymmetric tensor, equal to 0 if two indices are equal, and equal to the signature of the permutation $123 \mapsto ijk$ if indices are distinct (i.e., $\varepsilon_{1,2,3} = \varepsilon_{2,3,1} = \varepsilon_{3,1,2} = +1$ and $\varepsilon_{1,3,2} = \varepsilon_{3,2,1} = \varepsilon_{2,1,3} = -1$).
- θ: A scalar $\in [0,1]$, or a measurable function taking values in $[0,1]$, or an angle.
- λ: A scalar, real or complex, or an eigenvalue, or an element of $\Lambda \subset \mathbb{R}^p$.
- Λ: Reduced characteristic set ($\subset \mathbb{R}^p$) used in the *compensated compactness* theory (17.6).
- μ: Viscosity, or a Radon measure (for example, an H-measure).
- ν: Kinematic viscosity, or the unit exterior normal to an open set $\Omega \subset \mathbb{R}^N$ with Lipschitz boundary, or a Radon measure (for example, a Young measure).
- ϱ: Density of charge, or of mass.
- ρ_ε: Special regularizing sequence, with $\rho_\varepsilon(x) = \frac{1}{\varepsilon^N}\rho_1\big(\frac{x}{\varepsilon}\big)$ with $\varepsilon > 0$ and $\rho_1 \in C_c^\infty(\mathbb{R}^N)$ with $\int_{x\in\mathbb{R}^N} \rho_1(x)\,dx = 1$, and usually $\rho_1 \ge 0$.
- τ_h: Translation operator of $h \in \mathbb{R}^N$, acting on a function $f \in L^1_{loc}(\mathbb{R}^N)$ by $\tau_h f(x) = f(x-h)$ a.e. $x \in \mathbb{R}^N$.
- φ: A test function.
- χ: A characteristic function, i.e., taking only values 0 or 1.
- ψ: A test function, or when $\psi \in \mathbb{C}^4$ the function describing matter in the Dirac equation.
- ω or Ω: An open set of $\mathbb{R}^N$.
- $|\cdot|$: Absolute value of $\cdot$, or norm of $\cdot$ in $\mathbb{R}^N$ or in the Hilbert space H, or Lebesgue measure of the set $\cdot$.
- $\approx$: Is approximately.
- $\in$: Belongs to.
- $\notin$: Does not belong to.
- $\bar{\cdot}$: Closure of $\cdot$.
- $[\cdot,\cdot]$: Commutator, i.e., for operators A, B from a vector space into itself $[A,B] = A\,B - B\,A$.
- $\langle\cdot,\cdot\rangle$: Duality product.
  $\langle\langle\cdot,\cdot\rangle\rangle$: (Definition 30.1) for an H-measure μ, if $Q(x,\xi,U) = \sum_{i,j} q_{i,j}(x,\xi)\, U_i\overline{U_j}$, then $\langle\langle\mu, Q(x,\xi,U)\rangle\rangle = \sum_{i,j}\int_{\mathbb{S}^{N-1}} q_{i,j}\,d\mu^{i,j} \in \mathcal{M}(\Omega)$.
- $\exists$: There exists.
- $\forall$: For all.
- $\cap$: Intersection.
- $\mapsto$: Maps to.
- $||\cdot||$: Norm in V.
- $||\cdot||_*$: Dual norm in V'.
- $\perp$: Orthogonal to.
- ∇: Fréchet derivative (nabla operator).
- $||$: Parallel to
- $\partial\omega$ or $\partial\Omega$: The boundary of ω or Ω.

- $\{\cdot,\cdot\}$: Poisson bracket, i.e., for two smooth functions f, g on $\mathbb{R}^N \times \mathbb{R}^N$ (variable (x,ξ)) $\{f,g\} = \sum_i \frac{\partial f}{\partial \xi_i}\frac{\partial g}{\partial x_i} - \frac{\partial f}{\partial x_i}\frac{\partial g}{\partial \xi_i}$.
- $\cdot'$: Derivative of $\cdot$.
- $\to$: Converges to.
- $\rightharpoonup$: Converges weakly (or weakly $\star$) to.
- $(\cdot,\cdot)$: Scalar product in $\mathbb{R}^N$, or Hermitian product in $\mathbb{C}^N$.
- $\setminus$: Subtraction for sets, i.e., $A \setminus B$ is the set of points in A which do not belong to B.
- $\subset$: Subset of.
- Σ: Sum operator.
- $\otimes$: Tensor product.
- $\tilde{\cdot}$: Usually operator of extension by 0 outside an open set $\Omega \subset \mathbb{R}^N$.
- $\times$: Product of sets, or cross product (in $\mathbb{R}^3$), i.e., $(a \times b)_i = \sum_{j,k} \varepsilon_{i,j,k} a_j b_k$.
- $\cup$: Union.

References

1. AMIRAT Y., HAMDACHE K. & ZIANI A., Homogénéisation d'équations hyperboliques du premier ordre et application aux écoulements miscibles en milieu poreux, *Ann. Inst. H. Poincaré Anal. Non Linéaire* **6** (1989), no. 5, 397–417.
2. AMIRAT Y., HAMDACHE K. & ZIANI A., Étude d'une équation de transport à mémoire, *C. R. Acad. Sci. Paris Sér. I Math.* **311** (1990), no. 11, 685–688.
3. ARTSTEIN Z., Discrete and continuous bang-bang and facial spaces or: look for the extreme points, *SIAM Rev.* **22** (1980), no. 2, 172–185.
4. BABUŠKA I., Homogenization approach in engineering, *Computing methods in applied sciences and engineering (Second Internat. Sympos., Versailles,* 1975), Part 1, pp. 137–153. Lecture Notes in Econom. and Math. Systems, Vol. 134, Springer, Berlin, 1976.
5. BABUŠKA I., Homogenization and its application, *Mathematical and computational problems. Numerical solution of partial differential equations, III* (Proc. Third Sympos. (SYNSPADE), Univ. Maryland, College Park, Md., 1975), pp. 89–116, Academic Press, New York, 1976.
6. BENSOUSSAN A., LIONS J.-L. & PAPANICOLAOU G., *Asymptotic analysis for periodic structures*, Studies in Mathematics and its Applications 5, North-Holland, Amsterdam, 1978.
7. BERGMAN D., Bulk physical properties of composite media, *Homogenization methods: theory and applications in physics (Bréau-sans-Nappe, 1983)*, 1–128, Collect. Dir. Études Rech. Élec. France, 57, Eyrolles, Paris, 1985.
8. BIRKHOFF G., *A source book in classical analysis*, Birkhoff G. ed., with the assistance of Uta Merzbach. Harvard University Press, Cambridge, MA, 1973. xii+470 pp.
9. BOSTICK W.H., The morphology of the electron, *Internat. J. of Fusion Energy* **3** (1985), no. 1, 9–52.
10. BROWN W.F., *Micromagnetics*, Interscience, New York, 1963.
11. BURGERS J.M., A mathematical model illustrating the theory of turbulence, *Advances in Appl. Mech.* **1** (1948), 171–199.
12. CALDERÓN A.P., Commutators of singular integral operators, *Proc. Nat. Acad. Sci.* **53** (1965), 1092–1099.
13. CHENAIS D., Un ensemble de variétés à bord Lipschitziennes dans $\mathbb{R}^N$: compacité, théorème de prolongements dans certains espaces de Sobolev, *C. R. Acad. Sci. Paris Sér. A-B* **280** (1975), Aiii, A1145–A1147.
14. CHOQUET-BRUHAT Y., Solutions globales des équations de Maxwell–Dirac–Klein–Gordon (masses nulles), *C. R. Acad. Sci. Paris Sér. I Math.* **292** (1981), no. 2, 153–158.
15. CIORANESCU D. & MURAT F., Un terme étrange venu d'ailleurs, *Nonlinear partial differential equations and their applications, Collège de France Seminar*, vol. II (Paris 1979/1980), pp. 98–138, Res. Notes in Math., 60, Pitman, Boston, MA, 1982.

16. CIORANESCU D. & MURAT F., Un terme étrange venu d'ailleurs, II, *Nonlinear partial differential equations and their applications, Collège de France Seminar*, vol. III (Paris 1980/1981), pp. 154–178, Res. Notes in Math., 70, Pitman, Boston, MA, 1982.
17. COATS J., *High frequency asymptotics of antenna/structure interactions*, PhD thesis, University of Oxford, Oxford, 2002.
18. COIFMAN R. & MEYER Y., Au delà des opérateurs pseudo-différentiels, *Astérisque* **57** (1978), 1–185.
19. COIFMAN R., ROCHBERG R. & WEISS G., Factorization theorems for Hardy spaces in several variables, *Ann. of Math.* (2) **103** (1976), no. 3, 611–635.
20. COIFMAN R. & WEISS G., Extensions of Hardy spaces and their use in analysis, *Bull. Amer. Math. Soc.* **83** (1977), no. 4, 569–645.
21. DAVID G. & JOURNÉ J.L., Une caractérisation des opérateurs intégraux singuliers bornés sur $L^2(\mathbb{R}^N)$, *C. R. Acad. Sci. Paris, Sér. I, Math.* **296** (1983), 761–764.
22. DE GIORGI E. & SPAGNOLO S., Sulla convergenza degli integrali dell'energia per operatori ellitici del 2 ordine, *Boll. Un. Mat. Ital.* (4) **8** (1973), 391–411.
23. DE SIMONE A., Ph.D. thesis, University of Minnesota, Minneapolis, MN, 1992.
24. DONOGHUE, W.F., *Monotone matrix functions and analytic continuation*, Die Grundlehren der mathematischen Wissenschaften, Band 207, Springer-Verlag, New York, 1974. v+182 pp.
25. EHRENPREIS L., On the theory of kernels of Schwartz, *Proc. Amer. Math. Soc.* **7** (1956), 713–718.
26. ENÉ H. & SANCHEZ-PALENCIA E.E., Equations et phénomènes de surface pour l'écoulement dans un modèle de milieu poreux, *Jour. de Mécanique*, **14** (1975), 73–108.
27. FEFFERMAN C., Recent progress in classical Fourier analysis, *Proc. I.C.M. Vancouver*, 1974, pp. 95–118.
28. FEYNMAN R., LEIGHTON R.B. & SANDS M., *The Feynman lectures on physics: the definitive and extended edition*, 3 vols, Addison-Wesley, 2005.
29. FEYNMAN R. & LEIGHTON R., *Surely, you're joking, Mr. Feynman! Adventures of a curious character*, Hutchings E. ed., W W Norton, 1985.
30. FRANCFORT G.A. & MURAT F., Homogenization and optimal bounds in linear elasticity, *Arch. Rational Mech. Anal.* **94** (1986), 307–334.
31. FRANCFORT G.A. & MURAT F., Optimal bounds for conduction in two-dimensional, two-phase, anisotropic media, in *Nonclassical continuum mechanics*, Knops R.J., Lacey A.A. eds. (Durham, 1986), pp. 197–212, London Mathematical Society Lecture Note Series 122, Cambridge University Press, Cambridge, 1987.
32. FRANCFORT G., MURAT F. & TARTAR L., Fourth-order moments of nonnegative measures on S^2 and applications, *Arch. Rational Mech. Anal.* **131** (1995), no. 4, 305–333.
33. GÅRDING L. & LIONS J.-L., Functional analysis, *Nuovo Cimento* (10) **14** (1959) supplemento, 9–66.
34. GIBBON E., *The history of the decline and fall of the Roman Empire*, vol.1, 459 pp., vol. 2, 492 pp., vol. 3, 414 pp., Elibron Classics series, Adamant Media, Boston, MA, 2000. (Replica of 1787 edition J. J. Tourneisen, Basel.)
35. GHOUILA-HOURI A., Sur la généralisation de la notion de commande d'un système guidable, *Rev. Française Informat. Recherche Opérationnelle* **1** (1967), no. 4, 7–32.
36. GUTIÉRREZ S., *Laminations in linearized elasticity and a Lusin type theorem for Sobolev spaces*, PhD thesis, Carnegie Mellon University, Pittsburgh, PA, 1997.
37. GUTIÉRREZ S., Laminations in linearized elasticity: the isotropic non-very strongly elliptic case, *J. Elasticity* **53** (1998/99), no. 3, 215–256.

38. HASHIN Z. & SHTRIKMAN S., A variational approach to the theory of effective magnetic permeability of multiphase materials, *J. Applied Phys.* **33** (1962), 3125–3131.
39. HOPF E., The partial differential equation $u_t + uu_x = \varepsilon u_{xx}$, *Comm. Pure Appl. Math.* **3** (1950), 201–230.
40. HÖRMANDER L., *The analysis of linear partial differential operators. I. Distribution theory and Fourier analysis*, Grundlehren der Mathematischen Wissenschaften, 256. Springer-Verlag, Berlin, 1983. ix+391 pp. *The analysis of linear partial differential operators. II. Differential operators with constant coefficients*, Grundlehren der Mathematischen Wissenschaften, 257. Springer-Verlag, Berlin, 1983. viii+391 pp. *The analysis of linear partial differential operators. III. Pseudodifferential operators*, Grundlehren der Mathematischen Wissenschaften, 274. Springer-Verlag, Berlin, 1985. viii+525 pp. *The analysis of linear partial differential operators. IV. Fourier integral operators*, Grundlehren der Mathematischen Wissenschaften, 275. Springer-Verlag, Berlin, 1985. vii+352 pp.
41. ISAACS R., *Differential games. A mathematical theory with applications to warfare and pursuit, control and optimization*, John Wiley & Sons, Inc., New York, 1965. xvii+384 pp.
42. JOSEPH D., *Stability of fluid motions* I, II, Springer Tracts in Natural Philosophy, Vol. 27, 28, Springer-Verlag, Berlin, 1976.
43. KARLIN S., *Mathematical methods and theory in games, programming and economics. Vol. I: Matrix games, programming, and mathematical economics, Vol. II: The theory of infinite games*, Addison-Wesley Publishing Co., Inc., Reading, MA, 1959. Vol. I, x+433 pp. Vol. II, xi+386 pp.
44. KELLER J.B., A theorem on the conductivity of a composite medium, *J. Mathematical Phys.* **5** (1964), 548–549.
45. KNUTH D.E., *The TeXbook*, Addison-Wesley, Reading, MA, 1984. x + 483 pp.
46. KORÁNYI A., Note on the theory of monotone operator functions, *Acta Sci. Math.* **16** (1955), 241–245.
47. LANDAU L.D. & LIFSHITZ E.M., *Électrodynamique des milieux continus*, Mir, Moscou, 1969. *Electrodynamics of continuous media*, Pergamon Press, 1960.
48. LAX P.D., Shock waves and entropy, *Contributions to Nonlinear Functional Analysis*, Zarantonello E.H. ed., pp. 603–634, Academic Press, New York, 1971.
49. LAX P.D., MORAWETZ C.S. & PHILLIPS S., Exponential decay of solutions of the wave equation in the exterior of a star-shaped obstacle, *Comm. Pure Appl. Math.* **16** (1963), 477–486.
50. LAX P.D. & PHILLIPS R.S., *Scattering theory*, Pure and Applied Mathematics, Vol. 26 Academic Press, New York, 1967. xii+276 pp.
51. LIONS J.-L., *Contrôle optimal de systèmes gouvernés par des équations aux dérivées partielles*, avant propos de P. Lelong, Dunod, Paris; Gauthier-Villars, Paris, 1968, xiii+426 pp. *Optimal control of systems governed by partial differential equations.* Translated from the French by S. K. Mitter. Die Grundlehren der mathematischen Wissenschaften, Band 170 Springer-Verlag, New York, 1971. xi+396 pp.
52. LIONS J.-L., *Quelques méthodes de résolution des problèmes aux limites non linéaires*, Dunod; Gauthier-Villars, Paris, 1969. xx+554 pp.
53. LIONS J.-L., Some remarks on variational inequalities, 1970 Proc. Internat. Conf. on Functional Analysis and Related Topics (Tokyo, 1969), pp. 269–282, University of Tokyo Press, Tokyo.
54. LIONS J.-L., Asymptotic behaviour of solutions of variational inequalities with highly oscillating coefficients, *Applications of methods of functional analysis to problems in mechanics*, Germain P., Nayrolles B. eds (Joint Sympos., IUTAM/IMU, Marseille, 1975), pp. 30–55. Lecture Notes in Math., 503. Springer, Berlin, 1976.

55. LIONS J.-L. & MAGENES E., *Problèmes aux limites non homogènes et applications.* Vol. 1, Travaux et Recherches Mathématiques, No. 17 Dunod, Paris, 1968. xx+372 pp. *Non-homogeneous boundary value problems and applications.* Vol. I, translated from the French by P. Kenneth. Die Grundlehren der mathematischen Wissenschaften, Band 181, Springer-Verlag, New York, 1972. xvi+357 pp.
56. LIONS J.-L. & PEETRE J., Sur une classe d'espaces d'interpolation, *Inst. Hautes Etudes Sci. Publi. Math.* **19** (1964), 5–68.
57. GUPTA M. & SWAMI NIKHILANANDA, *The gospel of Sri Ramakrishna*, Ramakrishna-Vivekananda Center, New York, 1942. 1062 pp.
58. MANDEL J., *Cours de mécanique des milieux continus*, 2 vol. Gauthier-Villars, Paris, 1966. vol.1, 456 pp., vol. 2, 391 pp.
59. MARINO A. & SPAGNOLO S., Un tipo di approssimazione dell'operatore $\sum_{ij} D_i\Big(a_{ij} D_j\Big)$ con operatori $\sum_j D_j(b\, D_j)$, *Ann. Scuola Norm. Sup. Pisa Cl. Sci.* (3) **23** (1969), 657–673.
60. MASCARENHAS L., A linear homogenization problem with time dependent coefficient, *Trans. AMS* **281** (1984), no. 1, 179–195.
61. MCCONNELL W., On the approximation of elliptic operators with discontinuous coefficients, *Ann. Scuola Norm. Sup. Pisa Cl. Sci.* (4) **3** (1976), no. 1, 123–137.
62. MEYERS N., An L^p-estimate for the gradient of solutions of second order elliptic divergence equations, *Ann. Scuola Norm. Sup. Pisa* (3) **17** (1963), 189–206.
63. MILTON G., Modelling the properties of composites by laminates, in *Homogenization and effective moduli of materials and media*, Ericksen J.L., Kinderlehrer D., Kohn R.V., Lions J.-L. eds., pp. 150–174, IMA Volumes in Mathematics and its Applications, 1, Springer, Berlin, 1986.
64. MILTON G.W., *Theory of composites*, Cambridge monographs in applied and computational mathematics (No. 6), Cambridge University Press, Cambridge, 2002. xxviii+719 pp.
65. MORAWETZ C.S., The limiting amplitude principle, *Comm. Pure Appl. Math.* **15** (1962), 349–361.
66. MORAWETZ C.S., RALSTON J.V. & STRAUSS W.A., Decay of solutions of the wave equation outside nontrapping obstacles, *Comm. Pure Appl. Math.* **30** (1977), no. 4, 447–508. Correction *Comm. Pure Appl. Math.* **31** (1978), no. 6, 795.
67. MORTOLA S. & STEFFÉ S., Unpublished report, Scuola Normale Superiore, Pisa.
68. MORTOLA S. & STEFFÉ S., A two-dimensional homogenization problem, *Atti Accad. Naz. Lincei Rend. Cl. Sci. Fis. Mat. Natur.* (8) **78** (1985), no. 3, 77–82.
69. MURAT F., Un contre-exemple pour le problème du contrôle dans les coefficients, *C. R. Acad. Sci. Paris, Sér. A-B* **273** (1971), A708–A711.
70. MURAT F., Théorèmes de non-existence pour des problèmes de contrôle dans les coefficients, *C. R. Acad. Sci. Paris, Sér. A-B* **274** (1972), A395–A398.
71. MURAT F., H-convergence, Séminaire d'analyse fonctionnelle et numérique, *Université d'Alger*, 1977-78. English translation MURAT F. & TARTAR L., H-convergence, *Topics in the mathematical modelling of composite materials*, 21–43, Progr. Nonlinear Differential Equations Appl., 31, Birkhäuser, Boston, MA, 1997.
72. MURAT F., Compacité par compensation, *Ann. Scuola Norm. Sup. Pisa Cl. Sci.* (4), **5** (1978), 489–507.
73. MURAT F., Compacité par compensation II, *Proc. Intern. Meeting on Recent Methods in Nonlinear Analysis (Rome, 1978)*, 245–256, Pitagora, Bologna, 1979.
74. MURAT F., Compacité par compensation: condition nécessaire et suffisante de continuité faible sous une hypothèse de rang constant, *Ann. Scuola Norm. Sup. Pisa Cl. Sci.* (4) **8** (1981), no. 1, 69–102.

75. MURAT F. & TARTAR L., Calcul des variations et homogénéisation, *Homogenization methods: theory and applications in physics* (Bréau-sans-Nappe, 1983), 319–369, Collect. Dir. Etudes Rech. Elec. France, 57, *Eyrolles, Paris*, 1985. English translation: Calculus of variations and homogenization, *Topics in the mathematical modelling of composite materials*, 139–173, Prog. Nonlinear Differential Equations Appl., 31, Birkhäuser, Boston, MA, 1997.
76. NESI V., Quasiconformal mappings as a tool to study certain two-dimensional G-closure problems, *Arch. Rational Mech. Anal.* **134** (1996) no. 1, 17–51.
77. PAPANICOLAOU G.C. & VARADHAN S.R.S., Diffusion in regions with many small holes, *Stochastic differential systems* (Proc. IFIP-WG 7/1 Working Conf., Vilnius, 1978), pp. 190–206, Lecture Notes in Control and Information Sci., 25, Springer, Berlin, 1980.
78. PONTRYAGIN L.S., BOLTYANSKII V.G., GAMKRELIDZE R.V. & MISHCHENKO E.F., *The mathematical theory of optimal processes*, translated from the Russian by Trirogoff K.N.; Neustadt L.W. ed., Interscience Publishers John Wiley & Sons, Inc., New York, 1962. viii+360 pp.
79. RAUCH J. & TAYLOR M., Electrostatic screening, *J. Mathematical Phys.* **16** (1975), 284–288.
80. RENARDY M., HRUSA W.J. & NOHEL J.A., *Mathematical problems in viscoelasticity*, Pitman Monographs and Surveys in Pure and Applied Mathematics, 35. Longman Scientific & Technical, Harlow; John Wiley & Sons, Inc., New York, 1987. x+273 pp.
81. SANCHEZ-PALENCIA E.E., Solutions périodiques par rapport aux variables d'espaces et applications, *C. R. Acad. Sci. Paris, Sér. A-B* **271** (1970), A1129–A1132.
82. SANCHEZ-PALENCIA E.E., Equations aux dérivées partielles dans un type de milieux hétérogènes, *C. R. Acad. Sci. Paris, Sér. A-B* **272** (1971), A1410–A1413.
83. SANCHEZ-PALENCIA E.E., *Nonhomogeneous media and vibration theory*, 398 pp., Lecture Notes in Physics, Vol. 127, Springer-Verlag, Berlin, 1980.
84. SCHWARTZ L., *Étude des sommes d'exponentielles réelles*, Actualités Sci. Ind., no. 959. Hermann et Cie., Paris, 1943. 89 pp.
85. SCHWARTZ L., *Théorie des distributions*, Publications de l'Institut de Mathématique de l'Université de Strasbourg, No. IX–X. Nouvelle édition, entièrement corrigée, refondue et augmentée. Hermann, Paris, 1966. xiii+420 pp.
86. SCHWARTZ L., *Analyse mathématique*. I, II, Hermann, Paris, 1967. xxix+830 pp., xxiv+475+21+75 pp..
87. SERRIN J., A symmetry problem in potential theory, *Arch. Rational Mech. Anal.* **43** (1971), 304–318.
88. SIMON L., On G-convergence of elliptic operators, *Indiana Univ. Math. J.* **28** (1979), no. 4, 587–594.
89. SPAGNOLO S., Sul limite delle soluzioni di problemi di Cauchy relativi all'equazione del calore, *Ann. Scuola Norm. Sup. Pisa* (3) **21** (1967), 657–699.
90. SPAGNOLO S., Sulla convergenza di soluzioni di equazioni paraboliche ed ellitiche, *Ann. Scuola Norm. Sup. Pisa* (3) **22** (1968), 571–597; *errata, ibid.* (3) **22** (1968), 673.
91. SPAGNOLO S., Convergence in energy for elliptic operators, *Numerical solution of partial differential equations, III* (Proc. Third Sympos. (SYNSPADE), Univ. Maryland, College Park, Md., 1975), pp. 469–498. Academic Press, New York, 1976.
92. TARTAR L., Convergence d'opérateurs différentiels, *Atti Giorni Analisi Convessa e Applicazioni* (Roma, 1974), 101–104.
93. TARTAR L., Problèmes de contrôle des coefficients dans des équations aux dérivées partielles, *Control theory, numerical methods and computer systems modelling* (Internat. Sympos., IRIA LABORIA, Rocquencourt, 1974), pp. 420–426. Lecture Notes in Econom. and Math. Systems, Vol. 107, Springer, Berlin,

1975. English translation: MURAT F. & TARTAR L., On the control of coefficients in partial differential equations, *Topics in the mathematical modelling of composite materials*, 1–8, Prog. Nonlinear Differential Equations Appl., 31, Birkhäuser, Boston, MA, 1997.

94. TARTAR L., Quelques remarques sur l'homogénéisation, *Functional analysis and numerical analysis (Tokyo and Kyoto,* 1976), 469–481, Japan Society for the Promotion of Science, Tokyo, 1978.
95. TARTAR L., Homogénéisation en hydrodynamique, in *Singular perturbations and boundary layer theory,* Lyon 1976, Lecture Notes in Mathematics 594, pp. 474–481, Springer, Berlin, 1977.
96. TARTAR L., Nonlinear constitutive relations and homogenization, *Contemporary developments in continuum mechanics and partial differential equations*, De La Penha G.M., Medeiros L.A. eds. (Proc. Internat. Sympos., Inst. Mat., Univ. Fed. Rio de Janeiro, Rio de Janeiro), pp. 472–484, North-Holland Math. Stud., 30, North-Holland, Amsterdam, 1978.
97. TARTAR L., Estimations de coefficients homogénéisés, *Computing methods in applied sciences and engineering* (Proc. Third Internat. Sympos., Versailles, 1977), I, pp. 364–373. Lecture Notes in Math., 704, Springer, Berlin, 1979. English translation: Estimations of homogenized coefficients, *Topics in the mathematical modelling of composite materials*, 9–20, Prog. Nonlinear Differential Equations Appl., 31, Birkhäuser, Boston, MA, 1997.
98. TARTAR L., Compensated compactness and applications to partial differential equations, *Nonlinear analysis and mechanics: Heriot–Watt Symposium, Vol. IV*, pp. 136–212, Res. Notes in Math., 39, Pitman, Boston, MA, 1979.
99. TARTAR L., Solutions oscillantes des équations de Carleman, *Goulaouic–Meyer–Schwartz Seminar, 1980–1981*, Exp. No. XII, 15 pp., École Polytech., Palaiseau, 1981.
100. TARTAR L., Compacité par compensation: résultats et perspectives, *Nonlinear partial differential equations and their applications. Collège de France Seminar, Vol. IV* (Paris, 1981/1982), 350–369, Res. Notes in Math., 84, Pitman, Boston, MA, 1983.
101. TARTAR L., Estimations fines des coefficients homogénéisés, *Ennio De Giorgi colloquium*, Krée P. ed. (Paris, 1983), 168–187. Res. Notes in Math., 125, Pitman, Boston, MA, 1985.
102. TARTAR L., Remarks on homogenization, *Homogenization and effective moduli of materials and media*, Ericksen J.L., Kinderlehrer D., Kohn R.V., Lions J.-L. eds. (Minneapolis, Minn., 1984/1985), 228–246, IMA Vol. Math. Appl., 1, Springer, New York, 1986.
103. TARTAR L., Remarks on oscillations and Stokes equation, *Macroscopic modelling of turbulent flows*, Frisch U., Keller J.B., Papanicolaou G., Pironneau O. eds., pp. 24–31, Lecture Notes in Physics 230, Springer, Berlin, 1985.
104. TARTAR L., Nonlocal effects induced by homogenization, *Partial differential equations and the calculus of variations, Essays in honor of Ennio De Giorgi*, II, 925–938, Birkhaüser, Boston, 1989.
105. TARTAR L., H-measures, a new approach for studying homogenisation, oscillations and concentration effects in partial differential equations, *Proc. Roy. Soc. Edinburgh Sect. A* **115** (1990), no. 3-4, 193–230.
106. TARTAR L., H-measures and small amplitude homogenization, *Random media and composites* (Leesburg, VA, 1988), 89–99, SIAM, Philadelphia, PA, 1989.
107. TARTAR L., Memory effects and homogenization, *Arch. Rational Mech. Anal.* **111** (1990), no. 2, 121–133.
108. TARTAR L., Some remarks on separately convex functions, *Microstructure and phase transition*, 191–204, IMA Vol. Math. Appl., 54, Springer, New York, 1993.

109. TARTAR L., On mathematical tools for studying partial differential equations of continuum physics: H-measures and Young measures, *Developments in partial differential equations and applications to mathematical physics* (Ferrara, 1991), 201–217, Plenum, New York, 1992.
110. TARTAR L., Beyond Young measures, in Microstructure and phase transitions in solids (Udine, 1994), *Meccanica* **30** (1995), no. 5, 505–526 (issue dedicated to Jerrald ERICKSEN).
111. TARTAR L., An introduction to the homogenization method in optimal design, *Optimal shape design*, Tróia, Portugal, 1998, 47–156, Cellina A., Ornelas A. eds., Lecture Notes in Math., Vol. 1740, Fondazione C.I.M.E. Springer-Verlag, Berlin: Centro Internazionale Matematico Estivo, Florence, 2000.
112. TARTAR L., Mathematical tools for studying oscillations and concentration: from Young measures to H-measures and their variants, *Multiscale problems in science and technology*, Dubrovnik, Croatia, 2000, 1–84, Antonić N., van Duijn C.J., Jäger W., Mikelić A. eds., Challenges to Mathematical Analysis and Perspectives, Springer, 2002.
113. TARTAR L., On homogenization and Γ-convergence, in *Homogenization 2001*, Proceedings of the First HMS2000 International School and Conference on Homogenization. Naples, Complesso Monte S. Angelo, June 18–22 and 23–27, 2001, Carbone L., De Arcangelis R. eds., 191–211, Gakkotosho, Tokyo, 2003.
114. TARTAR L., Faut-t-il croire aux dogmes, en physique et ailleurs?, unpublished text, written in the fall of 2005 in memory of Abdelhamid ZIANI.
115. TARTAR L., Compensation effects in partial differential equations, Rendiconti – Accademia Nazionale delle Scienze detta dei XL – *Memorie di Matematica e Applicazioni* 123° (2005), vol. XXIX, fasc. 1, pagg. 395–454. Issue in memory of Luigi AMERIO.
116. TARTAR L., *An introduction to Navier–Stokes equation and oceanography*, 271 pp., Lecture Notes of Unione Matematica Italiana, Vol. 1, Springer, Berlin, 2006.
117. TARTAR L., *An introduction to Sobolev spaces and interpolation spaces*, 248 pp., Lecture Notes of Unione Matematica Italiana, Vol. 3, Springer, Berlin, 2007.
118. TARTAR L., Qu'est-ce que l'homogénéisation?, *Portugaliae Mathematica*, **64** (2007), no. 4, 389–444.
119. TARTAR L., *From hyperbolic systems to kinetic theory, a personalized quest*, 309 pp., Lecture Notes of Unione Matematica Italiana, Vol. 6, Springer, Berlin, 2008.
120. TRUESDELL C. & MUNCASTER R., *Fundamentals of Maxwell's kinetic theory of a simple monatomic gas, treated as a branch of rational mechanics*, Academic Press, New York, 1980.
121. UCHIYAMA A., On the compactness of operators of Hankel type, *Tohoku Math. J.* (2) **30** (1978), no. 1.
122. WEINBERGER H.F., Remark on the preceding paper of Serrin, *Arch. Rational Mech. Anal.* **43** (1971), 319–320.
123. YOUNG L.C., Generalized curves and the existence of an attained absolute minimum in the calculus of variations, *C. R. Soc. Sci. Lett. Varsovie, Classe III* **30** (1937), 212–234.
124. YOUNG L.C., Generalized surfaces in the calculus of variations, I, II, *Ann. of Math.* (2) **43** (1942), 84–103, 530–544.
125. YOUNG L.C., *Lectures on the calculus of variation and optimal control theory*, foreword by Wendell H. Fleming, W. B. Saunders Co., Philadelphia, 1969. xi+331 pp.
126. ZARANTONELLO E.H., The closure of the numerical range contains the spectrum, *Bull. Amer. Math. Soc.* **70** (1964), 781–787, and *Pacific J. Math.* **22** (1967), no. 3, 575–595.

Index

Editor in Chief: Franco Brezzi

Editorial Policy

1. The UMI Lecture Notes aim to report new developments in all areas of mathematics and their applications - quickly, informally and at a high level. Mathematical texts analysing new developments in modelling and numerical simulation are also welcome.

2. Manuscripts should be submitted to
   Redazione Lecture Notes U.M.I.
   umi@dm.unibo.it
   and possibly to one of the editors of the Board informing, in this case, the Redazione about the submission. In general, manuscripts will be sent out to external referees for evaluation. If a decision cannot yet be reached on the basis of the first 2 reports, further referees may be contacted. The author will be informed of this. A final decision to publish can be made only on the basis of the complete manuscript, however a refereeing process leading to a preliminary decision can be based on a pre-final or incomplete manuscript. The strict minimum amount of material that will be considered should include a detailed outline describing the planned contents of each chapter, a bibliography and several sample chapters.

3. Manuscripts should in general be submitted in English. Final manuscripts should contain at least 100 pages of mathematical text and should always include
   - a table of contents;
   - an informative introduction, with adequate motivation and perhaps some historical remarks: it should be accessible to a reader not intimately familiar with the topic treated;
   - a subject index: as a rule this is genuinely helpful for the reader.

4. For evaluation purposes, please submit manuscripts in electronic form, preferably as pdf- or zipped ps- files. Authors are asked, if their manuscript is accepted for publication, to use the LaTeX2e style files available from Springer's web-server at
   ftp://ftp.springer.de/pub/tex/latex/svmonot1/ for monographs
   and at
   ftp://ftp.springer.de/pub/tex/latex/svmultt1/ for multi-authored volumes

5. Authors receive a total of 50 free copies of their volume, but no royalties. They are entitled to a discount of 33.3 % on the price of Springer books purchased for their personal use, if ordering directly from Springer.

6. Commitment to publish is made by letter of intent rather than by signing a formal contract. Springer-Verlag secures the copyright for each volume. Authors are free to reuse material contained in their LNM volumes in later publications: A brief written (or e-mail) request for formal permission is sufficient.

FSC
www.fsc.org
MIX
Papier | Fördert gute Waldnutzung
FSC® C083411

Kolibri 360 GmbH
Ferdinand-Jühlke-Straße 7
99095 Erfurt, Deutschland
produktsicherheit@kolibri360.de